6th International Conference on Advances in Solidification Processes (ICASP-6)

IOP Conference Series: Materials Science and Engineering
Volume 1274

Le Bischenberg, France
19 – 24 June 2022

ISBN: 978-1-7138-7140-8
ISSN: 1757-8981

Printed from e-media with permission by:

Curran Associates, Inc.
57 Morehouse Lane
Red Hook, NY 12571

Some format issues inherent in the e-media version may also appear in this print version.

Institute of Physics
Dirac House, Temple Back
Bristol BS1 6BE UK

Phone: 44 1 17 929 7481
Fax: 44 1 17 920 0979

techtracking@iop.org

Additional copies of this publication are available from:

Curran Associates, Inc.
57 Morehouse Lane
Red Hook, NY 12571 USA
Phone: 845-758-0400
Fax: 845-758-2633
Email: curran@proceedings.com
Web: www.proceedings.com

TABLE OF CONTENTS

IOP Conf. Series: Materials Science and Engineering 1274 (2023) 011001 doi:10.1088/1757-899X/1274/1/011001

6th International Conference on Advances in Solidification Processes

20-24 June 2022, Le Bischenberg, France

The ICASP conference series is a preeminent forum for researchers to discuss the latest advances and future directions in **solidification science**. The 6th edition was held 20–24 June 2022 at "Le Bischenberg", in Bischoffsheim, on the Alsace Wine Route and at the foot of the Vosges in the east of France.

ICASP-6 covered all aspects of **research and applications** related to **solidification**. It was a great opportunity for researchers and engineers from academia and industry **to exchange on recent advances** and **to discuss outstanding challenges** in the field of solidification. There were 25 parallel sessions dealing with the following topics:

- Nucleation and grain refinement
- Interfaces
- Microstructure formation during solidification and remelting
- Dendritic growth and dendrite fragmentation
- Eutectic, peritectic, monotectic growth
- Rapid solidification
- Melting
- Stress and deformation, hot tearing, shrinkage, and porosity defects
- Multiphase flow and heat & mass transfer, mushy zone rheology, macrosegregation
- Thermophysical properties, thermodynamics of solidification and melting
- Additive manufacturing and welding
- Casting processes: ingot casting, shape casting, continuous/semi-continuous casting (steel and nonferrous alloys), remelting processes (ESR, VAR,...), etc.
- Multiscale modeling methods and through-process modeling
- Experimental and characterization methods

From 200 abstract received, the final program included 130 oral presentations (including 4 plenaries, 9 keynotes and 11 invited talks) and over 30 poster presentations.

The conference was also the occasion to give tribute to Gabriel Faivre (1944-2020), in a session dedicated to eutectics. This honorary session was organized by Silvère Akamatsu and Sabine Bottin-Rousseau, from the Institut des Nanosciences de Paris.

After the peer-review procedure, 60 articles were included in the proceedings. Each article was reviewed, heavily relying on the members of the International Scientific Committee or senior authors.

The editors and the conference chairmen gratefully acknowledge the International Scientific Committee for promoting the conference, proposing topics of plenary lectures, and reviewing the papers. The organizers thank the sponsors for supporting this event and the Université de Lorraine for support in the organization.

Julien Zollinger Jacob Kennedy Miha Založnik

julien.zollinger@univ-lorraine.fr jacob.kennedy@univ-lorraine.fr miha.zaloznik@univ-lorraine.fr

Institut Jean Lamour, CNRS – Université de Lorraine, Nancy, France

IOP Conf. Series: Materials Science and Engineering 1274 (2023) 011001 doi:10.1088/1757-899X/1274/1/011001

Sponsors

Chairmen

Miha Založnik	Institut Jean Lamour	France
Julien Zollinger	Institut Jean Lamour	France

Organizing Committee

Charles-André Gandin	Cemef MINES ParisTech	France
Jakob Olofsson	Jönköping University	Sweden
Silvère Akamatsu	Institut des Nanosciences de Paris	France
Sabine Bottin-Rousseau	Institut des Nanosciences de Paris	France
Jacob Kennedy	Institut Jean Lamour	France
Jean-Philippe Tinnes	Institut Jean Lamour	France
Bernard Rouat	Institut Jean Lamour	France
Abdelhalim Chirouf	Institut Jean Lamour	France
Lucie Gutman	Institut Jean Lamour	France
Victor Moudin	Institut Jean Lamour	France
Ilia Ushakov	Institut Jean Lamour	France
Ivan Cazic	Institut Jean Lamour	France

IOP Conf. Series: Materials Science and Engineering 1274 (2023) 011001 doi:10.1088/1757-899X/1274/1/011001

ICASP-6 International Scientific Committee

Christoph Beckermann	University of Iowa	USA
David Browne	University College Dublin	Ireland
Steve Cockcroft	University of British Columbia	Canada
Janin Eiken	Access	Germany
Zhongyun Fan	Brunel University	UK
Charles-André Gandin	MINES ParisTech	France
Amber Genau	University of Alabama	USA
László Gránásy	Wigner Research Center for Physics	Hungary
Ulrike Hecht	Access	Germany
Hani Henein	University of Alberta	Canada
Wanqi Jie	Northwestern Polytechnical University	China
Jacques Lacaze	University of Toulouse	France
Peter D. Lee	University College London	UK
Andreas Ludwig	University of Leoben	Austria
Mohammed M'hamdi	SINTEF	Norway
Nathalie Mangelinck	Aix-Marseille University	France
André Phillion	McMaster University	Canada
Markus Rettenmayr	Friedrich Schiller University	Germany
Božidar Šarler	University of Ljubljana	Slovenia
Yasushi Shibuta	The University of Tokyo	Japan
Damien Tourret	IMDEA Materials Institute	Spain
Peter Voorhees	Northwestern University	USA
Menghuai Wu	University of Leoben	Austria
Shoumei Xiong	Tsinghua University	China
Hideyuki Yasuda	Kyoto University	Japan
Miha Založnik	University of Lorraine	France
Gerhard Zimmermann	Access	Germany
Julien Zollinger	University of Lorraine	France

Peer Review Statement

All papers published in this volume have been reviewed through processes administered by the Editors. Reviews were conducted by expert referees to the professional and scientific standards expected of a proceedings journal published by IOP Publishing Publishing.

- **Type of peer review:** Single Anonymous
- **Conference submission management system:** Morressier
- **Number of submissions received:** 64
- **Number of submissions sent for review:** 63
- **Number of submissions accepted:** 60
- **Acceptance Rate (Submissions Accepted / Submissions Received × 100):** 93.8
- **Average number of reviews per paper:** 1
- **Total number of reviewers involved:** 29
- **Contact person for queries:**
  Name: Julien Zollinger
  Email: julien.zollinger@univ-lorraine.fr
  Affiliation: University of Lorraine - Institut Jean Lamour

Medium Range Ordering in liquid Al-based alloys: towards a machine learning approach of solidification

P Jarry[2,1] and N Jakse[1]

[1] Université Grenoble Alpes, CNRS, Grenoble INP, SIMaP, F-38000 Grenoble, France.
[2] C-TEC, Parc Economique Centr'alp, 725 rue Aristide Bergès, CS10027, Voreppe 38341 cedex, France.

Abstract. Ab initio molecular dynamic simulations of liquid Al93Cr7 and Al83Zn10Cr7 alloy have evidenced the presence of an icosahedral short range order (iSRO) which develops into icosahedral medium range order (iMRO) as the melt is undercooled. This atomic arrangement accounts for the presence of Dynamic Heterogeneities characterized by Al fast-dynamics regions and Cr-rich slow-dynamics regions. Characterisation of the medium range order was carried out by a direct connectivity approach. However, given the small size of the simulation (256 atoms), such characterisation remains partial. In order to better describe both iMRO formation and more dilute alloys closer to industrial compositions, a new modelling strategy has been initiated to allow in the long term for large-scale atomic-level simulations. Molecular Dynamics (MD) of million to billion atoms may indeed lead to meaningful results. Exploitation of such large amounts of MD-generated big data can be carried out by means of Machine Learning (ML) tools which provide relevant and powerful analysis methods. An unsupervised ML approach based on topological descriptors using persistent homology concepts is proposed to reveal the structural features of atomic arrangements without *a priori* knowledge on the studied system. This approach has been applied so far to pure Al melts. Both translational and orientational orderings are thus evidenced together with nucleation pathways, whose revealed features are beyond the hypotheses of the Classical Nucleation Theory.

1. Introduction

Metallic liquids may reach very large undercoolings despite similarities of the liquid and crystal in their density and coordination number as was pointed out in the seminal work of Turnbull [1]. The root cause for this general fact is that the local order that develops in the liquid phase is not compatible with crystalline order [2]. The necessary atomic rearrangement for nucleation requires a large driving force in the absence of heterogeneous nuclei [3]. The conditions for the formation of a stable embryo and for its ability to grow are governed by the diffusion of chemical species, thus by its efficiency in altering atomic arrangements in the liquid phase. Conversely, diffusion is strongly influenced by the local order in the liquid phase. Upon cooling in the metastable undercooled state, the liquid becomes progressively more viscous and the dynamics changes significantly, exhibiting Arrhenius-to-non-Arrhenius crossovers of diffusion, viscosity and structural relaxation time, the occurrence of dynamic heterogeneities (DH) and the breakdown of the Stokes–Einstein (SE) relation [4–6]. Understanding the underlying physics remains a major challenge to characterize order in disorder and to establish structure-dynamics relations from the experimental and simulation point of view. Recently the discovery of the possibility for icosahedral order in certain systems to promote nucleation [7] instead of inhibiting it has disrupted some received ideas based on the representation provided by small size AIMD simulations. Progress in the understanding of these phenomena in undercooled melts might partly be based on the ability to perform larger MD simulations than those achievable by AIMD but with a similar accuracy as well as to analyze the structural and dynamic properties from the huge amount of data generated by these large-scale simulations. The present contribution aims at describing recent developments based on Machine Learning (ML) tools loaned from Artificial Intelligence [8]. Section 2 describes the main issues in the structure and dynamics aspects of solidification mechanisms mostly in the case of Al-based alloys. Then, the construction of interatomic potentials by supervised Machine Learning training is illustrated in Section 3 for pure Aluminium. Section 4 reports recent developments of an unsupervised ML technique founded on persistent homology concepts enabling us to reveal structural features without *a priori*. Finally, in Section 5 we draw some conclusions and perspectives.

2. Impact of Icosahedral Short and Medium Range Order in alloy solidification

Recently the observation of Σ3 orientation relationships between chains of equiaxed grains in non-inoculated alloys: Al-Zn-Cr [7], Au-Cu-Ir [9], Ni-based alloys [10], has revealed the so far undetected possibility for fcc crystals to co-nucleate from adjacent facets of icosahedral motifs, which crystallographically leads to their above-mentioned mutual twin orientation relationships. Such an observation looked paradoxical, as enhanced diffusion was understood to be the mechanism that could destroy iSRO and allow for the big rearrangement leading to crystalline order. Whereas in the present case one had concomitantly strong iSRO promoted by Cr, hence low solute diffusivity, and iSRO-aided crystal nucleation instead of having an iSRO-inhibited nucleation! Interestingly, the AIMD study of Al-Cr and Al-Zn-Cr melts [11], **Figure 1**a, has revealed substantial decoupling of Cr diffusion from the diffusion of Al and Zn components by a factor of 3 at the liquidus temperature (1000 K).

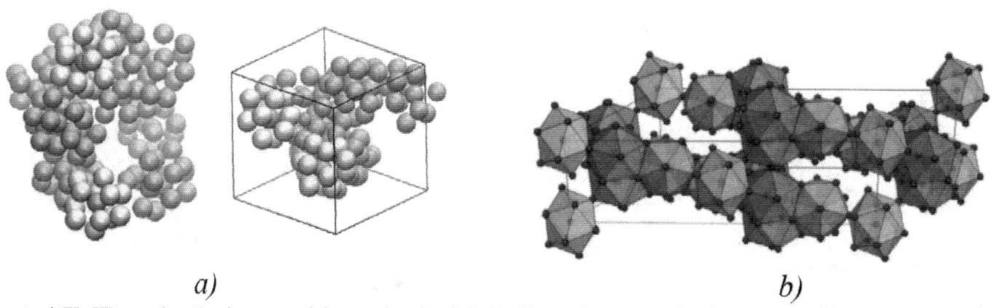

a) *b)*

Figure 1: *a)* AIMD calculations of icosahedral Medium Range Order in undercooled Al-Cr and Al-Zn-Cr melts at 800 K, after [11]. Cr atoms are drawn in blue, Zn atoms in grey, and Al atoms in pink; Al atoms in two typical Cr-centred icosahedra are highlighted in yellow for visual purpose. Left: $Al_{80}Cr_{20}$; *right*: $Al_{83}Zn_{10}Cr_7$, but local composition in the slow dynamics region is found to be $Al_{85}Zn_3Cr_{12}$; *b)* monoclinic lattice unit of the θ-$Al_{45}Cr_7$ crystal re-drawn [7] with the atomic coordinates of [12]. While Al atoms are shown in blue, Cr atoms at the centre of icosahedral clusters are shown in red.

Cr additions promote the formation of a heterogeneous local ordering characterized by a strong five-fold symmetry (icosahedral short-range order (iSRO)) *around Cr atoms*. In the undercooled melt, AIMD evidences the extension of iSRO to icosahedral medium-range order (iMRO) (**Figure 1**a), a most probable precursor of the θ-Al_7Cr phase (**Figure 1**b) which exhibits the pseudo-icosahedral symmetry of the Al_4Cr icosahedral phase, i.e. 12 atoms located at the vertices of deformed icosahedra (either 12Al or 11Al + 1Cr or 10Al + 2Cr) whose centres are occupied by a Cr atom [13]. Although the size of the AIMD simulation is clearly too small to allow for the actual description of the peritectic reaction leading to the formation of θ-Al_7Cr, it evidences a spinodal decomposition of the liquid phase into two Dynamic Heterogeneities (DH), one with fast diffusion in the Al-richer DH and one with slow diffusion in the Cr-rich DH characterised by iMRO, which appears to be the liquid forerunner of that peritectic reaction.

Conditions for iMRO-mediated nucleation:
The appearance of iMRO gives rise to a non-Arrhenian temperature dependence of diffusivity and viscosity, a breakdown of the Stokes-Einstein relation, an observation to be related with diffusion anomalies already reported for tetrahedrally coordinated systems and correlated with entropy anomalies [14]. This particular case of self-nucleation therefore appears to be, in some systems, a path of minimum entropy production governing "what is kinetically preferred within the range of thermodynamic possibilities" (Turnbull): on the one hand icosahedral embryos minimise the interfacial energy with the liquid phase and on the other hand $\Sigma3$ orientation relationships minimise the grain boundary energy between typically 5 grains growing from the same nucleus, leading to an overall minimisation of the interfacial energy cost of nucleation and early growth. Such a decrease in the interfacial energy cost and the reduced value of the free energy difference should lead to a corresponding decrease in the embryo critical size and has indeed been shown to dramatically increase the computed nucleation rate in the case of the Al-Mn system [15]. For this kind of self-nucleation mechanism to be active it is therefore reasonable to assume that a) no heterogeneous crystalline nucleus should be present; b) the iSRO must be promoted, which is generally the case around Transition Metals in Al alloys, such as in the case of Cr in Al melts; and is also promoted at higher undercoolings (at least in pure metals), which means that rapid solidification conditions will promote unravelling of this kind of self-nucleation mechanism; c) a separation must happen between a DH with strong iSRO serving as a nucleant precursor and a DH with low iSRO serving as crystalline precursor; e) lattice parameters between the icosahedral motif and the fcc phase should match; f) icosahedral embryos must form before the fcc phase or at least simultaneously, which means that the alloy composition should be such that primary intermetallic compounds (IMC) can form; g) finally the primary IMC must be peritectic, because its reaction with liquid Al will produce fcc-Al by incorporating the peritectic element (like Cr) into the solid solution, therefore increasing its freezing temperature, stabilising it and creating the constitutional undercooling necessary for further growth: a scheme proposed long ago by Maxwell and Hellawell [16]; contrariwise, nucleation of fcc-Al on a primary eutectic IMC is improbable when back diffusion of eutectic solutes into the forming Al solid solution is sufficient to lead to solutal remelting of the fcc embryo, preventing epitaxial attachment: in the latter case, fcc-nucleation must therefore occur separately, directly within the Al-rich DH. Whether this is possible or not will govern the bifurcation between crystallisation or amorphisation. Such a view is compatible with the fact that the best glass formers correspond to deep eutectics.
Let us add that diffusion not only governs nucleation, but also growth. A review and a tentative chart of the role played by the liquid Al alloy atomic arrangement in selection mechanisms during growth of the solid phases has recently been published [17].

3. Construction of interatomic potentials by supervised Machine Learning training for pure Al

The need to upscale in comparison with AIMD simulation, from typically a few hundreds of atoms over picoseconds up to typically 10^6 atoms and microseconds, more relevant to tackle solidification phenomena, has led to deriving interatomic potentials with semi-empirical functional forms that average out or otherwise model electronic degrees of freedom. A typical approach of current use for metallic materials of interest here are the Embedded-Atom Model (EAM) [18] and the modified EAM [19]. However, impressive progress was made in the last decade in the designing of potentials from *ab initio* calculations using supervised ML methods [8]. Various methods have been used, ranging from simple linear regression (LR) methods to highly non-linear regression methods using High Dimensional Neural Networks (HDNN) or Kernel Regression (KR) [20]. In general, the ML potentials reach an accuracy close to the *ab initio* calculations from which the database was formed. Among designed ML potentials obtained by training on a data base derived from AIMD, very few fully account for both the liquid and crystalline phases (which is called transferability): they are often limited to a good prediction of the melting point. This is due to the fact that building an AIMD database represents a huge computational effort and therefore is often limited to the narrow domain of states useful for the considered study. For solidification phenomena, a strategy very recently put forward consists in constructing a dataset obtained by sampling phase space trajectories generated by AIMD. Liquid states for temperatures above and below the melting point as well as underlying crystalline states from very low temperature (typically 10 K) up to melting are included. This way, time fluctuations are obtained by appropriate sampling of AIMD, so as to embrace the conditions of ergodicity (and eventual ergodicity breakdown) and therefore capture the dynamics and thermodynamics of the system, regardless of coordination numbers and diffusion properties.

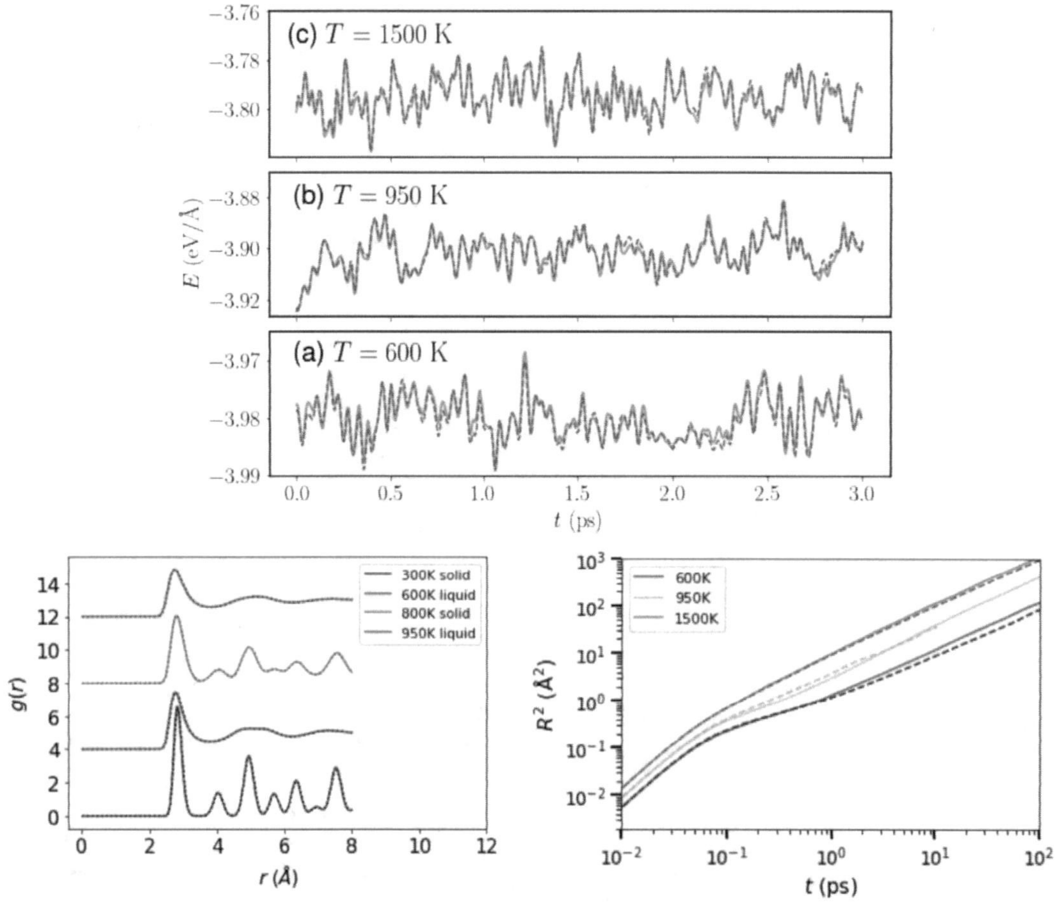

Figure 2: Comparison between AIMD and MD with the ML potential. *Top panel*: Energy per atom as a function of time at T = 600 K (a) in the undercooled region, T = 950 K (b) in the vicinity of the melting point, and at 1500 K (c) far above the melting point at zero pressure. *Lower left panel*: Pair-correlation function for low temperature liquid and solid states at room pressure. *Lower right panel*: Mean-square displacement for low temperature liquid and undercooled states at room pressure. The solid lines correspond to the HDNN potential and the dashed lines with corresponding colours are the AIMD simulations.

In our case 20% of the AIMD data were retained to create validation sets. They were used to monitor the mean squared error on the validation data during the learning process to detect overfitting. The learning process is terminated when the validation error starts to increase, thus preventing overfitting. Simultaneously, a least squared error norm regularization was performed to reduce the model complexity, which is an additional ingredient to prevent overfitting. This approach is illustrated here for pure aluminium [21]. The choice was made to train a HDNN on solely the energies of an AIMD database containing 24300 configurations collected as described above, on the basis of Behler-Parrinello (BP) descriptors [22]. Note that AIMD simulations for the database were performed with the Vienna Ab initio Simulation Package (VASP) [23] within the Local Density Approximation (LDA) functional which provides a good description of the liquid structure and self-diffusion coefficient. **Figure 2** illustrates the excellent match between AIMD and MD simulations on 10^4 atoms using the ML potential in terms of energy fluctuations, structure (through the pair-correlation function) as well as diffusion (by means of the mean-square displacement over the whole range of temperatures and pressures explored by the dataset). It must be noted that, as mentioned in [21], trajectories and calculated properties at temperatures other than those used in the training set were also in excellent agreement with AIMD. Another indication of the very satisfactory physical description obtained with MD using the HDNN potential is the good prediction of latent and specific heats. Moreover, early stages of crystallization were further investigated on a scale much larger than that of AIMD, which resulted in successful capture of homogeneous nucleation mechanisms with an accuracy close to that of *ab initio* [21].

4. Toward unsupervised ML for topological characterisation and nucleation description

To identify translational and orientational orderings during homogeneous nucleation in large-scale MD simulations, an unsupervised learning approach was developed through persistent homology (PH) within the topological data analysis (TDA). In a recently published article [24] we successfully assessed the accuracy of PH against classical methods, such as the common neighbour analysis (CNA) [25], the Steinhardt parameters [26] or the polyhedra template matching (PTM) [27] on known structures. CNA is a very robust method which detects small structural building blocks (bonded pairs) and gives access to a characterization in terms of global proportions of these pairs in the simulation box. The PTM is a powerful method to detect first coordination shell polyhedra that match given or preconceived polyhedra (template). Our methodology, which associates the PH and the GMM clustering is able to sort out any local structure up to the second neighbour shell without preconceived idea (i.e. in an unsupervised way) and therefore goes beyond these classical methods. The local atomic environment of each atom (called the local structure) is explored by sweeping spheres expanding from each atom, with concatenation rules consisting in filling tetrahedra when the spheres contact or overlap: the elemental bricks of polytetrahedral arrangements are therefore present by construction in the PH algorithm. The PH is thus used as a translational and rotational invariant descriptor to encode the local structures required for the clustering method, namely the Gaussian Mixture Models (GMM), which predicts features relevant for atom clusters identification and crystallography. Applied to the MD numerical experiment of nucleation performed on one million atoms along a isotherm close to the Time-Temperature-Transformation (TTT), this method allowed us to clearly capture nucleation events without *a priori*, as seen in **Figure 3**. As expected in pure Al as well as more generally for other metals, nucleation occurred away from the regions with stronger five-fold symmetry [28] which is consistent with Frank's argument [2]. Moreover, translational and orientational orderings take place simultaneously in emerging embryos.

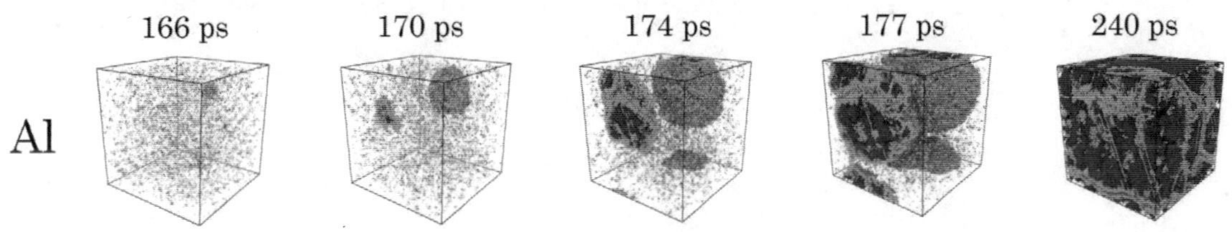

Figure 3: Snapshots of the molecular dynamics simulations for Al with 1 million atoms, during isothermal nucleation at different times for temperatures close to the nose of the TTT. Only clusters with fcc crystalline orderings are drawn, after [24].

5. Summary, conclusions and perspectives

As evidences of the role of the liquid structure on nucleation are being more and more unravelled, and as diffusion of species in the liquid phase is identified as a key parameter in alloy design, especially, but not only, in rapid solidification, quantum engineering of liquid alloys becomes a promising and strategic research avenue at the condition that full advantage be taken from Machine Learning tools. In the present contribution we have exposed the adopted methodology for upscaling atomistic simulations of liquid metals without losing the accuracy of *ab initio* molecular dynamics. Use of supervised or unsupervised ML is very promising to build empirical potentials from *ab initio* data sets and to identify the topological features of atomic arrangements through the liquid-to-solid phase change. The complete approach has been successfully applied, down to the nucleation stage, to pure Al, albeit at a still modest scale for the sake of proving the concept. Next steps will of course concentrate on Al-based alloys, in order to properly account, by means of large MD simulations on more complex alloy systems, for what AIMD has begun to teach us on very small ones, in terms of the compromise between chemical and icosahedral short range order, incipient iMRO, decoupling of solutes diffusion through the formation of dynamic heterogeneities and bifurcation between self-nucleation and amorphisation.

References

[1] Turnbull D. Kinetics of Solidification of Supercooled Liquid Mercury Droplets. J Chem Phys 1952;20:411–24.
[2] Frank FC. Proc R Soc Math Phys Eng Sci 1952;215.
[3] Sosso GC, Chen J, Cox SJ, Fitzner M, Pedevilla P, Zen A, et al. Crystal Nucleation in Liquids: Open Questions and Future Challenges in Molecular Dynamics Simulations. Chem Rev 2016;116:7078–116.
[4] Royall CP, Williams SR. The role of local structure in dynamical arrest. Phys Rep 2015;560:1–75.
[5] Binder K, Kob W. Glassy Materials and Disordered Solids: An Introduction to Their Statistical Mechanics. Revised. WORLD SCIENTIFIC; 2011.
[6] Jug G, Loidl A, Tanaka H. On the structural heterogeneity of supercooled liquids and glasses. EPL Europhys Lett 2021;133:56002. .
[7] Kurtuldu G, Jarry P, Rappaz M. Influence of Cr on the nucleation of primary Al and formation of twinned dendrites in Al–Zn–Cr alloys: Can icosahedral solid clusters play a role? Acta Mater 2013;61:7098–108.
[8] Schmidt J, Marques MRG, Botti S, Marques MAL. Recent advances and applications of machine learning in solid-state materials science. Npj Comput Mater 2019;5:83.
[9] Zollinger J, Rouat B, Guyon J, Pillai SK, Rappaz M. Influence of Ir Additions and Icosahedral Short Range Order (ISRO) on Nucleation and Growth Kinetics in Au-20.5Wt Pct Cu-4.5Wt PctAg Alloy. Metall Mater Trans A 2019;50:2279–88.
[10] Cazic I, Zollinger J, Mathieu S, El Kandaoui M, Plapper P, Appolaire B. New insights into the origin of fine equiaxed microstructures in additively manufactured Inconel 718. Scr Mater 2021;195:113740.
[11] Pasturel A, Jakse N. Influence of Cr on local order and dynamic properties of liquid and undercooled Al–Zn alloys. J Chem Phys 2017;146:184502.
[12] Cooper MJ. The structure of the intermetallic phase θ(Cr-Al). Acta Crystallogr 1960;13:257–63.
[13] Audier M, Durand-Charre M, Laclau E, Klein H. Phase equilibria in the Al-Cr system. J Alloys Compd 1995;220:225–30.
[14] Dyre JC. Perspective: Excess-entropy scaling. J Chem Phys 2018;149:210901.
[15] Battezzati L, Antonione C, Marino F. Some thermodynamic and kinetic aspects of icosahedral phase nucleation in AI-Mn. J Mater Sci 1989;24:2324–30.
[16] Maxwell I, Hellawell A. An analysis of the peritectic reation with particular reference to Al-Ti Alloys. Acta Metall 1975;23:901.
[17] Jarry P, Rappaz M. Recent advances in the metallurgy of aluminium alloys. Part I: Solidification and casting. Comptes Rendus Phys 2018;19:672–87.

[18] Daw MS, Baskes MI. Embedded-atom method: Derivation and application to impurities, surfaces, and other defects in metals. Phys Rev B 1984;29:6443–53.
[19] Baskes MI. Modified embedded-atom potentials for cubic materials and impurities. Phys Rev B 1992;46:2727–42.
[20] Goryaeva AM, Maillet J-B, Marinica M-C. Towards better efficiency of interatomic linear machine learning potentials. Comput Mater Sci 2019;166:200–9.
[21] Jakse N, Sandberg J, Saliou A, Jarry P, Devijver E, Voigtmann T, et al. Interatomic machine learning potentials for aluminium: application to solidification phenomena. ArXiv220101370 Cond-Mat Physics 2022.
[22] Behler J, Parrinello M. Generalized Neural-Network Representation of High-Dimensional Potential-Energy Surfaces. Phys Rev Lett 2007;98:146401.
[23] Kresse G, Furthmüller J. Efficiency of ab-initio total energy calculations for metals and semiconductors using a plane-wave basis set. Comput Mater Sci 1996;6:15–50.
[24] Becker S, Devijver E, Molinier R, Jakse N. Unsupervised topological learning for identification of atomic structures. Phys Rev E 2022;105:045304.
[25] Honeycutt JDana, Andersen HC. Molecular dynamics study of melting and freezing of small Lennard-Jones clusters. J Phys Chem 1987;91:4950–63.
[26] Steinhardt PJ, Nelson DR, Ronchetti M. Bond-orientational order in liquids and glasses. Phys Rev B 1983;28:784–805.
[27] Larsen PM, Schmidt S, Schiøtz J. Robust structural identification via polyhedral template matching. Model Simul Mater Sci Eng 2016;24:055007.
[28] Becker S, Devijver E, Molinier R, Jakse N. Unsupervised topological learning approach of crystal nucleation. Sci Rep 2022;12:3195.

IOP Conf. Series: Materials Science and Engineering 1274 (2023) 012002 doi:10.1088/1757-899X/1274/1/012002

A 3D characterization of equiaxed grains induced by mediated nucleation in additively manufactured Inconel 718

I Cazic[1,2], **M Engstler**[3], **J Zollinger**[2], **J Ghanbaja**[2], **M El Kandaoui**[1] and **B Appolaire**[2]

[1] Institut de Soudure, 4 bd Henri Becquerel, Espace Cormontaigne, Yutz, France
[2] Institut Jean Lamour, Université de Lorraine, Campus ARTEM, Allée André Guinier, F-54011 Nancy, France
[3] Universität des Saarlandes, Department of Materials Science and Materials Engineering, Saarbrücken Campus, Germany

Email: julien.zollinger@univ-lorraine.fr

Abstract. In additive manufacturing of Ni-based alloy IN718, what causes Icosahedral Short-Range Order (ISRO) or iQC mediated nucleation of equiaxed grains is still not understood. In order to have a better comprehension of the microstructure formation, a more comprehensive characterization both on grain structure and chemical heterogeneities must be achieved. In this work, a scanning electron microscope equipped with a focused ion beam (FIB) allowed to perform a 3D EDS chemical analysis and EBSD orientation mapping of the volume in an equiaxed grain zone. After the 3D reconstruction of the equiaxed grains has been performed, grain-to-grain orientation analysis allowed to analyze multiple twin boundaries orientation relationships in a volume containing hundreds of grains. Chemical mapping using EDS allowed to identify TiC inclusions from which a five-fold symmetry axis seem to emerge, indicating a potential nucleation site. Such areas were analyzed in transmission electron microscopy (TEM) to investigate the different present phases. The nucleation sequence and potential origin of mediated nucleation are further discussed, opening new prospects for grain refinement in additively manufactured nickel-based alloys.

1. Introduction

Ni-based alloy Inconel 718 has attracted many interests in additive manufacturing (AM) due to its good processability and high tensile properties up to 650 °C [1]. New challenges such as grain refinement have led to reach for columnar to equiaxed grain transition through controlling the processing conditions [2]. Equiaxed grains have been observed for many processes and different energies in literature [3]. Recently, even finer equiaxed grain zones have been encountered in processed Inconel 718 using coaxial laser and wire technology (LMD-W) [4]. The presence of such equiaxed grains when using LMD-W was demonstrated previously to originate from Icosahedral Short-Range Order (ISRO) mediated nucleation mechanism; through grain-to-grain orientation analysis using electronic backscattered diffraction (EBSD) that allows to identify multiple-twinning relationships between grains sharing common ⟨110⟩ directions that display five-fold symmetries [3]. So far, ISRO mediated nucleation mechanism inducing equiaxed grain growth in Inconel 718 is not explained by a chemical addition to the alloy; but is rather attributed to metastable short-range configurations in the liquid due to rapid remelting during processing [3]. Nevertheless, lack of results on heterogeneous ordering in the liquid for this alloy; and especially in the coaxial laser wire process conditions, does not permit to conclude on what exactly causes ISRO in this case. The purpose of this paper is to give new elements for the understanding of the ISRO origin, by characterizing the equiaxed microstructure of the as-built alloy.

2. Experimental methods and analysis

Samples of 150*30*10 mm were manufactured using Inconel 718 commercially available welding wire of 1.2 mm diameter in an inert atmosphere with a Precitec CoaxPrinter laser head fixed on an ABB IRB 6640-185 robot. The laser source is an IPG 10 kW Yb^{3+} - doped fibre solid-state laser with an operating wavelength of 1070 nm, providing a continuous laser beam. The bidirectional scanning strategy is shown in Figure 1, which consist of alternating the scanning direction along x axis at each layer. The increment along z axis between layers remains constant and is fixed to 0.8 mm. The laser power was fixed to 2000 W, the advance speed to 1.2 m/min and wire feeding rate to 2 m/min. The samples were cut in the x-z plane, z being the building direction and y the laser scanning direction, as shown in Figure 1.

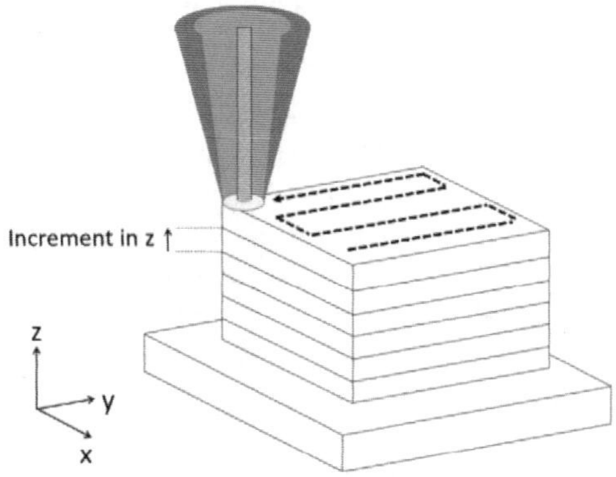

Figure 1. Scanning strategy during processing of the alloy

For this study, a focused ion beam scanning electron microscope Thermo Fisher Helios Xe-PFIB equipped with an EDAX Velocity EBSD camera and EDAX Octane EDS system was used. Serial sections of 100 nm were taken in a volume of 60*60*60 µm in the equiaxed grain zone. At each slice, both SE and BSE images were taken. Every second slice, an EDS and EBSD 2D mapping with 300*300-pixel resolution was realized at 20 kV, in total of 300 EDS and EBSD images. From these images, the 3D volume can be reconstructed, and Figure 2 (a) displays the 3D EBSD maps, that allows to identify each grain orientation of the sample. The Figure 2 (b) exhibits the population of Titanium carbides (TiC) identified by EDS mapping in the volume, which volume fraction is 0.8%.

Figure 2. 3D reconstruction of (a) EBSD unique grain colouring mapping of grains (b) EDS mapping of Titanium carbides in red colour.

TEM characterization was performed using a Jeol ARM200F FEG corrected in source and image, with a resolution in Scanning TEM (STEM) with a voltage of 200 kV. The microscope is equipped with both EDS detector and an Electron Energy loss spectroscope (EELS). Thin foils were extracted in equiaxed grain regions and polished using a Helios Nanolab 600i electronic microscope using FIB technique. SE and BSE imaging for characterizing the microstructure for searching extraction zones were performed using a Zeiss Gemini SEM500 equipped with an InLens detector.

3. Results and Discussion

3.1 Characterization and analysis of the equiaxed grains

One of the 2D EBSD images is shown in Figure 3 (a) where five grains present multiple-twin relationships. The corresponding ⟨110⟩ pole figure of the grains 1-2-3-4-5 is displayed in Figure 3 (b), that shows the common direction ⟨110⟩ of all five grains (circle in red), where the red arcs represent the common {111} planes between each pair of grains. The common {111} plane of each pair of grain is rotated by 70.5°, which is compatible with a 5-fold of a decahedron and was demonstrated previously to be a 5-fold symmetry axis [3]. The grains 1-5 present a rotation around the common direction ⟨110⟩ of 7 to 8°, which was already reported as a near-twin [3-5].

From the identified 5-fold symmetry axis, the grains 1-2-3-4-5 and the twin grain boundaries are represented in Figure 4 (a) and (b) in 3D. The figure 4 (c) shows that the 5-fold symmetry axis is connected to the surface of the TiC. In Figure 4 (d), the grains 1-2-3-4-5 are connected to the surface of the TiC inclusion as well and it seem they have grown up from the surface of the inclusion. The presence of TiC (fcc) can act as a nucleation site for various alloys [6], yet its relationship with the 5-fold symmetry axis of the Ni fcc grains is not possible, due to its NaCl structure type. The presence of an intermediary phase on the surface of the TiC that allows the fcc grains to nucleate and grow respecting the 5-fold symmetry is a more realistic hypothesis. In this case, an intermediary phase or a quasicrystal that would possess polygons in its structure such as icosahedron could be present and allow ISRO mediated nucleation in local clusters during rapid melting and re-melting conditions. This result requires more characterization of the TiC interface with the surrounding grains to find what population of phases is present.

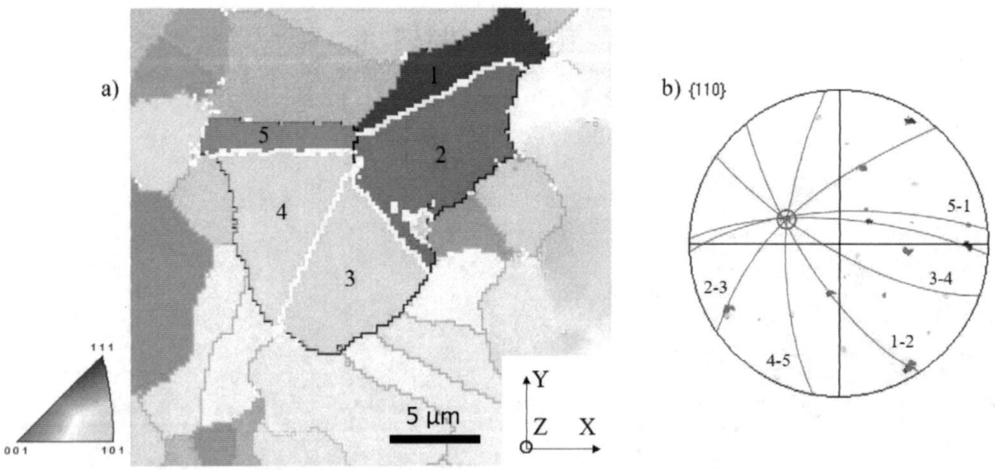

Figure 3. (a) 2D EBSD IPF (Z) map of selected 5 nearest-neighbour grains (1-2-3-4-5) with multiple-twin orientation relationships, taken from the equiaxed zone. Twin grain boundaries are displayed in white. (b) Pole figure ⟨110⟩ of grains from 1 to 5 sharing a common ⟨110⟩ direction, circled in red.

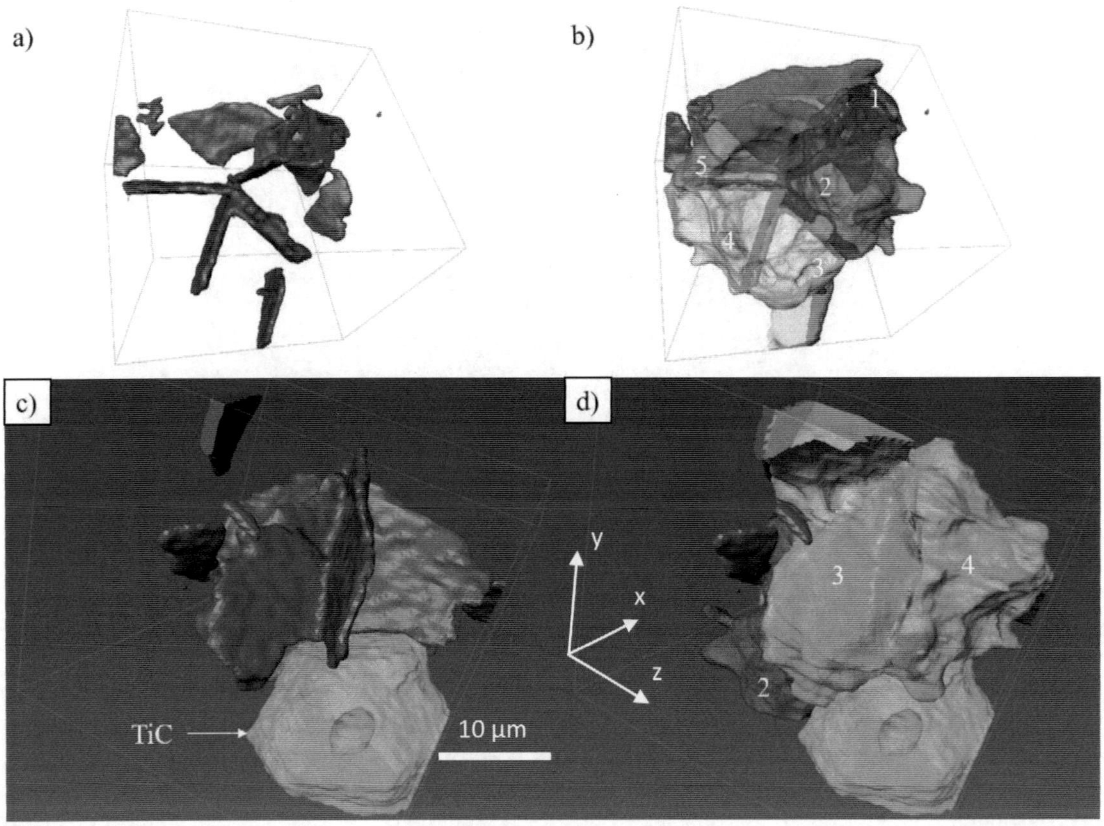

Figure 4. 3D representation of (a) twin grain boundaries constructing the identified 5-fold symmetry axis, in dark grey (b) grains 1-2-3-4-5 with their corresponding colours and twins in between (c) twin boundaries originating from a TiC particle, with an Al oxide at the centre of the particle (d) grains 1-2-3-4-5 connected to the TiC.

3.2 Investigation of phases around TiC particles

The Figure 5 shows an SE image using an InLens where a TiC inclusion identified by EDS is present in an equiaxed grain region surrounded by various growing phases. The oxide at the centre of this particle is enriched in Al and Mg but its structure could not be determined only by EDS. Nano sized precipitates were also observed being spread out randomly in the matrix, with undetermined chemical composition. Such population of phases were not observed in BSE or SE electronic imaging modes, nor observed in literature, due to the spot size and high voltage using BSE or SE classic detectors. To investigate at a lower scale for understanding what phases are present, transmission microscopy (TEM) is a more precise characterization tool.

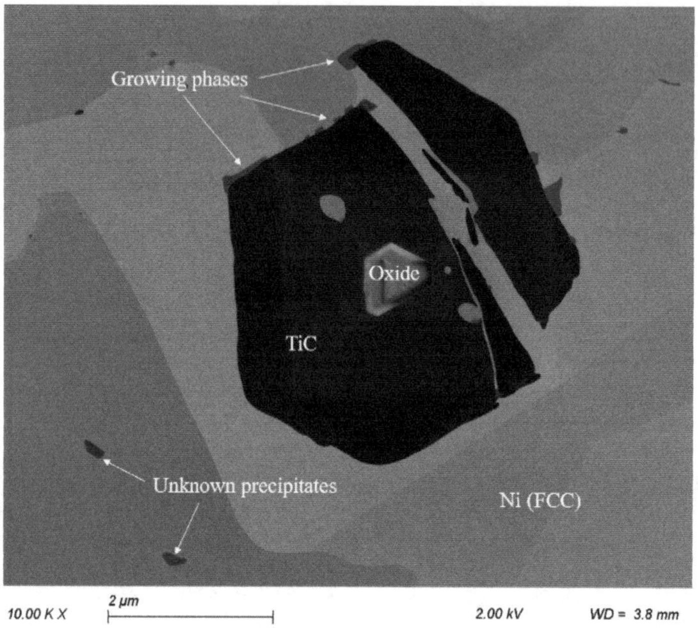

Figure 5. InLens SE imaging of a TiC in the equiaxed grain region. Various unknown phases are growing on the surface of the TiC particle that is cracked. Some undetermined nano sized precipitates are present randomly in the matrix.

The Figure 6 displays a TEM micrograph of the microstructure surrounding a TiC inclusion in the equiaxed region. Each of the phases reported in Figure 6 were identified with diffraction patterns and chemical composition using EDS. The TiC has nucleated on an Al_2MgO_4 oxide, and various phases seem to nucleate and grow on the surface of the TiC particle, such as NbC carbides and precipitates (~200 nm) that are in darker contrast in Figure 6. NbC growth on TiC has been already reported [6] and is expected due to similar lattice parameters. The resulting diffraction pattern of one of such precipitates also shows presence of the matrix, that is traduced by more intense and spaced diffraction spots. The smaller size of the diffraction network of the precipitate indicates that the lattice parameter of the precipitate is much higher than the matrix. The zone axis is not yet determined, and surprisingly the precipitate contains a close to Laves phase Ni_2Nb C14 chemical composition (% at.) such as 18.76 % Nb - 45.22 % Ni - 15.35 % Fe - 18.38 % Cr - 2.29 % Ti - 0.68 % Al content. Smaller nano-precipitates (~20 nm) were observed but are not reported in this study due to their size, with notable presence of 6.62 % of Mg, also at the interface between the TiC and the matrix.

These observations show chemical heterogenous zones during solidification around the TiC particles in the molten bath acting as nucleation sites for potentially unexpected phases where local equilibrium allows stable conditions for growth. The link between these phases on TiC with ISRO mediated nucleation is still not clear since such phases could dissolve and promote local nucleation and growth of the fcc phase and needs to be evaluated.

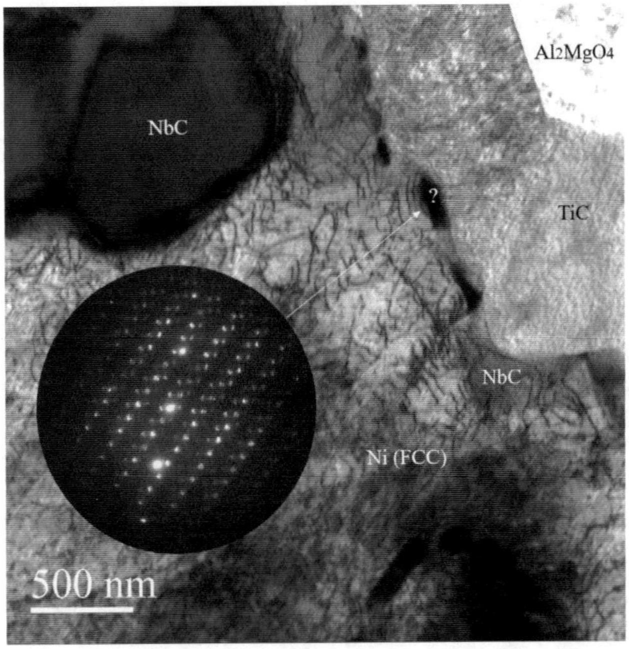

Figure 6. TEM micrograph of a TiC particle and the Ni matrix. An oxide Al2MgO4 is present at the centre of the TiC particle. Niobium carbides (NbC) are present in the matrix and on the surface of the TiC. Various unknown precipitates (dark contrast) are present at the surface of the TiC inclusion, with a diffraction pattern of an indicated unknown precipitate.

4. Conclusion

The use of 3D-EBSD and EDS characterization permitted to identify a 5-fold symmetry axis between nearest-neighbour grains and was found to originate from a TiC particle surface. Presence of various phases are reported using TEM characterization and undetermined precipitates on the TiC surface. The formation of such phases with high lattice parameter is the experimental proof of chemical heterogeneous zones in the liquid around the TiC particle impacting equiaxed grain growth. The TiC inclusions characterized in the equiaxed grain region seem to play a role in the formation of various phases that could potentially impact the ISRO mediated nucleation phenomenon and therefore grain refinement in Inconel 718. More in-depth investigation is ongoing to determine the crystalline structure of the present precipitates and their link to grain refinement.

Acknowledgements

This work was funded by INTERREG VA “Grande Région” funding program “FAFil” No. 050-4-08-126 and by the ANRT France under the CIFRE Ph.D. fellowship 2018-1630.

References

[1] Baufeld, B. 2012 *J. Mater. Eng. Perform.* 21, no 7, p. 1416-1421
[2] Dehoff, R. R. 2015 *Materials Science and Technology* 31, no 8, p. 931-938
[3] Cazic I, Zollinger J, Mathieu S, El Kandaoui M, Plapper P and Appolaire B. 2021 *Scripta Materialia* 195, p. 113740
[4] Bambach M, Sizova I, Kies F and Haase C 2021 *Additive Manufacturing* 47, p. 102269
[5] Kurtuldu G, Jarry P, Rappaz M 2013 *Acta Materialia* 61, 7098-7108
[6] Shen M Y, Tian X J, Liu D, Tang H B, Cheng X 2018 *Journal of Alloys and Compounds* 734, 188-195

Twinned dendrites in CrB-structured intermetallic NiGd

T Niersbach*, **M Kolbe, M Becker and F Kargl**

Institut für Materialphysik im Weltraum, Deutsches Zentrum für Luft- und Raumfahrt (DLR), 51170 Köln, Germany
* Corresponding author: till.niersbach@dlr.de

Abstract. The formation of twinned dendrites was investigated in orthorhombic CrB-structured NiGd. A NiGd droplet embedded in borosilicate glass-flux was processed by differential scanning calorimetry. A small undercooling of ≈ 20 °C was reached and a cross-section of the solidified sample was analyzed with EBSD. A twinned dendrite is found, that exhibits a twin boundary in its center, both parts sharing a <110> direction and a misorientation angle of 40° between its two parts. The present findings are in agreement with model predictions for CrB-structured alloy solidification based on the investigation of deeply undercooled and homogeneously nucleating NiZr. The present investigation on NiGd serves as a first important step to validate the model's predictions. Due to the observation of twinned dendrites at small undercoolings, the present study may further serve as a first-step to extend the model to heterogeneous growth regimes.

1. Introduction

Twinned dendrites are a common growth feature observed in different fcc-alloys [1,2,3] as well as in intermetallic phases with orthorhombic CrB-type structure such as NiZr [4]. Twinned dendrites are characterized by a common <110> growth direction split in their center by a twin plane and usually have a growth competition advantage over non-twinned dendrites [5]. For both types of alloys, the growth of twinned dendrites has been associated with the presence of a quasicrystalline nucleation core that either serves as template for epitaxial growth of fcc grains or directly as the center of nucleation. In the latter case, a homogeneously nucleated quasicrystal, based on an icosahedron, was proposed to be the starting point for the continuous growth of NiZr into a ten-fold twinned dendritic microstructure. Quasicrystalline nucleation is common in metallic melts [6] and Kobold et al. showed that at undercoolings of $\Delta T = 300$ K the system solidifies from homogeneous nucleation [7]. Evidence of the ten-fold symmetry was obtained through the *in-situ* observation of a decagonally shaped heat front propagation over the surface of an electrostatically levitated NiZr droplet, and subsequent microstructural analysis revealing ten domains of twinned dendrites.

Based on these observations the Hornfeck-Kobold-Kolbe (HKK) growth model was proposed that connects a quasicrystalline nucleation with a multiple twinned dendritic growth structure [4]. The growth features a primary dendrite that follows the <001> growth direction and ten secondary dendrites that span orthogonally to the primary dendrite along the <110> direction. Due to high growth velocities and low nucleation frequencies in highly undercooled melts, the dendrites continued to grow through the whole sample volume [8]. The center of each dendrite consists of a <110> twin boundary so that each grain is rotated to its neighbors by 36°. This angle corresponds to twice the angle of the face diagonal of the NiZr unit cell's (001), the (a-b) plane:

$$\varphi = 2 * \arctan\frac{a}{b} (\approx 36° \text{ for NiZr}) \; [8]$$

with a and b being the lattice parameters. The ten-fold structure in the NiZr system is therefore based on the lattice parameter ratio a/b of the CrB-type unit cell of NiZr. The orthorhombic CrB-structure composes eight atoms in a base centered unit cell (space group type Cmcm, No. 63). Most important is the fact, that the unit cell structure extends across the twin boundary, so that for an atom there will be low energetic difference between a position in the bulk crystal or near the twin boundary. The particular lattice parameters differ in each alloy of this structure type and therefore suggest that the proposed mechanism is possible in other CrB-type alloys as well, but with a different rotation angle φ. With the focus on the NiGd intermetallic alloy, the lattice parameters for this system (a = 0.3779(3) nm, b = 1.0367(5) nm, c = 0.4213(4) nm [9]) result in an expected angle of 40° between two neighboring grains. A nine-fold symmetry would therefore be expected macroscopically, as well as dendritic structures that exhibit this angle in their central twinning boundary. $Ni_{50}Gd_{50}$ is a congruently melting (T_m = 965 °C) intermetallic phase which is stable upon solidification down to room temperature [10]. To test whether the HKK growth mechanism is also possible in the NiGd system, it was examined for the presence of twinned dendrites. The finding of twinned dendrites would be an important observation, as it is a prerequisite for the possibility of the discussed growth mechanism in CrB-structured alloys.

2. Experimental

A spherical sample with a diameter of 3 mm was alloyed by arc-melting high purity pieces of elemental Ni (99.995 %, Alfa Aesar) and Gd (99.999 %, Alfa Aesar) under a protective argon atmosphere. The elements themselves, as well as the arc-melting facility, were handled within a glove box to reduce the exposure to oxygen as much as possible. Oxygen and water contents inside the glove box were maintained below 0.1 ppm. The sample was molten three times to ensure a homogeneous composition. The mass loss from arc-melting was found to be negligible for this system. The composition of the here used sample differs from the intermetallic $Ni_{50}Gd_{50}$ composition. According to the phase diagram [10], the NiGd phase solidifies first for the chosen composition of $Ni_{44}Gd_{56}$, followed by the eutectic reaction NiGd-$NiGd_3$ at a temperature of 704 °C. The final solidification at the eutectic temperature increases the visibility of the primary NiGd dendrites using microscopy methods.

Power differential scanning calorimetry (DSC) was used in this work to melt and solidify the sample and to detect any possible unknown phase transitions. With the intention to achieve a high undercooling of the melt, the alumina crucible was filled with borosilicate glass shards in the center of which the sample was positioned. It was ensured that the sample had no direct contact to the crucible. The glass fluxing ideally creates a melt embedding for the sample upon melting and prevents or at least reduces oxidization and reactions with the crucible material. Additionally, the glass embedding provokes diffusion of remaining oxides from the sample into the glass. For the measurement a reference crucible was placed thermally isolated next to the crucible with the sample. The mass of both crucibles and the content is known, and a temperature profile is determined. The electrical power that is needed to keep both crucibles at the same temperature is plotted over the actual temperature. Positive values for the required power indicate endothermic reactions and negative values indicate to exothermic reactions.

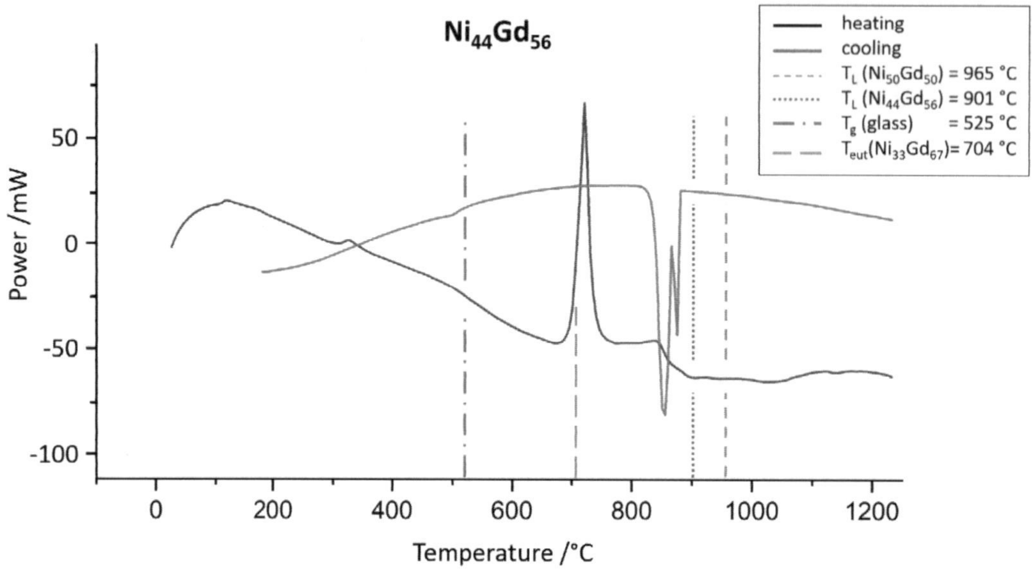

Figure 1. Required power, to keep both crucibles at the same temperature, plotted over the predetermined temperature range of the DSC experiment. Visible are the heating and the cooling curve for $Ni_{44}Gd_{56}$. The glass transition temperature of borosilicate glass, and the liquidus temperatures of NiGd and $Ni_{44}Gd_{56}$, as well as the eutectic temperature are marked.

Figure 1 shows the required power over the temperature range to keep both crucibles at the same temperature. The high spike of the heating curve at around 700 °C marks the beginning of the melting process and corresponds to the eutectic temperature of 704 °C. The cooling curve shows a high negative double spike beginning sharply at 880 °C. This would imply an undercooling of 21 °C relative to a liquidus temperature for $Ni_{44}Gd_{56}$ of 901 °C. However, different than expected, the microstructure analysis revealed the existence of the two phases NiGd and Ni_2Gd. The presence of the Ni_2Gd phase indicates a change of composition into a Ni-rich alloy. A reasonable explanation for this change is a reaction of the sample with the borosilicate glass that led to a diffusive flow of Gd into the glass. Taking such a compositional change into account the undercooling has to be reevaluated. The expected composition lies between $Ni_{50}Gd_{50}$ and $Ni_{53}Gd_{47}$. The eutectic temperature for $Ni_{53}Gd_{47}$ is 899 °C. Despite the compositional change, a small undercooling between 20 °C ($Ni_{53}Gd_{47}$) and 85 °C ($Ni_{50}Gd_{50}$) can be expected. Contrary to expectations, this graph shows an exothermic baseline trend during heating and an endothermic baseline trend during cooling. This could be traced back to the glass embedding, as experiments, where the crucible was filled only with glass, showed this behavior as well. The exact cause for it remains unknown.

After the experiment the sample was prepared for the microstructure analysis. It was embedded and a cross-section was polished with 0.05 µm alumina suspension and finalized with a basic 0.05 µm SiO_2 suspension. The sample was analyzed with a Leo 1530 VP scanning electron microscope. Microscopy images were taken with a forward scatter detector (FSD). Electron backscatter diffraction (EBSD) scans were executed with a Nordlys 2 detector from Oxford hkl and analyzed with the Channel 5 software.

3. Results and Discussion

Figure 2 shows an FSD image of a section of the processed sample. It shows the microstructure in this sample separated into the NiGd dendritic structure and a second interdendritic Ni_2Gd phase. The yellow rectangle highlights a twinned dendrite, which is examined closer in figure 3 using EBSD analysis. Figure 3 shows the dendrite with its <110>, <100>, <010> and <001> inverse pole figures (IPF). In the EBSD image the Ni_2Gd phase (light gray) and another, differently orientated NiGd grain (dark gray) are colored dark to emphasize the twinned dendrite. The dendrite is split in its center by a straight grain boundary. It has a misorientation angle of approximately 40° around the <001> axis and is highlighted in red (misorientation angle: 38° < red < 42°). This is visualized as well in the graph in the lower left of figure 3. From the IPF plots it can be seen that both sides of the dendrite, purple and light green, have the same <001> direction, as both share the same spot in the IPF plot. Their respective <100> and <010> directions are rotated to each other by 40°. Since the upper hemisphere of the pole figure makes up 180°, the <100> directions seem to be rotated by 140° to each other (180° - 40° = 140°). The <110> inverse pole figure additionally proves that both grains share a common <110> direction. This is the $[\bar{1}10]_p$ direction for the purple colored grain and the $[110]_g$ direction for the green colored grain. This shared direction is highlighted by an orange circle. Both grains' <010> directions are inclined by 20° to their respective side, with respect to the <110> grain boundary direction. These results match with the observations in the NiZr system that led to the HKK-model.

It can be observed in figure 2 that the secondary arms of the dendrite are symmetrically on both sides of the twin boundary. Therefore, it can be reasonably assumed that the grain boundary plane is nearly orthogonal to the viewing direction. The red line of the grain boundary in the EBSD image has been transferred to the stereographic projection of the IPF plot (upper right in figure 3, red dotted line). Basing on this red dotted line, a plane nearly orthogonal to the viewing direction has been selected (gray dashed line). We observed that the $[\bar{1}10]_p$, $[110]_g$ and the common [001] lie within this plane. A precise determination of the geometrical orientation of the grain boundary is difficult. A well-known method is to remove the surface of the sample and follow the boundary into the bulk of the sample. After gentle polishing and removing a layer of about 10 μm, we concluded that the above described plane is a good selection. Further experiments have to prove this result.

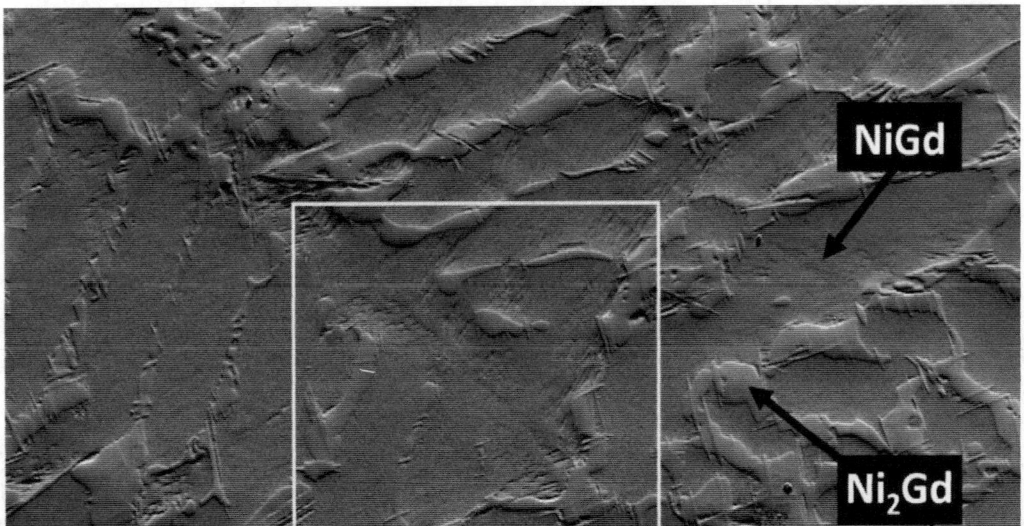

Figure 2. Forward scatter detector image taken from the Ni-Gd sample after DSC processing. It can be seen that two phases are present, separated from each other. The dendritic structure consists of the intermetallic NiGd phase and is enclosed by the interdendritic Ni_2Gd phase. Marked in yellow is the area that was further investigated by EBSD in figure 3.

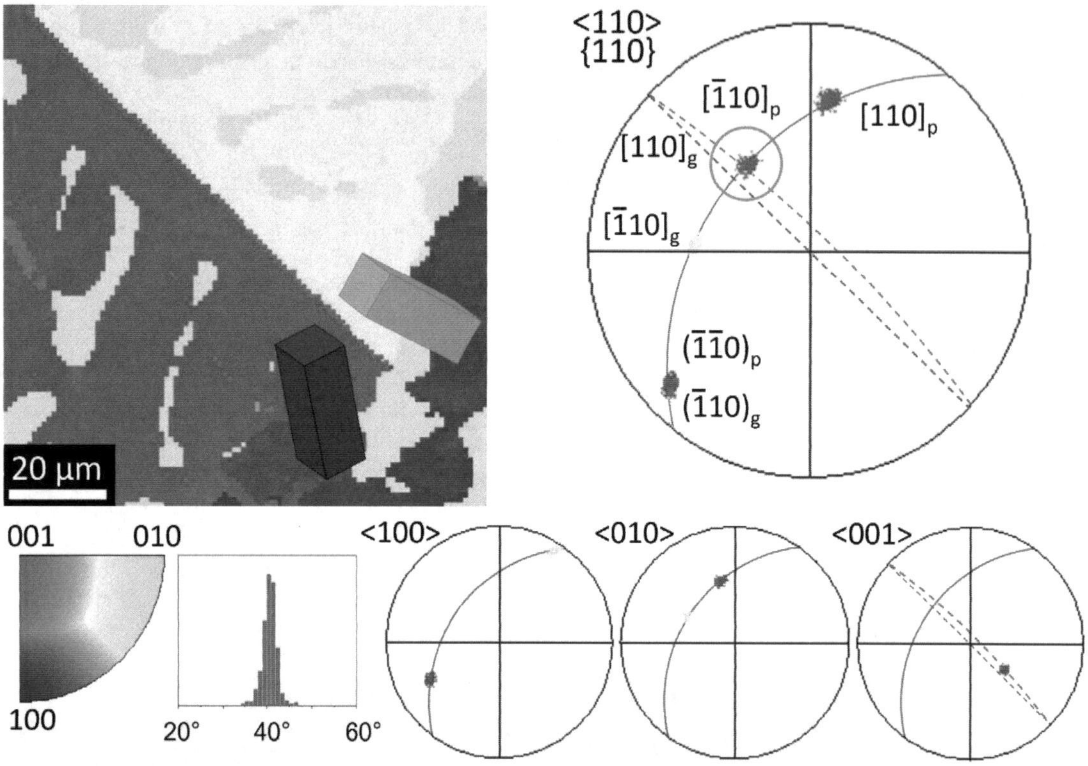

Figure 3. EBSD image of the twinned dendrite from figure 2 with the respective IPF plots and a qualitative mapping of the grain boundary angle. The grain orientation is mapped in the depicted IPF color scheme. The dendrite itself is separated in its central line by a 40° grain boundary, as marked in red in the EBSD image and shown in the mapping in the lower left. Both dendrite parts exhibit the same <001> direction and are rotated by 40° around the <001> axis, which can be directly seen in the <010> IPF plot. Moreover, both grains share one <110> direction (marked by the orange circle as the data points overlap each other). The dotted red line is the projection of the grain boundary visible in the EBSD image and the gray dashed line indicates the twin boundary plane that corresponds to the $(\bar{1}\bar{1}0)_p$ and $(\bar{1}10)_g$ planes of the purple and green grain, respectively. The Ni_2Gd phase is pictured in light gray. The dark gray area shows a differently orientated NiGd grain.

4. Conclusion

In the present work a twinned dendrite in the NiGd system was analyzed using EBSD. The dendrite is split in its center by a straight grain boundary. The twin relationship is characterized by a 40° rotation around the common <001> axis. As a consequence, the twin grains also share one common <110> direction. The grain boundary plane is spanned by the <110> and [001] direction common to both grains. The two corresponding <010> directions enclose the common <110> direction and are inclined by 20° to both sides. The angle of 20° is also present in the NiGd unit cell, and it corresponds to the face diagonal of the (001) (a-b) plane. A twin boundary construction on the basis of the unit cell geometry is therefore possible and is in line with the microscopic observations. This type of twinned dendrite was analyzed to be a requirement for the Hornfeck-Kobold-Kolbe model for CrB-type alloys [4]. Our findings show that the proposed quasicrystalline growth mechanism might be also possible in NiGd. In contrast to the observations made in the NiZr system, on which the HKK-model is based, only low undercooling was reached for NiGd in this work. It may suggest that the model could also be valid for the case of heterogeneous nucleation. Further studies with samples at higher undercoolings are planned to look for other evidence, such as multiple twinning events, to better understand the nucleation and growth mechanism in NiGd.

References

[1] Henry S, Rappaz M and Jarry P 1998 *Metall. Mater. Trans. A* **29** 2807–17
[2] Salgado-Ordorica M A and Rappaz M 2008 *Acta Mater.* **56** 5708–18
[3] Kurtuldu G, Sicco A and Rappaz M 2014 *Acta Mater.* **70** 240–248
[4] Hornfeck W, Kobold R, Kolbe M, Conrad M and Herlach D 2018 *Nat Commun* **9** 4054
[5] Salgado-Ordorica M, Phillion A and Rappaz M 2013 *Metall. Mater Trans A* **44** 2699–2706
[6] Schenk T, Holland-Moritz D, Simonet V, Bellissent R and Herlach D 2002 *Phys. Rev. Lett.* **89** 075507
[7] Kobold R, Kolbe M, Hornfeck W and Herlach D 2018 *J. Chem. Phys.* **148** 114502
[8] Kobold R 2016 *Crystal Growth in Undercooled Melts of Glass Forming Zr-based Alloys* (Dissertation, Ruhr-Universität Bochum)
[9] Romaka V V, Romaka L, Horyn A, Rogi P, Stadnyk Y, Melnychenko N, Orlovskyy M and Krayovskyy V, 2016 *J. Sol. State Chem.* **239** 145–152
[10] Xu G, Cui Y, Fei H, Zhang L, Zheng F, Liu L and Jin Z 2012 *Int. J. Mater. Res.* **103** 1179-87

Acknowledgments

The work was partially funded by ESA (European Space Agency) within ESA MAP LIPHASE contract no. 4200015005/01/NL/SH. We gratefully thank Nuria Navarrete Argiles and Carolina Villada Vargas from the Institut für Materialphysik im Weltraum, DLR, for handling the DSC.

Characterisations of the Al-Mn intermetallic phases formed under pulse magnetic fields solidification

K Xiang[1, *], S Huang[1] and J Mi[1]

[1] *Department of Engineering, University of Hull, East Yorkshire, HU6 7RX, UK*
* Corresponding author: k.xiang-2021@hull.ac.uk

Abstract. In this research, the pulse electromagnetic fields were applied during the solidification of an Al-8Mn alloy. The changes of the microstructures due to different magnetic pulses applied were characterised using scanning electron microscopy, energy dispersive X-ray spectroscopy and X-ray tomography. The results show that without the magnetic field, the Al_4Mn phases are thin plates in 3D space. At a magnetic field of 0.8T, the Al_4Mn phases are transformed into a columnar morphology companied by fine and fragmented Al_6Mn particles. At the magnetic flux density of 1.2T, the size of Al_4Mn phases become more refined with their orientation gradually aligned towards the direction of the magnetic field, revealing that the pulse electromagnetic fields have a strong effect on the evolution of the Al-Mn phases in Al-8Mn alloy under solidification.

1. Introduction

In metal alloys, metallic compounds (i.e., intermetallic phases) are an essential structural constituent for strengthening. Controlling the nucleation and growth dynamics of intermetallic phases to achieve the desired structure, size, morphology, and distribution has been always one of the important research directions in materials science and engineering disciplines [1]. In addition to the conventional chemical methods, i.e., adding different chemical elements into an alloy to tailor the intermetallic phases, physical based external field methods, e.g. electrical current, magnetic fields, ultrasonic waves, etc [1- 4] are also very effective in term of controlling the growth dynamics of different types of intermetallic phases through the solidification processes. Most external field methods avoid the need of using other specific chemicals, hence they are generic and applicable to most alloy systems, offering a certain degree of environmental benefit [1, 4, 5]. For example, Manuwong et al. [2] studied the solidification of Al alloys under electromagnetic pulses and found the Lorentz force generated inside the melts could disrupt the growth of Al dendrites and Al_2Cu phases, demonstrating a great application potential in industry.

Al-based alloys are widely used in automotive, buildings, transportation, and packaging industry due to their good mechanical property, excellent corrosion resistance, formability, and recyclability [6]. Mn is an alloying element in Al alloys. Adding 1.0 to 1.5 wt% into an Al alloy is a common practice to provide strength and toughness [7]. The common Al-Mn intermetallic compounds are Al_6Mn, Al_4Mn, and $Al_{11}Mn_4$ (especially when the Mn content is higher than 1.6 wt%). They are often brittle intermetallics, initiating cracks and deteriorating the toughness [8, 9]. The damaging effects are closely linked to the structure, size, morphology, and distribution of these Al-Mn intermetallic compounds [10, 11]. Thus, many attempts have been made to optimise their structure and geometrical characteristics [12, 13]. To date, however, there has been a very limited report on how pulse electromagnetic fields can be used to control and manipulate the microstructural evolution of different types of those Al-Mn intermetallics during the solidification processes. In this work, we used an Al-8Mn(wt%) alloy as the experimental alloy, and a bespoke PEMF apparatus to generate magnetic fields of different magnetic flux densities inside the alloys during solidification. We then used Scanning Electron Microscopy (SEM), Energy Dispersive X-ray Spectroscopy (EDS), and X-ray tomography techniques to characterize the resultant microstructures, revealing the correlation between the change of size/morphologies and the applied magnetic pulses.

2. Experiment

An Al-8Mn(wt%) alloy was used in the experiment. It was made by melting pure Al (99.96 wt%) and pure Mn (99.85 wt%) together using an electrical resistance furnace. Firstly, a pure Al ingot of 230 grams was melted at ~700 °C for 2 hours inside an alumina crucible (the crucible inner surface was coated with a boron nitride spray). Then, fine particles (1-3 μm in diameter, 20 grams) of pure Mn were added into the melt and held at ~850 °C for 7 hours. During the melting operation, stirring of the melt was carried out every half hour to homogenize the melt temperature and speed up the diffusion of the solid Mn particles into the Al melt until all solid Mn was completely dissolved into the Al melt and a homogenous Al-8Mn was achieved. Finally, the melt was drawn uphill into a quartz tube (Φ 10 mm × 150 mm) connected to a counter-gravity casting apparatus [14] to solidify as cylindrical cast bars. The negative pressure applied was 7 KPa.

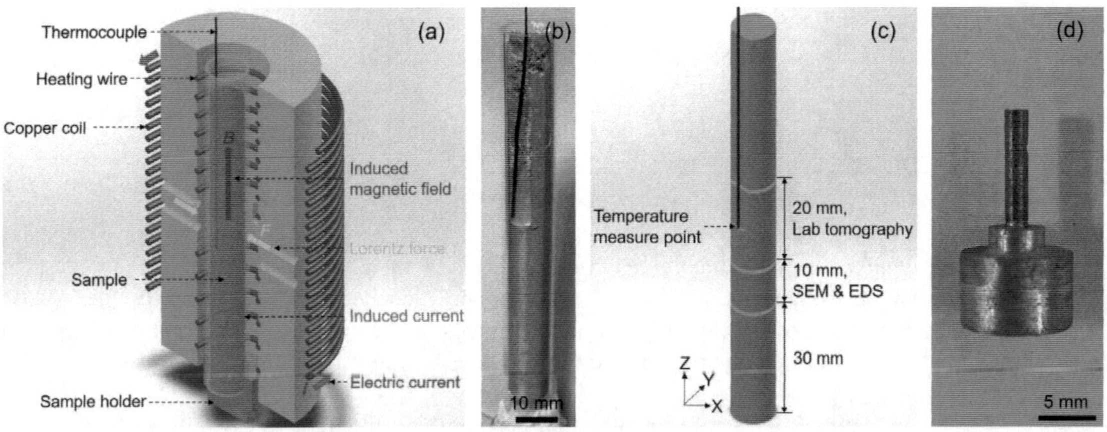

Figure 1 (a) A schematic of the pulse electromagnetic filed apparatus. (b) The experimental sample state before PEMF process. (c) A schematic of the sample state after PEMF process. (d) The sample machined into geometry for tomography scan.

Before the PEMF process, these cast bars were machined into a certain size (Φ 9 mm × 100 mm) until they can fit into the new tubes. Then, the round bar surface (5cm away from the top) was filed off for a few mm to allow a K-type thermocouple to be placed between the sample and the quartz tube as shown in figure 1b. Finally, the tube was put into the centre of the electric resistance furnaces inside the copper coil as shown in figure 1a, and heated up at ~800 °C for 30 mins to melt the alloy completely. When the furnace was turned off, different discharging voltages with a pulse at a frequency of 1 Hz were applied to create a PEMF inside the melt continuously during solidification until the temperature decreased below ~500°C, as shown in figure 2a. Experimental parameters used in the PEMF process are tabulated in Table 1 below, and a more detailed experimental procedure can be found elsewhere [14].

Two cylinders with 20 mm and 10 mm in length respectively were machined off from the cast bar along the Z-axis (figure 1c). Then, the X-Y cross-section of the 10 mm cylinder was polished for SEM observation, and the 20 mm cylinder was further machined into a size of Φ 2 mm × 10 mm for tomography scan, as shown in figure 1d. The 2D morphology and quantitative elements information were analysed by SEM (Cambridge Stereoscan 360 and ZEISS EVO-60) and EDS (Oxford X-MAX 80). Lab-based X-ray tomography was carried out on the samples using an X-Tek HMX 160 μCT scanner and the parameters for tomography acquisition are listed in detail in Table 2. The open-source ImageJ software was applied to pre-process the tomography stacks (including adjusting brightness, contrast, and Trainable Weka Segmentation [16]), and Avizo 3D v2021.1 (FEI, France) was used to perform the final segmentation and visualization.

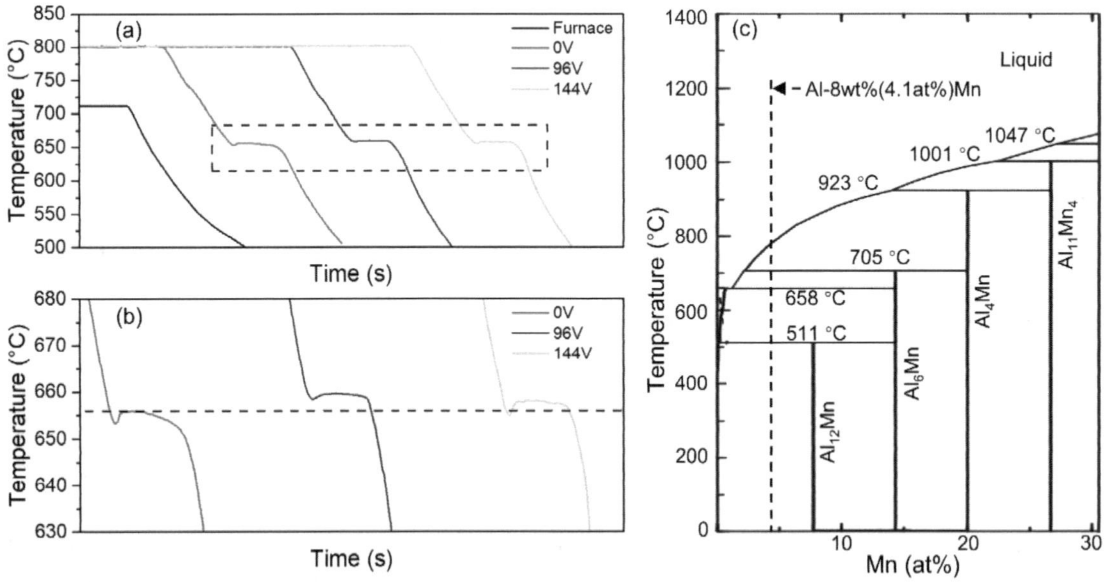

Figure 2 (a) The temperature variation versus time for different samples during the PEMF process. (b) magnified observation of the boxed region in (a) and the dashed line indicates the phase transformation temperature. (c) Local enlarged equilibrium solidification phase diagram of the Al-Mn binary alloy [17] and the dashed line indicates the alloy composition used in this experiment.

Table 1. The experimental parameters of the PEMF process.

Sample	Discharging Voltage (V)	Magnetic Flux Intensity (T)	Phase Transformation Temperature (°C)	Cooling Rate (°C/s)
Al-8Mn-0T	0	0	653.4	0.78
Al-8Mn-0.8T	96	0.8	658.7	0.84
Al-8Mn-1.2T	144	1.2	654.9	0.87

Table 2. The parameters used for the tomography acquisition.

Experimental Parameters	Detailed information
X-ray beam	Polychromatic radiation
X-ray tube energy	135 kV
X-ray tube current	115 μA
Effective magnification	35×
Effective pixel size	4.5 μm
Exposure time	500 ms
Field-of-view	824 × 767
No. of projection	1538

3. Results and Discussion

Figure 2b shows the degree of undercooling recorded in the three samples during solidification. According to the dashed line denoted in the Al-Mn binary alloy phase diagram (figure 2c) [17], the Al_4Mn phases are precipitated from the melt when the temperature drops below the liquidus, followed by peritectic transformation (Liquid + $Al_4Mn \leftrightarrow Al_6Mn$) at 705 °C and eutectic transformation (Liquid $\leftrightarrow$ Al + Al_6Mn) at 658 °C. When the temperature further decreases, peritectoid transformation (Al + $Al_6Mn \leftrightarrow Al_{12}Mn$) occurs at 511°C. However, the $Al_{12}Mn$ phase is meta-stable and tends to transform back to the equilibrium Al_6Mn phase under elevated temperature [13].

The typical 2D morphologies of Al-8Mn alloys under different magnetic flux intensities are presented in figure 3. As displayed in figure 3a, many elongated streaks are formed in the Al matrix when no magnetic field is applied, and most of them almost intersect but don't quite join together, which seems like being cut off or impeded by the other streaks while growing. Their average width and length are measured to be 52.2 μm and 2388.8 μm, respectively. EDS analysis results show that the Mn and Al contents of these streaks were 19.56 and 80.44 at%, respectively, which confirms that they all belong to the primary Al_4Mn phases. Figure 3d is a magnified observation of the boxed region in figure 3a, indicating that these streaks are fairly weak in their nature - the thermal stress inside has caused ruptures and tears throughout aligned towards the longitude of the streaks, significantly decreasing the strength of the entire alloy. Besides, it can also be seen that some tiny clusters with most widths < 1 μm are dispersedly distributed around the Al matrix, which are confirmed to be Al_6Mn eutectic phases by EDS.

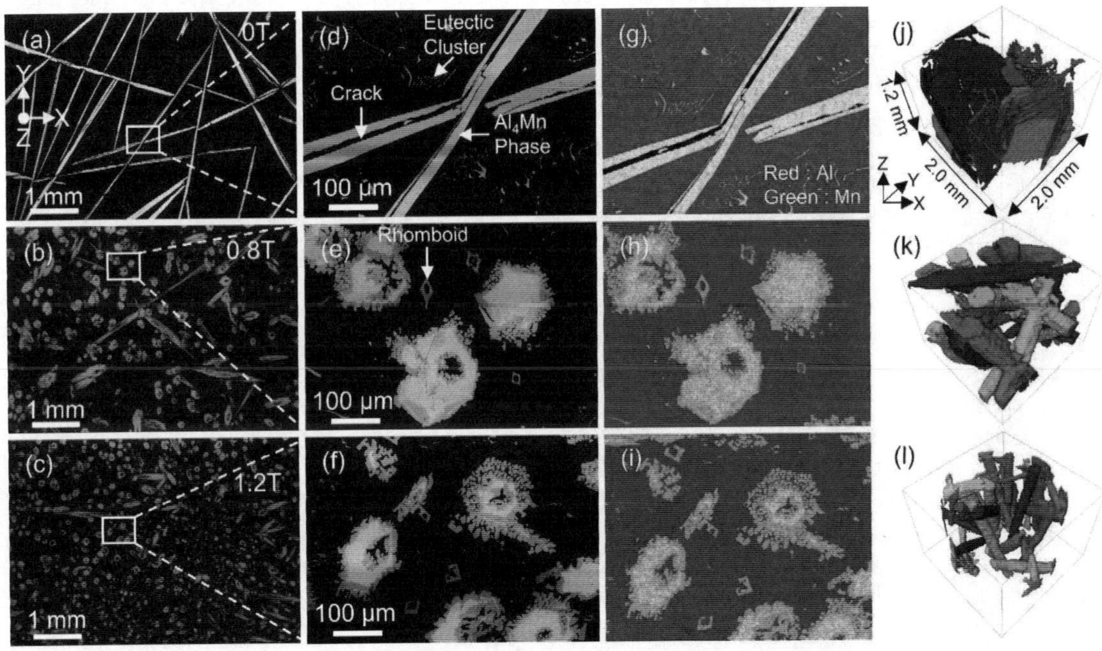

Figure 3 (a), (b) and (c) Typical SEM images for the Al-8Mn-0T, Al-8Mn-0.8T and Al-8Mn-1.2T samples, respectively; (d), (e) and (f) magnified images corresponding to the boxed area in (a), (b), and (c), respectively. (g), (h) and (i) EDS element mapping for Al and Mn elements in the selected regions. (j), (k), and (l) Typical 3D rendering of the Al_4Mn phases in the selected volumes of Al-8Mn-0T, Al-8Mn-0.8T, and Al-8Mn-1.2T samples, respectively. (For interpretation of the references to colour in this figure legend, the reader is referred to the web version of this article.)

When a magnetic field of 0.8T is applied, figure 3b reveals that two distinct morphologies appear in the matrix– short strips and hexagonal crystals with a hole formed inside, showing a similar feature to that found in [9]. The average grain size of the hexagonal crystals is measured to be 165.6 μm. In addition, Figure 3e shows that some edges of the hexagons are fragmented into a mass of fine particles with sizes less than 9.0 μm. With the magnetic flux intensity increasing to 1.2T, these hexagonal crystals are further refined to a grain size of 113.3 μm on average, and these fine particles surrounding the hexagonal crystals show a tendency of epitaxial diffusion growth, demonstrating a remarkable grain refinement. Besides, a careful observation (figures 3e and 3f) shows that these hexagonal crystals have a sandwich type of contrast with bright in the middle and grey around the edge. Combined with the EDS results, it shows that the bright area contains higher Mn content (15.7 at% Mn) and lower Al content (84.3 at%) than that of the grey area (12.8 at% Mn and 87.5 at% Al). Therefore, it indicates that the bright area should be Al_4Mn phases, while the grey area should be Al_6Mn phases formed by the peritectic reaction during solidification. Apart from these, several incomplete or hollow rhomboid crystals (containing 11.8 at% Mn and 88.2 at% Al) are also dispersed inside the matrix, which is similar to the Al_6Mn phases formed under high static magnetic fields found by Li et al [18] and Luo et al [19].

Figure 3j displays that the true 3D structure (morphology) of the Al_4Mn phases in Al-8Mn-0T is a plate, and their lateral lengths and thicknesses are close to the lengths and widths of those streaks observed in the 2D section (figure 3d). In addition, these plates are intersected with each other, causing a rupture during solidification. This observation is similar to that found by Puncreobutr et al [20]. They confirmed that the impingement and branching should be the main interaction mechanism for the Fe-rich intermetallics in Al alloys during solidification. Figures 3k and 3i reveal that after the application of the magnetic fields, the 3D structure of the Al_4Mn phases is no longer a plate-like, but a hollow columnar morphology, which means that those strips and hexagonal grains in figures 4b and c should be the same morphology in 3D characterization, but just different orientations. Besides, it's worth noting that with the increase of magnetic flux intensity, the Al_4Mn phases show an orientational tendency with their longitudinal direction gradually aligned towards the Z-axis, i.e. the direction of the magnetic field. Li et al [19] pointed out that this should be closely related to the magnetocrystalline anisotropy of the Al-Mn phase and the intensity of magnetic flux. The above analyses suggest that the pulse magnetic fields have a strong effect on the solidification evolution of the Al_4Mn phases. Therefore, the future work will be focused on using synchrotron X-ray imaging (radiography and tomography) techniques as we did in other alloy systems in the past few years [21-23] to character in situ the evolution of the Al-Mn phase during solidification under PEMF process.

4. Conclusions

The characteristics of the primary Al_4Mn phases solidified under different magnetic flux intensities were studied mainly using SEM, EDS, and X-ray tomography. Without the magnetic field, the Al_4Mn phases show a plate-like morphology in 3D space. With the application of a magnetic pulse of 0.8T, the Al_4Mn phases are transformed into a columnar morphology, surrounded by fined and fragmental Al_6Mn particles. When the magnetic flux intensity is increased to 1.2T, the size of Al_4Mn phases tend to be further refined to 113.3 μm with their longitudes gradually aligned towards the direction of the magnetic field, revealing that the pulse electromagnetic fields have a strong effect on the evolution of the Al-Mn phase in Al-8Mn alloy under solidification.

Acknowledgements

The author would like to acknowledge the PhD Scholarship awarded by the Chinese Scholarship Council (Funding No. 202108500033 for Mr. Kang Xiang at the University of Hull, UK from Sep 2021 to Sep 2025), Mr. Garry Robinson and Mr. Timothy S Dunstan for support on Electron microscopy analyses, and Ms. Sue Taft for support on X-ray tomography acquisition.

References

[1] Eskin D G and Mi J 2018 Solidification Processing of Metallic Alloys Under External Fields *Springer Series in Materials Science 273, Springer Nature Switzerland AG, Cham Switzerland.*

[2] Manuwong T, Zhang W, Kazinczi P L, Bodey A J, Rau C and Mi J 2015 *Metall. Mater. Trans. A* **46** 2908-15.

[3] Wang B, Tan D, Lee T L, Khong J C, Wang F, Eskin D, Connolley T, Fezzaa K and Mi J 2018 *Acta Mater.* **144** 505-15.

[4] Zhang Z, Wang C, Koe B, Schlepütz C M, Irvine S and Mi J 2021 *Acta Mater.* **209** 116796.

[5] Wang F, Eskin D, Mi J, Connolley T, Lindsay J and Mounib M 2016 *Acta Mater.* **116** 354-63.

[6] Birol Y 2008 *Scripta Mater.* **59** 611-4.

[7] McAlister A J and Murray J L 1987 *J. Phase Equilib.* **8** 438-47.

[8] Chen Z, Hou Y, Xie B and Zhang Q 2020 *Materials* **13** 2388.

[9] Juarez-Islas J A, Warrington D H and Jones H 1989 *J. Mater. Sci.* **24** 2076-80.

[10] Darling K A, Roberts A J, Armstrong L, Kapoor D, Tschopp M A, Kecskes L J and Mathaudhu S N 2014 *Mater. Sci. Eng., A* **589** 57-65.

[11] Kang H, Li X, Su Y, Liu D, Guo J and Fu H 2012 *Intermetallics* **23** 32-8.

[12] Cao C, Chen D, Ren J, Shen J, Meng L and Liu J 2019 *Mater. Lett.* **255** 126535.

[13] Fan Y and Makhlouf M M 2017 *Mater. Sci. Eng., A* **69** 1-7.

[14] Tan D, Lee T L, Khong J C, Connolley T, Fezzaa K and Mi J 2015 *Metall. Mater. Trans. A* **46** 2851-61.

[15] Du W 2018 *PhD Thesis, University of Hull* chapter 3.

[16] Arganda-Carreras I, Kaynig V, Rueden C, Eliceiri KW, Schindelin J, Cardona A and Seung H S 2017 *Bioinformatics* **33** 2424-6.

[17] Asgar-Khan M and Medraj M 2009 *Mater. Trans.* **50** 1113-22.

[18] Li L, Zhao Z, Zuo Y, Zhu Q and Cui J 2013 *J. Mater. Res.* **28** 1567-73.

[19] Luo D, Guo J, Yan Z and Li T 2009 *Rare Metals* **28** 302-8.

[20] Puncreobutr C, Phillion A B, Fife J L, Rockett P, Horsfield A P and Lee P D 2014 *Acta Mater.* **79** 292-303.

[21] Wang F, Eskin D, Mi J, Wang C, Koe B, King A, Reinhard C and Connolley T 2017 *Acta Mater.* **141** 142-53.

[22] Koe B, Abraham C, Bailey C, Greening B, Small M, Connolley T and Mi J 2020 *HardwareX* **7** e00104.

[23] Wang F, Eskin D, Connolley T, Wang C, Koe B, King A, Reinhard C and Mi J *Mater. Lett.* **213** 303-5.

IOP Conf. Series: Materials Science and Engineering 1274 (2023) 012005 doi:10.1088/1757-899X/1274/1/012005

β grain refinement during solidification of Ti-6Al-4V in Wire-Arc Additive Manufacturing (WAAM)

J. R. Kennedy[1,2], A.E Davis[1], A.E. Caballero[3], E.J. Pickering[1], P.B. Prangnell[1]

[1]University of Manchester, Department of Materials, Manchester, United Kingdom
[2] Now at Institut Jean Lamour, Nancy, France
[3] Welding Engineering and Laser Processing Centre, Cranfield University, Bedfordshire, United Kingdom

Abstract. Constructing titanium aerospace parts by near-net-shape processing has the potential to greatly reduce cost and lead time, one method for this is Wire-Arc Additive Manufacturing (WAAM). Conventional WAAM processing with the most common Ti alloy, Ti-6Al-4V, results in solidification by epitaxial growth from previously deposited layers and a structure dominated by columnar β grains which are heavily <001> fibre textured and cm's in scale. In order to prevent these large grains from forming, while maintaining deposition parameters, the solidification conditions were modified by the additions of particles to the melt; either using inoculant, TiN particles, or the solutal growth restrictor, Y, also added as elemental powder that dissolved in the melt. The powder particles were added by adhering them to the deposited tracks to avoid the costs of manufacturing new wires. With TiN inoculants the morphology of β grains was completely modified to equiaxed grains averaging 300 µm in diameter. Y additions narrowed the columnar grains from 1-2mm to 100-300 µm. Y also induced a change to equiaxed grains, late in solidification, in the region which was remelted by subsequent deposition. However, Yttria particles were found to have formed interdendritically with an interconnected skeletal morphology. High-resolution EBSD analysis showed both TiN and yttria particles exhibit specific orientation relationships with the solidified β grains, which were confirmed experimentally.

1. Introduction

Titanium alloys are known for their good strength/density ratio as well as corrosion/oxidation resistance [1], and as such are one of the most used alloy systems in aerospace applications [2]. However, Ti wrought parts have high associated costs due to the extensive processing and machining requirements needed for their production [3]. Additive manufacturing (AM) technologies have been proposed as a solution [4]; as they provide a potential production route to produce near-net-shape parts without the requirement for dies or extensive machining [5]. A variety of these technologies have shown great promise and have been used to produce Ti-6Al-4V (Ti-64) samples [6–8], as it is the most commonly used Ti alloy, not only in aerospace but biomedical applications. Among these technologies most are limited to the production of relatively small parts in a confined build volume; however, Wire-Arc Additive Manufacturing (WAAM) is particularly well suited for the production of large parts due to its high deposition rate and less constrained build envelope [9]. A significant drawback to the implementation of these processes is the formation of large, heavily textured, β grains during solidification which can negatively impact the properties of finished products [10]. A successful approach has been implemented here to limit the size and texture of these β grains by using conventional solidification technologies, namely the addition of inoculants or solutal growth restricting elements during deposition, to directly influence solidification and prevent their formation.

2. Background

WAAM differs from more common AM processes, as it uses an electric arc as the heat source and wire as the feedstock material, which is deposited in a layerwise fashion by feeding the wire into a melt pool [11]. A major drawback of this process is the formation of β grains with a columnar morphology, preferentially aligned with the build direction, which cause mechanical anisotropy in finished parts [12,13]. The columnar β grains grow across the boundaries between added layers as they are formed by epitaxial growth of the grains at the fusion boundary of the previously deposited layers. As Al and V, the main alloying elements in Ti-64 do not partition strongly during solidification [14] and thus do not contribute significantly to solutal growth restriction, or constitutional undercooling, alloy additions to increase solutal growth restriction is attractive. Strongly partitioning elements are segregated into the liquid, slowing the growth of the columnar front while also promoting nucleation by increasing the constitutional undercooling [15]. Elements which have previously been investigated for this purpose in Ti alloys fabricated using AM technologies include Be [16], B [17], La [18], Fe or Ni [19]. Y additions to Ti-64 have resulted in an order of magnitude refinement of the β grain size, attributed to both the solutal effect of Y on growth restriction as well as the potential heterogeneous nucleation of β on yttria during solidification [20]. Y has also been observed to change bead morphology and fluid flow during arc welding, while still refining the grains by solutal growth restriction [21]. WAAM is an attractive processing route for Y additions as the molten processing should result in the formation of beneficial yttria particles which remain small due to the short processing time and pin grain growth. Inoculation is a method of grain refinement based on introducing potent nucleation sites into a melt, either by direct powder additions or via precipitation reactions [22]. The effectiveness of a particular inoculant is difficult to determine as processing conditions play a large role and an inoculant which is ineffective in one set of conditions may be highly effective in another [23]. Previous studies on inoculation of Ti during WAAM focused on La based inoculants [24–26]. ZrN additions have also been proposed for Ti alloys where TiN forms in the melt with recorded orientation relationships (Nishiyama-Wasserman and Kurdjumov-Sachs [27]). Analysis of the Ti-N binary phase diagram suggests that this dissolution/precipitation reaction requires nitrogen contents in excess of 5 wt% [28]. Direct addition of TiN particles to act as inoculants may then be a promising inoculation process for Ti-64 during WAAM.

3. Materials and Methods

Four 20-layer high walls were built on identically sized Ti-64 substrates using the same build parameters and Ti-64 wire. In order to modify the solidification of β grains Y and TiN particles were added to the melt by adhering them to the top of each of the last 10 layers deposited. This was achieved by premixing the particles with a polyurethane adhesive, which was then brushed on to the wall during production directly after the layer was deposited, further details of which can be found in [29–31]. Metallographic preparation (to 0.25 μm colloidal silica), was carried out on cross sections for analysis. Final bulk compositions are given in Table 1 obtained by inductively coupled plasma (ICP) and gas emission infra-red absorption fusion (LECO) tests on regions from the top 10 layers of each sample. All three modified samples show slight increases in C content, the Ti64-Y also shows an increase in Y and O with an addition level of 0.76 wt% Y, while Ti64-TiN has an increase of N, corresponding to a calculated addition level of 0.26 wt% TiN particles [31].

Table 1: Bulk compositions (wt%) measured from the top 10 layers of each build

wt%	Ti	Al	V	Fe	O	N	C	Y
Ti64	89.43	6.22	3.95	0.20	0.17	0.01	0.02	0
Ti64-P	89.51	6.13	3.95	0.19	0.16	0.01	0.05	0
Ti64-Y	88.60	6.11	3.90	0.21	0.22	0.02	0.07	0.76
Ti64-TiN	89.58	6.10	3.78	0.22	0.19	0.07	0.08	0

The β structure which forms during solidification is not directly observable at room temperature in Ti-64 as there is a solid state β to α transformation which occurs during cooling which follows the BOR (i.e. $\{110\}_{\beta}//\{0001\}_{\alpha}$ and $<111>_{\beta}//<11\bar{2}0>_{\alpha}$) which allows the β grain structure to be reconstructed [32]. Here this was done using the methodology of Davies and Wynne [33,34] from α EBSD orientation maps of the sample cross sections acquired with a FEG-SEM operated at 20 kV indexed with Oxford instruments (OI) Aztec and analysed with Channel 5.

4. Results and Discussion

Reconstructed β grain EBSD orientation maps showing the full ND-TD cross sections for all four samples are provided in Fig. 1, where a) consists entirely of Ti-64 and in b) - d) the top 10 layers were produced with just adhesive, Y or TiN additions respectively. The bottom parts of each build, produced without additives, consist of a typical WAAM microstructure [8,12,35]. This structure continues in both Ti64 and Ti64-P similarly, indicating that the addition of the polyurethane has little effect on the solidification of β grains. The demarcation between the solely Ti-64 material built before Y particle additions were made can be clearly seen in the orientation maps as the grain structure changes significantly. Below where additions were used the β microstructure consists of mm scale grains, primarily oriented with a common <100> direction near the ND axis. This texture continues into the regions with Y additions, as can be seen in the pole figures shown in Figure 1. The columnar grains in the front of the wall appear tilted in the TD, which is reflected in the {001} pole figure. In contrast, the sample with TiN additions shows a drastic transition from the columnar grains in the lower Ti-64 only region of the wall to significantly refined, equiaxed grains, with a reduction in texture strength, immediately on addition of the TiN particles. In the refined upper-region of the sample, no orientations are seen to be strongly favoured. The width of the grains was seen to be reduced significantly by the addition of both TiN and Y powder to the build. Near the base of the wall, each sample showed similar behaviour with larger grain widths of roughly 1 mm. In the case of the Ti64 and Ti64-P builds, both showed a similar behaviour with the average grain width remaining above 1mm in the top ten layers for both samples. This indicates that any grain refinement observed in the other samples can be attributed solely to the Y and TiN powder additions. Both the Ti64-Y and Ti64-TiN samples show a similar reduction in grain widths, however morphologically we can see from Figure 1 that while the grains remain columnar in the Ti64-Y sample they change to equiaxed in the Ti64-TiN build. This reduction occurs somewhat gradually at a height of less than 10mm, due to the remelting which occurs during deposition of subsequent layers, which results in a stepwise change in the dilution when applying composition changes during deposition [36].

4.1 Ti-64 with Y additions

The top, Y modified region of the Ti64-Y sample showed significant columnar grain refinement compared to the baseline Ti64 and Ti64-P samples. The texture of the β grains is reduced in magnitude, but remains favoured towards a <001>β//ND texture. Alongside this columnar refinement, the structure also contained second phase precipitates not found in conventional Ti-64. These particles exhibited a skeletal structure; one such example of this morphology is shown in Figure 2. The particles were confirmed to be Y_2O_3, yttria, by the EBSD phase map which successfully indexed them. It is also interesting to note from the orientation map that the clusters of particles appear in the map with the same colour-orientation, indicating that they are not independent particles but rather a single inter-connected particle with a 3D morphology that has been bisected during sectioning and metallography [31]. These particles were not uniformly distributed, but rather showed an obvious relationship with the β grain structure.. In Figure 2 it can be seen that they also closely follow the low-angle grain boundaries seen in the β map. This yttria banding can be seen directly from the first layer deposited with Y additions [31]. However, in contrast to the Ti64 and Ti64-P samples the length of the columnar grains was limited to two to three layers, with the exception of the large grain on the left of the sample in Figure 1.. Thus, Y additions have successfully changed the solidification of Ti-64 during WAAM from being dominated by epitaxial growth to a combination of epitaxial growth with limited classical nucleation and growth.

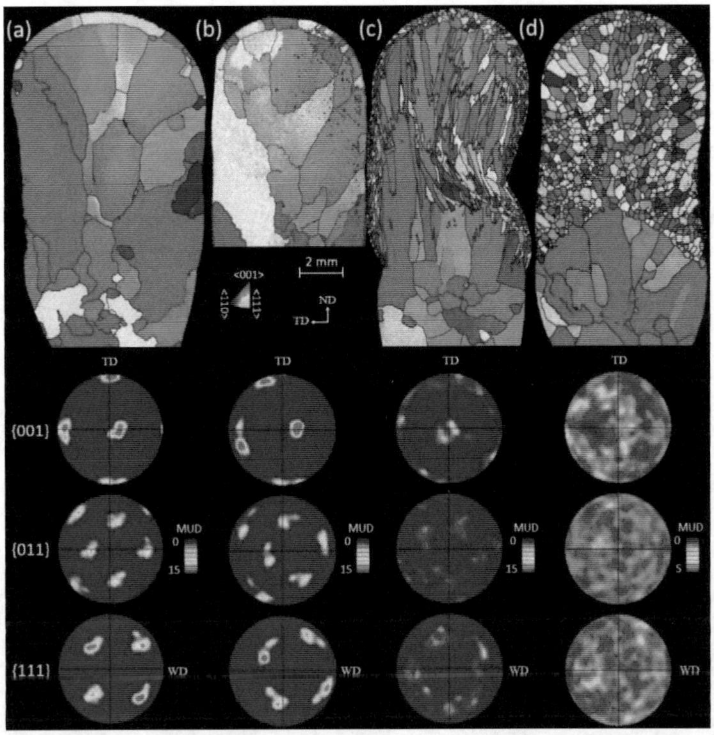

Figure 1: Reconstructed IPF-ND β EBSD orientation maps and pole figures for the; (a) Ti64 baseline, (b) Ti64-P (adhesive only), and (c) Ti64-Y, and (d) Ti64-TiN modified samples.

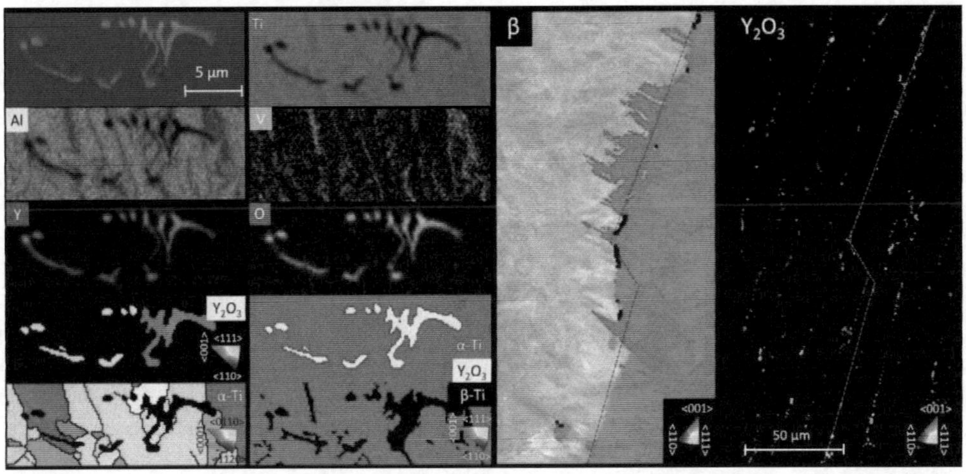

Figure 2: Chemical and EBSD, IPF//ND orientation and phase maps of the inter-dendrtitic yttria second phase particles found in the sample deposited with yttrium powder.

4.2 Ti-64 with TiN additions

Similar to the Ti64-Y sample the Ti64-TiN sample also showed significant grain refinement, but this was accompanied by the columnar grains being fully replaced by a finer ~300 μm equiaxed microstructure. In contrast to the Ti64-Y sample the <001>β//ND texture was also eliminated. The Ti64-TiN build did not exhibit a homogeneous distribution of particles, but rather the particles were primarily located in clusters. These heterogeneously distributed particles were determined to be TiN, which survived contact with the melt. In Figure 3 it can be seen that EBSD analysis indexed the particles as TiN with differing orientations between most of the particles, and the chemical analysis indicates that the particle-rich regions are lean in Al, but rich in N. This clustering is likely the result of the application method of mixing the particles with polyurethane adhesive prior to application to the wall. This is evidenced in Figure 3 by the higher than nominal levels of carbon within the clusters. The solidification behaviour during deposition of the Ti64-TiN sample has thus changed from the primarily epitaxial growth mechanism to favour a classical inoculation related, nucleation and growth mechanism, where the TiN particles were sufficiently effective to act as heterogenous nucleation sites for β-Ti [27,30], which then grow and block the epitaxial growth from previous layers.

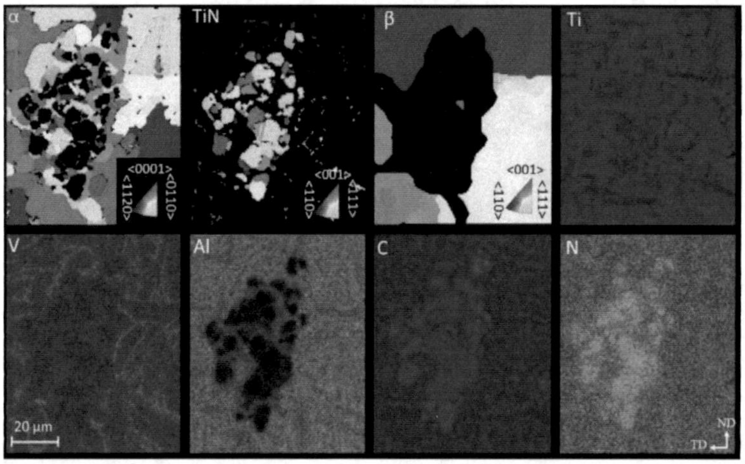

Figure 3: Chemical and IPF//ND orientation maps of a typical TiN cluster.

4.3 Particle-Matrix Orientation Relationships

Using the EBSD orientation data obtained previously it is possible to plot the frequency of the misorientation angles between the Ti and TiN or Y_2O_3 particles, as shown in Figure 4, with respect to the BOR. As the β to α solid state transformation obeys the BOR plotting the misorientations of the particles relative to the BOR from α is analogous to comparing the particle and β orientations. Both the yttria and TiN particles deviate strongly from random, indicating that preferential orientation relationships are present. The TiN particles are preferentially misoriented from the BOR by 43°, which corresponds to the Kurdjumov-Sachs (KS) OR; $\{111\}_{fcc}//\{1\bar{1}0\}_{bcc}$, $<\bar{1}10>_{fcc}//<\bar{1}11>_{bcc}$. This OR exhibits an atomic misfit of only 6% between TiN and β-Ti64 in the temperature range of solidification [30], which is likely to be why it is preferred and leads to effective grain refinement. In contrast, the Yttria-Ti misorientation distribution shows three distinct peaks, indicating that three separate orientation relationships are possible. Proposed ORs between Ti and Yttria include; $[010]_{\beta}(002)_{\beta}//[110]_{yttria}(004)_{yttria}$ by Zhang et al. [20], with a 45° misorientation and $<111>_{\beta}\{0\bar{1}1\}_{\beta}//<110>_{yttria}\{1\bar{1}0)\}_{yttria}$ by Wang et al. [37], with a 35.3° misorientation neither of which can account for all three peaks seen in the misorientation distribution here. A further $\{111\}_{\beta}<112>_{\beta}//\{011\}_{yttria}<011>$

yttria OR which has a 30° misorientation has also been proposed alongside the other potential ORs [31]. Thus, all three misorientation peaks can be correlated with potential ORs. This indicates that when the yttria formed inter-dendritically it was probably nucleated collaboratively with the solid β, and conversely yttria particles released in the melt can be responsible for the nucleation of β, which helps explain the refinement of the columnar grains in the Ti64-Y build.

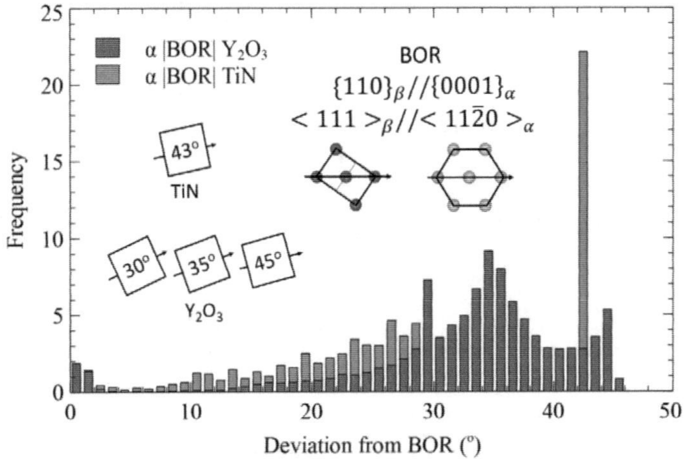

Figure 4: Frequency plot of the deviation between the orientation of the yttria and TiN particles from BOR with α-Ti, obtained from EBSD maps.

5. Conclusions

Two potential methods of grain refining Ti-64 during WAAM were investigated; inoculation via TiN particles, and solutal growth restriction via the addition of elemental Y. Both methods were successful at refining the solidification structure. The Y additions were responsible for refinement of the β grains to an average width of 300 μm, however they retained a columnar grain structure with a large aspect ratio, extending through multiple deposited layers. The grain lengths were, however, reduced compared to unmodified Ti-64, which has been attributed to limited nucleation of β from yttria particles, which formed interdendritically during solidification and were released into the melt during subsequent deposition. The Y additions did result in a CET during solidification, although this was achieved in the top region of the melt tracks which were subsequently remelted during deposition. Three potential ORs were observed between the yttria particles and β. The TiN additions were much more effective in inducing a CET and gave rise to refined equiaxed grains with an average size of 300 μm and random texture. The TiN particles were found to obey the KS OR with the β, and are suitable inoculants for Ti-64 WAAM. Both particles additions show potential for application to WAAM processing. However, while sufficient to demonstrate a proof of concept, the experimental method used for applying the powder additions are not well suited for industrial application and further study is required to develop a suitable addition method that leads to more efficient powder dispersal in the melt pool.

Acknowledgements
The authors are appreciative of the EPSRC programme grant NEWAM (EP/R027218/1) and the Innovate UK project Open Architecture Additive Manufacturing (OAAM) for supporting aspects of this research. The authors acknowledge the use of equipment within Advanced Metals Processing and Characterisation themes of the Henry Royce Institute for Advanced Materials, funded through EPSRC grants EP/R00661X/1, EP/S019367/1, EP/P025021/1 and EP/P025498/1. P.B. Prangnell is also grateful to the Royal Academy of Engineering, UK, and Airbus for supporting his research through the Airbus-University of Manchester Centre for Metallurgical Excellence. J.R. Kennedy is grateful to

ArcelorMittal, ArcelorMittal: Industeel, CEA, EDF and Framatome for supporting his research through the Industrial chair <<Solidification>> at Institut Jean Lamour.

References

[1] Welsch G, Boyer R and Collings E W 1993 *Materials Properties Handbook: Titanium Alloys* (ASM International)
[2] Lütjering G and Williams J C 2007 *Titanium* (Springer Science & Business Media)
[3] Shokrani A, Al-Samarrai I and Newman S T 2019 Hybrid cryogenic MQL for improving tool life in machining of Ti-6Al-4V titanium alloy *Journal of Manufacturing Processes* **43** 229–43
[4] Duda T and Raghavan L V 2016 3D Metal Printing Technology *IFAC-PapersOnLine* **49** 103–10
[5] Sames W J, List F A, Pannala S, Dehoff R R and Babu S S 2016 The metallurgy and processing science of metal additive manufacturing *International Materials Reviews* **61** 315–60
[6] Alfaify A Y, Hughes J and Ridgway K 2018 Critical evaluation of the pulsed selective laser melting process when fabricating Ti64 parts using a range of particle size distributions *Additive Manufacturing* **19** 197–204
[7] Nicoletto G, Konečná R, Frkáň M and Riva E 2018 Surface roughness and directional fatigue behavior of as-built EBM and DMLS Ti6Al4V *International Journal of Fatigue* **116** 140–8
[8] Zhao H, Ho A, Davis A, Antonysamy A and Prangnell P 2019 Automated image mapping and quantification of microstructure heterogeneity in additive manufactured Ti6Al4V *Materials Characterization* **147** 131–45
[9] Marinelli G, Martina F, Ganguly S and Williams S 2019 Development of Wire + Arc additive manufacture for the production of large-scale unalloyed tungsten components *International Journal of Refractory Metals and Hard Materials* **82** 329–35
[10] Ayed A, Valencia A, Bras G, Bernard H, Michaud P, Balcaen Y and Alexis J 2020 Effects of WAAM Process Parameters on Metallurgical and Mechanical Properties of Ti-6Al-4V Deposits *Advances in Materials, Mechanics and Manufacturing* Lecture Notes in Mechanical Engineering ed F Chaari, M Barkallah, A Bouguecha, B Zouari, M T Khabou, M Kchaou and M Haddar (Cham: Springer International Publishing) pp 26–35
[11] Martina F, Mehnen J, Williams S W, Colegrove P and Wang F 2012 Investigation of the benefits of plasma deposition for the additive layer manufacture of Ti–6Al–4V *Journal of Materials Processing Technology* **212** 1377–86
[12] Colegrove P A, McAndrew A R, Ding J, Martina F, Kurzynski P and Williams S 2016 System Architectures for Large Scale Wire + Arc Additive Manufacture 10th international conference on trends in welding research (tokyo, japan) p 5
[13] Williams S W, Martina F, Addison A C, Ding J, Pardal G and Colegrove P 2016 Wire + Arc Additive Manufacturing *Materials Science and Technology* **32** 641–7
[14] Bermingham M J, McDonald S D, Dargusch M S and StJohn D H 2008 Grain-refinement mechanisms in titanium alloys *Journal of Materials Research* **23** 97–104
[15] Rappaz M and Thévoz P H 1987 Solute diffusion model for equiaxed dendritic growth: Analytical solution *Acta Metallurgica* **35** 2929–33
[16] Bermingham M J, McDonald S D, StJohn D H and Dargusch M S 2009 Beryllium as a grain refiner in titanium alloys *Journal of Alloys and Compounds* **481** L20–3
[17] Roy S, Suwas S, Tamirisakandala S, Miracle D B and Srinivasan R 2011 Development of solidification microstructure in boron-modified alloy Ti–6Al–4V–0.1B *Acta Materialia* **59** 5494–510
[18] Liu M, Liu S, Chen W, Chen C, Lv Y, Zhang X, Lei P, Lin Y and Zhou K 2019 Effect of trace lanthanum hexaboride on the phase, grain structure, and texture of electron beam melted Ti-6Al-4V *Additive Manufacturing* **30** 100873
[19] Welk B A, Taylor N, Kloenne Z, Chaput K J, Fox S and Fraser H L 2021 Use of Alloying to Effect an Equiaxed Microstructure in Additive Manufacturing and Subsequent Heat Treatment of High-Strength Titanium Alloys *Metall Mater Trans A* **52** 5367–80

[20] Zhang D, Qiu D, Gibson M A, Zheng Y, Fraser H L, Prasad A, StJohn D H and Easton M A 2020 Refining prior-β grains of Ti–6Al–4V alloy through yttrium addition *Journal of Alloys and Compounds* **841** 155733

[21] Nordin M C, Edwards G R and Olson D L The Influence of Yttrium Microadditions on Titanium Weld Metal Cracking Susceptibility and Grain Morphology 11

[22] Greer A L 2003 Grain Refinement of Alloys by Inoculation of Melts *Philosophical Transactions: Mathematical, Physical and Engineering Sciences* **361** 479–95

[23] Wang T, Chen Z, Fu H, Gao L and Li T 2012 Grain refinement mechanism of pure aluminum by inoculation with Al–B master alloys *Materials Science and Engineering: A* **549** 136–43

[24] Bermingham M J, McDonald S D and Dargusch M S 2018 Effect of trace lanthanum hexaboride and boron additions on microstructure, tensile properties and anisotropy of Ti-6Al-4V produced by additive manufacturing *Materials Science and Engineering: A* **719** 1–11

[25] Bermingham M J, StJohn D H, Krynen J, Tedman-Jones S and Dargusch M S 2019 Promoting the columnar to equiaxed transition and grain refinement of titanium alloys during additive manufacturing *Acta Materialia* **168** 261–74

[26] Mereddy S, Bermingham M J, StJohn D H and Dargusch M S 2017 Grain refinement of wire arc additively manufactured titanium by the addition of silicon *Journal of Alloys and Compounds* **695** 2097–103

[27] Qiu D, Zhang D, Easton M A, St John D H and Gibson M A 2018 Refining As-cast β-Ti Grains Through ZrN Inoculation *Metall and Mat Trans A* **49** 1444–9

[28] H.A. Wriedt, J.L. Murray 1986 *Binary alloy phase diagrams. Vol. 3.* (ASM)

[29] Kennedy J R, Davis A E, Caballero A, Garner A, Donoghue J, Williams S, Zollinger J, Bouzy E, Pickering E J and Prangnell P B 2020 Isomorphic grain inoculation in Ti-6Al-4V during additive manufacturing *Materials Letters: X* **8** 100057

[30] Kennedy J R, Davis A E, Caballero A E, Williams S, Pickering E J and Prangnell P B 2021 The potential for grain refinement of Wire-Arc Additive Manufactured (WAAM) Ti-6Al-4V by ZrN and TiN inoculation *Additive Manufacturing* **40** 101928

[31] Kennedy J R, Davis A E, Caballero A E, Byres N, Williams S, Pickering E J and Prangnell P B 2022 β Grain refinement by yttrium addition in Ti-6Al-4V Wire-Arc Additive Manufacturing *Journal of Alloys and Compounds* **895** 162735

[32] Gey N and Humbert M 2002 The calculation of a parent grain orientation from inherited variants for approximate (b.c.c.–h.c.p.) orientation relations *Journal of Applied Crystallography* **35** 401–5

[33] Davies P S, Wynne B P, Rainforth W M, Thomas M J and Threadgill P L 2011 Development of Microstructure and Crystallographic Texture during Stationary Shoulder Friction Stir Welding of Ti-6Al-4V *Metall and Mat Trans A* **42** 2278–89

[34] Davies P S W 2009 An investigation of microstructure and texture evolution in the Near-α titanium alloy timetal 834.

[35] Ho A, Zhao H, Fellowes J W, Martina F, Davis A E and Prangnell P B 2019 On the origin of microstructural banding in Ti-6Al4V wire-arc based high deposition rate additive manufacturing *Acta Materialia* **166** 306–23

[36] Kennedy J R, Davis A E, Caballero A E, White M, Fellowes J, Pickering E J and Prangnell P B 2021 Microstructure transition gradients in titanium dissimilar alloy (Ti-5Al-5V-5Mo-3Cr/Ti-6Al-4V) tailored wire-arc additively manufactured components *Materials Characterization* **182** 111577

[37] Wang X, Zhang L-J, Ning J, Li S, Zhang L-L and Long J 2021 Hierarchical grain refinement during the laser additive manufacturing of Ti-6Al-4V alloys by the addition of micron-sized refractory particles *Additive Manufacturing* **45** 102045

Effect of Eu and P additions with Ta grain refiner on the solidification microstructure of Al-7Si-0.3Mg alloys

I Spacil[1], P Schumacher[1,2] and J Li[1,*]

[1] Institute of Casting Research, Montanuniversität Leoben, A-8700 Leoben, Austria
[2] Austrian Foundry Research Institute, A-8700 Leoben, Austria
[*] E-mail: jiehua.li@unileoben.ac.at

Abstract. Al-7Si-0.3Mg based alloys have been widely used. Nevertheless, at least three issues during melt treatment (grain refinement of Al grains, modification of eutectic Si and refinement of eutectic grain size) need to be further explored. In this contribution, we focus on these three melt treatments. For grain refinement of Al grains, the stochiometric Al-2.2Ti-1B grain refiner rod together with 0.12 wt. % Ta was used. For modification of eutectic Si, Eu up to 500 ppm was used. The best modification can be obtained with 200 ppm Eu. By further increasing Eu up to 500 ppm, modification level becomes worse due to the formation of Al_2Si_2Eu. For the refinement of eutectic grain size, a combined addition of Eu (up to 500 ppm) and P (up to 40 ppm) was used. With increasing P concentration, the modification of eutectic Si became worse. The size of eutectic Si increases, the shape factor of eutectic Si decreases. When the P concentration is higher, more Eu is needed for a well-modified eutectic Si. This investigation provides a novel concept for melt treatment of Al-Si-Mg based alloys.

1. Introduction

Light weight and thin walled designs are widely used in the automotive industry to reduce the energy consumption. The hypoeutectic Al-Si-Mg alloys fulfill that requirement due to its excellent casting properties and high strength to weight ratio. Mechanical properties of Al-Si-Mg alloys depend on the chemical composition, solidification rate, soundness and melt treatment [1]. Prior to casting, necessary melt treatments such as grain refinement of Al grains, modification of eutectic Si and refinement of eutectic grain size are employed to improve mechanical properties.

A smaller and uniformly distributed grain size of primary phase (α-Al) through grain refinement improves castability (mainly fluidity and resistance to hot tearing) by mass flow and mechanical properties (especially elongation) by fine distribution of Fe containing intermetallic phases. Grain refinement is achieved by the addition of potent substrates (TiB_2, NbB_2, etc.) for heterogenous nucleation which are active at a low undercooling during solidification [2]. Al-5Ti-1B grain refiners are commonly employed for grain refinement of Al-Si-Mg alloys due to the following facts: (i) all B is bounded into stable TiB_2 with a stochiometric weight ratio (2.2:1), (ii) excess Ti forms Al_3Ti, (iii) free Ti enhances the grain refinement via increasing growth restriction (Q). On the other hand, free Ti interacts with Si and forms titanium silicide (Al, Ti, Si) so-called τ phase. The τ phase is formed on surface of TiB_2, therefore the formation of Al_3Ti 2DC (two-dimensional compound) can be prevented. The Al_3Ti 2DC is identical to bulk Al_3Ti formed in its peritectic reaction. The former compound acts as nucleation sites for α-Al below Ti concentration for bulk Al_3Ti to be present. The interaction between Ti and Si on the interface of TiB_2 can cause a so-called Si poisoning when the Si concentration is higher than 2 wt. % [3, 4]. Most of Al-Si-Mg based alloys contain more than 7 wt. % Si, thus overcoming or at least reducing the negative Si poisoning effect is crucial to obtain high mechanical properties of Al-Si-Mg based alloys in the foundry industry. Recently [5, 6], the solute Ta was implemented to reduce the grain size in Al-Si-Mg casting alloys via formation of Al_3Ta on the basal plane of TiB_2. To date, there is a lack of available data to prove no interaction of Ta with Si or other solutes. This paper provides new insights of using solute Ta as grain refiner.

The second melt treatment influencing the Al-Si-Mg properties, particularly ductility, is modification of eutectic Si. Grain refinement in combination with modification of eutectic is reported to have beneficial effects on mechanical properties [7]. Si is transformed from large brittle plates to a fine fibrous coral-like morphology. Chemical modification is achieved via addition of certain modifying element (i.e. Na [8], Sr [7]). Other reported elements (i.e. Yb [9]) can only refine the eutectic Si. To date, three modification mechanisms are accepted: the impurity-induced twinning (IIT) [10], the twin plane re-entrant edge (TPRE) [11], and the poisoning of TRPE mechanism [12]. Neither well-accepted mechanism can describe all observations regarding the morphology transformation of eutectic Si and the refinement of eutectic grain size. The eutectic Si nucleates and growths on the AlP phase. AlP facilitates the nucleation, thus increasing the nucleation temperature (T_N). Addition of Na or Sr in the melt reduces the amount of AlP by formation of Na_3P and Sr_3P_2 compounds and the remaining modifier modifies the eutectic Si to a fibrous coral-like morphology. When the potent AlP is removed, the nucleation temperature decreases. However, neither Na_3P nor Sr_3P_2 are efficient potent substrates for eutectic grains, therefore the eutectic grain size becomes very coarse (up to 1 mm) [13].

On the other hand, it has been reported [14] that Eu could be a suitable modifying element among the group of rare elements. Further, EuP was proposed according to the lattice mismatch and experimentally verified to be potent substrate for eutectic Si [13], which is in contrast to Na_3P and Sr_3P_2. Although the nucleation potency of EuP may be not as efficient as AlP, EuP seems to be beneficial for nucleation of not only eutectic Si but also eutectic grains.

Conventional Al-Si-Mg based casting alloys contain a P concentration in excess of 10 ppm but due to impurity accumulation and a recycling process much higher P is observed. There is still a lack of a detailed investigation on controlling solidification behaviour of Eu in combination with a higher P content. This paper is aimed to elucidate solidification behaviour of Al-7Si-0.3Mg based alloys with a controlled addition of Ta (0.12 wt. %), Eu (up to 500 ppm) and P (up to 40 ppm). The thermal analysis (Quick-Cup method) and middle lab-scale (50 kg) melting for gravity die casting were performed to obtain quantitative data for a better understanding of microstructural changes.

2. Casting experiment

Four Al-7Si-0.3Mg alloys with the addition of 200 ppm Eu and 10 ppm P (Alloy 1), with the addition of 500 ppm Eu and 10 ppm P (Alloy 2), with addition of 200 ppm Eu and 40 ppm P (Alloy 3), and with the addition of 500 ppm Eu and 40 ppm P (Alloy 4) were prepared. The presence of 10 ppm P is directly from the standard commercial purity (CP) Al ingots. In the present study, the concentrations are in wt. %. Each alloy was prepared with an amount of 50 kg in the electrical resistance furnace. Firstly, the CP Al ingots (99.8 %) were melted. Secondly, Si and Mg were alloyed using CP Si (99.97 %) and CP Mg (99.8 %). Impeller degassing using a degassing medium of industrial purity Ar (99.5 %) and a gas flow of 5 l/min for 15 minutes was applied. After degassing, the melt was kept at 760 °C for 10 min to let the remaining purging gas to leave the melt and to separate the inclusions produced by degassing. Samples for density-index (DI) measurement were taken from the degassed and skimmed melt and evaluated immediately after the solidification. If the DI was less than 1.5 %, the Al-2.2Ti-1B grain refiner rod with an exact stochiometric Ti/B weight ratio of 2.2:1 was added, the grain refiner contained only TiB_2. Solute Ta (0.12 wt. %) was added using CP Al-10Ta master alloy. CP Al-5Eu master alloy was added for the modification of eutectic Si. The P was alloyed using CP Al-3P master alloy.

The melting temperature was kept at 760 °C for additional 5 minutes to ensure the full dissolution of master alloys. Prior to casting, the melt was stirred to ensure homogeneous distribution potent TiB_2 particles and thereby to avoid fading by sedimentation. Samples were casted into BN coated and preheated (at 180 °C) diez mould. 20 ppi (pores per inch) filters were used to remove the possible non-metallic inclusions. Typical cooling rate of 4 °C/s are observed which are similar to TP-1 test [15]. The Quick-Cup method was performed to obtain thermal analysis data. Spark analysis samples were also cast. Main alloying elements (i.e. Si, Mg, Fe, Ti, P) were measured using spark analysis (spark emission spectrometer (OBLF VeOS)). The Ta and Eu concentrations were measured using inductive coupled plasma optical emission spectrometer (ICP-OES).

Samples for microstructure characterization were prepared using a standard metallographic process (grinding using SiC paper up to 1200 mesh and polishing using 3 µm diamond solution and OPS). To clearly observe the α-Al grains electro polishing (Barker's etching) was performed. Optical microscope (Zeiss AXIO) in a polarized mode was used to characterize solidification microstructure. MIA (Multiple Image alignment) was used to assemble in total 16 images from the centre of the etched samples together as one large image with the aim to obtain an overview for grain size measurement. To observe the eutectic grain size, modified Murakami reagent (MMR) etching was implemented using a mixture of 240 ml H_2O, 40 g NaOH and 20 g $K_3[Fe(CN)_6]$. Deep etching was implemented using a 15 % HCl to analyze the 3D morphology of eutectic Si by removing the primary phase (α-Al). The grain size, second dendrite arm spacing (SDAS) and eutectic grain size were measured and evaluated using the ImageJ software. The grain size was measured according to ASTM E112 standard. The shape factor (SF) was calculated using equation (A.1, see appendix A) to quantify the shape of Si particles. If the particle is spherical, the SF is close to one, otherwise SF decreases. SEM (FEI Quanta 200 and JSM-7200) was used to determine the 3D morphology of eutectic Si and intermetallic phases (Al_2Si_2Eu), respectively.

3. Results and discussions

3.1. Thermal analysis

Table 1 shows thermal analysis data of Al-7Si-0.3Mg based alloys with additions of Eu and P. The nucleation temperature (T_N), the minimum temperature (T_{min}) and the growth temperature (T_G) were measured. The T_N indicates the starting point of nucleation. The recalescence (ΔT) is extracted from $T_G - T_{min}$ [16].

In Alloys 1, 3, 4, no recalescence of α-Al was detected. In Alloy 2, the recalescence of α-Al was observed to be about 0.6 K. This indicates that the heterogenous nucleation of α-Al becomes worse and a larger grain size may be expected in Alloy 2. The T_{min} of α-Al is not significantly changed for all investigated alloys.

At the eutectic region, the highest T_N (576.4 °C) and T_{min} (571.7 °C) was observed in Alloy 3 (Table 1). Alloy 3 also demonstrates a smaller recalescence and a shift of the T_{min} to a higher temperature. A higher T_N of eutectic phase causes a shorter growth time of eutectic phase. The measured recalescence for Alloy 1, Alloy 2 and Alloy 4 strongly indicates that the Eu addition could deplete or remove the AlP, which plays a role as nucleation site for eutectic Si. Except for Alloy 3, T_G is shifted to a lower temperature, thus it can be reasonably assumed that Eu also hinders the growth of eutectic Si via poisoning TRPE or IIT [17]. The T_{min} depression (of eutectic growth temperature) indicates a well modified structure. It is attributed to the poisoning of AlP by modifier (Eu), indicating that the nucleation occurs at lower temperatures. The results from thermal analysis measurement still need to be supported by further microstructural analysis.

In the present work, the T_N represents only the nucleation temperature of eutectic Si. The T_N of α-Al is not included because of its difficulty to detect the start of the nucleation due to the impurity effect.

Table 1. Thermal analysis data of nucleation and growth of eutectic Si.

	T_N (°C)	T_{min} (°C)	T_G (°C)	$T_{eq}{}^{a} - T_N$ (K)	ΔT (K)
Alloy 1	571.4	564.9	568.8	5.6	3.9
Alloy 2	572.0	565.0	566.2	5.0	1.2
Alloy 3	576.4	571.7	572.2	0.6	0.5
Alloy 4	570.7	567.4	569.7	6.3	2.3

[a] Equilibrium temperature equals to 577 °C.

3.2. As-cast microstructure

Figure 1 shows the α-Al dendritic structure in four alloys with different concentrations of Eu and P observed by optical microscopy (OM). The addition of stochiometric grain refiner and solute (0.12 wt. % Ta) result in a successful grain refinement. The grain size and SDAS of four alloys are plotted in figure 2 (a). The measured grain sizes of α-Al of four alloys are less than 200 μm. Small grain size is attributed to effective grain refinement and high cooling rate from gravity die casting (4 °C/s). SDAS measurement was challenging because the grain size is small and therefore dendrites are not strongly pronounced. Alloy 3 shows the lowest SDAS value of about ~18.8 μm. It is well-accepted that modifier has no significant effect on α-Al grain size. The grain size measurement results confirm that solute Eu has no negative effect on grain refinement of primary α-Al phase. It is consistent with recent experiments [5] performed in the identical alloy using Sr (200 ppm) to modify the eutectic Si. Interestingly, the grain size shows a slightly decreasing tendency with increasing Eu and / or P concentration.

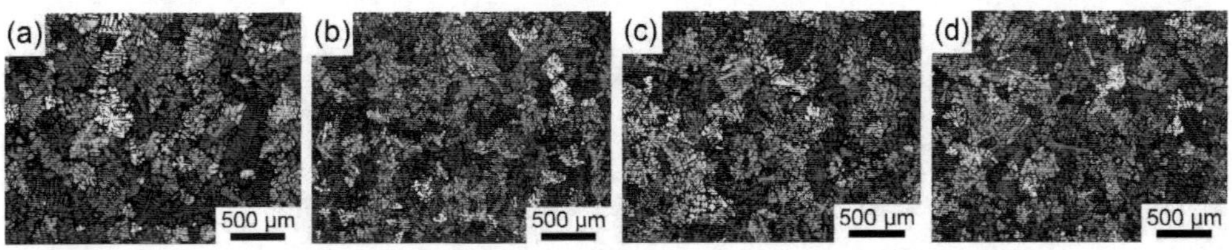

Figure 1. Optical microscopy image of Alloy 1 (a), Alloy 2 (b), Alloy 3 (c), and Alloy 4 (d) observed in a polarized mode after Barker's etching.

Figure 2 (b) shows the eutectic grain size and the shape factor of eutectic Si. The eutectic grain size was measured after the MMR (section Experiment) by OM. Alloy 1 shows a very coarse eutectic grain size (~1600 μm). With the addition of 500 ppm Eu (Alloy 2) the measured eutectic grain size is approximately 1800 μm. The smallest eutectic grain size is measured in Alloy 3 (~1100 μm). This result and thermal analysis data (table 1) indicate an easier nucleation of eutectic Si which is attributed to a greater amount of EuP particles with increasing P content. Therefore, the addition of 200 ppm Eu in combination with 40 ppm P is reasonably assumed to be insufficient to obtain a well-modified as-cast structure despite its decrease in eutectic grain size. A further increase of Eu concentration up to 500 ppm in combination with 40 ppm P shows an increase in eutectic grain size approximately to 1500 μm compared with Alloy 3. It is clear that with higher P content the nucleation of eutectic Si becomes easier. However, a simultaneous increase of Eu is required in order to obtain desired modifying effect.

The SF is not significantly changed for 200 ppm Eu and 500 ppm Eu addition (~0.85). A slight increase of eutectic grain size and a decrease of SF in Alloy 2 indicates that the best modification of eutectic Si can be obtained with addition of only 200 ppm Eu. A significant decrease of SF (0.66) in addition of 200 ppm Eu and 40 ppm P indicates a poor modification effect of eutectic Si. With a further increase of Eu up to 500 ppm, the SF increases to 0.78. This result again supports the observation that the modification effect becomes worse and more Eu is needed with increasing P concentration.

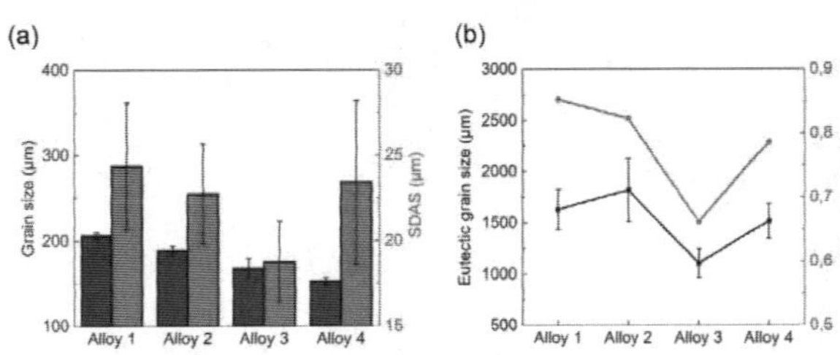

Figure 2. Grain size of α-Al and SDAS measured after Barker's etching (a) and eutectic grain size measured after MMR etching and shape factor of eutectic Si (b).

3.3. Scanning electron microscopy (SEM)

The effect of Eu and P on solidification microstructure was further investigated using SEM. SEM images (figure 3) show the 3D morphology of eutectic Si at a high magnification after deep etching. Figure 3 (a), (c), (d) show a fine and fully fibrous eutectic structure. A significant decrease in nucleation undercooling (5.6 K) and increase in recalescence (3.9 K) in Alloy 1 strongly indicates an interaction between Eu and P, resulting in a well-modified structure (figure 3 (a)). However, the eutectic structure of Alloy 3 (in figure 3 (c)) shows a plate like morphology. The size of eutectic plates is also much larger in comparison to a fine branching of modified eutectic Si observed in other alloys. This indicates that the addition of 200 ppm Eu is not sufficient to transform all potent AlP particles to EuP nucleation sites and to provide sufficient Eu for modification. Increasing P content and its negative effect on modification must be compensated by increasing amount of added modifier (Eu).

Figure 3. SEM image demonstrating the 3D morphology of eutectic Si at a high magnification in Alloy 1 (a), Alloy 2 (b), Alloy 3 (c), and Alloy 4 (d) after deep etching.

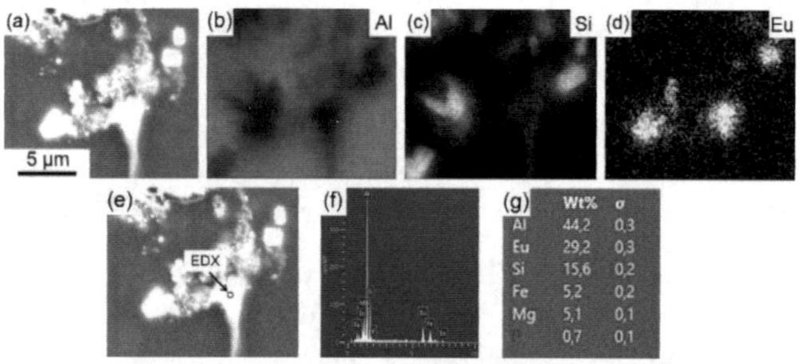

Figure 4. SEM image and EDS mapping of intermetallic phase observed at the grain boundaries (a), (b), (c), (d) and EDX analysis of intermetallic particle (e), (f), (g) in Alloy 2 (500 ppm Eu). Note, the sample was investigated prior to etching.

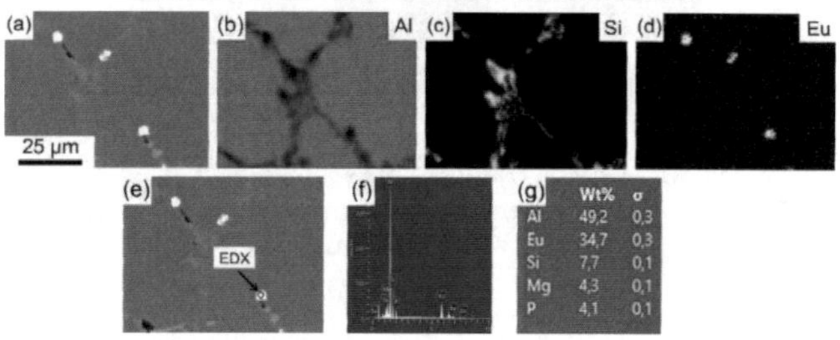

Figure 5. SEM image and EDS mapping of intermetallic phase observed at the grain boundaries (a), (b), (c), (d) and EDX analysis of intermetallic particle (e), (f), (g) in Alloy 4 (500 ppm Eu and 40 ppm P). Note, the sample was investigated prior to etching.

Figures 4 and 5 show Eu-rich intermetallic phases observed at the grain boundaries of as-cast samples in alloys containing 500 ppm Eu and 500 ppm Eu in combination with 40 ppm P, respectively. SEM-mapping (figures 4 (a), (b), (c), (d) and 5 (a), (b), (c), (d)) shows the main elements incorporated in intermetallic particles. EDX measurement was implemented to identify chemical composition of observed particles. The results from SEM-mapping and EDX analysis indicates the presence of Al_2Si_2Eu phase which is related to higher Eu concentrations. Al_2Si_2Eu is believed to be not responsible for the growth of eutectic Si [16]. Noteworthy, the EDX analysis also detects the trace P element incorporated into Al_2Si_2Eu. Within increasing P, more P was measured within Al_2Si_2Eu (figure 5 (g)) presumably in a EuP particle.

Using Ta to overcome Si poisoning, Eu to modify eutectic Si, and P to refine eutectic grains appears to be a promising methodology to achieve a well-controlled microstructure. Further investigations are still required to explore the effect of solutes (Ta, Eu, P) on heat treatment and mechanical properties.

4. Conclusion

The solidification microstructure of Al-7Si-0.3Mg based alloys with controlled addition of Eu up to 500 ppm and P up to 40 ppm was investigated using thermal analysis, OM and SEM in a gravity die casting condition. The main conclusions are listed as follows:

1. A significant grain refinement (~200 μm) in Al-7Si-0.3Mg using a stochiometric grain refiner (Al-2.2Ti-1B) and solute Ta (0.12 wt. %) with controlled additions of Eu and P was achieved. Interestingly, the grain size also tends to slightly decrease with increasing concentration of Eu (up to 500 ppm) and / or P (up to 40 ppm).
2. The presence of 200 ppm Eu addition is proposed to perform best modification effect in Al-7Si-0.3Mg based alloys. The presence of 200 ppm Eu combined with 40 ppm P leads to an unmodified (plate-like) microstructure with a finer eutectic grain size and a smaller shape factor value.
3. The addition 500 ppm Eu results in a slightly worse modification by consuming Eu to form pre-eutectic Al_2Si_2Eu phase. The presence of 500 ppm Eu combined with 40 ppm P shows a well-modified structure with a refined eutectic grain size (1100 μm). The modification of eutectic Si becomes worse with increasing P (up to 40 ppm). Therefore, more Eu is required to obtain a well-modified structure. Ratio of Eu and P to achieve a fibrous morphology and a refined eutectic grain size need to be further investigated.
4. No interaction of Ta and Si has been observed, indicating a promising melt treatment to obtain a well-controlled solidification microstructure.

Acknowledgments
The financial support from Austrian Science Fund (FWF) (P 32378-N37) and BMBWF (KR 06/2020) is gratefully acknowledged.

Appendices

Appendix A. Shape factor calculation

To obtain SF according to equation (A.1), the perimeter and area of Si particles were obtained from optical microscope image using ImageJ software.

$$SF = \frac{1}{\sum_{i=1}^{n} A_i}\left(\sum_{i=1}^{n} A_i \frac{2\cdot\sqrt{\pi A_i}}{P_i}\right) \quad \text{(A.1)}$$

where A_i is the area of Si particle, P_i is the perimeter of Si particle, n is the number of particles.

References

[1] Paray F and Gruzleski J E 1992 *Cast Metals* **5** 187–98
[2] Greer A L, Bunn A M, Tronche A, Evans P V and Bristow D J 2000 *Acta Mater.* **48** 2823–35
[3] Qiu D, Taylor J A, Zhang M-X and Kelly P M 2007 *Acta Mater.* **55** 1447–56
[4] Gröbner J, Mirković D and Schmid-Fetzer R 2005 *Mater. Sci. Eng. A* **395** 10–21
[5] Spacil I, Holec D, Schumacher P and Li J 2022 *Solid State Phenom.* **327** 54–64
[6] Pammer M, Li J and Schumacher P 2021 *Gießerei-Rundschau* **68** 15–20
[7] Kori S A, Murty B S and Chakraborty M 2000 *Mater. Sci. and Eng. A* **283** 94–104
[8] Pacz A 1921 U.S Patent No.1387900
[9] Li J H, Suetsugu S, Tsunekawa Y and Schumacher P 2013 *Metall. Mater. Trans. A* **44** 669–81
[10] Lu S-Z and Hellawell A 1987 *Metall. Mater. Trans. A* **18** 1721–33
[11] Hamilton D R and Seidensticker R G 1960 *J. Appl. Phys.* **31** 1165–8
[12] Day M G and Hellawell A 1968 *Proc. Math. Phys. Eng. Sci.* **305** 473–91
[13] Li J H, Ludwig T H, Oberdorfer B and Schumacher P 2018 *Int. J. Cast Met. Res.* **31** 319–31
[14] Nogita K, McDonald S D and Dahle A K 2004 *MATER. TRANS.* **45** 323–6
[15] Aluminum Association 2012 *Standard test procedure for aluminum alloy grain refiners 2012* (Washington DC: Aluminum Association)
[16] Li J H, Wang X D, Ludwig T H, Tsunekawa Y, Arnberg L, Jiang J Z and Schumacher P 2015 *Acta Mater.* **84** 153–63
[17] Li J, Hage F, Wiessner M, Romaner L, Scheiber D, Sartory B, Ramasse Q and Schumacher P 2015 *Sci. Rep.* **5** 13802

Revealing atomic structure evolution of an Al-1.5Fe alloy in the liquid state using X-ray total scattering and empirical potential structure refinement

S Huang[1*], L Qin[1], J Zhao[1], K Xiang[1], S Luo[2], S Michalik[3], J Mi[1]

[1] *Department of Engineering, University of Hull, Cottingham Road, Hull, HU6 7RX, UK*
[2] *School of Materials Science and Engineering, Hefei University of Technology, Hefei, 230009, Anhui, PR China*
[3] *Diamond Light Source Ltd., Harwell Science and Innovation Campus, Didcot, Oxfordshire OX11 0DE, UK*
* Corresponding author: s.huang-2018@hull.ac.uk

Abstract. We used *in-situ* real-time synchrotron X-ray total scattering and empirical potential structure refinement method to study the evolution of atomic structure of an Al-1.5Fe alloy melt from 750°C to 630 °C. Detailed data normalization and analyses for the total scattering data were presented in order to achieve sensible atomic structure simulation using the empirical potential structure refinement (EPSR) method with constraints of R-factor is $<10^{-3}$. Reliable construction of 3D atomic structure of Al-Fe melt in the temperature range reveals that more Al atoms coordinate Fe atoms and the Fe-centered clusters get more compact with the temperature decreases. More systematic work linking the liquid structure to the nucleation structure is under way.

1. Introduction

Phase changes of materials driven by different thermodynamic forces are universal in nature. Understanding of the underlying governing laws and the time-evolving dynamics have been at the centre of materials science and condensed-matter physics for centuries. For a man-made material or an alloy system, a precise understanding is critical for us to design, develop and optimize existing and new materials [1-4].

Iron is a common impurity in commercial aluminium alloys, which can be easily picked up in the manufacturing process and recycling processes of Aluminium and its alloys [5, 6]. The presence of a small quantity of Fe can induce the formation of hard and brittle intermetallic iron-rich phases which have deleterious effects on castability, machinability and mechanical properties [7]. To date, the full potential of this composite material is yet to be fulfilled, in part due to the lack of a complete picture of the transition from Al-Fe liquid state to its primary phase, although some work has been performed to study the local structure of Al-Fe melt. For instance, Qin et al [8] implemented a wide angle X-ray diffraction investigating the changes of the pre-peak in S(Q) with increased Fe content. They suggested that the origin of the pre-peak is the Fe-Fe correlation in Al-rich alloys and the Al-Al correlation in Fe-rich alloys, but a mixture of Fe-Fe and Al-Al correlation around the composition of FeAl alloy. Il'inskii et al [9] observed that the variation of the component concentrations in Fe-Al alloys causes the change of volume fraction of each cluster type while retaining the atomic composition and arrangement within the clusters. Despite these progresses, the focus of present studies has mainly been on the element effect of Al and Fe on the local structure of Al-Fe melt. Clearly, insight into the dynamics and evolution of Al-Fe alloys melt from the perspective of temperature dependence is still lacking [10].

Here, we report a systematic work to probe the temperature dependence of evolution of atomic structure of Al-1.5wt% Fe (weight percentage) alloys melt cooled from 750 to 630°C, by means of synchrotron X-ray total scattering (SXTS) and empirical potential structure refinement (EPSR) method.

2. 3D atomic reconstruction

This work mainly consists of 3 parts: SXTS experiment, standard data normalization by GudrunX and 3D atomic structure construction by EPSR. The *in-situ* SXTS experiment was carried out at the beamline I12 of Diamond Light Source (DLS), UK, with use of a bespoke furnace (Fig. 1(a)). To remove background information or separate out the effects of self-absorption, polarization, fluorescence and Compton scattering, GudrunX was used for standard corrections, by which we extracted the total structure factor S(Q) and pair distribution function $G(r)$. Lastly, evolution of 3D atomic structure of the matter was reconstructed by substituting S(Q) and $G(r)$ into EPSR simulation.

2.1 Sample preparation and SXTS experiment

All of as-cast Al-1.5wt%Fe alloys samples were machined into cylindrical bars with a diameter of 3mm and length of 25mm. In a SXTS experiment, with set-up in Fig.1 (a), a monochromatic X-ray of 100.03 KeV (0.12361 Å) was directed to a slit of a bespoke electric resistance furnace and illuminated onto the alloys sample inserted. The assembly of as-prepared sample capsulated in a quartz capillary (diameter: 3.5mm, thickness: 0.1mm) sitting on a ceramic post was enclosed inside 17 mm-diameter quartz tube. The loaded sample was heated from room temperature at a rate of 41.5 °C/min till 750 °C, with a dwell time of 30 min for ensuring it in fully melting, followed by a cooling of 2.72 °C /min back to room temperature. As the sample is heated and cooled in a programmed manner, the X-ray scattering patterns were imaged onto a flat panel detector (Pilatus CdTe 2M, 5×1679 pixels with 172× 172 μm2 per pixel). 30s exposure time was used for each pattern. In the measurement, the sample-to-detector distance (center-to-center) was set at 408.02 mm, and for high Q range (up to 20Å^{-1}), the detector was intentionally positioned off-center to the beam. Calibration and conversion of the 2D patterns into 1-D intensity profiles were carried out using DAWN [11].

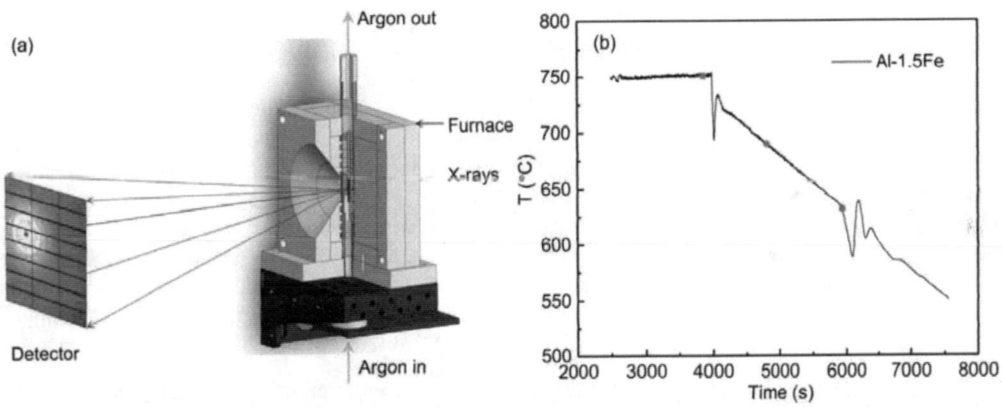

Fig. 1. (a) A schematic of the experimental setup in the synchrotron X-ray total scattering experiment. (b) The cooling profile of Al-1.5Fe melt in the synchrotron X-ray total scattering experiment, where the red dots point to the selected temperatures (750, 690 and 630°C) for the subsequent EPSR simulations.

2.2 Data normalization using GudrunX

The synchrotron X-ray total scattering data was normalized using GudrunX, based on the Krogh-Moe and Normand method [12-14]. The measured X-ray total scattering intensity is a sum of counterparts of sample and background:

$$I^{meas}(Q) = a(Q)I^{sample}(Q) + bI^{bk}(Q) \quad (1)$$

where $Q = 4\pi\, sin(\theta)\,/\lambda$ is the scattering momentum, $a(Q)$ is the transmission factor which can be deduced according to information about the beamline instrument and materials involved, b is the scale factor for background correction, $I^{sample}(Q)$ is the scattering intensity of the sample, and $I^{bk}(Q)$ represents the background scattering intensity in absence of sample of interest.

$I^{sample}(Q)$ is regularly parsed into three parts, including the coherent scattering intensity $I^{coh}(Q)$, the incoherent scattering intensity $I^{incoh}(Q)$ and the multiple scattering intensity $I^{mul}(Q)$, with linear additivity as:

$$NI^{sample}(Q) = I^{coh}(Q) + I^{incoh}(Q) + I^{mul}(Q) \tag{2}$$

in which N is the normalization factor:

$$N = \frac{-2\pi^2 \rho_0 + \int_0^{Q_{max}} Q^2 \frac{I^{incoh}(Q) + \sum_i c_i f_i^2(Q)}{(\sum_i c_i f_i(Q))^2} dQ}{\int_0^{Q_{max}} Q^2 \frac{I^{sample}(Q)}{(\sum_i c_i f_i(Q))^2} dQ} \tag{3}$$

Where ρ_0 is the average atomic density of the sample and Q_{max} is the maximum scattering vector. c_i and f_i the concentration and scattering form factor of i_{th} element respectively. According to the Krogh-Moe-Norman method [13, 14], the structure factor, $S(Q)$, can be calculated by Eq.(4):

$$S(Q) = \frac{I^{coh}(Q) - \left(\sum_i c_i f_i^2(Q) - \left(\sum_i c_i f_i(Q)\right)^2\right)}{(\sum_i c_i f_i(Q))^2} \tag{4}$$

Then, the experimental distribution function, $G(r)$, can be obtained by a Fourier transformation of the structure factor

$$F(r) = \int_0^{Q_{max}} Q(S(Q) - 1)\sin(Qr)\, dQ \tag{5}$$

And the pair distribution function, $g(r)$, can be calculated by Eq.(6):

$$g(r) = 1 + \frac{F(r)}{4\pi r \rho_0} \tag{6}$$

In this work, the f_i of each element can be calculated using the analytic functions. Also, the $I^{Incoh}(Q)$ can be computed using the analytic approximation. The $I^{Mul}(Q)$ is generally neglected in X-ray scattering study since it is usually small. Then for this normalization procedure, three important parameters, Q_{max}, ρ_0 and b, have to be determined. In this paper, the above three parameters were optimized following a refinement procedure proposed by Kaplow et al [15] and Eggert et al [16], and the optimization process was performed using a dedicated MATLAB script Shifeng et al [17] developed.

2.3 EPSR simulation

EPSR simulation allows us to extract the partial pair distribution functions (PDFs) and reconstruct the 3D atomic structure information based upon the SXTS data of liquid or amorphous solids. Starting with and referring to a refined empirical interatomic potential, EPSR executes to produce the simulated structural arrangement by stepwise moving the positions of the atoms against those defined in the empirical information, and it only ceases when the best possible agreement with the measured structure factors is reached [18].

The reference potential of an arbitrary atom pair (atom α and β) is described by a combination of Lennard-Jones and Coulomb potentials:

$$U_{\alpha\beta} = 4\varepsilon_{\alpha\beta}\left[\left(\frac{\sigma_{\alpha\beta}}{r}\right)^{12} - \left(\frac{\sigma_{\alpha\beta}}{r}\right)^{6}\right] + \frac{1}{4\pi\varepsilon_0}\frac{q_\alpha q_\beta}{r}, \tag{7}$$

where ε_0 is the permittivity of free space, $\varepsilon_{\alpha\beta}$ is the well depth parameter, $\sigma_{\alpha\beta}$ is the range parameter, and q_α and q_β are Coulomb charges of atoms α and β. The mixing rule suggests $\varepsilon_{\alpha\beta} = \sqrt{\varepsilon_\alpha \varepsilon_\beta}$ and $\sigma_{\alpha\beta} = (\sigma_\alpha + \sigma_\beta)/2$.

In the simulation, a cubic box with a side length of 72.10 Å and containing 20000 atoms in total (Al: Fe=19854:146 in Al-1.5Fe melt) is built towards the correct density. The Lennard-Jones parameters used for modelling the Al, Fe atoms are listed in Table 1. The operation details of EPSR were provided in [19]. The simulation was performed at the University of Hull supercomputing cluster, Viper. It takes approximately 5 hours of computing time (~5000 iterations) to complete a typical simulation with a satisfactory R-factor of <10^{-3}.

Table 1. The Lennard-Jones potential parameters used in EPSR modelling [20], package version 25.

Atoms	Epsilon (kJ/mol)	Sigma (Å)	Charge
Al	0.32677	2.5000	0
Fe	0.43180	2.4775	0

3. Results and discussion

Fig. 2 (a) and (b) show the structure factor [F(Q)], and PDF [G(r)] obtained from the SXTS experiments and the corresponding EPSR modelling at different temperatures for Al-1.5Fe melt. The simulated results matched the experimental data very well. There is no obvious difference in F(Q) and G(r) of Al-1.5Fe under different temperatures. To understand the coordination environment of Fe atoms, the partial PDF of Al-Fe pair was extracted (see in Fig.2 (c)). As the temperature decreased from 750 °C to 630 °C, the 1st peak position of $g(r)_{Al-Fe}$ moved from ~2.37Å to ~2.34 Å, and the intensity increased from ~2.227 to ~2.323. The cooling process also witnesses an increase of the coordination number (CN) of Fe atoms in their first shell (with a cutting distance of 3.7 Å), in which the CN increase from 10.712 to 10.774. To get a clear view of the temperature-dependent evolution of Fe-centred clusters, all the Fe atoms and the Al atoms in their first shell were extracted from the simulation box at the specified temperatures (see in Fig.2 (e)). Most of the atoms in the first shell of Fe atoms are Al atoms, and some of the Fe-cantered clusters also connect each other to form medium-range order.

The temperature-induced changes on the first peak position of Al-Fe pair, the coordination number of Fe atoms, and 3D distribution of Fe-centred clusters show that the local coordination environment of Fe atoms changes with the temperature decrease. Our previous study reported the shrink behaviour of Al-Sc pair during cooling; we proposed it would result in a more compact short-range order in the melt [3]. Although the Fe-centred clusters are not compatible with the long-range packing FCC Al_3Fe primary phase, the increase of coordination number of Fe atoms from 10.71 to 10.77 shows a trend of more Al atoms appear in the coordination environment of Fe atoms. There are four different types of medium-range ordered structures (MROs) formed by sharing the Al-Al bond, vertex, edge, and face between the neighbouring clusters, which is of great importance in understanding the formation of long-packing clusters, while how these medium-range orders differ under different temperature still needs further structure analysis. More systematic analyses are underway to link the atomic structure in the liquid state to those at the nucleation stage of the different types of Fe-containing intermetallic phases, which are of paramount importance for developing a processing strategy for removing the Fe impurities from the Al melt, which has attracted worldwide interest in the Al research and industry.

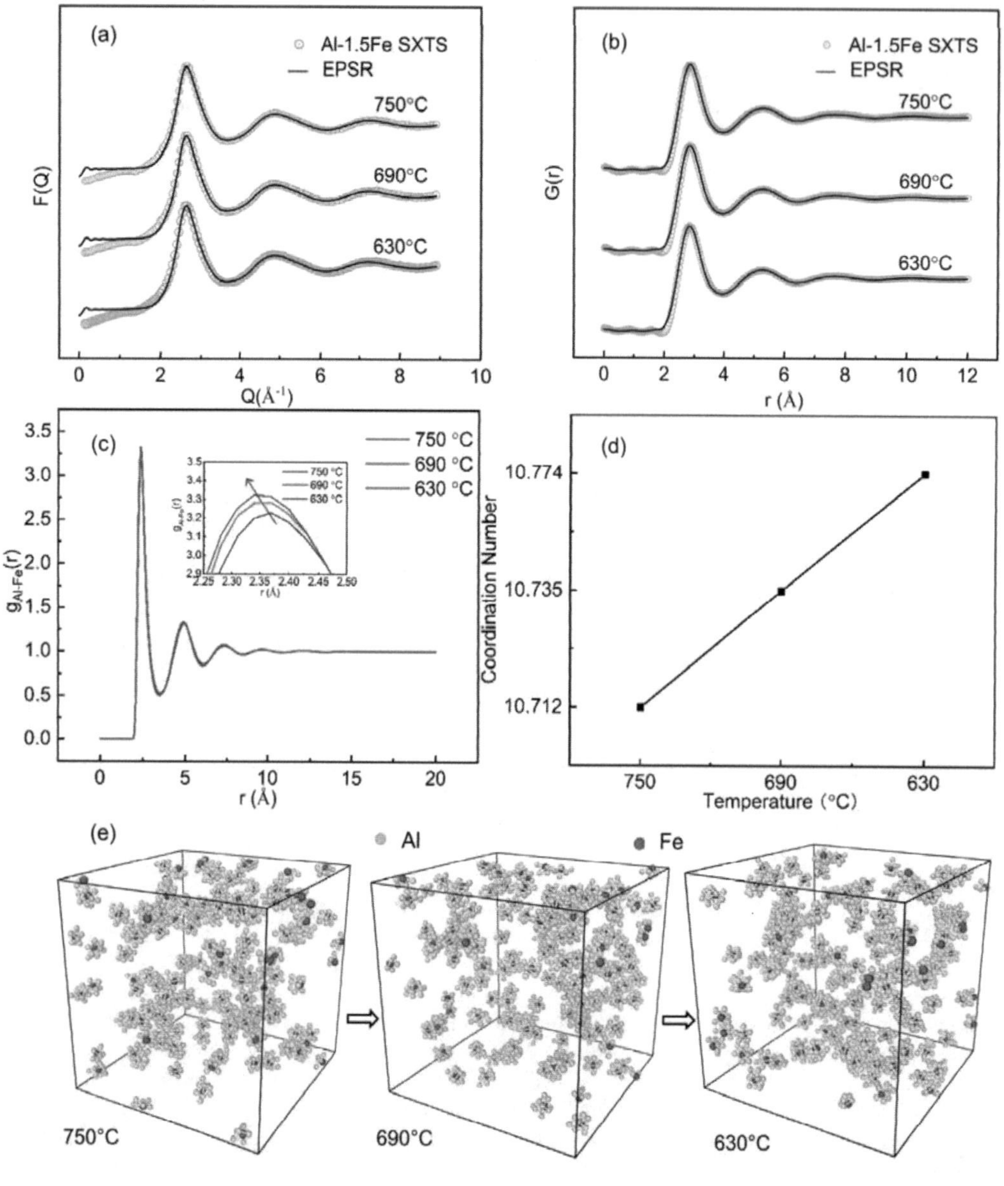

Fig. 2. The SXTS experimental (open circles) and EPSR simulated (solid lines) results of (a) structure factor F(Q) and (b) total PDFs G(r) for the liquidus Al-1.5Fe alloys melt at the targeted temperatures. (c) The partial PDFs of the Al-Fe pair in the Al-1.5Fe melt at the targeted temperatures. (d) The coordination number of Fe-Al pair as a function of temperature in the Al-1.5Fe melt. (e) Atomic structure reconstructed via EPSR modelling for the Fe-centred Al atom clusters at (left) 750 °C, (middle) 690 °C and (right) 630 °C.

4. Conclusion

The combined synchrotron X-ray total scattering and EPSR simulation allowed us to reveal the atomic structure of Al-1.5Fe and its evolution in the temperature range from 750 to 630°C, during which more Al atoms coordinate Fe atoms and the Fe-centered clusters get more compact in the melt. More systematic analyses are underway to link the atomic structure in the liquid state to those at the nucleation stage of the different types of Fe-containing intermetallic phases.

5. Acknowledgements

The authors would like to acknowledge the synchrotron X-ray beam time allocation by beamline I12 of Diamond Light Source (experimental No. MG27571-3), and the University of Hull Supercomputer, Viper. Huang and Xiang would also like to acknowledge the financial support from the China Scholarship Council (CSC) for their PhD study at University of Hull.

References

[1] Reichert H, Klein O, Dosch H, Denk M, Honkimäki V, Lippmann T and Reiter G 2000 *Nature.* **408** 839-41.

[2] Guerdane M, Teichler H and Nestler B 2013 *Phys. Rev. Lett.* **110** 086105.

[3] Huang S, Luo S, Qin L, Shu D, Sun B, Lunt A J, Korsunsky A M and Mi J 2022 Scripta Mater. **211** 114484

[4] Eskin D G and Mi J 2018 Solidification Processing of Metallic Alloys Under External Fields *Springer Series in Materials Science 273, Springer Nature Switzerland AG, Cham Switzerland.*

[5] Zhao Y, Du W, Koe B, Connolley T, Irvine S, Allan P, Schlepütz C M, Zhang W, Wang F and Eskin D G 2018 *Scripta Mater.* **146** 321-26.

[6] Li Z, Qin L, Guo B, Yuan J, Zhang Z, Li W and Mi J 2022 *Acta Metallurgica Sinica (English Letters)* **35** 115-23.

[7] Terzi S, Taylor J, Cho Y, Salvo L, Suéry M, Boller E and Dahle A 2010 *Acta Mater.* **58** 5370-80.

[8] Qin J, Bian X, Sliusarenko S I and Wang W 1998 *J. Phys. Condens. Matter.* **10** 1211.

[9] Il'Inskii A, Slyusarenko S, Slukhovskii O, Kaban I and Hoyer W 2002 *Mater. Sci. Eng. A* **325** 98-102.

[10] Yang J, Zhang J, Dai Y, Ma J, Li F, Bian F, Mi J and Sun B 2015 *Computational Materials Science* **109** 41-8.

[11] Filik J, Ashton A W, Chang P C, Chater P A, Day S J, Drakopoulos M, Gerring M W, Hart M L Magdysyuk O V, Michalik S, Smith A. Tang C, Terrill N J, Wharmby M T and Wilhelm H 2017 *J. Appl. Crystallogr.* **50** 959-66.

[12] Soper A K 2011 *Swindon, UK: Science & Technology Facilities Council.* 2011-3.

[13] Krogh-Moe J 1956 *Acta Crystallogr.* **9** 951-3.

[14] Norman N 1957 *Acta Crystallogr.* **10** 370-3.

[15] Kaplow R, Strong S and Averbach B 1965 *Phys. Rev.* **138** 1336.

[16] Eggert J H, Weck G, Loubeyre P and Mezouar M 2002 *Phys. Rev. B* **65** 174105.

[17] Luo S, Khong J C, Daisenberger D, Huang S, McMillan P F and Mi J 2022 Phys. Rev. B. **105** 064203

[18] Soper A K 2005 *Phys. Rev. B* **72** 104204.

[19] Soper A K 1996 *Chem. Phys.* **202** 295-306.

[20] Magomedov M N 2006 *High Temp.* **44** 513-29.

The cooling rate effects on the Nb-B inoculant potential of refining grains and dendritic scale in 6201 alloy

F E Saldanha[1], S M A Sousa[2], G L Gouveia[2], J E Spinelli[1,2,3]

[1] Department of Materials Engineering, Federal University of São Carlos UFSCar, 13565-905 - São Carlos, SP, Brazil.
[2] Federal University of São Carlos, Graduate Program in Materials Science and Engineering, 13565-905 - São Carlos, SP, Brazil.
[3] Author to whom any correspondence should be addressed.

Nb-B refiners are thought to be more stable than Ti-based refiners, becoming promising inoculants for Al alloys. However, the combined effect of adding NbB on dendritic scale and grain size (GS) with a wide range of solidification cooling rates remains unknown. It is well known that the shape of the α -Al phase is determined by a balance between inoculating components content/type and cooling rate. The inoculation of Al alloys with Ti-based inoculants, for example, significantly lowers GS while having no significant effect on secondary dendritic arm spacing (SDAS). Increased cooling rate, on the other hand, results in grain refinement, but only to a limited extent. Because of a combination of appropriate mechanical properties and suitable electrical conductivity, the 6201 alloy is widely employed in the transmission and distribution of electricity. Despite being a thermomechanically treated alloy, it goes through a melting and solidification process first, which entail research on this level of processing. In order to assess the impact of adding NbB to the 6201 alloy's refinement level (either macro or microscopic), directional solidification (DS) and centrifugal casting (CC) processes were utilized to obtain samples for examination. For the DS samples, cooling rates were determined, whereas for the CC samples, they were approximated for two alloys: 6201 (reference) and 6201-0.5wt.%Nb-0.1wt.%B alloys. Comparing the refined and non-refined alloy results, the GS of the DS 6201-NbB alloy was much smaller (about 13 times) across a range of cooling rates from 0.5 K/s to 20 K/s, whereas only 3 times smaller GS values were detected for CC samples at higher cooling rates. For particular cooling rates of 2.5 K/s and 240 K/s, the GS/SDAS ratios of values related to slow and fast cooling were 6.9 and 6.2, demonstrating that the cooling rate sensitivity of both GS and SDAS was similar in the NbB-inoculated alloy.

1. Introduction

The use of inoculants in Al alloys castings has been practiced for a long time [1,2]. Despite its limited chemical stability, the Al-Ti-B system has been employed in the most well-known commercial inoculants. For Al alloys with high Si concentration, their utilization is inefficient [3,4], with different solutions being investigated over decades. In contrast, several studies have indicated that, in addition to its chemical stability, Nb-B inoculant is effective in refining Al alloys [4-7]. Moreover, the use of Nb in various applications, such as superalloys, thin films, medical implants, Ti and Al alloys, Cu alloys and superconductors, and ceramic and electrolytic capacitors, has expanded significantly in recent years [8-11].

Owing to promising results with refining via Nb-B, greater stability in the molten bath and the commercial possibility of expanding the range of Nb uses, there has been a rising interest in integrating Nb into Al alloys. Further research with alloys of various chemistries and varying solidification conditions is required for this integration and usage as a refiner to be really successful in the form of the Nb-B inoculant. Understanding the GS vs. SDAS relationships is also a significant task to be accomplished as stated by Easton and Davidson [12]. In their study, they detected, for instance, the development of α-Al cells for GS/ SDAS less than 3 in the TiB_2-inoculated 1050 and 2014 alloys. As such, GS/SDAS ratio can be very useful to tailor micromorphologies in refined Al alloys.

The inoculant's efficiency verification for a range of cooling rates, which means for diverse casting techniques, such as sand casting, permanent mold casting, and high-pressure die casting, is also noteworthy. Following Nb–B inoculation, it is critical to evaluate resulting microstructure characteristics.

The experimental study reported in this paper was meant to solve the previously enumerated problems, and as a result, evaluate the applicability of Nb–B inoculation in commercial 6201 alloy investigated over a wide range of cooling rates. The focus was on examining the GS and SDAS resulting from two important processes: centrifugal casting and directional solidification.

2. Experimental procedure

The 6201 and 6201+NbB molten alloys were directionally solidified in vertical cylindrical bipartite ingot molds coated with a refractory material (to minimize radial heat losses). K-type thermocouples were distributed along with the height of the ingot molds from the water-cooled bases, in order to obtain data on temperature as a function of time (5Hz) and position in relation to the bases. When the molten alloys reached liquidus temperatures + 5% at the thermocouples closest to the ingot's bases, water flow was activated against the external parts of the bottom, promoting upward directional solidification.

In centrifugal casting, the same alloys were processed in a hermetic chamber fixed perpendicularly to a rotation axis. The inductively molten alloys were driven into the Cu mold cavity by the force resulting from the rotation of the chamber (400 rpm). Samples at high solidification cooling rates with 3 mm thick plate geometry were produced. Both solidification systems can be seen in Figure 1.

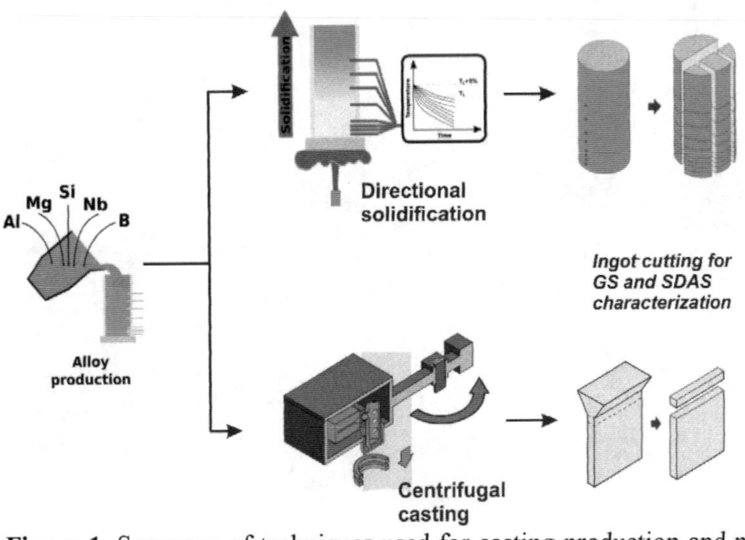

Figure 1. Sequence of techniques used for casting production and microstructural characterization of the 6201 and 6201-NbB alloys.

The samples obtained under different processing conditions were polished and etched with a concentrated Keller's solution (2 ml HF + 3 ml HCl + 5 ml HNO3 + 75 ml H2O) for 60s to reveal the dendritic arrangement and etched with 3% HF in water for 90 seconds to reveal the grain arrangement. After acquiring images by using an optical microscope, the intercept technique was used to measure the SDAS [13], while the Heyn linear intercept method [14] was used to determine the GS.

3. Results and discussions

The thermal profiles of both alloys in Figure 2 depict the cooling down processes recorded during DS. Each line represents temperature x time registration by using thermocouple in a strategic position (P) across the castings. Several samples were extracted throughout the casting length at P of 5, 10, 15, 20, 25, 45, 70, and 90 mm from the water-cooled bottom for metallography investigations. Some optical micrographs inserted in Figure 2 may illustrate the dendritic dimension variations owing to very distinct kinetic conditions associated with the different positions along the casting length. Each position corresponds to a distinct cooling rate, which lowers as the solidification front progresses to the top of the casting. The size of the cells or dendrites at each site reflects this. Microstructures get coarser for positions further away from the bottom of the casting. Because some temperature x time pairs reflect the passing of the solidification front, two optical images of each alloy in two different positions were added on the level of the α-Al temperature and related times.

Even though the focus here is the dendritic array, it can be seen that the α-Al dendrites formed for the 6201-Nb-B alloy start to miss their characteristic shape at P of 20 mm. At these regions and at lower positions towards the casting basis, α-Al cells start to prevail. This occurrence appears to be a transition from reverse cells to dendrites as will be discussed later regarding to the CC results as well. Theoretical and experimental solidification investigations show that the increase in $\dot{T}$ resulted in a change of the micromorphology of the solid/liquid interface from planar to cellular and subsequently to dendritic, implying a direct effect of solidification kinetics on the morphology of the solidification interface. At highly increased $\dot{T}$, the dendritic front transformed into a cellular and planar front, defining a reverse transition that produced the supposed high-cooling rate cells [15].

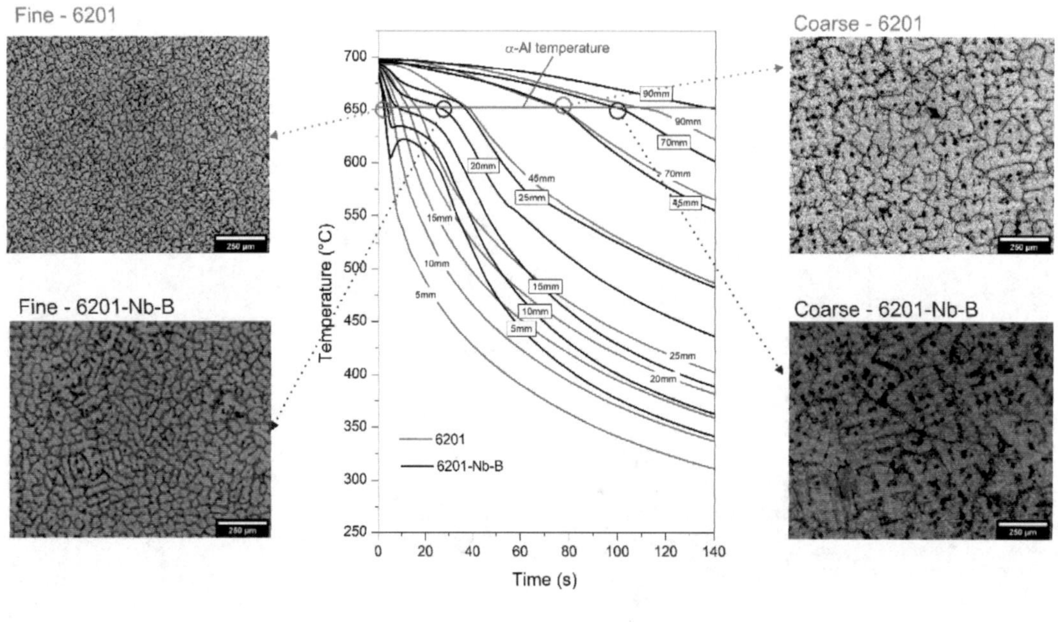

Figure 2. Experimental cooling curves for the DS 6201 and 6201-NbB alloys. Positions shown in mm refer to thermocouples' locations from metal/mold interface. Optical micrographs of some of these positions have been inserted.

The thermal data obtained permitted accurate determination of the slope of the experimental cooling curves (the cooling rate $\dot{T}$ = dT/dt) immediately after the passage of the *liquidus* isotherm by each thermocouple position along the length of the castings. The approximated $\dot{T}$ related to the centrifuged cast samples was determined using extrapolation from the SDAS-scaling relation found for the non-inoculated DS samples. The determined $\dot{T}$ for the CC process was roughly 240 K/s. Further studies remain necessary to improve this technique of extrapolating solidification cooling rate.

The grain size evolutions following Nb-B inoculation in the DS and CC samples can be seen in Figure 3. According to the GS and SDAS measurements in Figure 4 and considering the two different solidification regimes, i.e., slow and fast cooling, approximately 13 times of GS reduction was observed owing to the Nb-B inoculation for a range of fairly low cooling rates during DS. In contrast, for a high solidification rate during CC, only 3 times GS reduction was attained due to the Nb-B addition. This may be expected owing to the high severity of heat extraction rate at the solidification front during CC, which may tailor the alloy's solidification more decisively regardless its modification. The same has also been observed by other authors [5-7].

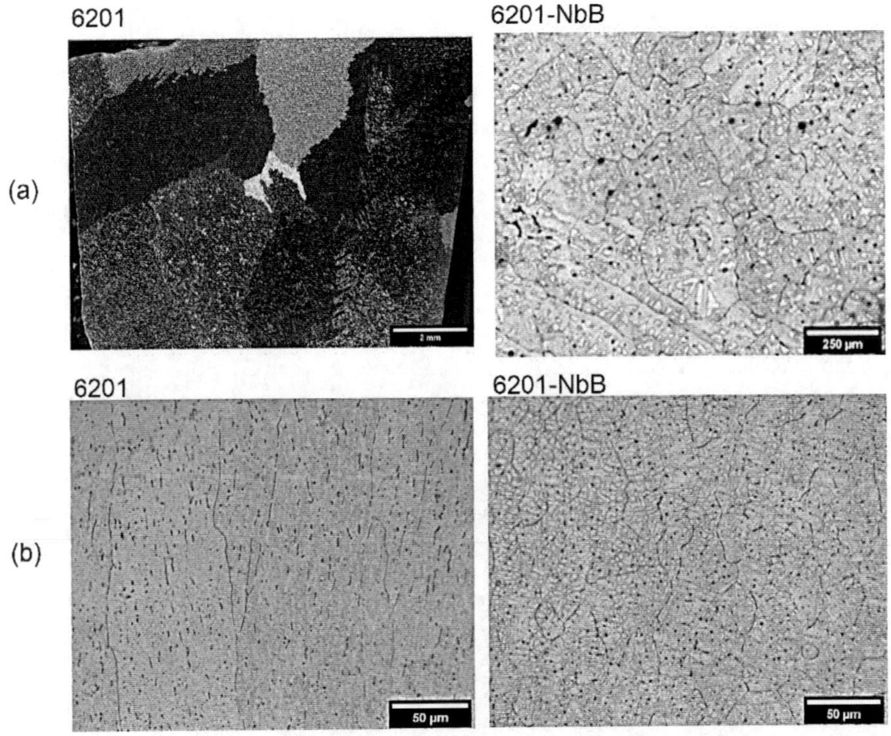

Figure 3. Optical microstructures evidencing the α-Al grain dimensions along the 6201 alloys: (a) transient DS (slow solidification) and (b) CC in copper mold (rapid solidification).

By examining Figure 4, it is possible to establish useful proportions between GS and SDAS, i.e., GS/SDAS ratios associated with low and high cooling rates. Understating these ratios may be useful for tailoring mechanical properties of Al alloys. On the one hand, some authors [16] argue that Al alloys with refined grains, which exhibit higher density of grain boundaries, are generally stronger than coarse-grained ones. On the other hand, some research works [17] affirmed that the SDAS is the dominant factor affecting mechanical properties of Al alloys.

The present results demonstrated the following GS/SDAS ratios for the Nb-B inoculated alloy: 6.9 [2.5 K/s] and 6.2 [~240 K/s] and the following ones for the 6201 alloy: 97.0 [2.5 K/s] and 20.5 [~240 K/s]. The cooling rate sensitivity of both microstructure features becomes quite similar when the alloy is inoculated with Nb-B. As a result, when evaluating refined grains of alloys containing Nb-B for high cooling rate processes, it is also critical to evaluate the impacts of SDAS on the mechanical properties.

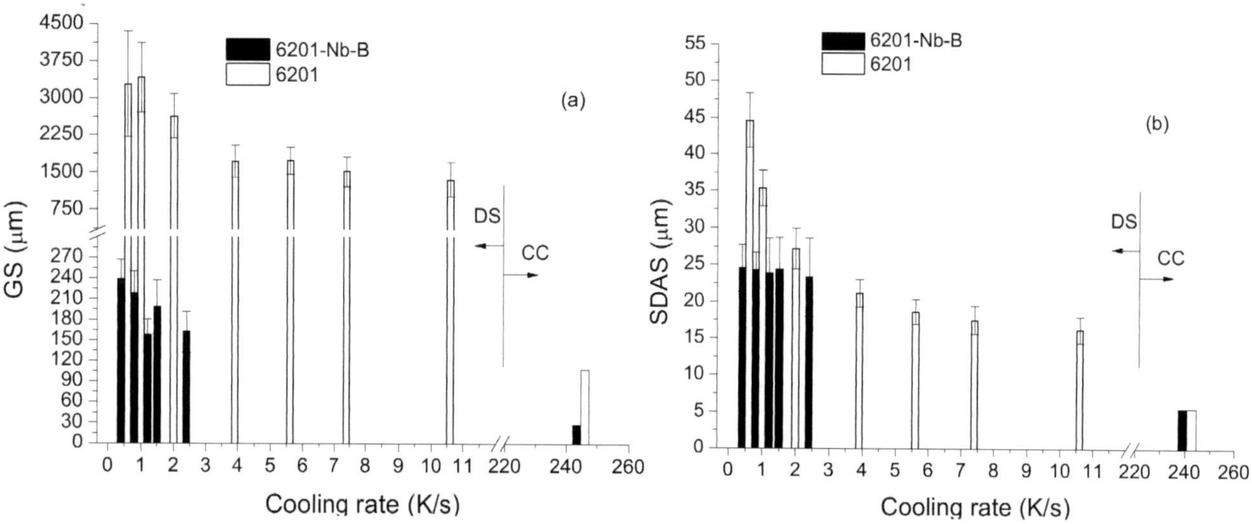

Figure 4. Cooling rate-microstructure variations for the 6201 and 6201-NbB alloys: (a) GS and (b) SDAS.

Another interesting result refers to the growth of high-cooling rate cells in the Nb-B inoculated CC samples as circled in Figure 5. For Ti-inoculated Al samples, as demonstrated by Easton and Davidson [12], the growth of cells may be favored for GS/SDAS lower than 3. However, the current findings indicate that for Nb-B in 6201 alloy, this growth may occur at a ratio of roughly 6. This means that for a same SDAS a coarser grain may allow the growth of cells.

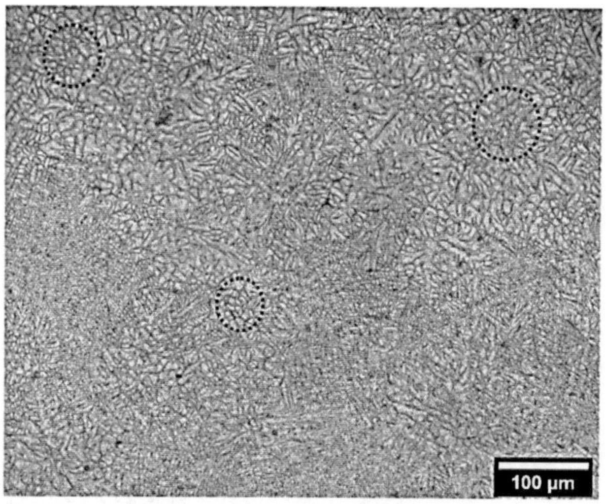

Figure 5. Mixture of cells and dendrites in the CC 6201-NbB alloy.

4. Conclusions

Nb-B refinement in the 6201-Nb-B alloy was demonstrated to be effective even at high cooling rates of 10^2-10^3 °C/s. In contrast to the findings in the literature about cell growth in Ti-B-modified alloys, Nb-B permitted cells emerging for coarser grains. The GS/SDAS ratio of the Nb-B inoculated alloy as a function of slow and fast solidification conditions was maintained relatively constant. In the case of the alloy without inoculation, this ratio was 5 times higher for the slow cooled samples. Therefore, the cooling rate sensitivity of both GS and SDAS was similar for the Nb-B-inoculated sample.

References

1. Sigworth GK and Kuhn TA 2008 *ASM International* 1256
2 Liu Z 2017 *Metall. Mater. Trans. A* **48** 4755
3 Quested TE, Dinsdale AT and Greer AL 2006 *Mater. Sci. Technol.* **22** 1126
4 Li Y, Jiang Y, Liu B, Luo Q, Hu B, Li Q 2021 *J. Mater. Sci. Technol.* **65** 190
5 Bolzoni L, Nowak M, and Hari Babu N 2015 *Mater. Des.* **66** 376
6 Bolzoni L and Hari Babu N 2019 *J. Alloys Compd.* 152807
7 Bolzoni L and Hari Babu N 2019 *J. Mater. Res. Technol.* **8** 5631
8 Rowe CED 1999 *Tantalum-Niobium Int. Study Cent. Bull.* **97** 2
9 Graham R, Sutherlin R, Wah C 2001 *in: Niobium Sci. Technol.,* Orlando, US 337
10 Alves AR, Coutinho AR 2015 *Mater. Res.* **18** 106
11 Grill R, Gnadenberger A 2006 *Int. J. Refract. Metals Hard Mater.* **24** 275
12 Easton M, Davidson C and StJohn D 2011 *Mater. Trans.* **52** 842
13 Gündüz M and Çadirli E 2002 *Mater. Sci. Eng. A* **327** 167
14 ASTM International ASTM E112-13: 2014
15. Trivedi R, Sekhar JA, Seetharaman V 1989 *Metall. Trans. A.* **20** 769
16. Birol, Y 2013 *Mater. Sci. Eng. A* **559** 394
17. Ghassemali E, Riestra M, Bogdanoff T, Kumar BS, Seifeddine S 2017 i*n: ICTP Cambridge, UK* 17

Acknowledgments
The authors acknowledge FAPESP (grant 2019/23673-7) and CNPq. This study was financed in part by the Coordenação de Aperfeiçoamento de Pessoal de Nível Superior - Brasil (CAPES) - Finance Code 001.

IOP Conf. Series: Materials Science and Engineering 1274 (2023) 012009 doi:10.1088/1757-899X/1274/1/012009

Large-scale phase-field simulations for dendrite growth: A review on current status and future perspective

T Takaki

Faculty of Mechanical Engineering, Kyoto Institute of Technology, Matsugasaki, Sakyo-ku, Kyoto 606-8585, Japan

takaki@kit.ac.jp

Abstract. The current status of large-scale phase-field (PF) simulations for dendrite growth is reviewed by focusing on the study conducted by our group. The discussion includes the competitive growth of multiple columnar dendrites, dendrite growth with liquid flow and solid motion, permeability prediction, and cross-scale simulations using the PF method. All PF simulations introduced here were executed using a graphics processing unit (GPU) or a GPU supercomputer to significantly accelerate the PF simulations. Finally, the future perspectives of large-scale dendrite-growth PF simulations are discussed.

1. Introduction

Dendrite is a typical growth morphology formed during alloy solidification [1, 2]. The growth of multiple columnar and equiaxed dendrites forms the final solidification structures that determine the properties of the casting products. Previous studies on dendrite growth have been summarized in some remarkable reviews [3-5]. Numerical simulations of dendrite growth are essential [6] to better understand dendrite growth and optimize solidification structures and processes.

The phase-field (PF) method is the most accurate model for expressing dendrite growth. This method has the following advantages: interface tracking is not necessary to express interface migration, the curvature effect is naturally included in the model, complicated and smooth interface morphologies can be expressed, and the PF equation can be easily discretized using the common finite difference method [7-13]. In 1990, Kobayashi [14] successfully described dendrite growth using the PF method. Surprisingly, he has already demonstrated three-dimensional (3D) dendrite-growth simulations [15]. Thereafter, the PF method was extended to alloy solidification [16, 17]. To date, quantitative PF models [18-23], in which the results do not depend on the interface thickness, have been generally used for alloy solidification. However, the PF method is computationally expensive because it is a diffused interface model. Thus, dendrite-growth simulations using the PF method were limited to two-dimensional (2D) or one 3D dendrite until around 2010.

To circumvent this issue, substantial numerical efforts have been made for performing PF simulations. The adaptive mesh refinement (AMR) method has been applied to PF simulations ever since the emergence of the PF method [24-27]. In PF simulations using AMR, fine grids are arranged only around the interface, and coarse grids are used in the bulk regions. Thus, AMR is effective when the interface fraction is small, whereas the computational efficiency of AMR becomes worse than that of the uniform mesh when the interface fraction is large. In dendrite-growth PF simulation, the AMR method is highly efficient when the arm spacing is significantly wider than the curvature radius of the dendrite tip. However, AMR coding is complicated and time consuming, sacrificing one of the advantages of PF method, which is easy to code using simple discretization methods such as the finite difference method. Recently, open AMR libraries have become increasingly available. Parallel computing using multiple cores is a common approach for accelerating PF simulations [28]. A scheme combining the AMR method and parallel computing was also employed to accelerate PF simulations [29-31].

In 2011, Yamanaka et al. conducted the first dendrite-growth PF simulations using a graphics processing unit (GPU) [32]. In the same year, we won the Gordon Bell Prize, which is one of the most prestigious awards in high-performance computing, together with a group at the Tokyo Institute of Technology. We achieved a large-scale dendrite-growth PF simulation in 4,096 × 1,024 × 4,096 grid points in parallel computing using 512 GPUs (NVIDIA Tesla M2050) of the GPU-rich supercomputer TSUBAME2.0 at the Tokyo Institute of Technology [33]. In 2013, we conducted the PF simulation of dendrite growth in 4,096 × 4,104 × 4,096 grid points using 768 GPUs in TSUBAME2.0 [34]. This PF simulation was four times larger than our previous simulations. At present, single-GPU computing has become common in the PF community [6, 20, 35, 36], although parallel computing using multiple GPUs is still limited to a small number of groups. Recently, Sakane et al. enabled AMR dendrite-growth PF simulations through parallel computing using multiple GPUs [37]. A GPU is composed of thousands of cores, which is different from the central processing unit (CPU) that contains only a few cores. Moreover, a GPU enables massive parallelism, which is suitable for repetitive computing. The PF equation can be discretized using a time-explicit scheme, and the dendrite-growth PF simulation requires many arithmetic operations in a single step. This indicates that the dendrite-growth PF simulation has a highly efficient coalesced memory access and high arithmetic intensity, which is suitable for GPU and parallel-GPU computing. GPU M2050 (NVIDIA Tesla M2050), which was installed on TSUBAME2.0 (November 2011–October 2013), had single-precision 1.03 TFlops with 448 CUDA cores, while A100 (NVIDIA Tesla A100), which is a high-end GPU of 2022, has single-precision 19.5 TFlops with 6,912 cores. In the past decade, the computational speed of a GPU has increased by approximately 20 times. Such increase in computational speed has not been achieved in the CPU. In addition, the theoretical performance of A100 is almost equal to that of the fastest supercomputer around 2001. In future, we expect further performance improvements in GPU computing.

In this article, the current status of large-scale PF simulations for dendrite growth is reviewed, with a focus on the simulation results of our group [38, 39]. Herein, we discuss the competitive growth of multiple columnar dendrites (Section 2), dendrite growth with liquid flow and solid motion (Section 3), permeability prediction (Section 4), and cross-scale simulations using the PF method (Section 5). Finally, future perspectives of large-scale PF simulations are discussed.

2. Competitive growth of multiple columnar dendrites

Solidification microstructures are formed through the competitive growth of a large number of dendrites. Selection behavior among columnar dendritic grains has been well known as the Walton and Chalmers (WC) model [40]. In this model, favorably oriented (FO) grains, with a smaller angle between the preferred growth direction and heat flow direction, continue to grow by overgrowing unfavorably oriented (UO) grains with larger angles. In contrast to the WC selection model, unusual selection behaviors were observed at the converging grain boundary (GB) in bi-crystal systems, where the UO grain keeps growing by overgrowing the FO grain [41]. The mechanism of the unusual selection was clarified via 2D PF simulations [36, 42, 43], in which the lateral migration of the FO dendrite at the GB is the key to the unusual selection. In addition, unusual selection was clearly shown in the in situ observation in a thin film of a transparent alloy [44]. The 2D competitive behaviors at converging and diverging GBs have been investigated in detail in several studies [45-51]. Competitive growth with liquid flow is also an interesting and challenging topic [52, 53].

Because the PF simulations of competitive growth between columnar grains require a relatively long time in a multiple-grain system, the computational cost is high, even in 2D problems. Therefore, 3D simulations of competitive growth are a challenging topic even at present. We performed large-scale 3D PF simulations of the competitive growth among multiple dendrites, as shown in figure 1. In the competitive growth of the bi-crystal system shown in figure 1(a), we confirmed that unusual selection can occur even in 3D problems, whereas in 3D, the dendrite interaction at the GB is significantly more complicated than that in 2D [54, 55]. Although some 3D PF simulations for bi-crystal competition have been conducted, their computational domain size is thought to be insufficient [56, 57]. Figure 1(b) shows the result of polycrystalline competitive growth [58]. We found that unusual selection occurs locally and temporarily; however, FO grains grow preferentially by following the WC model. In other words, unusual selection is remarkable only in a special situation such as a bi-crystal. Thus, it was concluded that competitive growth following the WC model commonly occurs during the polycrystalline solidification [59, 60]. In the simulations shown in figures 1(c) and (d) [61, 62], we investigated an array of primary arms in a single crystal [63], which is important for microsegregation and permeability. It was concluded that the hexagonal array is dominant for columnar dendrites/cells, including dendrites growing in the temperature gradient direction [61] and those with reasonably tilted angles [62].

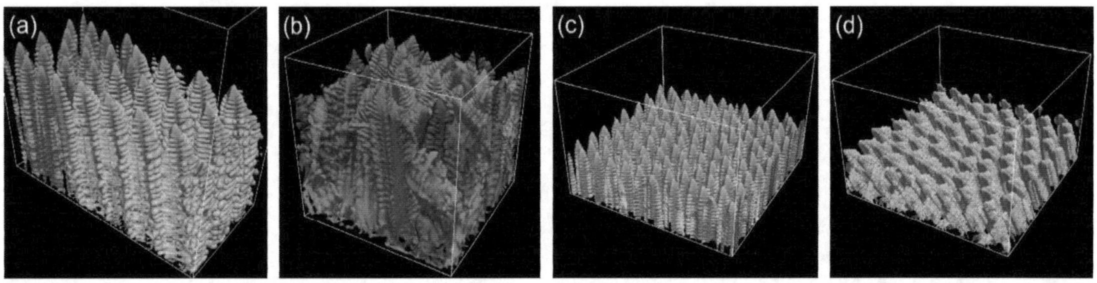

Figure 1. Columnar dendrite morphologies of an Al–3wt%Cu alloy for a (a) bi-crystal (768 × 1,536 × 1536 grid points) [64], (b) polycrystal ($1{,}024^3 + 512^2 \times 256$ grid points) [58], and (c)(d) single crystal ($1{,}536^2 \times 512 + 768^2 \times 256$ grid points) [62] computed by large-scale PF simulations using a GPU supercomputer. The grid size Δx was set to $\Delta x = 0.75$ μm. In (b-d), coarse grids with a size of $2\Delta x$ were used for the upper-half region with only the liquid phase to reduce computational cost, although only the fine grid region is visualized in (b).

High accuracy evaluation of competitive growth is still challenging; therefore, such evaluations have been limited to thin films [65, 66] and cellular structures [35, 67]. Furthermore, there have been no simulations of competitive growth among dendrites with a large primary arm spacing and developed secondary arms. Although the frozen temperature approximation [68] has been widely used in directional solidification simulations, considering the thermal effect is also important and challenging [69, 70]. High-performance computing can help to clarify these phenomena and issues.

3. Dendrite growth and semi-solid deformation with liquid flow and solid motion

Solidification of alloys is a complicated multi-physics phenomenon including solid–liquid phase transformation, transport of solute and heat, liquid flow, solid motion, and solid deformation. PF modeling has been evolving from purely diffusive conditions to conditions with liquid flow [71], solid motion [72-76], and solid deformation [77, 78]. As expected, the computational cost remarkably increases when additional phenomena are considered.

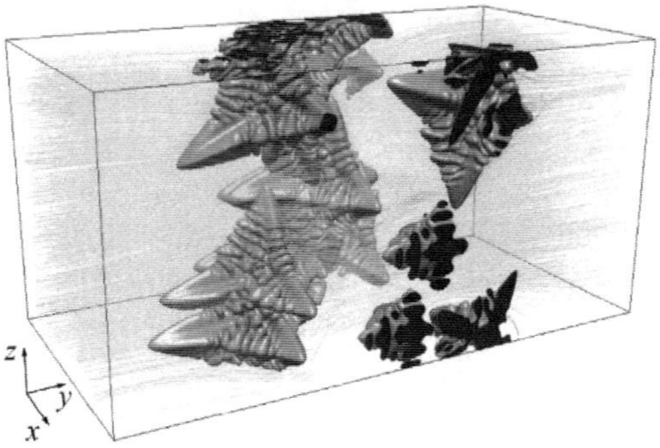

Figure 2. Snapshot of a large-scale PF–LB simulation of the growth of multiple equiaxed dendrites with forced convection [79]. The PF–LB simulation was conducted for 1,536 × 3,072 × 1,536 grid points by parallel computing using 512 GPUs (NVIDIA Tesla K20X) on the TSUBAME2.5. The simulation time (wall time) for 90,000 steps was approximately 17 h.

The lattice Boltzmann (LB) method [80] for liquid flow is a suitable scheme for parallel computing because its equations can be solved explicitly without solving the Poisson equation, which is required when directly solving the Navier–Stokes equations. Thus, recently, the LB method has been actively used along with the PF equation, referred to as the PF–LB model, to simulate dendrite growth under liquid flow conditions [81, 82]. We enabled large-scale PF–LB simulations using a GPU supercomputer [79]. Figure 2 shows the growth of fixed multi-equiaxed dendrites under forced convection using 512 GPUs (NVIDIA Tesla K20X) of the supercomputer TSUBAME2.5 [79]. This PF–LB GPU parallel scheme was applied to inclined dendrites with forced convection [83] and dendrite growth with natural convection [84, 85]. In these simulations, we used the single-relaxation time LB model, in which the kinetic viscosity ν is related to the relaxation time τ as $\nu = \Delta t c^2 (\tau - 0.5)/3$ with time increment Δt and discrete particle velocity c. Considering $c = \Delta x/\Delta t$, $\nu = \Delta x^2 (\tau - 0.5)/(3\Delta t)$. As τ should be close to 1, setting $\tau = 1$ results in $\nu = \Delta x^2/(6\Delta t)$. Assuming the grid size $\Delta x = 1$ μm and time increment $\Delta t = 10$ μs, we obtain $\nu = 1.67 \times 10^{-8}$ m²/s, which is 1/60 – 1/20 of the actual kinetic viscosity of Al–Cu alloys [86]. To achieve simulations with the actual ν of an Al–Cu alloy, we need a significantly smaller time increment, making the computation much more expensive. Recently, we introduced the two-relaxation time LB model, which allows for a significantly larger relaxation time [87]. In addition, the LB simulation became stable upon introducing interpolated bounce-back at the solid–liquid interface [87].

In simulations with liquid flow, we usually need a significantly larger computational domain than that for purely diffusive conditions, and AMR is very effective under such conditions. Guo et al. have been actively developing parallel–AMR [88], and recently applied it to dendrite growth under thermosolutal convection [89]. We developed a multi-GPU parallel–AMR scheme [37] for purely diffusive PF simulations, in which octree block decomposition and dynamic load balancing were incorporated in the code [90]. Block decomposition enables effective computing in GPUs [91] and dynamic load balancing is necessary for large-scale computing [90]. We are currently developing a multi-GPU parallel–AMR scheme for the PF–LB model. Figure 3 shows an example of dendrite growth with natural convection of Al–3wt%Cu using a multi-GPU parallel–AMR simulation. The domain size was set to $256 \times 4096 \times 8192 \Delta x^3$ with $\Delta x = 0.391$ μm. Using uniform grids with a size of Δx requires 8.59×10^9 grid points, while the number of grid points used in the simulation shown in figure 3(b) was 3.96×10^8, i.e., the reduction ratio in the number of grid points was approximately 95%. Although this is still an expensive computation using 64 GPUs (NVIDIA Tesla P100) of the supercomputer TSUBAME3.0, we could significantly reduce the computational cost because it required 256 GPUs when using uniform grids with a size of Δx. Currently, we are trying to further improve the computational efficiency of the multi-GPU parallel–AMR [92].

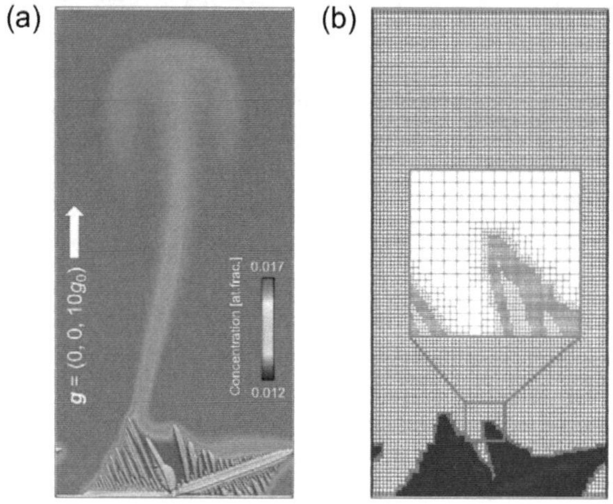

Figure 3. Multi-GPU parallel–AMR PF–LB simulation of dendrite growth with natural convection of Al–3wt%Cu: (a) Solute concentration distribution and (b) block decomposition at step number 3.8×10^5 ($\Delta t = 7.266 \times 10^{-6}$ s). The block is decomposed in to 8^3 grids. Parallel computing using 64 GPUs (NVIDIA Tesla P100) of the supercomputer TSUBAME3.0 was performed for a domain with dimensions of $0.1 \times 1.6 \times 3.2$ mm^3. The number of grid points is 3.96×10^8 which is 0.046 times that of $256 \times 4096 \times 8192$ grid points when using uniform grids. The simulation time for 3.8×10^5 steps was approximately 16.3 h.

The motion of equiaxed grains is one of the major factors that causes macrosegregation [93, 94]. Recent studies have enabled dendrite-growth PF simulations with the motion of multiple dendrites [95-98]. We have also enabled such simulations [99] as well as large-scale simulations with a large number of mobile equiaxed dendrites in 2D [100]. More recently, we enabled it in the frame of a multi-phase-field (MPF) model [101] using a double-obstacle potential [102, 103]. This model can accurately express a series of solidification processes, from dendrite growth with motion to grain growth after vanishing of the liquid phase [101]. Figure 4 shows a simulation example for the growth of a large number of dendrites with solid motion followed by grain growth [101]. This MPF–LB model can be applied to semi-solid deformations [101, 104]. Figure 5 shows a simulation example using the MPF–LB model [101] of the semi-solid shear deformation with a solid fraction f_s of 0.78. Unlike the simulations using the discrete element method [105, 106], our MPF–LB model can simulate semi-solid deformation containing arbitrarily shaped grains with a solid–liquid phase transformation. The shear band formation due to Reynolds dilatancy, which is a characteristic phenomenon in semi-solid deformation [107, 108], can be clearly seen in figure 5. Parallel computing using 16 GPUs (NVIDIA Tesla P100) of the supercomputer TSUBAME3.0 were performed for the simulations shown in figures 4 and 5; because the simulations shown in these figures are still limited to 2D, we expect that the 3D simulations will be performed in the future.

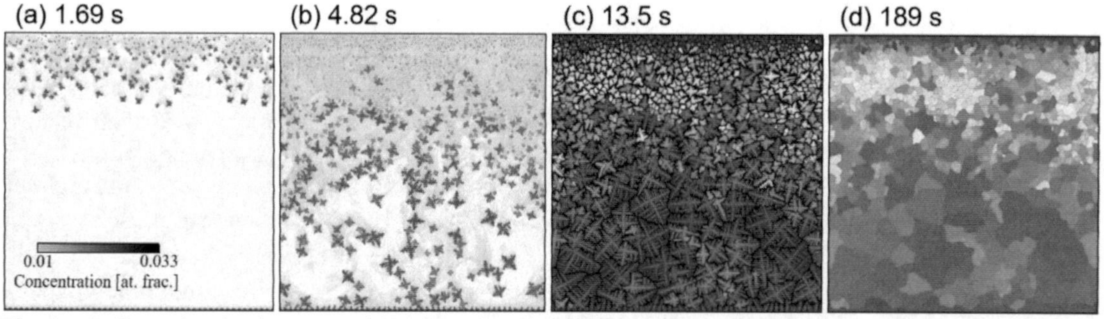

Figure 4. Growth of a large number of equiaxed dendrites with sedimentation and subsequent grain growth of Al–3wt%Cu. Parallel computing using 16 GPUs was performed for a domain with dimensions of 7.37^2 mm^2 ($16{,}384^2$ grid points) [101]. The color indicates the solid grain number.

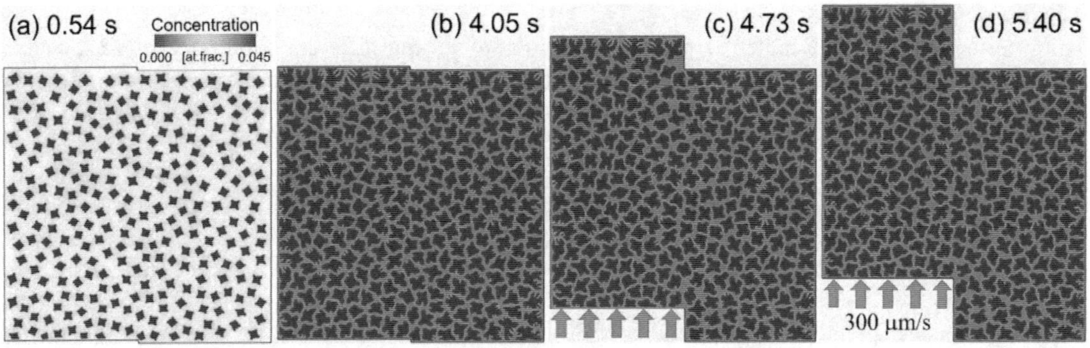

Figure 5. Growth of equiaxed dendrites (a, b) and subsequent shear deformation (c, d) of a semi-solid state of Al–3wt%Cu. Solid fraction f_s at the end of solidification (b) was $f_s = 0.78$. Parallel computing using 16 GPUs was conducted for a domain with dimensions of 1.843^2 mm^2 ($4{,}096^2$ grid points). The simulation time for 4×10^5 steps ($\Delta t = 1.35 \times 10^{-5}$ s) was approximately 9.6 h.

4. Permeability prediction

Permeability is an important parameter used in simulations to predict macrosegregation [109-111]. Permeability predictions via computational fluid dynamics [112-114] or flow experiments using sample models produced by a 3D printer [115-117] for experimentally obtained dendritic structures have been conducted. These experiments were conducted as alternative methods for direct experimental measurement of permeability. However, numerical predictions of both the microstructures and interdendritic liquid flow are promising for the systematic evaluation of permeability. Permeability predictions for dendritic structures computed by PF simulations have been reported [118-121] owing to improvements in computer performance. However, these dendrite structures were not realistic as only one columnar dendrite or a small number of equiaxed dendrites was considered.

In our previous studies, we enabled permeability prediction using large-scale PF and LB simulations for both normal and parallel flows to columnar dendritic structures [122, 123]. Because of the large-scale LB simulations for columnar structures with multiple dendrites, as shown in figure 1(c), it was concluded that a regular hexagonal array of columnar dendrites is appropriate as a permeability prediction alternative to a columnar structure with multiple dendrites, and the Kozeny–Carman equation [124] with k_C = 9 (normal flow) and 3 (parallel flow) can express the dimensionless permeability for the flow through columnar structures [122, 123]. Other advantages of this permeability prediction method are that the distribution of the PF variable can be directly utilized to express the dendrite structures in the LB simulation and the periodic boundary conditions can be applied to the LB simulation. Currently, we are investigating the permeability of columnar dendrites with more developed secondary arms [125], tilted columnar dendrites (figure 1(d)), polycrystalline columnar dendrites (figure 1(b)), and equiaxed structures. Figure 6 shows a simulation example of the interdendritic liquid flow for a polycrystalline columnar structure computed by the PF simulation.

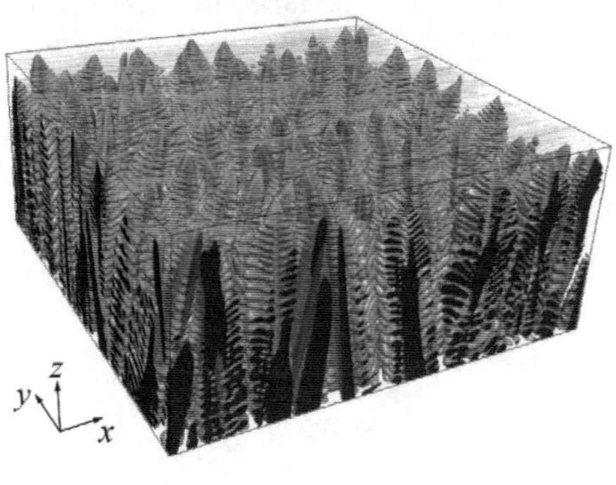

Figure 6. Snapshot of a large-scale LB simulation of interdendritic liquid flow for a polycrystalline columnar structure computed by the PF method. The PF simulation was conducted for $2{,}048^2 \times 1{,}536$ grid points, and the LB simulation was conducted for $2{,}048^2 \times 896$ grid points by removing an upper domain with only liquid. The LB simulation was performed using 224 GPUs (NVIDIA Tesla K20X) on the TSUBAME2.5.

5. Cross-scale simulations

Owing to the improved computer performance, simulations using different numerical models, such as molecular dynamics (MD) and PF methods, which have been conventionally used in different spatiotemporal conditions, are becoming feasible in the same spatiotemporal scales. For example, large-scale MD simulations using parallel-GPU computing exhibited solidification structures [126, 127] and reproduced grain growth [128, 129]. In such situations, the concept of conventional multi-scale modeling has been changing gradually, and we demonstrated a new criterion for multi-scale modeling [38]. The next step of conventional multi-scale modeling may be 'cross-scale modeling', where different methods can be used in the same spatiotemporal manner; in other words, different methods can be integrated.

One promising method that integrates different numerical and experimental methods is data assimilation, which was originally developed for meteorology and oceanography [130-132]. Koyama suggested coupling data assimilation and the PF method in materials science [133]. He stated that *data assimilation is the process by which observations are incorporated into a computer model of a real system.* In other words, we can predict the unknown parameters and/or conditions through the integration. In the PF simulations of material microstructure evolution, the lack of material parameters is a serious issue. Yamanaka has been conducting pioneering work in introducing data assimilation into the PF simulations [134-137]. Ohno et al. developed a novel inference method using an ensemble Karman filter (EnKF) of the solid–liquid interfacial parameters of a pure metal by coupling PF and MD simulations [138, 139].

Figure 7 presents an illustration of data assimilation using the EnKF. In the EnKF, we can use nonlinear system models, such as PF equations, by introducing an ensemble approximation, where N simulations are carried out simultaneously, and the probability distribution function of the state is derived approximately. Figure 7(a) is an image of the dendritic morphological change in the N PF simulations over time $t = t^1_{obs} \sim t^2_{obs}$, and figure 7(b) shows the changes in an unknown parameter ε, which is inferred in this example, during time $t = 0 \sim t^4_{obs}$. At time $t = 0$, the N PF simulations start from different initial conditions, where, as shown in figure 7(b), different values of ε are set for N simulations based on a Gaussian distribution. Then, the N PF simulations were performed until $t = t^1_{obs}$, where the first observation data are available. The numerical simulation stage is called forecasting. During forecasting, N state vectors $\mathbf{x}_t^{(i)}$ ($i = 1$–N) are updated by following time evolution equations of each field. For example, the state vector $\mathbf{x}_t^{(i)}$ can be set as $\mathbf{x}_t^{(i)} = (\boldsymbol{\phi}_t^{(i)}\ \mathbf{u}_t^{(i)}\ \varepsilon^{(i)})^{\mathrm{T}}$ for isothermal solidification of a binary alloy [19] with the vectors of the PF variable $\boldsymbol{\phi}_t^{(i)}$ and dimensionless supersaturation $\mathbf{u}_t^{(i)}$ at all grid points and an unknown parameter $\varepsilon^{(i)}$. At $t = t^1_{obs}$, the state vectors $\mathbf{x}_t^{(i)}$ are modified by the Bayesian inference theory, or the EnKF, based on the observation data; in other words, the simulation results of the dendrite morphology are modified to be close to the observation results. This modification stage is called filtering. Subsequently, the forecasting was performed until $t = t^2_{obs}$. As explained above, data assimilation consists of two stages: forecasting and filtering. The unknown parameters $\varepsilon^{(i)}$ are expected to gradually approach the true value, as shown in figure 7(b). Although the unknown parameters $\varepsilon^{(i)}$ do not change during forecasting in the example of figure 7(b), the other components of $\mathbf{x}_t^{(i)}$, such as the PF variable and dimensionless supersaturation, change. In the present example, only one unknown value was used, whereas multiple unknown parameters could be inferred simultaneously. In the twin experiment, which was used to validate data assimilation systems of polycrystalline grain growth, we showed simultaneous inference of 30 unknown parameters [140]. We expect actual simultaneous inference of GB properties, which has anisotropy depending on the misorientation between grains and GB orientation, by combining large-scale MPF grain growth simulations [141-143] and in situ observations via data assimilation.

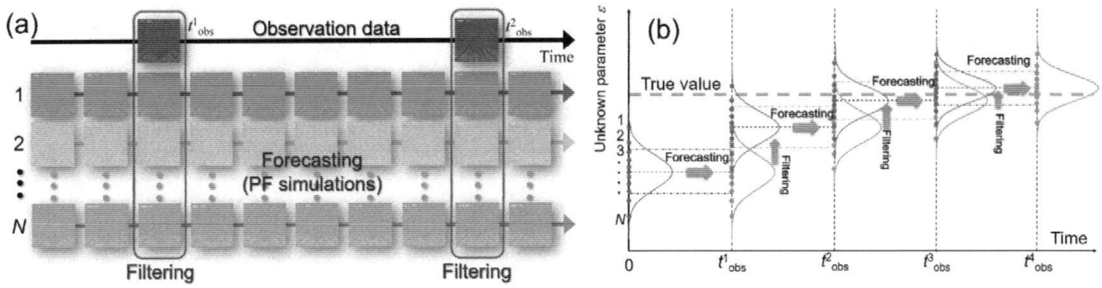

Figure 7. Illustration of data assimilation using the EnKF. Changes in (a) dendrite morphology and (b) an unknown parameter ε during data assimilation. Forecasting (numerical simulation to update the state vector) and filtering (modification of the state vector) are repeated in sequential data assimilation.

In the EnKF, a Gaussian distribution is expressed by an ensemble approximation using N discrete values, and hundreds of simulations should be performed simultaneously, as shown in figure 7. Therefore, in 3D problems, the computational cost becomes much more expensive than that for 2D. To reduce N, Yamanaka et al. introduced a local EnKF to the dendrite-growth PF simulation [135], and, to accelerate data assimilation in 3D, they performed parallel computing [134]. We are currently attempting to develop a method that integrates the PF simulations and time-resolved in situ X-ray observations [144, 145]. In our previous estimation to reproduce X-ray imaging in thin samples by the PF simulation, we required approximately 5×10^9 grid points [38]. Thus, to integrate the PF simulations and in situ observations, a high-performance computing and efficient data assimilation scheme are required.

6. Conclusions and perspectives

Herein, I briefly reviewed the current status of large-scale PF simulations for dendrite growth, with a focus on the results obtained by our group on competitive growth of multiple columnar dendrites, dendrite growth with liquid flow and solid motion, permeability prediction, and cross-scale simulation using the PF method. All the PF simulations introduced here were executed using a GPU or GPU supercomputer.

Solidification is a multi-physics phenomenon; several phenomena such as solid–liquid phase transformation, transport of solute and heat, liquid flow, and solid motion need to be represented simultaneously in simulations. However, it is still a significant challenge to reproduce the solidification multi-physics phenomenon in 3D in the framework of the PF method, and the application of high-performance computing is indispensable. Even if such multi-physics simulations become possible, a full-scale PF simulation of the actual casting scale will not be possible in the future. In such a case, multi-scale modeling that utilize results of systematic dendrite-scale PF simulations in higher-scale simulations is considered to be a good research direction. The permeability prediction introduced here is an example of such a multi-scale modeling, and it is expected that an effective permeability model that considers process history can be constructed through systematic permeability evaluations using the PF and LB methods. Furthermore, as introduced in the section of cross-scale simulation, PF simulations incorporating experimental observations are one of the major future research directions, and the development of such a technology is indispensable for understanding dendrite solidification in more detail. It is expected that numerical simulations will not only reproduce experiments, but also enable numerical simulation-driven material and process design by integrating with data science, such as machine learning and data assimilation. I believe that further development of PF models, computational schemes, and high-performance computing will be important in the future of solidification science.

Acknowledgments

This review was supported by Dr. S. Sakane (Kyoto Institute of Technology) and Prof. T. Aoki (Tokyo Institute of Technology). The simulation results presented herein were obtained in collaboration with Prof. M. Ohno (Hokkaido University), Prof. Y. Shibuta (University of Tokyo), Prof. H. Yasuda (Kyoto

University), Dr. S. Sakane (Kyoto Institute of Technology), and Prof. T. Aoki (Tokyo Institute of Technology). The recent studies introduced in this paper were supported by KAKENHI Grant-in-Aid for Scientific Research (A) 20H00217. Large-scale simulations conducted using the GPU supercomputer TSUBAME at the Tokyo Institute of Technology were supported by HPCI and JHPCN.

References

[1] Kurz W and Fisher D J 1989 *Fundamentals of Solidification* (Switzerland: Trans Tech Publications Ltd.)

[2] Dantzig J A and Rappaz M 2016 *Solidification: 2nd Edition - Revised & Expanded*: CRC Press LLC)

[3] Asta M, Beckermann C, Karma A, Kurz W, Napolitano R, Plapp M, Purdy G, Rappaz M and Trivedi R 2009 *Acta Mater.* **57** 941-71

[4] Kurz W, Fisher D J and Trivedi R 2019 *Int. Mater. Rev.* **64** 311-54

[5] Kurz W, Rappaz M and Trivedi R 2021 *Int. Mater. Rev.* **66** 30-76

[6] Karma A and Tourret D 2016 *Curr. Opin. Solid State Mater. Sci.* **20** 25-36

[7] Chen L Q 2002 *Annu. Rev. Mater. Sci.* **32** 113-40

[8] Moelans N, Blanpain B and Wollants P 2008 *Calphad* **32** 268-94

[9] Singer-Loginova I and Singer H M 2008 *Rep. Prog. Phys.* **71**

[10] Steinbach I 2009 *Modell. Simul. Mater. Sci. Eng.* **17** 1-31

[11] Wang Y and Li J 2010 *Acta Mater.* **58** 1212-35

[12] Takaki T 2014 *ISIJ Int.* **54** 437-44

[13] Tourret D, Liu H and Llorca J 2022 *Prog. Mater. Sci.* **123** 100810

[14] Kobayashi R 1993 *Physica D* **63** 410-23

[15] Kobayashi R 1994 *Exp. Math.* **3** 59-81

[16] Wheeler A A, Boettinger W J and McFadden G B 1992 *Phys. Rev. A* **45** 7424-39

[17] Warren J A and Boettinger W J 1995 *Acta Metall. Mater.* **43** 689-703

[18] Karma A and Rappel W J 1998 *Phys. Rev. E* **57** 4323-49

[19] Ohno M and Matsuura K 2009 *Phys. Rev. E* **79** 031603

[20] Ohno M 2012 *Phys. Rev. E* **86** 051603

[21] Ohno M, Takaki T and Shibuta Y 2016 *Phys. Rev. E* **93** 012802

[22] Ohno M, Takaki T and Shibuta Y 2017 *Phys. Rev. E* **96** 033311

[23] Ohno M 2020 *ISIJ Int.* **60** 2745-54

[24] Braun R J and Murray B T 1997 *J. Cryst. Growth* **174** 41-53

[25] Provatas N, Goldenfeld N and Dantzig J 1999 *J. Comput. Phys.* **148** 265-90

[26] Lan C W, Chang Y C and Shih C J 2003 *Acta Mater.* **51** 1857-69

[27] Takaki T, Fukuoka T and Tomita Y 2005 *J. Cryst. Growth* **283** 263-78

[28] George W L and Warren J A 2002 *J. Comput. Phys.* **177** 264-83

[29] Guo Z and Xiong S M 2015 *Comput. Phys. Commun.* **190** 89-97

[30] Gong T Z, Chen Y, Cao Y F, Kang X H and Li D Z 2018 *Comput. Mater. Sci.* **147** 338-52

[31] Greenwood M, Shampur K N, Ofori-Opoku N, Pinomaa T, Wang L, Gurevich S and Provatas N 2018 *Comput. Mater. Sci.* **142** 153-71

[32] Yamanaka A, Aoki T, Ogawa S and Takaki T 2011 *J. Cryst. Growth* **318** 40-5

[33] Shimokawabe T, Aoki T, Takaki T, Yamanaka A, Nukada A, Endo T, Maruyama N and Matsuoka S 2011 *Proceedings of 2011 SC - International Conference for High Performance Computing, Networking, Storage and Analysis* 1-11

[34] Takaki T, Shimokawabe T, Ohno M, Yamanaka A and Aoki T 2013 *J. Cryst. Growth* **382** 21-5

[35] Tourret D, Debierre J M, Song Y, Mota F L, Bergeon N, Guérin R, Trivedi R, Billia B and Karma A 2015 *Phys. Rev. E* **92** 042401
[36] Tourret D and Karma A 2015 *Acta Mater.* **82** 64-83
[37] Sakane S, Takaki T and Aoki T 2022 *Materials Theory* **6** 3
[38] Shibuta Y, Ohno M and Takaki T 2018 *Adv. Theory Simul.* **1** 1800065
[39] Shibuta Y, Ohno M and Takaki T 2015 *JOM* **67** 1793-804
[40] Walton D and Chalmers B 1959 *Trans. Metall. AIME* **215** 447-57
[41] Zhou Y Z, Volek A and Green N R 2008 *Acta Mater.* **56** 2631-7
[42] Li J, Wang Z, Wang Y and Wang J 2012 *Acta Mater.* **60** 1478-93
[43] Takaki T, Ohno M, Shimokawabe T and Aoki T 2014 *Acta Mater.* **81** 272-83
[44] Yu H, Li J, Lin X, Wang L and Huang W 2014 *J. Cryst. Growth* **402** 210-4
[45] Takaki T, Ohno M, Shibuta Y, Sakane S, Shimokawabe T and Aoki T 2016 *J. Cryst. Growth* **442** 14-24
[46] Guo C, Li J, Yu H, Wang Z, Lin X and Wang J 2017 *Acta Mater.* **136** 148-63
[47] Tourret D, Song Y, Clarke A J and Karma A 2017 *Acta Mater.* **122** 220-35
[48] Pineau A, Guillemot G, Tourret D, Karma A and Gandin C A 2018 *Acta Mater.* **155** 286-301
[49] Guo C, Takaki T, Sakane S, Ohno M, Shibuta Y and Mohri T 2020 *Int. J. Heat Mass Transfer* **160** 120196
[50] Lee J, Ohno M, Shibuta Y and Takaki T 2021 *J. Cryst. Growth* **558** 126014
[51] Dorari E, Ji K, Guillemot G, Gandin C-A and Karma A 2022 *Acta Mater.* **223** 117395
[52] Takaki T, Sakane S, Ohno M and Shibuta Y 2019 *Modell. Simul. Mater. Sci. Eng.* **27** 054001
[53] Laxmipathy V P, Wang F, Selzer M and Nestler B 2021 *Comput. Mater. Sci.* **186** 109964
[54] Yang C, Liu L, Zhao X, Wang N, Zhang J and Fu H 2013 *J. Alloys Compd.* **578** 577-84
[55] Hu S, Yang W, Cui Q, Huang T, Zhang J and Liu L 2017 *Mater. Charact.* **125** 152-9
[56] Guo C, Li J, Wang Z and Wang J 2018 *Mater. Des.* **151** 141-53
[57] Guo C, Li J, Wang Z and Wang J 2020 *Comput. Mater. Sci.* **176** 109531
[58] Takaki T, Sakane S, Ohno M, Shibuta Y, Aoki T and Gandin C-A 2018 *Materialia* **1** 104-13
[59] Gonzales F and Rappaz M 2008 *Metall. Mater. Trans. A* **39** 2148-60
[60] Gandin C A, Rappaz M, West D and Adams B L 1995 *Metall. Mater. Trans. A* **26** 1543-51
[61] Takaki T, Sakane S, Ohno M, Shibuta Y, Shimokawabe T and Aoki T 2016 *Acta Mater.* **118** 230-43
[62] Takaki T, Sakane S, Ohno M, Shibuta Y and Aoki T 2022 *Comput. Mater. Sci.* **203** 111143
[63] Jacobi H and Schwerdtfeger K 1976 *Metall. Trans. A* **7** 811-20
[64] Takaki T, Sakane S, Ohno M, Shibuta Y, Shimokawabe T and Aoki T 2016 *ISIJ Int.* **56** 1427-35
[65] Clarke A J, Tourret D, Song Y, Imhoff S D, Gibbs P J, Gibbs J W, Fezzaa K and Karma A 2017 *Acta Mater.* **129** 203-16
[66] Song Y, Akamatsu S, Bottin-Rousseau S and Karma A 2018 *Phys. Rev. Mater.* **2** 053403
[67] Mota F L, Bergeon N, Karma A, Trivedi R and Debierre J-M 2020 *Phys. Rev. E* **102** 032803
[68] Echebarria B, Folch R, Karma A and Plapp M 2004 *Phys. Rev. E* **70** 061604
[69] Song Y, Tourret D, Mota F L, Pereda J, Billia B, Bergeon N, Trivedi R and Karma A 2018 *Acta Mater.* **150** 139-52
[70] Zhang A, Du J, Yang J, Guo Z, Wang Q, Jiang B, Pan F and Xiong S 2020 *Phys. Rev. E* **102** 043313
[71] Beckermann C, Diepers H J, Steinbach I, Karma A and Tong X 1999 *J. Comput. Phys.* **154** 468-96

[72] Do-Quang M and Amberg G 2008 *J. Comput. Phys.* **227** 1772-89
[73] Medvedev D, Varnik F and Steinbach I 2013 *Procedia Comput. Sci.* **18** 2512-20
[74] Rojas R, Takaki T and Ohno M 2015 *J. Comput. Phys.* **298** 29-40
[75] Takaki T, Rojas R, Ohno M, Shimokawabe T and Aoki T 2015 *IOP Conf. Ser.: Mater. Sci. Eng.* **84** 012066
[76] Sakane S, Takaki T, Ohno M and Shibuta Y 2019 *Comput. Mater. Sci.* **164** 39-45
[77] Yamaguchi M and Beckermann C 2013 *Acta Mater.* **61** 4053-65
[78] Ren J-k, Chen Y, Cao Y-f, Xu B, Sun M-y and Li D-z 2021 *J. Mater. Sci.* **56** 12455-74
[79] Sakane S, Takaki T, Rojas R, Ohno M, Shibuta Y, Shimokawabe T and Aoki T 2017 *J. Cryst. Growth* **474** 154-9
[80] Chen S and Doolen G D 1998 *Annu. Rev. Fluid Mech.* **30** 329-64
[81] Miller W, Succi S and Mansutti D 2001 *Phys. Rev. Lett.* **86** 3578-81
[82] Medvedev D and Kassner K 2005 *Phys. Rev. E* **72**
[83] Sakane S, Takaki T, Ohno M, Shibuta Y, Shimokawabe T and Aoki T 2018 *J. Cryst. Growth* **483** 147-55
[84] Takaki T, Rojas R, Sakane S, Ohno M, Shibuta Y, Shimokawabe T and Aoki T 2017 *J. Cryst. Growth* **474** 146-53
[85] Takaki T, Sakane S, Ohno M, Shibuta Y and Aoki T 2020 *Comput. Mater. Sci.* **171** 109209
[86] Konstantinova N Y, Popel' P S and Yagodin D A 2009 *High Temp.* **47** 336-41
[87] Sakane S and Takaki T 2021 *Comput. Mater. Sci.* **186** 110070
[88] Zhang X, Kang J, Guo Z, Xiong S and Han Q 2018 *Comput. Phys. Commun.* **223** 18-27
[89] Zhang A, Jiang B, Guo Z, Du J, Wang Q, Pan F and Xiong S 2021 *Adv. Theory Simul.* **4** 2000251
[90] Watanabe S, Aoki T and Takaki T 2020 *Parallel Comput.* **97** 102647
[91] Schive H-Y, Tsai Y-C and Chiueh T 2010 *Astrophys. J., Suppl. Ser.* **186** 457-84
[92] Sakane S, Takaki T, Ohno M, Shibuta Y and Aoki T 2019 *Modell. Simul. Mater. Sci. Eng.* **27** 054004
[93] Ludwig A, Wu M and Kharicha A 2015 *Metall. Mater. Trans. A* **46** 4854-67
[94] Wu M, Ludwig A and Kharicha A 2019 *Metals* **9** 229
[95] Liu X, Tang F, Zhao W, Cai J and Wei Y 2022 *Comput. Mater. Sci.* **204** 111182
[96] Ren J-k, Chen Y, Cao Y-f, Sun M-y, Xu B and Li D-z 2020 *J. Mater. Sci. Technol.* **58** 171-87
[97] Meng S, Zhang A, Guo Z and Wang Q 2020 *Comput. Mater. Sci.* **184** 109784
[98] Rátkai L, Pusztai T and Gránásy L 2019 *Npj Comput. Mater.* **5** 113
[99] T. Takaki, R. Sato, R. Rojas, M. Ohno and Shibuta Y 2018 *Comput. Mater. Sci.* **147** 124–31
[100] Sakane S, Takaki T, Ohno M, Shibuta Y and Aoki T 2020 *Comput. Mater. Sci.* **178** 109639
[101] Yamanaka N, Sakane S and Takaki T 2021 *Comput. Mater. Sci.* **197** 110658
[102] Steinbach I and Pezzolla F 1999 *Physica D* **134** 385-93
[103] Miyoshi E, Takaki T, Ohno M and Shibuta Y 2020 *ISIJ Int.* **60** 160-7
[104] Yamanaka N, Sakane S and Takaki T 2020 *IOP Conf. Ser.: Mater. Sci. Eng.* **861** 012067
[105] Su T C, O'Sullivan C, Nagira T, Yasuda H and Gourlay C M 2019 *Acta Mater.* **163** 208-25
[106] Su T C, O'Sullivan C, Yasuda H and Gourlay C M 2020 *Acta Mater.* **191** 24-42
[107] Gourlay C M and Dahle A K 2007 *Nature* **445** 70-3
[108] Nagira T, Yokota H, Morita S, Yasuda H, Yoshiya M, Gourlay C M, Sugiyama A, Uesugi K and Umetani K 2013 *ISIJ Int.* **53** 1195-201
[109] Schneider M C and Beckermann C 1995 *Int. J. Heat Mass Transfer* **38** 3455-73
[110] Reddy A V and Beckermann N C 1997 *Metall and Materi Trans B* **28** 479-89

[111] Ahmad N, Rappaz J, Desbiolles J-L, Jalanti T, Rappaz M, Combeau H, Lesoult G and Stomp C 1998 *Metall. Mater. Trans. A* **29** 617-30
[112] Madison J, Spowart J, Rowenhorst D, Aagesen L K, Thornton K and Pollock T M 2010 *Acta Mater.* **58** 2864-75
[113] Madison J, Spowart J E, Rowenhorst D J, Aagesen L K, Thornton K and Pollock T M 2011 *Metall. Mater. Trans. A* **43** 369-80
[114] Puncreobutr C, Phillion A B, Fife J L and Lee P D 2014 *Acta Mater.* **64** 316-25
[115] Khajeh E and Maijer D M 2010 *Acta Mater.* **58** 6334-44
[116] Khajeh E and Maijer D M 2011 *Acta Mater.* **59** 4511-24
[117] Berger R, Apel M, Laschet G, Jessen W, Schröder W, Wipperfürth J, Austermann J and Hopmann C 2021 *Metals* **11** 1895
[118] Natsume Y, Takahashi D, Kawashima K, Tanigawa E and Ohsasa K 2013 *ISIJ Int.* **53** 838-47
[119] Ludwig A, Kharicha A, Hölzl C, Domitner J, Wu M and Pusztai T 2014 *Eng. Anal. Bound. Elem.* **45** 29-35
[120] Böttger B, Haberstroh C and Giesselmann N 2016 *JOM* **68** 27-36
[121] Berger R, Apel M and Laschet G 2021 *Materialia* **15** 100966
[122] Takaki T, Sakane S, Ohno M, Shibuta Y and Aoki T 2019 *Acta Mater.* **164** 237-49
[123] Mitsuyama Y, Takaki T, Sakane S, Shibuta Y and Ohno M 2020 *Acta Mater.* **188** 282-7
[124] Carman P C 1956 *Flow of gases through porous media* (New York: Academic Press)
[125] Mitsuyama Y, Takaki T, Sakane S, Shibuta Y and Ohno M 2020 *IOP Conf. Ser.: Mater. Sci. Eng.* **861** 012029
[126] Shibuta Y, Oguchi K, Takaki T and Ohno M 2015 *Scientific Reports* **5** 13534
[127] Shibuta Y, Sakane S, Miyoshi E, Okita S, Takaki T and Ohno M 2017 *Nat. Commun.* **8** 10
[128] Okita S, Miyoshi E, Sakane S, Takaki T, Ohno M and Shibuta Y 2018 *Acta Mater.* **153** 108-16
[129] Shibuta Y, Sakane S, Miyoshi E, Takaki T and Ohno M 2019 *Modell. Simul. Mater. Sci. Eng.* **27** 054002
[130] Evensen G 1994 *Journal of Geophysical Research: Oceans* **99** 10143 62
[131] Rodell M, Houser P R, Jambor U, Gottschalck J, Mitchell K, Meng C J, Arsenault K, Cosgrove B, Radakovich J, Bosilovich M, Entin J K, Walker J P, Lohmann D and Toll D 2004 *Bull. Amer. Meteor.* **85** 381-94
[132] Houtekamer P L and Zhang F 2016 *Monthly Weather Review* **144** 4489-532
[133] Koyama T, Tsukada Y and Kawai Y 2015 *PTM 2015 - Proceedings of the International Conference on Solid-Solid Phase Transformations in Inorganic Materials 2015* 831-2
[134] Takahashi K and Yamanaka A 2021 *Comput. Mater. Sci.* **190** 110296
[135] Yamanaka A and Takahashi K 2020 *Mater. Today Commun.* **25** 101331
[136] Yamanaka A, Maeda Y and Sasaki K 2019 *Mater. Des.* **165** 107577
[137] Sasaki K, Yamanaka A, Ito S-i and Nagao H 2018 *Comput. Mater. Sci.* **141** 141-52
[138] Ohno M, Oka Y, Sakane S, Shibuta Y and Takaki T 2020 *Phys. Rev. E* **101** 052121
[139] Nagatsuma Y, Ohno M, Takaki T and Shibuta Y 2021 *Nanomaterials* **11** 2308
[140] Miyoshi E, Ohno M, Shibuta Y, Yamanaka A and Takaki T 2021 *Mater. Des.* **210** 110089
[141] Miyoshi E, Takaki T, Ohno M, Shibuta Y, Sakane S, Shimokawabe T and Aoki T 2017 *Npj Comput. Mater.* **3** 25
[142] Miyoshi E, Takaki T, Ohno M, Shibuta Y, Sakane S, Shimokawabe T and Aoki T 2018 *J. Mater. Sci.* **53** 15165-80

[143] Miyoshi E, Takaki T, Sakane S, Ohno M, Shibuta Y and Aoki T 2021 *Comput. Mater. Sci.* **186** 109992
[144] Yasuda H, Morishita K, Nakatsuka N, Nishimura T, Yoshiya M, Sugiyama A, Uesugi K and Takeuchi A 2019 *Nat. Commun.* **10** 3183
[145] Yasuda H, Kawarasaki T, Tomiyori Y, Kato Y and Morishita K 2019 *IOP Conf. Ser.: Mater. Sci. Eng.* **529**

IOP Conf. Series: Materials Science and Engineering 1274 (2023) 012010 doi:10.1088/1757-899X/1274/1/012010

The Effects of Melt Pool Geometry and Scan Strategy on Microstructure Development During Additive Manufacturing

Alexander F Chadwick[1] **and Peter W Voorhees**[2]

Department of Materials Science and Engineering, Northwestern University, 2220 Campus Drive, Evanston, IL 60208 USA.

Abstract. During the powder bed fusion additive manufacturing (AM) process, hundreds of grains may solidify simultaneously within the melt pool, depending upon the processing conditions and the grain size of the base plate and powder. Thus, a small variation in the shape of the melt pool or scan strategy may result in very different as-printed microstructures, since new grains may become more favourably aligned with the overall direction of growth. The microstructure affects the performance of the additively built component, and it is therefore critical to understand the interplay between the initial microstructure, material properties, heat source, and scan strategy on determining the as-printed microstructure. We examine columnar grain growth during AM using a phase-field model to perform simulations with single melt pools predicted by the cylindrically symmetric Rosenthal solution. Quantitative measures of the competitive grain growth are extracted from the simulation data. To aid in their understanding, we map these quantities onto visualizations of the microstructure and compare them to the inverse pole figure map, from which we find that the crystallographic directions of the grains that grow closest to the centre of the laser track may not be preferentially aligned with the scan direction.

1. Introduction

With additive manufacturing (AM) techniques such as powder bed fusion (PBF), a heat source locally produces a pool of molten metal that resolidifies as the heat source advances. As time evolves, the combined effects of heat flow, mass transport, and competitive grain growth lead to a rich multiphysics problem [1]. Additionally, the resulting microstructures are complex in nature, and it is often necessary to employ three-dimensional measurement techniques to accurately characterize and quantify features of interest [2,3]. Although many variables affect the as-printed microstructure, there is significant interest in the role of the melt pool geometry [3–7]. Changes in the melt pool in turn affect how well grains and grain boundaries are aligned with thermal gradients or the scan and build directions, to name a few examples, and the resulting effects accumulate over the entire AM build.

Recently, the authors proposed a phase-field model (PFM) of the PBF process [7], where we find that much of the solidification occurs while the grain boundaries are normal to the solid-liquid interface. By comparison, competitive grain growth due to kinetic anisotropy is only prevalent along the centre of the melt pool where the resulting variations in mobility and undercooling allow grains to become offset relative to their neighbours. A quantitative analysis found that a characteristic length scale relating the

[1] Email: alexander.chadwick@northwestern.edu
[2] Email: p-voorhees@northwestern.edu

change in grain volume and the initial solid-liquid interfacial area strongly correlates with how closely a grain tip approaches the centre of the laser track. Additionally, grains that had <100> crystallographic directions parallel to the normal direction of the solid-liquid interface later in the solidification process competed more effectively, as they were more likely to be favourably aligned with the melt pool boundary in the regime where kinetic anisotropy became significant. Subsequently, we have implemented a novel approach to reassign grains between the diffuse order parameters of the PFM that prevents grains from merging together into one feature [8]. This permits more grains and crystallographic orientations to be sampled in the resulting simulations. In the present work, simulations with grain remapping allow for an examination of how the analysed quantities from our previous study correlate with three-dimensional visualizations of the predicted microstructures, including typical experimental approaches such as inverse pole figure (IPF) mappings.

2. Model Description

In the present study, we employ the PFM of the PBF process proposed by Chadwick and Voorhees [7,8]. The total Helmholtz free energy of the system, F, is constructed as a functional of diffuse phase-field order parameters, their gradients, and the latent heat of fusion:

$$F = \int_V \left[W f_{MW}(\{\phi\}) + \frac{\kappa}{2}\sum_{i=0}^{N} |\nabla\phi_i|^2 + L\frac{T_{Liq} - T}{T_{Liq}} h(\{\phi\}) \right] dV, \tag{1}$$

where ϕ_i is the i-th order parameter, f_{MW} is a multiwell free energy with well height W, κ is the gradient energy coefficient, L is the latent heat of fusion, T_{Liq} is the liquidus temperature, T is the local temperature, and $h(\{\phi\})$ is an interpolation function between phases. Each order parameter evolves according to its own Allen-Cahn-type equation to minimize F [9]:

$$\frac{\partial\phi_i}{\partial t} = -M_i(\{\phi\}, \mathbf{n})\frac{\delta F}{\delta\phi_i}. \tag{2}$$

Here, t is time, $M_i(\{\phi\}, \mathbf{n})$ is the phase-field mobility that is a function of the values of $\{\phi\}$ and the solid-liquid interface normal vector, $\mathbf{n}$, and $\delta F/\delta\phi_i$ is the variation of F with respect to ϕ_i. The values of W, κ, and M_i are related to the interfacial energy, γ, the diffuse interface thickness, ζ, and the anisotropic solid-liquid interface mobility, $\mu(\mathbf{n})$. Consistent with cubic symmetry, we take

$$\mu(\mathbf{n}) = \mu_0(1 + \epsilon_4\{4(n_1^4 + n_2^4 + n_3^4) - 3\}), \tag{3}$$

where μ_0 is the modulus of the mobility, ϵ_4 is the strength of the anisotropy, and each n_i is a component of the normal vector in the crystal coordinate axis.

The evolution of each ϕ_i requires knowledge of the local temperature field. While it is possible to use any heat flow solution, here this is accomplished by coupling equation (2) to Rosenthal's solution for the moving point heat source [10],

$$T(X,\rho) = T_0 + \frac{Q}{2\pi\kappa_t(X^2+\rho^2)^{1/2}} exp\left[-\frac{v_P}{2\alpha}(X + (X^2+\rho^2)^{1/2})\right], \tag{4}$$

where X and ρ are the position parallel and normal to the laser track, respectively, T_0 is the ambient temperature, Q is the absorbed power, κ_T is the thermal conductivity, α is the thermal diffusivity, and v_P is the pulling velocity. The form of equation (4) recasts the Cartesian simulation frame into the moving cylindrical reference frame of the heat source, which simplifies the quantitative analysis of the simulation results. The values of X and ρ and are related to the simulation frame by $X = x - x_0 - v_P t$ and $\rho = (y^2 + z^2)^{1/2}$, respectively, where x, y, and z are the coordinates in the simulation frame and x_0 is the initial position of the laser at $t = 0$. As X varies in time, so too does T, and therefore a time-dependent driving force exists in equations (1) and (2). Equation (4) has cylindrical symmetry about the X axis: $\rho = 0$ when a point lies exactly on the laser track, and there is a unique value, ρ_{max}, that corresponds to the maximum width and depth of the melt pool for a given set of thermal and processing parameters. As grains solidify, traversing from $\rho = \rho_{max}$ to $\rho = 0$, the orientation of the grain tip continuously changes, exposing different crystallographic directions [7]. As a note, other thermal models are readily coupled to the PFM. However, the in-depth examination of the as-printed microstructures from these models is ongoing and is beyond the scope of the present work.

3. Results and Discussion

The numerical implementation of the model is identical to the previous studies and includes grain remapping to reduce the number of order parameters [7,8]. The model parameters, which are identical to those in [8], represent an analogue of 316L stainless steel and are listed in table 1. A rectilinear geometry with dimensions 192 μm × 76.8 μm × 38.4 μm in the x, y, and z directions, respectively, is generated that initially contains 3,169 grains with an average grain size of 6.7 ± 2.9 μm. A total of 2,241 grains remain in the microstructure after one pass of the laser is performed, including both the grains that interact with the melt pool and unmelted baseplate material. To perform a quantitative analysis on the as-printed microstructure, a filtering step is applied to remove (i) grains that touch either end of the simulation domain in the scan direction, (ii) grains that had not finished solidifying by the end of the simulation, and (iii) grains that did not increase in volume by at least 5 μm^3. This filter primarily removes unmelted grains and grains on the edges of the domain, although some exceptionally small grains that interact with the melt pool are also removed in this step. After filtering, 497 grains remain in the dataset.

Table 1. The values of the parameters employed in the phase-field simulation. If multiple references are noted, the employed value is chosen to be consistent with the range of literature values.

Material Parameters			Model Parameters	
Parameter	Value	Reference	Parameter	Value
L	1.3×10^{-9} J/μm³	[11,12]	v_P	1 m/s
T_{Liq}	1700 K	[11]	Q	25 W
γ	2.5×10^{-13} J/μm²	[12,13]	T_0	300 K
μ_0	2.2×10^5 μm/s K	[13]	x_0	38 μm
ϵ_4	0.30	[7]	Δx	0.24 μm
κ_T	2.7×10^{-5} W/μm K	[11]	$\zeta/\Delta x$	1.7
α	5.2×10^6 μm²/s	[11]	τ	1×10^{-7}

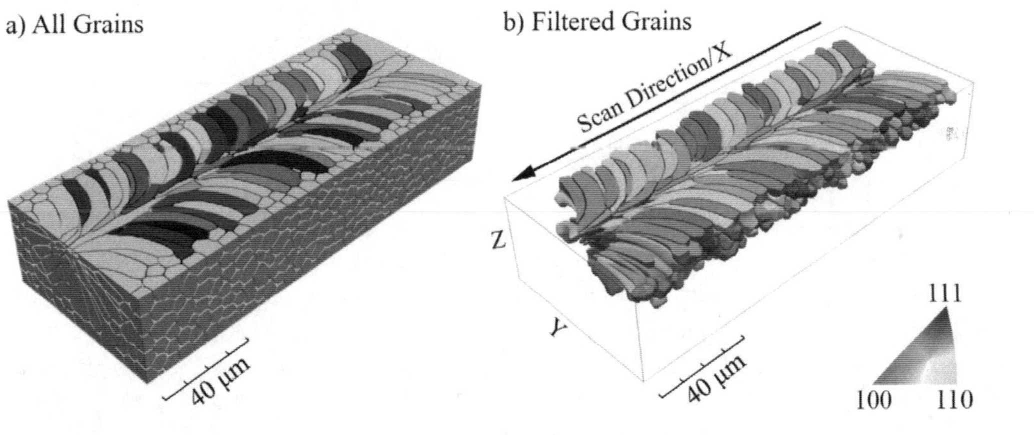

Figure 1. The as-printed microstructures predicted by the described phase-field model, with all grains included in (a) and only the filtered grains included in (b). The IPF colouring in (b) is with respect to the indicated scan direction along the X axis.

We present the final as-printed microstructures in figure 1. In figure 1a, the entire microstructure is shown; grains that satisfy the filter criteria are coloured according to a unique integer identifier and the excluded grains are coloured grey. In figure 1b, only the filtered grains are included, and these are coloured according to the IPF map with respect to the scan direction. While the IPF colouring is a convenient representation of the crystallography throughout the domain, the resulting figure does not necessarily reflect competitive grain growth effects. One might expect that red-coloured grains, which have a <100> crystallographic direction close to the scan direction, would be able to outcompete their neighbours during solidification, as according to equation (3) the <100> directions have the fastest mobility. However, the local velocity of the interface is related to the normal of the solid-liquid interface and not the crystallographic orientation of a grain with respect to the scan direction. Therefore, these grains are favourably aligned only when they have reached the centre of the laser track where the solid-liquid interface normal is parallel to the scan direction, and many grains that reach the centreline in figure 1b do not have a <100> crystallographic direction parallel to the scan direction.

To further examine the microstructural evolution, we perform an analysis similar to our previous study [7]. At every output of the simulation, each grain is analysed to determine its total volume, V, its solid-liquid interfacial area, A, and the position and temperature of the grain tip. Additionally, we calculate the average value of the solid-liquid interfacial mobility across the entire solid-liquid interfacial area of the grain, which is denoted $\langle\mu\rangle$. From the Cartesian coordinates of the grain tip, the radial distance of the grain tip from the laser axis, ρ_{tip}, can then be calculated. We assume that a grain begins solidifying once $\rho_{\mathrm{tip}} = \rho_{\mathrm{max}} = 26.8$ μm, and solidification ends once $A = 0$. The trends obtained by the analysis are plotted in figure 2.

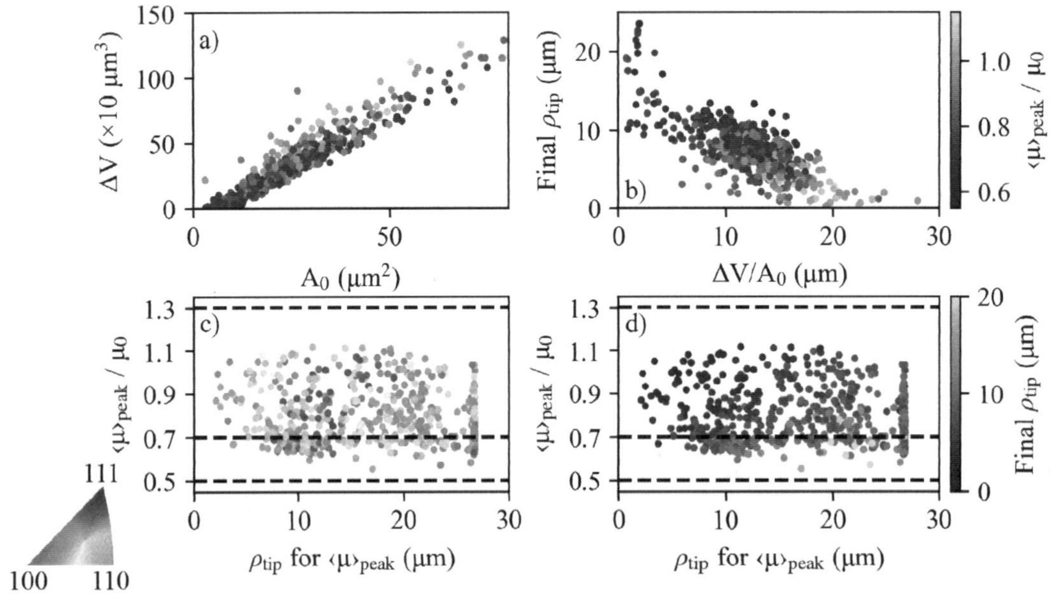

Figure 2. Scatterplots from the quantitative analysis of the filtered dataset. (a) The change in grain volume, ΔV, as a function of initial solid-liquid interfacial area, A_0, and the peak average interfacial mobility, $\langle\mu\rangle_{\mathrm{peak}}$. (b) The radial position of the grain tip, ρ_{tip}, when it detached as a function of $\Delta V/A_0$ and $\langle\mu\rangle_{\mathrm{peak}}$. (c) The value of $\langle\mu\rangle_{\mathrm{peak}}$ as a function of the value of ρ_{tip} where it was measured, coloured according to the IPF mapping of each grain with respect to the scan direction of figure 1. (d) The same as in (c), but each point is coloured by the final value of ρ_{tip} at the end of solidification. The value of $\langle\mu\rangle_{\mathrm{peak}}$ is normalized by μ_0 for clarity. From top to bottom, the dashed lines in (c) and (d) indicate the normalized values for <100>, <110>, and <111> directions, respectively.

In figure 2a, we observe that the change in volume of the grain from the start to the end of solidification, ΔV, is generally linear with respect to the solid-liquid interfacial area at the start of solidification, A_0. The scatter of the data is related to the maximum value of the average mobility, $\langle\mu\rangle_{\mathrm{peak}}$, with faster grains (yellow end of the colour scale) typically having larger values of $\Delta V/A_0$. Figure 2b shows the final value of ρ_{tip} when solidification ends for a grain as a function of the calculated $\Delta V/A_0$ and $\langle\mu\rangle_{\mathrm{peak}}$. Generally, the final radial position of the grain tip decreases with increasing $\Delta V/A_0$; therefore, the grains with larger $\langle\mu\rangle_{\mathrm{peak}}$ also tend to have ρ_{tip} closer to zero when they detach from the interface. Thus, faster grains tend to grow to the centre of the melt pool. Both of these results agree with the previous study [7], even with the increased number of grains and orientations in the dataset. In figure

2c and 2d, we plot the normalized value of $\langle\mu\rangle_{\mathrm{peak}}$ as a function of the value of ρ_{tip} where that value was measured. Since the normal to the interface changes with time, the value of $\langle\mu\rangle$ will also change with time—or equivalently with the position of the grain as shown. In 2c, the points are coloured according to the IPF colour of the grain, and we observe that many of the grains with their <100> crystallographic directions aligned along the scan direction experience $\langle\mu\rangle_{\mathrm{peak}}$ towards $\rho_{\mathrm{tip}} = \rho_{\mathrm{max}}$. This implies that the interfacial normal was oriented such that $\langle\mu\rangle_{\mathrm{peak}}$ occurs near ρ_{max}. However, a smaller fraction of these <100>-aligned grains experience $\langle\mu\rangle_{\mathrm{peak}}$ towards $\rho_{\mathrm{tip}} = 0$. Qualitatively, this is expected: a grain with a <100> direction parallel to the laser track is also likely to have a <100> direction normal to the laser track such that it can experience a peak in its mobility at either $\rho = \rho_{\mathrm{max}}$ or $\rho = 0$ if it grows to the centre of the laser track. Figure 2c also shows that higher-index orientations appear more likely to have their value of $\langle\mu\rangle_{\mathrm{peak}}$ occur between the extrema of ρ. Lastly, in figure 2d, the points are coloured according to their final value of ρ_{tip}, and we observe that grains that experience a higher $\langle\mu\rangle_{\mathrm{peak}}$ later in the solidification process (smaller values of ρ_{tip}) tend to grow closer to the centre of the laser track. As with figure 2a and 2b, the trend in 2d is consistent with the previous study [7].

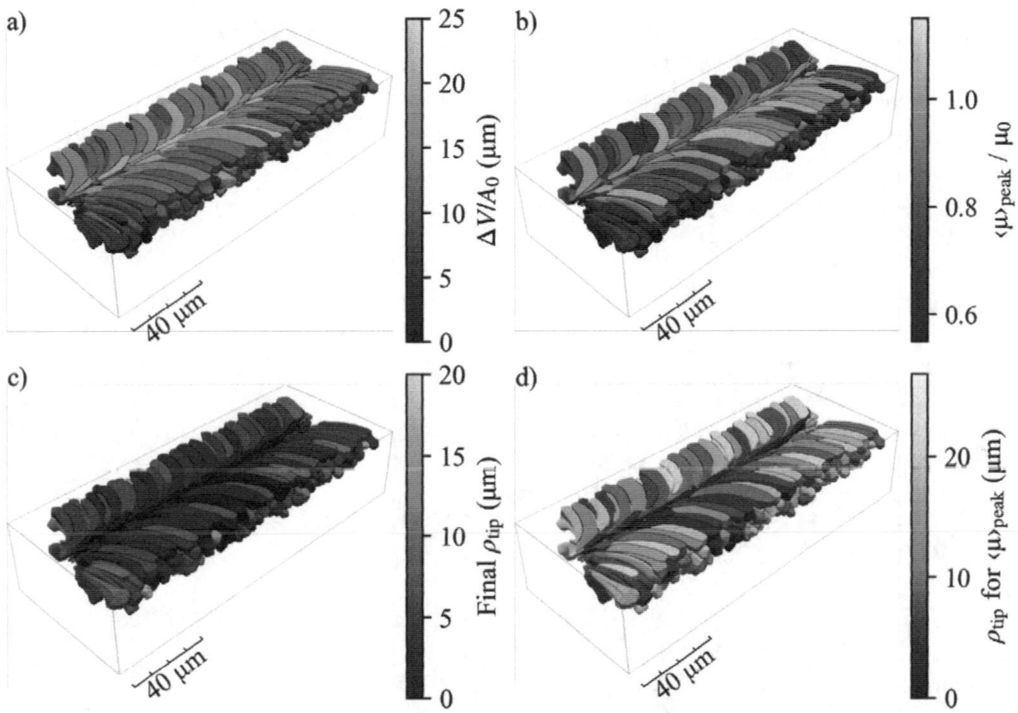

Figure 3. The filtered grains from figure 1, coloured according to (a) $\Delta V/A_0$, (b) the normalized $\langle\mu\rangle_{\mathrm{peak}}$, (c) the value of ρ_{tip} when solidification ended, and (d) the value of ρ_{tip} where $\langle\mu\rangle_{\mathrm{peak}}$ occurred.

The quantities in figure 2 may additionally be mapped onto the microstructure in figure 1b, which leads to the results in figure 3. Panes (a) through (d) plot $\Delta V/A_0$, the normalized $\langle\mu\rangle_{\mathrm{peak}}$, the final value of ρ_{tip} at the end of solidification, and the value of ρ_{tip} for $\langle\mu\rangle_{\mathrm{peak}}$ for each grain, respectively. In figure 3a, we find that the grains that reach the very centreline of the track are typically towards the upper end of the colour scale (larger $\Delta V/A_0$). While these grains do not necessarily have the largest values of $\langle\mu\rangle_{\mathrm{peak}}$ in figure 3b, visibly they are more towards the orange-yellow end of the spectrum than the purple-blue end, and thus they are experiencing somewhat more favourable mobilities. However, in figures 3c and 3d, many of the grains that visibly reach the centreline are towards the lowest values of

ρ_{tip}, both for their final radial distance from the laser track as well as where $\langle\mu\rangle_{\text{peak}}$ occurred, consistent with the trends in figure 2. When compared with figure 1b, we observe that many of the grains that compete effectively according to the quantitative analysis do not appear to have a <100> crystallographic direction aligned with the scan direction.

4. Conclusions

We performed a simulation study of single-track additively manufactured geometries using a phase-field model that allowed for remapping of grains between diffuse order parameters to significantly increase the number of sampled grains and crystallographic orientations. As in a previous study, the final radial distance of a grain tip, ρ_{tip}, is related to its characteristic length scale $\Delta V/A_0$. Additionally, grains that expose fast-growing <100> crystallographic directions near the centre of the melt pool compete more effectively than grains that expose <100> directions near the sides of the melt pool. We also mapped analysed quantities onto plots of the as-printed microstructure, from which we obtain several conclusions. (1) The inverse pole figure in the scan direction does not necessarily reflect the most kinetically favourable grains during a single laser scan. For the chosen thermal parameters, many grains that reach the laser track centreline have higher-index, or slow growth, directions aligned with the scan direction. This is reflected in plots of the as-printed microstructure as well as in the reduced-dimension analysis. (2) Grains along the laser track centreline, as expected, have small final values of ρ_{tip} at the end of the solidification. Thus, these grains also have large values of $\Delta V/A_0$ and experienced $\langle\mu\rangle_{\text{peak}}$ closer to the laser track than the grains that did not reach the centreline. (3) Although the grains that are visible along the centreline are in the upper half of the range of measured values of $\langle\mu\rangle_{\text{peak}}$, they do not necessarily have the largest value of $\langle\mu\rangle_{\text{peak}}$. In other words, they do not exclusively expose <100> crystallographic directions at their best-aligned location in the melt pool.

5. References

[1] Michopoulos J G, Iliopoulos A P, Steuben J C, Birnbaum A J and Lambrakos S G 2018 *Addit. Manuf.* **22** 784–99
[2] Rowenhorst D J, Nguyen L, Murphy-Leonard A D and Fonda R W 2020 *Curr. Opin. Solid State Mater. Sci.* **24** 100819
[3] Polonsky A T, Lenthe W C, Echlin M P, Livescu V, Gray G T and Pollock T M 2020 *Acta Mater.* **183** 249–60
[4] Liu J and To A C 2017 *Addit. Manuf.* **16** 58–64
[5] Agrawal A K, Meric de Bellefon G and Thoma D 2020 *Mater. Sci. Eng. A* **793** 139841
[6] Scipioni Bertoli U, MacDonald B E and Schoenung J M 2019 *Mater. Sci. Eng. A* **739** 109–17
[7] Chadwick A F and Voorhees P W 2021 *Acta Mater.* **211** 116862
[8] Chadwick A F and Voorhees P W 2022 *Int. J. Numer. Methods Eng.* 10.1002/nme.6966
[9] Allen S M and Cahn J W 1979 *Acta Metall.* **27** 1085–95
[10] Rosenthal D 1946 *Trans. Am. Soc. Mech. Eng.* **68** 849–66
[11] Kim C S 1975 *Thermophysical Properties of Stainless Steels* (Argonne, IL: Argonne National Laboratory)
[12] Ueno K and Shibuta Y 2020 *ISIJ Int.* **60** 2301–5
[13] Sun D Y, Asta M and Hoyt J J 2004 *Phys. Rev. B* **69** 174103

Acknowledgments

This work is based on research sponsored by the Office of Naval Research (ONR) under project number N00014-18-1-2782 P00002. This research was also supported in part through the computational resources and staff contributions provided for the Quest high performance computing facility at Northwestern University, which is jointly supported by the Office of the Provost, the Office for Research, and Northwestern University Information Technology. The authors thank Dr. David Rowenhorst and Dr. Kirubel Teferra at the United States Naval Research Laboratory for providing the functions to calculate the inverse pole figure colours.

IOP Conf. Series: Materials Science and Engineering 1274 (2023) 012011 doi:10.1088/1757-899X/1274/1/012011

Effects of melt pool flow on porosity levels in arc welding

F. Wu[1], K. V. Falch[1], M. Drakopoulos[2], W. U. Mirihanage[1,*]

[1] The University of Manchester, Manchester M24 4SA, UK
[2] Diamond Light Source, Didcot OX11 0DE, UK

Abstract. Arc welding is one of the widely used approaches for joining metals. During the arc welding process, an electric arc creates intense heat to fuse metals that forms the melt pool between the parts to be welded together. Intensive flow fields are observable within the fusion weld pools as a result of multiple driving forces. The flow patterns within the melt pool significantly determine the shape of the weld joint and other attributes of the solidified joint, such as microstructure and defects. Porosity is one of the solidification-related defects that can bring a detrimental impact. During the formation and solidification of weld pools, gas bubbles that are formed can be driven in or out from the pool by the flow. In this work, employing in situ synchrotron X-rays, we have observed how different flow conditions and air-liquid interface are contributed to retaining and releasing the gas bubbles that formed during the arc welding. The results suggest that underpinning driving forces, such as electromagnetic forces, act on molten metal to retain the pores inside the weld joints; but, gravity-driven effects can contribute to reduce the porosity, with appropriate process conditions.

1. Introduction

Arc welding processes, generating enough heat to melt metals through electricity, have a wide range of applications for modern energy and transports industries to produce critical components or joints [1-3]. Multiple physical and chemical phenomena caused by the interaction between heat sources and base materials change the shape, microstructures and mechanical features of final welds. The flow behaviour of the liquid metals or alloys plays a crucial role in dominating the heat and mass transfer during the phase transformations [4,5].

Previous researches have demonstrated that porosity on the welds is affected by the existence of hydrogen or air/gas from the processing modes. There are two types of microporosity formed during the fusion welding processes; shrinkage porosity results from shrinkage during the liquid metal solidification process, while gas porosity originates from trapped air or undissolved hydrogen. In contrast, insufficient metals feeding and high solubility of hydrogen results in the generation of macroporosity [6-8]. The negative impacts of porosity on the welds are reflected in severely deteriorating the weld products properties, however, this can be mitigated by controlling hydrogen content, air flow rate or the temperature of the environment according to the previously reported results [6]. Besides, thermocapillary force, arc forces, surface tension force and gravity-driven forces have been shown to change not only the flow conditions but also influence behaviour such as pores elimination in fusion-based welding and additive manufacturing processes [9-11].

However, the detailed in-situ information about interrelationship of molten metal flow and porosity formation are limited due to the instantaneity of fusion welding process and opacity of metallic materials, when conventional metallographic methods are used to observe the melt pools and weld joints. Synchrotron X-ray imaging has recently become an important way to study the dynamic processes, such as melting and solidification, in the welding and additive manufacturing of metallic alloys [10-16]. Notably, data collected from real-time X-ray experiments is critical for developing physically accurate simulation models of dynamic melt pools [17,18]. By combining in-situ synchrotron x-ray imaging experiments and optical microscopy results, in this contribution we analyse the influence of melt pool flow on porosity levels in arc welding processes under different process parameters.

2. Experimental

LM24/A380 alloy (Al-8%wt. Si alloy) with 25 mm height, 12 mm length and 6 mm thickness is set as the subject to provide fundamental melt pool related phenomenon under tungsten inert gas (TIG) welding process. All of the specimens were cleaned with sandpaper from grits up to 1200. The chemical composition of base metal is listed in Table 1. In-situ synchrotron X-ray imaging was carried out at I12 beamline - Diamond Light Source, UK to observe and record the melt pools formation and pores motion inside the melt pool. Transmission images were collected at 500 Hz frame rate using 60 keV monochromatic X-ray beam. Direct current (DC) and Alternative current (AC) modes with different welding currents and duration were set to obtain various melt pool evolution and provide general and fundamental phenomena that occur in arc welding. Optical microscopy was used for examine post process microstructure and porosity. All images are processed by ImageJ® [19]. Further experiment details can be found in refs. [10,11,20]. Fig.1 shows the schematic description of the experimental setup. In this experiment, the welding torch was stationary and fixed around 1 mm aboved the center of samples without filler metals.

Table 1. Chemical composition (wt %) of experimental samples.

LM24/A380	**Al**	**Si**	**Cu**	**Mg**	**Fe**	**Mn**	**Zn**	**Ni**
	Balance	8.9	3.2	0.2	0.6	0.23	2.4	0.03
	Zr	**Ti**	**V**	**Pb**	**Sn**			
	2.4	0.06	0.01	0.03	0.01			

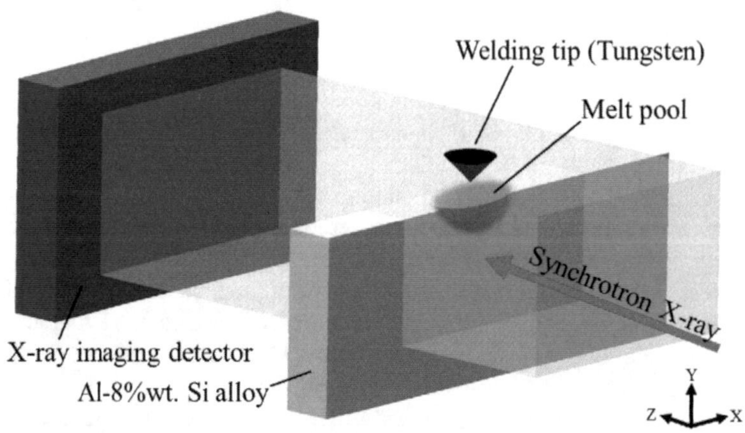

Figure 1. Schematic view of the experimental set-up for synchrotron X-ray imaging.

3. Results and discussion

The center surface height variation of the weld pool, according to the welding time and power input levels, compared with the original air-liquid interface is shown in Fig. 2. There was a similarity in the evolution features of the melt pool's liquid-air interface height despite different welding parameters. As the arc started, the center surface height of the melt pool increased and remained for a period of time before rapidly descending to the lowest point. Generally, the peak height of the center melt pool in the AC mode is almost twice that of DC mode. In contrast, the lowest point at the center of the melt pool is directly correlated with the energy input. DC mode with longer weld time or higher current can lead to lower center surface height as a result of the greater net heat/energy input.

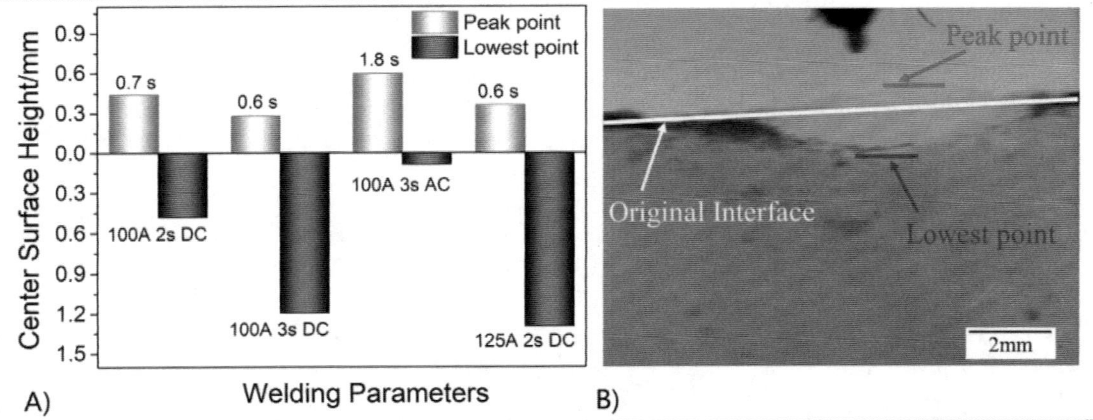

Figure 2. (A) Quantified extreme value of center surface height under different weld parameters. (B) A superimposed images showing weld pool center surface change (maximum vs. minimum) during the arc welding process.

Fig. 3 shows an example of pores motion during the welding process in the constructed images by superposing the extracted bubble positions based on the normal initial projection of 100 A DC, 2 s welding case. In the X-ray image, light colors areas represent the pores in the melt pool, the volume change caused by molten materials and the empty background (with shielding gas) . As shown in Fig.3.(A), the upward trajectory of the pores are significantly suppressed by the arc and surface tension gradient forces driven fluid flow. In the initial 2 s, the pores moved upward slowly and temporarily remained almost stationary below the surface of the molten pool until the arc is stopped. Here, the circulating fluid flow, either inward or outward, tends to retain pores in the molten pool. However, the pores moved up quickly and escaped from the melt pool once molten metal started to solidify, as indicated in Fig.3.(B). These bubbles appeared significant escaping tendency when arc stopped, as strong buoyancy effects pushed them out of the denser molten metal pool, which is consistent with the existing research [21].

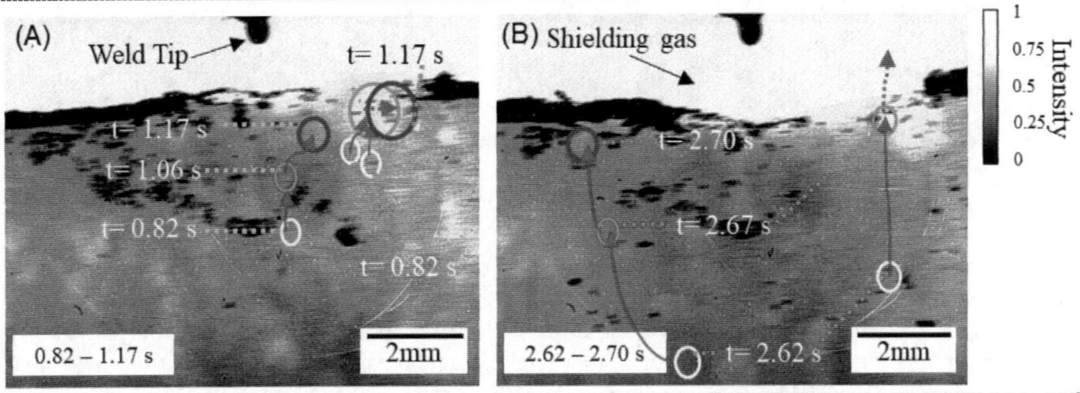

Figure 3. The movement of pores in the melt pool. (A) at the initial stage of the welding. (B) latter stage of welding. Weld current: 100 A DC with 2 s weld duration. The colour scale means the intensity of X-rays passed through the samples.

The average speed and speed histograms of the pore transport in the horizontal and vertical directions with and without the arc is presented in Fig. 4. As the arc started, the arc forces and the gradually increased surface tension forces accelerated the pores movement in the horizontal direction. On the contrary, the average speed of pores motion in the vertical direction increased about 40% as the arc stopped. Based on this difference, it can be discovered that pores with less density (compared with base materials) are more easily to be squeezed out from melt pools when gravity-driven forces become the dominant driving force of liquid metal. In addition, the movement speed in the vertical direction keeps greater than that in the horizontal direction (with or without arc). It can also be seen from the speeds histograms that the presence of arc has more significant effect on the vertical direction. This is because when the arc stopped, gravity as the main driving force has a greater impact on the vertical movement of bubbles.

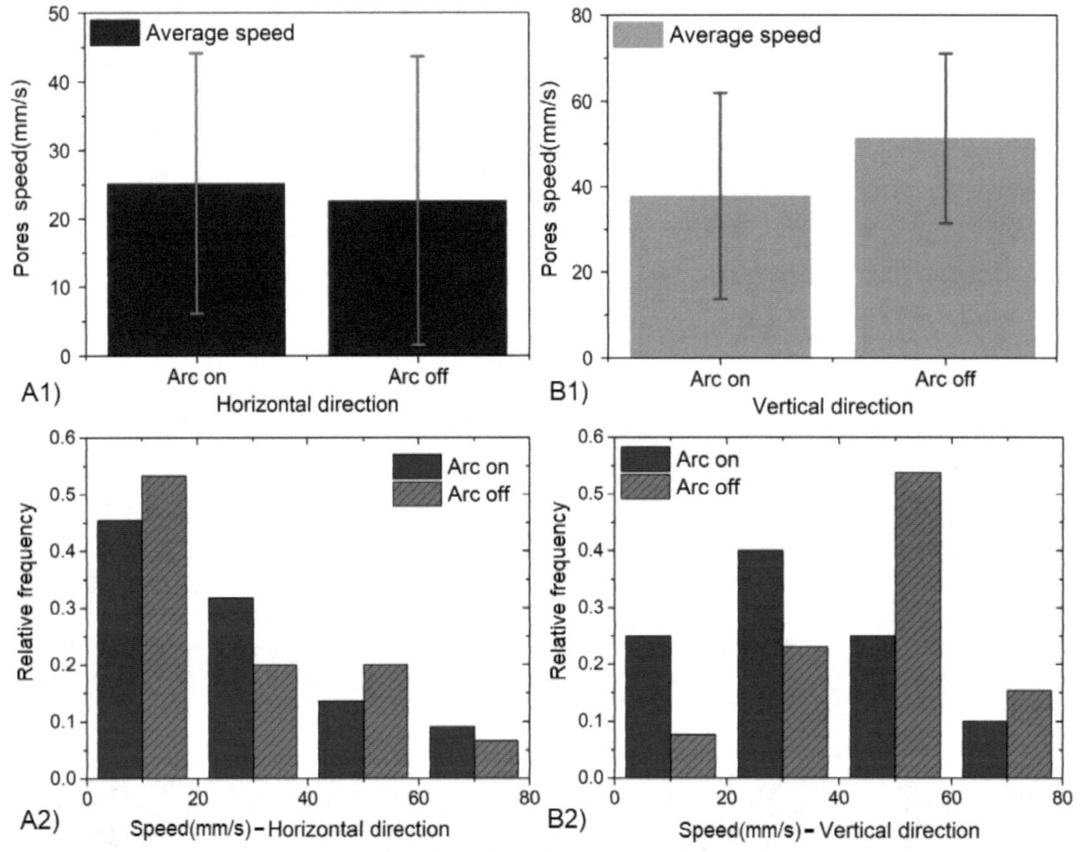

Figure 4. Average pores motion speed with arc on and off. (A1) in the horizontal direction. (B1) in the vertical direction. Pores motion speeds relative frequency histogram (A2) in the horizontal direction. (B2) in the vertical direction. Weld current:100 A DC with 2 s weld duration.

The porosity levels of the finial weld pools from different welding parameters and welding time are observed from optical microscopes. Fig.5 shows a representative micrograph of porosity distribution in the weld pool and porosity variation with different parameters. Under the same current and welding time, AC mode produced more porosity than the DC mode (about twice as much). Besides, increasing the welding time has a greater effect on reducing the porosity than turning up the current in the DC mode. According to Joule's Law, 125 A with 2 s weld time generated 2875 J heat through weld tip, while 3450 J heat are generated in the 100 A with 3 s weld time [22]. Therefore, more metal alloys are melted with the higher heat input, which increases the self-weight of a given volume of the melt pool (liquid/molten metal) and enhances the impact of gravity-driven forces on the fluid flow and pores motion. However, increased heat input and resulting molten metal volume can further complicate the interaction with atmosphere [23]. Thus, further work with wider experimental parameters and melt pool sizes are suggested to extract enhanced empirical relationships.

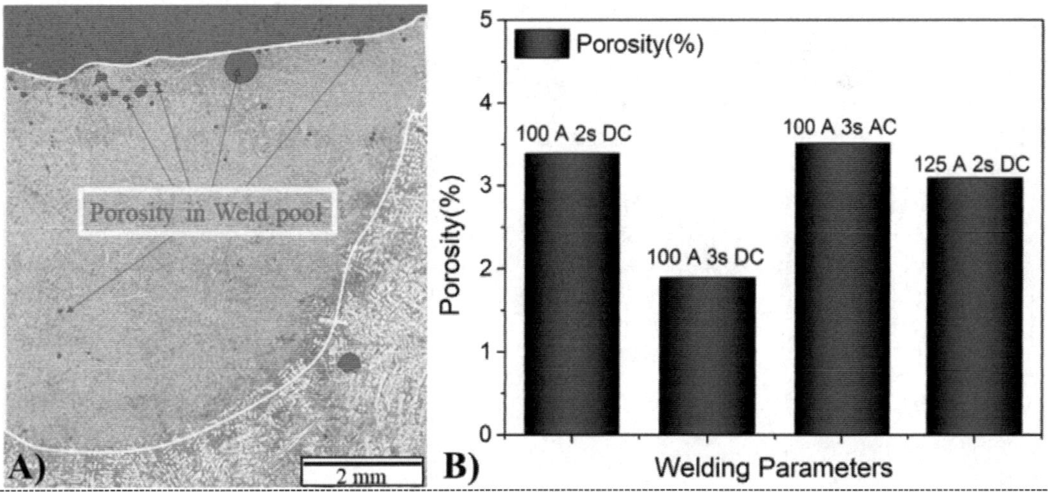

Figure 5. (A) an example of calculation method for porosity levels. The yellow line is the boundary of the weld pool. (B) The change of porosity levels of the weld pools with varying welding parameters..

4. Summary

In situ synchrotron X-ray imaging and ex situ optical microscopes is used to observe pores motion and final porosity levels occur in the arc welding. Preliminary analysis on gas bubbles motion with force regime changes, we can identify a simple propensity on formation of the gas porosity based on welding process parameters. When the volume of molten metal within the pool increases with the increasing power input or/and weld time, pores are more likely to release due to the domination of gravity-driven effects within the melt pool. Further analysis with wider experimental parameters may need to understand the exact limiting factors of the current observations.

Acknowledgments

This research work is supported by EPSRC(UK) grant EP/P02680X/1 and Diamond Light Source is acknowledged for granting the beamtime (EE EE20611).

References

[1] Zitouni, A. H., et al., 2020 Int. J. Interact. Des. Manuf 14 (1), 173
[2] IvánTabernero, et al., 2018 Procedia CIRP 68, 358
[3] Liu, A., et al., 2013 International Journal of Machine Tools and Manufacture 65: 1-7
[4] Li, H., et al., 2020. In 3rd International Conference on Electron Device and Mechanical Engineering, ICEDME 2020, Institute of Electrical and Electronics Engineers Inc. pp 486
[5] Xie, M. X., et al., 2019 Mater 12 (9)
[6] Ardika, R. D., et al 2021 In 26th International Conference on Fracture and Structural Integrity, IGF26 2021, C ed.; Elsevier B.V. Vol. 33, pp 171
[7] Mohandas, T., et al., 1999 Metall Mat Trans A Phys Metall Mat Sci 30 (13), 789
[8] Luan, R., et al., 2017 In 13th IEEE Conference on Automation Science and Engineering, CASE 2017, IEEE Computer Society, Vol. 2017-August, pp 548
[9] Hojjatzadeh, S. M. H., et al., 2019 Nat. Commun 10 (1)
[10] Wu, F., et al., 2021 International Journal of Heat and Mass Transfer 179, 121679

[11] Wu, F., et al., 2020 Materials & Design 190, 108534
[12] Zhao, C., et al., 2020 Sci 370 (6520), 1080
[13] Cunningham, R., et al., 2019 Sci 363 (6429), 849
[14] Mirihanage, W. U., et al., 2017 Acta Mater 68, 159
[15] Mirihanage, W. U., et al., 2015 In 14th International Conference on Modeling of Casting, Welding and Advanced Solidification Processes, MCWASP 2015, 1 ed.; Yasuda, H., (ed.) Institute of Physics Publishing, Vol. 84
[16] Guo, D., et al., 2021 Materials & Design 207, 109782
[17] Pequet, C., et al., 2002 Metall Mat Trans A Phys Metall Mat Sci 33 (7), 2095
[18] Khalajzadeh, V., et al., 2020 Metallurgical and Materials Transactions A 51 (5), 2239
[19] Schneider, C. A., et al.,2012 Nature Methods 9 (7), 671
[20] Drakopoulos, M., et al., 2015 J. Synchrotron Radiat 22, 828
[21] Md. R. U. Ahsan, et al., 2016 Science and Technology of Welding and Joining 21:3, 209-215
[22] Kawabe, N., et al., 2016 Weld World 60 383–392
[23] Y Naito et al., 2006 Welding International 20:12, 936-941

Effect of laser scanning velocity on the microstructural changes and mechanical properties of the H13 steel part treated by laser surface remelting

Jichang Xie[1],*, Ruifeng Di[2], Jishuai Li[3], Yanan Liu[4], Rija Nirina Raoelison[5], Mohamed Rachik[1]

[1] Laboratoire Roberval, UTC, Sorbonne universités, Université de Technologie de Compiègne, Centre de recherche Royallieu, CS60319, 60203 Compiègne Cedex, France
[2] Laser Technology Institute, Tiangong University, Tianjin 300387, China
[3] UCLouvain, Institute of Mechanics, Materials and Civil Engineering, IMAP, 1348, Louvain-la-Neuve, Belgium
[4] School of Mechatronics Engineering, Harbin Institute of Technology, Heilongjiang, Harbin 150001, China
[5] Université de Bourgogne Franche-Comté – UTBM, Laboratoire Interdisciplinaire Carnot de Bourgogne, UMR 6303 CNRS, 90100 Belfort, France

E-mail: jichang.xie@utc.fr

Abstract. This paper investigates the effect of laser scanning velocity on the structural changes and mechanical properties of a H13 steel part treated by laser surface remelting. A high energy density laser with a large laser beam was used to improve the processing efficiency. The room temperature phases are dominated by martensite, retained austenite, and carbides. The laser treatment effects under various exposition conditions are compared in terms of heat affected depth, surface topography, and mechanical properties. The strengthening mechanisms include grain refinement, dislocation and precipitation strengthening, which act simultaneously during the laser remelting process to confer a high hardness at the surface of the H13 steel workpiece. Due to its higher yield strength and lower elongation, failure in the remelting zone occurred at the grain boundaries of the dendrites and in between the heat affected zone (HAZ) and the parent metal (Base Metal). These results can support the mass production of surface hardened H13 steel moulds using a laser treatment (high laser density and large laser heat source) with apprehensible changes of properties.

1. Introduction

Thanks to the high energy density of the laser beam, laser-based processes for material processing have attracted much attention in additive manufacturing [1, 2], repair [3], joining [4], cutting [5] and surface treatment [6] or finishing [7] of metals. In comparison to laser-based additive manufacturing and repair, the laser surface remelting is an effective way to extend the components longevity without additional material. These benefits include continuous production, lack of mould and expensive equipment, simplicity and perspective of industrialization [8]. Due to the inherent high cooling rate of laser-based processes [9], the laser surface melting could enhance the mechanical properties by phase transformation (for example, austenite to martensite) and finer grain [10]. The produced metal materials have excellent mechanical performance that can be used in various industries.

In the recent decades, many materials and composites, particularly the H13 tool steel has been produced via this method [11-15] owing to its high toughness, fatigue and wear resistance under elevated temperature working conditions. These remelted H13 steel can increase in strength, fracture toughness, microhardness due to the existence of desirable ultrafine nanostructures. However, due to a limited energy density in the case of small laser spot, the processing efficiency is limited.

In this paper, a high energy density laser using a large beam diameter (5 mm) is utilized to study the treatment effect on H13 tool steel under various exposition conditions, i.e. under different laser scanning speeds. Herein, we investigate the microstructure development in H13 tool steel tensile specimens after the laser remelting. The microhardness distribution due to the laser treatment is discussed and correlated with the grain size, and the phase composition.

2. Material and methods

The annealed H13 tool steel was cut into tensile samples (Fig. 1). The H13 steel has a chemical composition (in wt.%) of 0.63% Si, 0.94% V, 4.87 % Cr, 0.44 % Mn, 1.13% Mo, and balance Fe.

A system RC-HLDM4000 (Raycham, China) was used for the laser remelting of H13 steel which includes a laser apparatus (Laserline 4000, Germany, with the wavelength of 1080 nm), and a KUKA KR 16-1610-2 robot. The experiment was carried out in an atmospheric environment using the argon as shielding gas and the rate was set to 15 ml/s. The distance between the coaxial laser head and the substrate was 45 mm. Based on our preliminary experiments, the laser power was set to 2500 W. The laser scanning speeds were set to 6 and 24 mm/s, respectively, and moves along the direction denoted in Fig. 1. In addition, the diameter of the laser spot is 5 mm. Considering that the lower scanning speed will introduce higher residual stress, for the low scanning speed (6 mm/s), the workpieces with the thickness of 15 mm were used to avoid distortions. For the scanning speed of 24 mm/s, the thickness of workpieces is 10 mm.

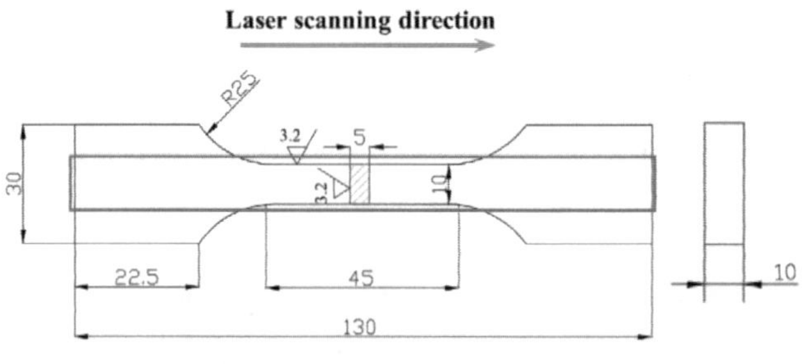

Fig. 1 Schematic of the remelted workpiece. The zone in the red rectangle is the single-track remelted zone.

All the testing specimens were cut from the gauge section in Fig. 1 and abraded using successive grades of 220 to 1200 grits of silicon carbide paper, and then polished on a velvet cloth using diamond suspension. Then, they were ultrasonically cleaned in distilled water and alcohol bath sequentially after etching (4% Nital solution for 3 min) for metallographic observation. The Keyence VHX-600 optical microscope was used to observe the fractured surface and cross-section of the remelted area.

Table 1 Samples identification and laser processing conditions.

Sample	Scanning speed (mm/s)	Power (W)	Thickness of sample (mm)
V6	6	2500	15
V24	24	2500	10

The microstructure was characterized by scanning electron microscopes (SEM) (ZEISS Sigma equipped with X-ray energy-dispersive spectrometer (EDS) and electron backscatter diffraction (EBSD, Oxford)). In addition to the previous mechanical polishing, an ion milling system (HITACHI IM4000, Japan) was used to further polish the specimens for EBSD observation.

A Vickers microhardness tester was used to evaluate the microhardness of the laser remelted samples. A series of indentations were performed along the depth orientation from the top surface of the samples, a 200 g load was applied for each point for 10 s.

3. Results and discussion

The cross-section of the remelted zone is shown in Fig. 2. The samples with different laser scanning speeds (6mm/s, 24mm/s) are defined as V6 and V24, respectively. The differences due to these exposition conditions are correlated to the heat/energy input used during the processing. As shown in Fig. 2a, the whole irradiated surface of the V6 sample is remelted and subsequently solidified. Due to the lower laser scanning speed, the input energy received per unit area is high, and the convection of the molten pool increased accordingly. Meanwhile, the liquid surface tension causes a shrinking of the molten pool from the edge to the center [16]. Consequently, the convexity is formed on the irradiated surface of the V6 sample. Based on the measurement, the thickness of the remelted zone in the V6 sample is 1.97 mm. The existence of dendrite (the details are shown in Fig. 3) is a typical microstructure within the melted zone due to an extremely high cooling rate (10^4-10^6 K/s) [17] and segregation [18]. Furthermore, the area beneath the partial remelted area is a heat affected zone (HAZ) with a thickness of 1.00 mm.

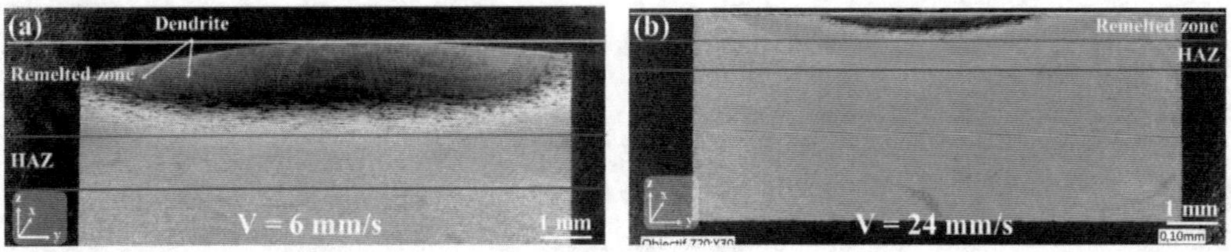

Fig. 2 Cross-section of both V6 and V24 specimens.

For the short exposition condition (V24, Fig. 2b), the irradiated surface is almost smooth, which means that the perturbation of the molten pool is marginal. The thickness of the remelted zone and the HAZ in the V24 sample are 0.53 and 0.49 mm, respectively.

The thickness and width of the two defined areas are reported in Table 2. For the two samples, the thickness decreases with the increase of laser scanning speed.

Table 2 Samples identification and laser processing conditions.

Sample	Thickness of remelted zone (mm)	Thickness of HAZ (mm)	Width of remelted zone (mm)
V6	1.97	1.00	10
V24	0.53	0.49	6.61

To further study the cross-section of the samples, the SEM and EBSD were used to observe the V6 sample. In Fig. 3a, we can clearly see an oxidation layer. With regard to the remelted zone, the microstructure shows the typical laser treated microstructure like the laser melting deposition but without the planar and equiaxed grain in the bottom and top surface, respectively [19]. According to the EBSD results, the remelted zone mainly exhibits the phases of martensite, austenite and Cr23C6 (Fig. 3b). In the upper part of the remelted zone (Fig. 3c), the grain structure is mainly cellular. In addition, there are dendrites in the center of the remelted zone (Fig. 3d). In the bottom area (Fig. 4e), the absence of planar grains may be attributed to the huge heat input (large laser spot and high laser power) and the homogeneity of the material.

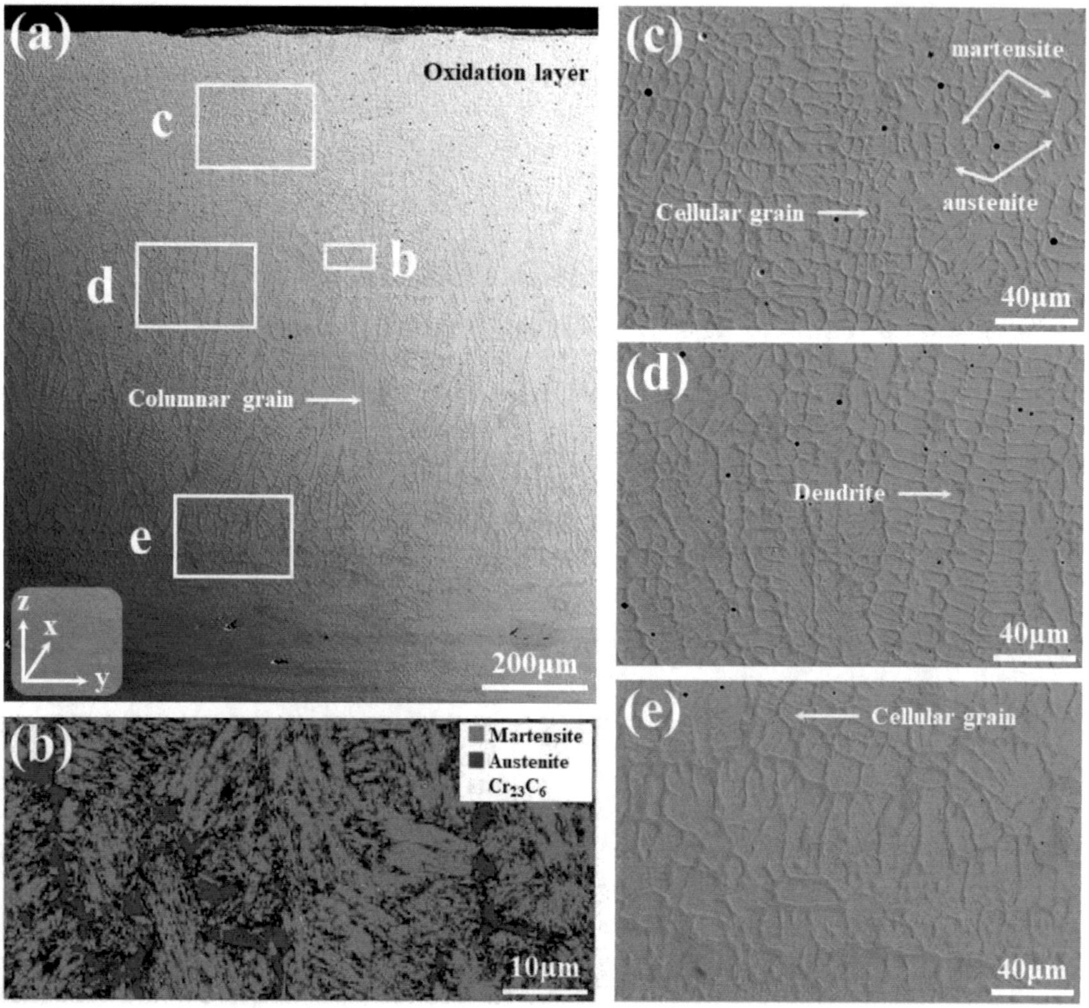

Fig. 3 Microstructure of V6 sample. (a) Cross-section captured by SEM. (b) Area b in panel e captured by EBSD. The red, blue, and yellow phases are martensite, austenite, and Cr23C6, respectively. The black pixels are the unindexed areas. (c-e) Area c-e in panel a.

Since the hardness is a key property for the H13 tool steel, the microhardness distribution along the depth direction and the corresponding strengthening mechanism are further investigated. Fig. 4a shows the microhardness distribution along the depth direction (Fig. 6b) in V6 sample. Combined with Table. 2, it shows that the remelted zone has a high hardness which is above 700 HV0.2. Fig. 4b shows the indentations along the center line of the cross section. The distance between the center of adjacent indentations is 50 μm.

It is well known that polycrystalline alloys with small grain sizes are stronger [20], and the contribution of grain boundary strengthening to the yield strength can be estimated by the Hall-Petch equation [21]:

$$\sigma = \sigma_0 + kd^{-1/2}$$

where σ0 is the friction resistance for dislocation movement within polycrystalline grains, k is the Hall-Petch slope, and d is the average grain size. Therefore, the EBSD were further used to measure the grain size along the certer line (similar to Fig. 4b). The average size of the grain at different depths in V6 sample was profiled in Fig. 4c. The comparison of Figs. 4a & c shows that the hardness decreases with the increase of grain size. Particularly, at a depth of ~1000 μm, the increase of grain size causes a lower hardness.

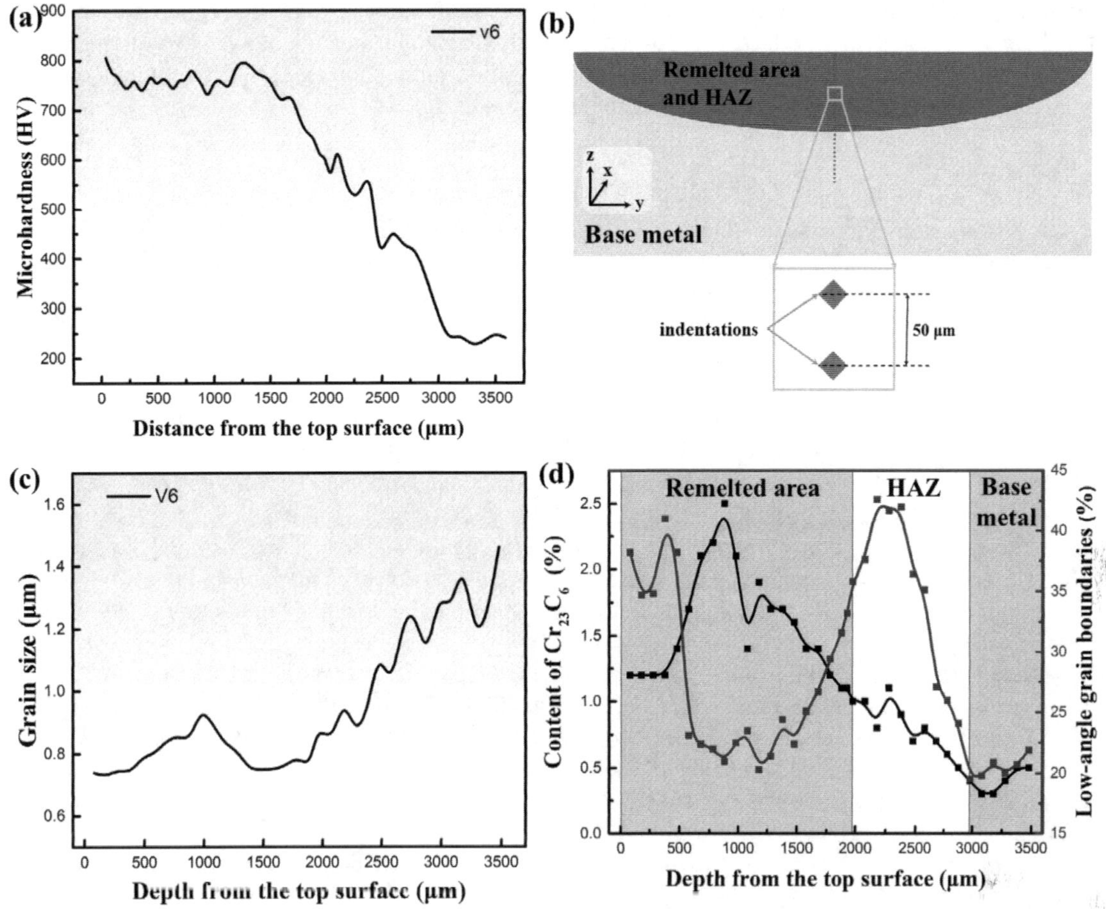

Fig. 4 The results tested along the depth direction in V6 sample: (a) the microhardness; (b) the sketch of the microhardness test process; (c) the average grain size; (d) the content of Cr23C6 and the low-angle grain boundaries (LAGB, ⩽ 15°). All grain sizes computed in this work are equivalent circle diameters.

It should be mentioned that the LAGBs are caused by low stacking fault energy, which results from the rapid solidification during laser processing [22]. The LAGBs are expected to act as a barrier to dislocation movement that increases the strength of materials [23] and leads to more strain hardening [24], as shown in Fig. 4d. The LAGB at the top area is around 37%, which is much higher than that at the center of the remelted zone (~ 20%) which leads to higher hardness. Similarly, the bottom of the remelted zone and the upper part of the HAZ also show a higher hardness.

Based on previous works, the dislocations are mainly stacked at the grain boundaries [25], which means that the finer grains are accompanied by more dislocations. The slip system interaction is a significant hardening mechanism. Particularly, the localized strain gradient is triggered by the phases in alloy with different hardness [26] and the dispersed precipitation (carbides) [27] will amplify this phenomenon. Therefore, intragranular carbides also play an important role in slip activation and development, because it could change the direction of slip bands or activate new slip systems thus resulting in local plasticity and giving rise to the residual stress field [25]. According to Fig. 4d, the high content of Cr23C6 in remelted zone also plays an important role in the hardness improvement.

4. Conclusions

In this work, H13 tool steel tensile workpieces were remelted by a large size Gaussian laser spot under different laser scanning speeds. The hardness of the remelted area in all samples is in the range of 700 to 850 $HV_{0.2}$. The grain structures in the remelted zone are mainly cellular, dendritic and columnar, and there is no evident interface in between the remelted zone and the HAZ. The strengthening mechanisms include the grain refinement and precipitation strengthening that act simultaneously to impart the high hardness of H13 steel during the laser remelting.

5. References

[1] DebRoy, T., et al., *Additive manufacturing of metallic components–process, structure and properties*. Progress in Materials Science, 2018. **92**: p. 112-224.

[2] Lu, H., et al., *Effects of laser shock peening on the hot corrosion behaviour of the selective laser melted Ti6Al4V titanium alloy*. Corrosion Science, 2021. **188**: p. 109558.

[3] Lai, Q., et al., *Investigation of a novel functionally graded material for the repair of premium hypereutectoid rails using laser cladding technology*. Composites Part B: Engineering, 2017. **130**: p. 174-191.

[4] Jiao, J., et al., *CFRTP and stainless steel laser joining: Thermal defects analysis and joining parameters optimization*. Optics & Laser Technology, 2018. **103**: p. 170-176.

[5] Sharma, A. and V. Yadava, *Experimental analysis of Nd-YAG laser cutting of sheet materials–A review*. Optics & Laser Technology, 2018. **98**: p. 264-280.

[6] Lu, J., et al., *High-performance integrated additive manufacturing with laser shock peening–induced microstructural evolution and improvement in mechanical properties of Ti6Al4V alloy components*. International Journal of Machine Tools and Manufacture, 2020. **148**: p. 103475.

[7] Lambiase, F., S. Genna, and C. Leone, *Laser finishing of 3D printed parts produced by material extrusion*. Optics and Lasers in Engineering, 2020. **124**: p. 105801.

[8] Rahmatabadi, D., et al., *Characterizing the elastic and plastic properties of the multilayered Al/Brass composite produced by ARB using DIC*. Materials Science and Engineering: A, 2019. **753**: p. 70-78.

[9] Gan, Z., et al., *Numerical simulation of thermal behavior and multicomponent mass transfer in direct laser deposition of Co-base alloy on steel*. International Journal of Heat and Mass Transfer, 2017. **104**: p. 28-38.

[10] Di Schino, A. and J.M. Kenny, *Grain refinement strengthening of a micro-crystalline high nitrogen austenitic stainless steel*. Materials Letters, 2003. **57**(12): p. 1830-1834.

[11] Chiang, K.-A. and Y.-C. Chen, *Laser surface hardening of H13 steel in the melt case*. Materials letters, 2005. **59**(14-15): p. 1919-1923.

[12] Jae-Ho, L., et al., *Laser surface hardening of AISI H13 tool steel*. Transactions of nonferrous metals society of China, 2009. **19**(4): p. 917-920.

[13] Karmakar, D.P., M. Gopinath, and A.K. Nath, *Effect of tempering on laser remelted AISI H13 tool steel*. Surface and Coatings Technology, 2019. **361**: p. 136-149.

[14] Oh, S. and H. Ki, *Prediction of hardness and deformation using a 3-D thermal analysis in laser hardening of AISI H13 tool steel*. Applied Thermal Engineering, 2017. **121**: p. 951-962.

[15] Meng, C., et al., *Effect of thermal fatigue on microstructure and mechanical properties of H13 tool steel processed by selective laser surface melting*. Metals, 2019. **9**(7): p. 773.

[16] Bertoli, U.S., et al., *In-situ characterization of laser-powder interaction and cooling rates through high-speed imaging of powder bed fusion additive manufacturing*. Materials & Design, 2017. **135**: p. 385-396.

[17] Lei, J., et al., *Comparative study on microstructure and corrosion performance of 316 stainless steel prepared by laser melting deposition with ring-shaped beam and Gaussian beam*. Optics & Laser Technology, 2019. **111**: p. 271-283.

[18] Krell, J., et al., *General investigations on processing tool steel X40CrMoV5-1 with selective laser melting*. Journal of Materials Processing Technology, 2018. **255**: p. 679-688.

[19] Lei, J., et al., *Enhanced corrosion and wear resistance properties of carbon fiber reinforced Ni- based composite coating by laser cladding.* Surface and Coatings Technology, 2018. **334**: p. 274-285.

[20] Kato, M., *Hall–Petch relationship and dislocation model for deformation of ultrafine-grained and nanocrystalline metals.* Materials Transactions, 2014. **55**(1): p. 19-24.

[21] Yuan, R., I.J. Beyerlein, and C. Zhou, *Coupled crystal orientation-size effects on the strength of nano crystals.* Scientific reports, 2016. **6**(1): p. 1-9.

[22] Chen, H., et al., *A novel approach to direct preparation of complete lath martensite microstructure in tool steel by selective laser melting.* Materials Letters, 2018. **227**: p. 128-131.

[23] Shanmugasundaram, T., E. Bouzy, and A. Chokshi, *Strengthening and weakening by repeated dynamic impact in microcrystals and nanocrystals.* Materials Science and Engineering: A, 2015. **639**: p. 97-102.

[24] Raabe, D., *Recovery and recrystallization: phenomena, physics, models, simulation.* Physical Metallurgy, 2014: p. 2291-2397.

[25] Zhao, C., et al., *A comparative assessment of iron and cobalt-based hard-facing alloy deformation using HR-EBSD and HR-DIC.* Acta Materialia, 2018. **159**: p. 173-186.

[26] Ashby, M., *The deformation of plastically non-homogeneous materials.* The Philosophical Magazine: A Journal of Theoretical Experimental and Applied Physics, 1970. **21**(170): p. 399- 424.

[27] Song, R., D. Ponge, and D. Raabe, *Improvement of the work hardening rate of ultrafine grained steels through second phase particles.* Scripta Materialia, 2005. **52**(11): p. 1075-1080.

Acknowledgments
This work was financially supported by the China Scholarship Council (Files No. 201808120131).

Grain structure predictions for metallic additive manufacturing processes

D Dreelan [1,2], A Ivankovic [1,2] and D J Browne[1,2]

[1] School of Mechanical & Materials Engineering, University College Dublin, Dublin 4, Ireland
[2] I-Form Research Centre for Advanced Manufacturing, Science Foundation Ireland

Abstract. Additive manufacturing has transformed the way we think about component fabrication. Generating a geometry in a layer-by-layer fashion presents many advantages over traditional subtractive methods, but also presents many challenges pertaining to the highly localised and energetic nature of the heat source. Since the material passes through multiple heating and cooling cycles throughout the build, some of which completely melt and erase the microstructure, a dynamic simulation is necessary to determine the grain structure that emerges. Grains are generally, but not exclusively, highly textured with columnar grains commonly spanning multiple layers. Fast, efficient and parallelised envelope cellular automata based models are used to simulate the nucleation and growth of the individual crystals that comprise the grain structure, with trade-offs being made between intra-grain detail and computational efficiency so that meso-scale simulations are possible. Simplified, but physically sound thermal models are used to predict the thermal conditions at the melt pool periphery, which are weakly coupled to the grain growth model. Dendrite tip kinetics models are used to determine alloy specific growth laws as a function of local undercooling. The effect of various processing parameters on as-solidified grain size, morphology and texture are investigated for aluminium alloys 3D printed by laser powder bed fusion.

1. Introduction

The advent of additive manufacturing (AM) marks the beginning of a new era in component fabrication. Unlike traditional techniques, components are built from the ground up by selectively melting metallic powder, fusing it to the solid metal underneath. Without the need to manoeuvre a tool to remove material from a homogenous bulk material, far more intricate and application optimised structures can be realised.

In powder bed fusion processes, such as selective laser melting (SLM) or electron beam melting (EBM), a highly localised, high powered and fast-moving radiative heat source is used to selectively melt regions of a bed of metallic powder. To achieve complete fusion with the underlying metal, be it with the solid metal build plate for the first layer, or with previously solidified layers for subsequent layers of the build, the melt pool typically penetrates some depth into the underlying metal. Due to the high intensity and velocity of the heat source, thermal conditions are typically favourable to columnar or cellular solidification regimes, with grains tending to become aligned with the heat source. Grains with crystallographic orientation that are most closely aligned with the local thermal gradient will grow preferentially and mechanically block grains that are less aligned. It is through this competitive growth that crystallographic texture emerges.

In virtually all engineering applications, with the exception of high temperature creep resistance, an equiaxed grain structure is desirable. Therefore, models that predict the onset of columnar-to-equiaxed transition (CET) are highly valuable. For CET modelling the most suitable modelling regime [1] is dependent on the spatial and temporal scale of the desired prediction. Of most practical importance are macro and meso scale models, which have the potential to simulate microstructure formation over several layers of an AM build, and even up to the scale of an entire part. At these scales, Hunt diagram based enthalpy models [2], cellular automata [3] and front tracking models [4,5] are most effective. Due to the high thermal gradient in AM processes, heterogenous nucleation of new equiaxed grains ahead of the columnar front is rare under standard processing conditions, so CET can be difficult, but not impossible to achieve. In the absence of CET columnar grains tend to span multiple layers, with each grain growing epitaxially from the microstructure of layer underneath.

Hence, in order to predict the final polycrystalline microstructure of an AM produced component, a dynamic simulation which keeps track of the evolution, and destruction of the grain structure during each heating and cooling cycle of the build.

2. Methods

The nucleation and competitive growth of each individual metallic grain in the microstructure is simulated here using cellular automata (CA) techniques. The defining characteristic of CA solidification models is that they are semi-discrete, with the state of a cell being one of a discrete number of possible states at any given moment. The model can be classified as sharp interface, as regions of solid are separated from regions of liquid by a single layer of interface cells.

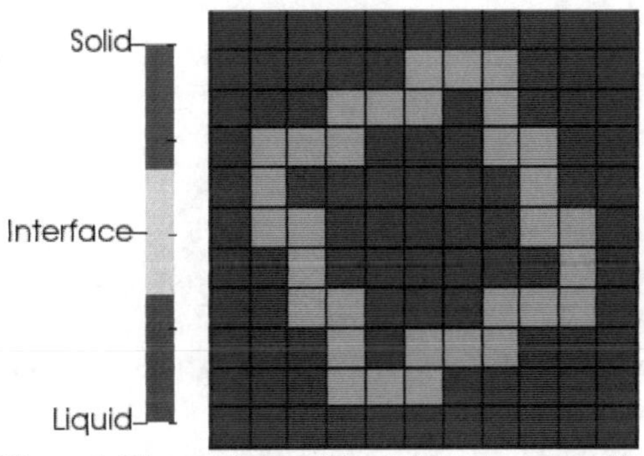

Figure 1: Discrete States of the Cellular Automata Algorithm

Rules are defined to govern the conditions under which these cell states may change. One such rule is that solid and liquid regions must always be separated by a single layer of interface cells. Another rule is that any cell whose cell centre temperature is higher than the liquidus temperature must be either liquid or interface, which is obvious by the definition of the liquidus temperature. The model is only semi-discrete, as growth is calculated continuously for interface cells, with many time steps occurring between state changes. This is where the major strength of CA models lies, as interface cells typically only make up a very small fraction of the total cells in the domain, allowing for much higher numbers of cells and consequently larger domains with higher numbers of grains to be simulated than would be possible by continuum-based methods such as phase field modelling [6] or molecular dynamics [7], particularly in 3D. The propagation of interface cells during solidification is managed by the classic DeCentred Square / Octahedron Algorithm (DCSA) [3]. Each interface cell has a uniform square or octahedron associated with it, for two and three dimensions respectively. The orientation of these shapes are defined by their Euler angles, with one component for 2D simulations, and three components for 3D. Apart from when a cell is nucleated, the centre of these shapes typically does not coincide with the cell centre of the interface cell to which it belongs, which greatly decreases the influence of mesh orientation on growth directions. Since the shapes are uniform, only a single value for the half diagonal length is needed to define their size. During solidification, these shapes grow. Eventually the apices of a shape will engulf the cell centre of a neighbouring cell, which is captured by it and inherits the orientation and grain identification number from the capturing cell. When a cell is captured, its state is changed from liquid to interface, and a new square/octahedron is initiated for that new interface cell. The starting size of this shape is determined by the specifics of the algorithm described in detail in the original paper [3]. The centre of the new shape is determined such that it is fully contained within the shape that captured it, with one of its apices coinciding with the apex that captured it.

The growth rate of the half diagonal of the squares/octahedra are governed by the KGT dendrite tip kinetics model [8], which assumes a parabolic dendrite tip operating at marginal stability. The model is highly non-linear and requires numerical methods to resolve, so curve fitting is used to obtain alloy specific growth law polynomials with respect to local thermal undercooling which are cheap to compute at each time step.

Heterogenous nucleation and subsequent CET, although rare for standard AM processing conditions and materials, is accounted for. At the start of the simulation, all possible nucleation sites are set and given a nucleation undercooling according to a distribution. Seed particles, which can arise from dendrite fragments or the intentional addition of grain refiners, are assumed to follow a Gaussian distribution of spherical diameters. From nucleation theory, the undercooling at which the growth of a spherical particle becomes thermodynamically favourable is inversely proportional to the diameter of the particle [9], and so nucleation undercoolings are also assumed to follow a Gaussian distribution. Special considerations are made during the choice of distribution parameters, so that they can effectively be applied to AM length scales. Initial microstructures can be generated in DREAM.3D [10], or by running the CA and nucleation model on an isothermal cooling simulation, with the latter being found to be several orders of magnitude faster and producing virtually identical results.

As already discussed, grain growth is assumed to be purely dependent on temperature through the use of dendrite tip kinetics models which inherently account for solute diffusion. The grain growth model is weakly coupled to a thermal model. An enthalpy based thermal model, using the Voller method for latent heat evolution [11], has been developed which supports various liquid fraction relationships including linear and Scheil which is most suited to rapid solidification as is the case in AM.

Reading of temperature histories presents powerful opportunities to increase the computational efficiency of the CA model. Since both the current and future thermal history are known and are sufficiently close to each other such that temperature can be assumed to be bounded over the time span, only cells that could possibly change state within the time span can be isolated, and the rest ignored. This is particularly pertinent to AM simulations as the heat source is highly localised, meaning that state transformations are occurring only in small regions of the domain at any given moment. Shared memory parallelisation using OpenMP is used to split the workload among multiple processing units, greatly increasing the speed and scalability of the model.

3. Case Study

We consider here a single pass of a laser over a 400μm × 1200μm × 200μm solid block of Al-4wt.%Cu alloy. The incident intensity of the laser follows a 2D Gaussian distribution, with a full width half max spot size of 25μm, a total power of 200W and a scan speed of 1000μm/s. The laser travels a total distance of 800μm, before the power is switched off and the molten metal is allowed to fully solidify. The starting equiaxed grain structure was generated in DREAM.3D [10], with a mean equivalent sphere diameter of 10μm. The F.C.C. Al-4wt.%Cu alloy was assumed to follow a Gulliver-Scheil liquid fraction relationship with respect to temperature. Material properties for the alloy are shown in table 1.

Table 1. Material properties of Al-4wt.%Cu

Parameter	Symbol	Unit	Value	Reference
Liquidus Temperature	T_l	°C	649	[12]
Solidus Temperature	T_s	°C	585	[12]
Solvent Freezing Temperature	T_f	K (°C)	660.37	[13]
Specific Latent Heat of Melting	L	J/kg	352733.7	[4,14]
Thermal Conductivity	κ	W/m.°C	100	[4,14]
Density	ρ	Kg/m^3	2800	[15]
Specific Heat Capacity	C_p	J/(kg.°C)	1071.4	[4,14]
Partition Coefficient	k	-	0.14	[12]
Liquidus Slope	m_l	°C/wt.%	-2.6	[12]
Diffusion Coefficient into Liquid	D_l	m^2/s	2.4×10^9	[12]
Gibbs-Thomson Coefficient	Γ	°C.m	2.36×10^{-7}	[12]
Dendrite Tip Selection Parameter	σ^*	-	0.025	[8]
Mean Nucleation Undercooling	ΔT_μ	°C	3.0	[16]
Nucleation Undercooling Standard Deviation	ΔT_σ	°C	0.5	[16]
Maximum Nucleation Site Density	n_{max}	sites/m^3	1×10^{10}	[16]

The growth law polynomial, as obtained by curve fitting to the KGT dendrite tip kinetics model is shown below.

$$V(\Delta T) = 2.27 \times 10^{-5}(\Delta T)^{3.053}$$

The alloy was preheated to 600K (326.85°C) and the bottom surface was held at this temperature throughout the simulation, with all other surfaces being cooled by convection with a heat transfer coefficient of 80 W/m^2 and radiation with an emissivity of 0.2 and temperature of 300K (26.85°C) for the surroundings.

A Gaussian distribution of activation undercoolings was assumed for all potential heterogenous nucleation sites, with a moderate mean nucleation undercooling of 3.0°C. Throughout the literature, common values for dilute Al-Cu alloys without the addition of grain refining inoculant particles range between 2.0°C [16] up to 28.84°C in some experimental measurements [17]. Shifting the activation criteria to lower thermal undercoolings would increase the propensity for CET to occur. As previously discussed, CET is rare under standard AM processing of aluminium alloys, so a small value was chosen to improve confidence in a prediction of a fully columnar grain structure.

4. Results

Figures 2 shows the grain structure evolution as predicted by the model for the Al-4wt.%Cu case study, as a function of time. Figure 3 shows the final as-printed grain structure.

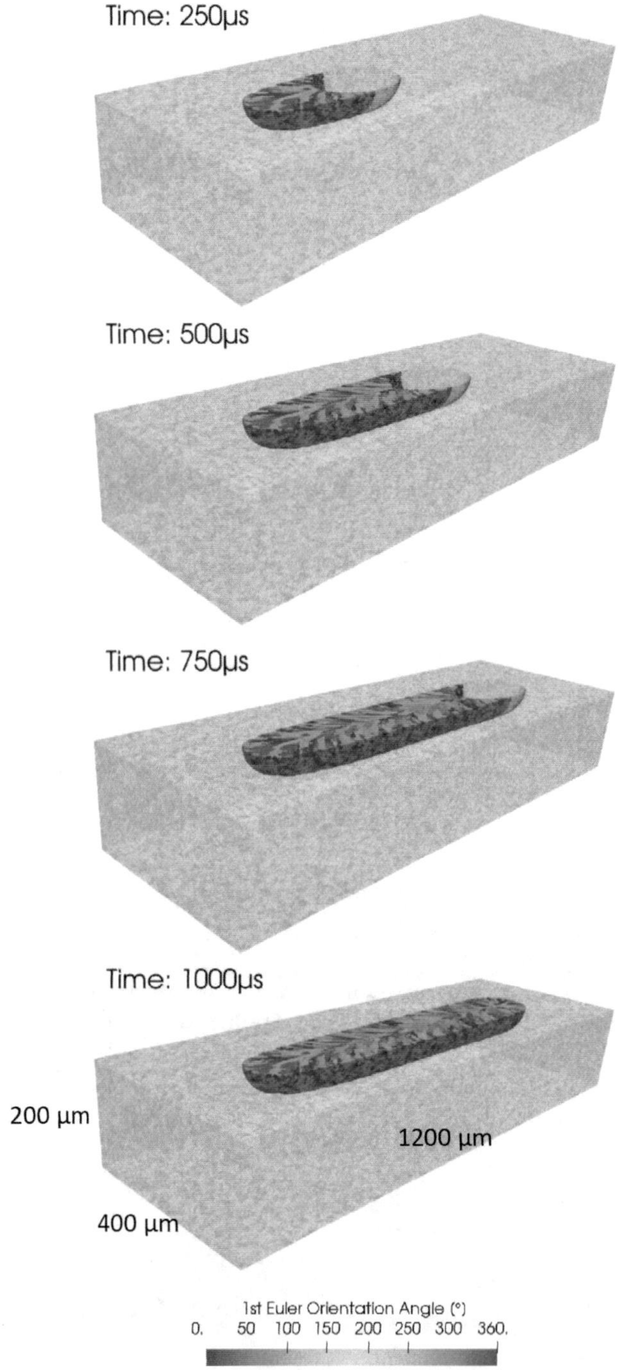

Figure 2: 3D view of single-track melting/solidification of a solid block of Al-4wt.%Cu. Snapshots at various times throughout the simulation. Cells that were melted at any stage during the simulation are highlighted so that the outer melt pool shape can be easily seen.

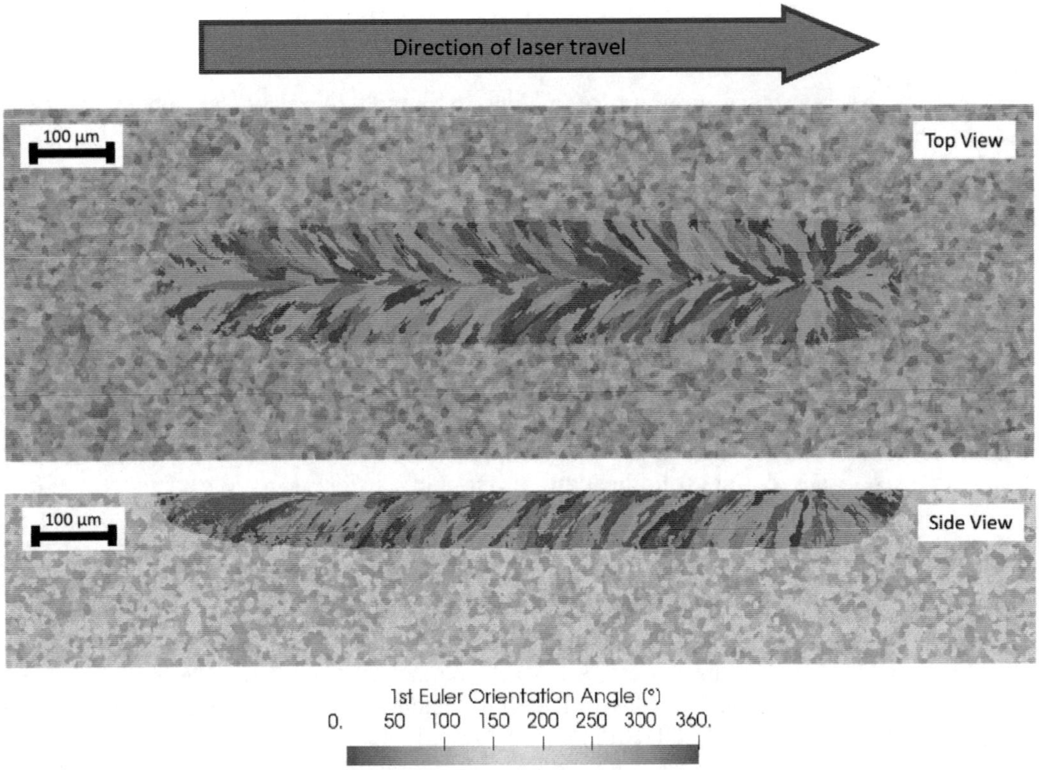

Figure 3: Grain structure prediction during the first layer of an Al-4wt.%Cu AM build. *Top:* View from top, along the axis of the laser heat source. *Bottom:* View from side perpendicular to the direction of laser travel, of a slice along the centre of the laser path. Colouring is based on the third Euler angle component, in degrees.

5. Discussion

Figure 2 shows the evolution of melt pool profile and grain structure at snapshots in time throughout the simulation. Cells that were have not yet been melted are shown translucently so that the melt pool morphology can be easily seen. It is seen that grains generally grow towards the centre of the laser heat source, in the opposite direction to heat flow.

As can be seen particularly in the side view of figure 3, crystallographic texture emerges as grains grow epitaxially from the existing grain structure at the periphery of the melt pool. The orientation of the columnar grains changes throughout solidification. In the early stages of the simulation, the thermal gradient is orientated towards the horizontal due to the rapid movement of the laser heat source and the material being relatively cool as not much energy had yet been supplied from the laser. During the middle of the simulation, the orientation of the thermal gradient becomes closer to vertical and reaches steady state. Once the laser reaches the end of its raster and the power is switched off, the thermal gradient changes direction with respect to the horizontal, pointing towards the centre of the last location of the laser spot. As can be seen, the orientation of the columnar grains matches the orientation of the local thermal gradient throughout all stages of the simulation, confirming that crystallographic texture tends to be aligned with the thermal gradient during solidification. As is common in AM processing of aluminium alloys, CET did not occur, and none of the potential nucleation sites set by the heterogenous nucleation model were activated.

The true value of grain structure predictions will be through the dynamic simulation of multiple layers of an AM build, so that how crystallographic texture, grain size and morphology develop. When, or if, CET occurs is also of great importance. Planned model developments to extend simulations to multiple layers, promises to address this. Work is on-going to validate these grain structure predictions, against EBSD imaging of aluminium alloys, and is expected to be completed in time to present at the conference.

Acknowledgments
This publication has emanated from research conducted with the financial support of Science Foundation Ireland under Grant number 16/RC/3872.

References

[1] Mirihanage WU, Dai H, Dong H, Browne DJ. Computational modeling of columnar to equiaxed transition in alloy solidification. *Advanced Engineering Materials*. 2013 Apr;15(4):216-29.
[2] Yang X, Barrett RA, Tong M, Harrison NM, Leen SB. Towards a process-structure model for Ti-6Al-4V during additive manufacturing. *Journal of Manufacturing Processes*. 2021 Jan 1;61:428-39.
[3] Gandin CA, Rappaz M. A 3D cellular automaton algorithm for the prediction of dendritic grain growth. *Acta Materialia*. 1997 May 1;45(5):2187-95.
[4] Browne DJ, Hunt JD. A fixed grid front-tracking model of the growth of a columnar front and an equiaxed grain during solidification of an alloy. *Numerical Heat Transfer, Part B: Fundamentals*. 2004 May 1;45(5):395-419.
[5] McFadden S, Browne DJ, Gandin CA. A comparison of columnar-to-equiaxed transition prediction methods using simulation of the growing columnar front. *Metallurgical and Materials Transactions A*. 2009 Mar 1;40(3):662-72.
[6] Sahoo S, Chou K. Phase-field simulation of microstructure evolution of Ti–6Al–4V in electron beam additive manufacturing process. Additive manufacturing. 2016 Jan 1;9:14-24.
[7] Kurian S, Mirzaeifar R. Selective laser melting of aluminum nano-powder particles, a molecular dynamics study. Additive Manufacturing. 2020 Oct 1;35:101272.
[8] Kurz W, Giovanola B, Trivedi R. Theory of microstructural development during rapid solidification. *Acta metallurgica*. 1986 May 1;34(5):823-30.
[9] Quested TE, Greer AL. Grain refinement of Al alloys: Mechanisms determining as-cast grain size in directional solidification. *Acta Materialia*. 2005 Oct 1;53(17):4643-53.
[10] Groeber MA, Jackson MA. DREAM. 3D: a digital representation environment for the analysis of microstructure in 3D. *Integrating materials and manufacturing innovation*. 2014 Dec;3(1):56-72.
[11] Voller VR, Swaminathan CR. ERAL Source-based method for solidification phase change. *Numerical Heat Transfer, Part B Fundamentals*. 1991 Jan 1;19(2):175-89.
[12] Rebow M, Browne DJ. On the dendritic tip stability parameter for aluminium alloy solidification. *Scripta Materialia*. 2007 Mar 1;56(6):481-4.
[13] Roeser WF, Lonberger ST. Methods of testing thermocouples and thermocouple materials. US Department of Commerce, National Bureau of Standards; 1958.
[14] Browne DJ. Modelling columnar and equiaxed growth (Doctoral dissertation, University of Oxford) 2003.
[15] Aluminum 2014 Alloy (UNS A92014) [Internet]. AZoM.com. 2022 [cited 18 January 2022]. Available from: https://www.azom.com/article.aspx?ArticleID=6614
[16] Zhu MF, Hong CP. A modified cellular automaton model for the simulation of dendritic growth in solidification of alloys. Isij International. 2001;41(5):436-45.
[17] Ahmadein M, Pustal B, Berger R, Subašić E, Bührig-Polaczek A. Grain nucleation parameters for aluminum alloys: Experimental determination and model validation. Metallurgical and Materials Transactions A. 2009 Mar;40(3):646-53.

Microstructural transition and grain selection across the wire/powder interface in hybrid additive manufacturing

I Ushakov[1], A Aaddi[1], J Wursthorn[2] and D Daloz[1]

[1] Université de Lorraine, IJL, F-54000 Nancy
[2] IREPA LASER, F-67400 Illkirch

Abstract. Hybrid additive manufacturing is a promising process to produce Nickel-based alloy complex parts and builds, respecting high deposition rates during wire deposition and geometrical precision by powder deposition in laser DED process. This combination of processes should allow producing large parts with geometrically complex structures, not achievable by other technologies. Nevertheless, hybrid transitions of this type are not well described in the literature, in terms of microstructure establishment and grain selection across the interface, geometrical form of interface, and their potential influence on mechanical properties. Microstructure development across the interface wire/powder will be presented in this work with interface characterization by optical microscopy and EBSD. Grains selection across several powder layers in the hybrid wall, microstructure analysis will be presented for both wire and powder depositions.

1. Introduction

Hybrid additive manufacturing (AM) is a promising process to produce Nickel-based alloy complex parts and builds, respecting high deposition rates during wire deposition and geometrical precision by powder deposition in laser directed-energy deposition (DED) process. While, laser DED process is quite well studied separately for wire and powder deposition and almost matured for some alloy families [1,2], this combination of processes leads to hybrid transitions, that are not well described in the literature. The closest attempt, combining oscillating selective laser melting (SLM) and powder DED in the case of Inconel 718 was recently reported [3]. In this work, microstructure and texture development through the interface have been studied, as well as the influence of heat treatment, microsegregation, composition, and nature of precipitates. The hybrid part was mechanically tested and showed that DED part was the weakest, because of δ-phase precipitation in the interdendritic spaces [4]. No similar data exists in the literature about wire-to-powder DED transition, which could be a critical point in AM of large-scale builds. In order to understand the behaviour of this particular zone, microstructure establishment, grain selection and their orientation, during wire/powder transition have been investigated and presented in this paper.

2. Materials and experimental methods

The hybrid AM DED wall was built in 2 steps (Figure 1). Firstly, the wire wall was built by successive deposition of 3 beads from commercially In625 wire of 1.6 mm diameter. Secondly, the powder wall was built on the top of the wire wall with a deposition of 9 powder beads (In625, granulometry of 45-105 μm). Deposition parameters are given in Table 1.

Table 1. Process parameters, used in study.

Parameter	Wire	Powder
Laser power (W)	6800	4500
Wire feeding rate ($mmin^{-1}$)	4.2	
Mass powder flux ($gmin^{-1}$)		52
Torch speed ($mmin^{-1}$)	0.5	1.5
Hatch spacing (mm)	6.4	2.6
Layer thickness (mm)	2.2	1.2
Heat input (Jmm^{-1})	816	180
Volume energy (Jmm^{-3})	58	57.7
Deposited mass (kgs^{-1})	3.06	3.12
Mass energy (Jkg^{-1})	8×10^{6}	1.44×10^{3}

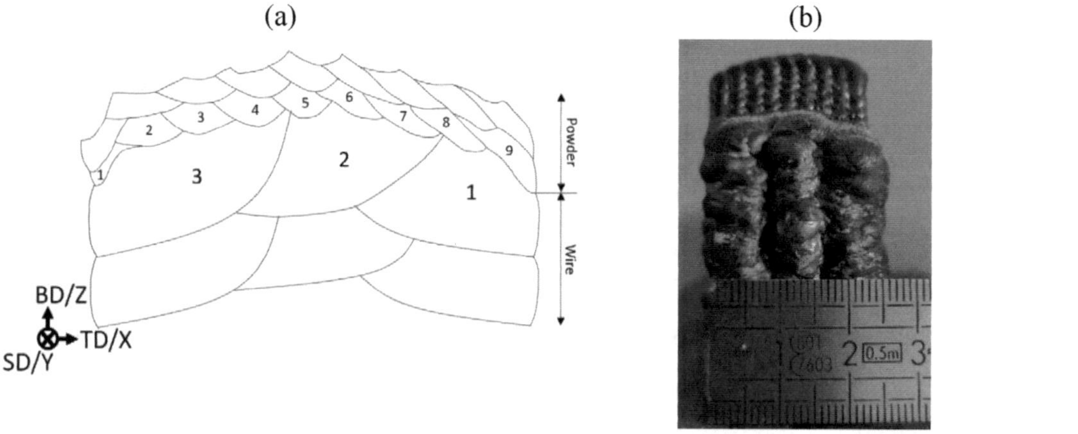

Figure 1. Schematic principle of the wire/powder deposition with a deposition sequence (a), Image of the wire/powder hybrid wall (b).

The microstructures were examined using optical microscopy (OM) and electron-backscatter diffraction (EBSD) mapping. Samples were cut in BD-TD plane (cross-section) for both wire and powder deposition. They were ground using 80-4000 abrasive papers, polished with diamond suspensions 3 µm and 1 µm. Etching for OM was performed using 95%HCl+5%H_2O_2 solution for In625. Final polishing with a colloidal silica suspension (OP-S) was applied before EBSD mapping, which has been performed with a SEM Quanta FEG 650, equipped with EBSD camera. Oxford HKL software was used for EBSD analysis.

3. Results

3.1. Wire wall structure

The representative macrostructure of the wire wall consisting of 3 central bead zones and 2 bead intersection zones is presented in figure 2. The microstructure of the wire wall has a bimodal shape. In this way, grain size was measured as a minor elliptical axis for two modes of distribution ($\bar{D}_1$ and $\bar{D}_2$), separately. After that, values have been averaged for each mode of grain size distribution. So, $\bar{D}_1$ characterizes more columnar dendritic grains of the central bead zone and $\bar{D}_2$ characterizes more columnar dendritic grains of the intersection bead zone. The central bead zone consists of long columnar grains (marked yellow in figure 2), which for a part grow through several layers, and for the rest new grains appear from nucleation events. This zone shows typical FCC Goss <001> grain orientation due to the direction of thermal gradient during solidification, with a grain size of $\bar{D}_1$(µm) = 22.9, $\bar{D}_2$(µm) = 330.5. Side beads have more inclined grains as a result of the direction of heat extraction during solidification. The intersection zone shows rare nucleation events with a prevalence of huge grains (marked blue in figure 2), growing epitaxially, but altering their grain dendritic growth direction from one deposited bead to another, which is typical for FCC material, because of <100> cubic symmetry [5]. Intersection zone presents mostly Cubic <001> orientation of big grains ($\overline{D}_2$(μm) = 408).

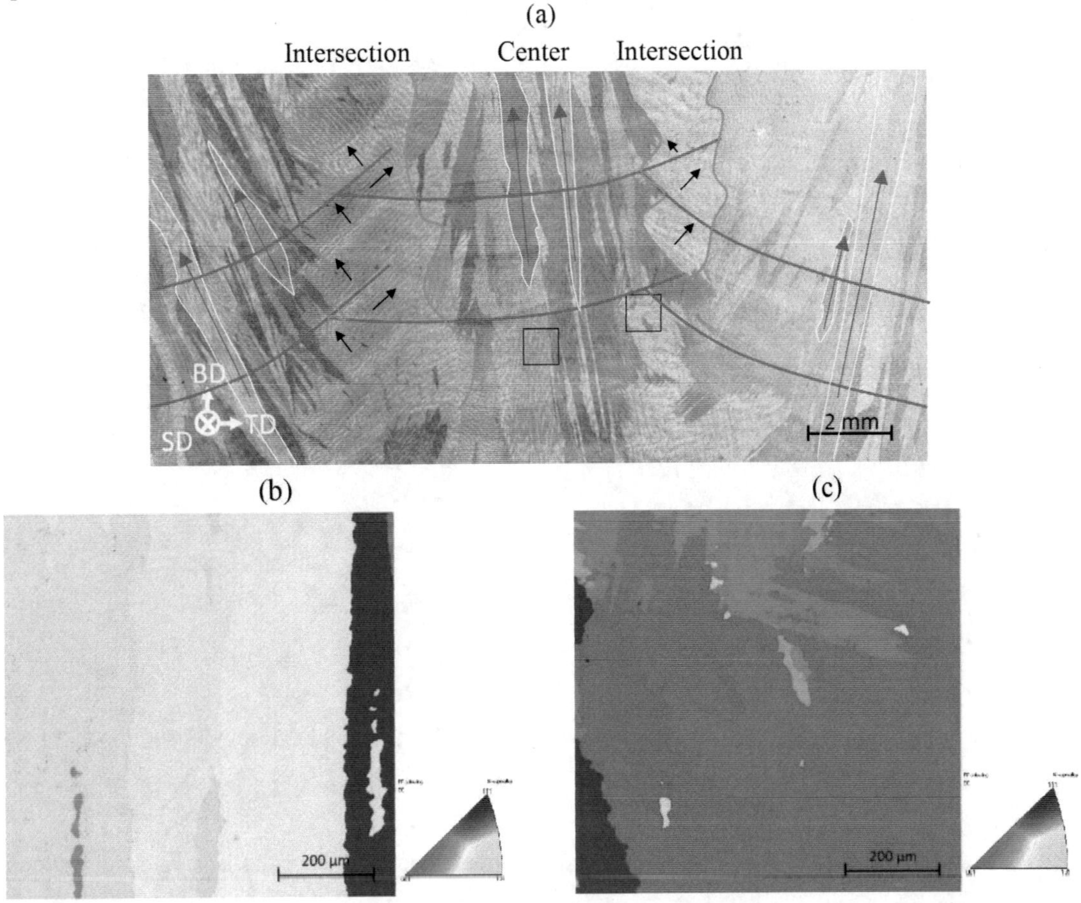

Figure 2. Optical macrostructure image (a); EBSD maps of the wire wall structure: the central bead zone (b), and the intersection bead zone (c).

3.2. Powder wall structure

The macrostructure of powder stationary deposition is presented in figure 3. In total, there are 9 central bead zones and 8 bead intersection zones. Macrostructure image shows similar to wire microstructure, but, on the contrary, huge columnar dendritic grains (marked blue in figure 3), growing mostly epitaxially, from bead intersection zone ($\overline{D}_2$(μm) = 531.5) dominate over small columnar dendritic grains (marked yellow at figure 3) of central bead zone. Grains of the central bead zone ($\overline{D}_2$(μm) = 203.7) have Cubic <001> grain orientation, and there is also germination of new grains during solidification. Huge grains of bead intersection zone has <111> Fibre parallel to SD orientation.

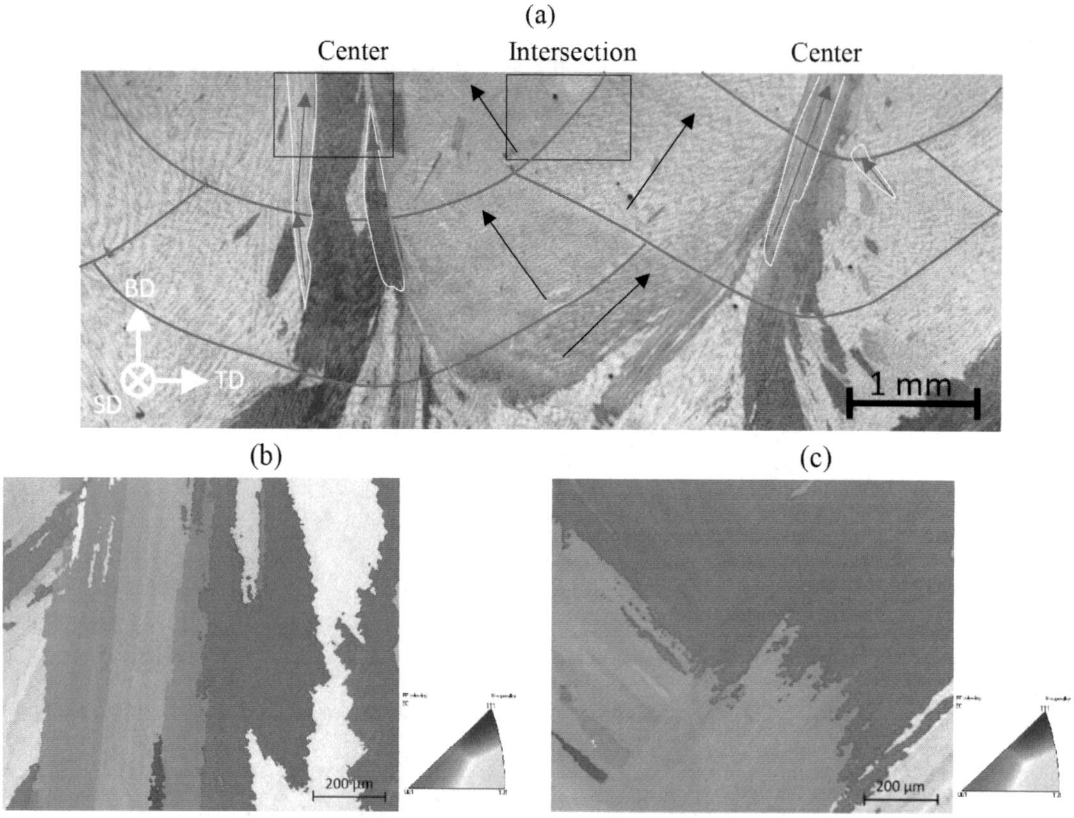

Figure 3. Optical macrostructure image (a); EBSD maps of the powder stationary regime: the central bead zone (b), and the intersection bead zone (c).

3.3. Wire/powder transition

Wire/powder transition, shown in figure 4, shows a complex macrostructure that can be decomposed in 3 main cases:

- Powder central zone deposited on wire central zone (figure 4(a)):

Columnar dendritic grains of wire central zone growing epitaxially through wire/powder interface and selected by successive powder layers to give Cubic <001> grain orientation in the powder central zone. This grain selection takes more than two layers. Grain germination occurs in this case.

- Powder intersection zone deposited on wire central zone (figure 4(b)):

Columnar dendritic grains of wire central zone growing epitaxially through wire/powder interface. The powder intersection zone should select a new Fibre <111> orientation for huge columnar dendritic grain, that forms later in this zone with successive powder layers. Instead of that, the powder intersection zone gives Cubic <001> orientation, selected by the central powder zone. As a result, this grain selection takes at least three powder layers. Grain germination occurs in this case, as well.

- Powder intersection zone deposited on wire intersection zone (figure 4(c)):

This case (and equivalent one: powder central zone deposited on wire intersection zone) gives big grains, growing epitaxially, with almost no grain germination. This grain selection is fast: one powder layer could be sufficient. With grain selection, big grains from this zone give directly Fibre <111> grain orientation in the powder intersection zone, which could even replace columnar grains of the powder central zone.

Further competition of these three different mechanisms gives uniform powder microstructure with the prevalence of huge columnar dendritic grains, selected by powder bead intersection zones. At least three powder layers are needed to delete wire microstructure in powder deposition.

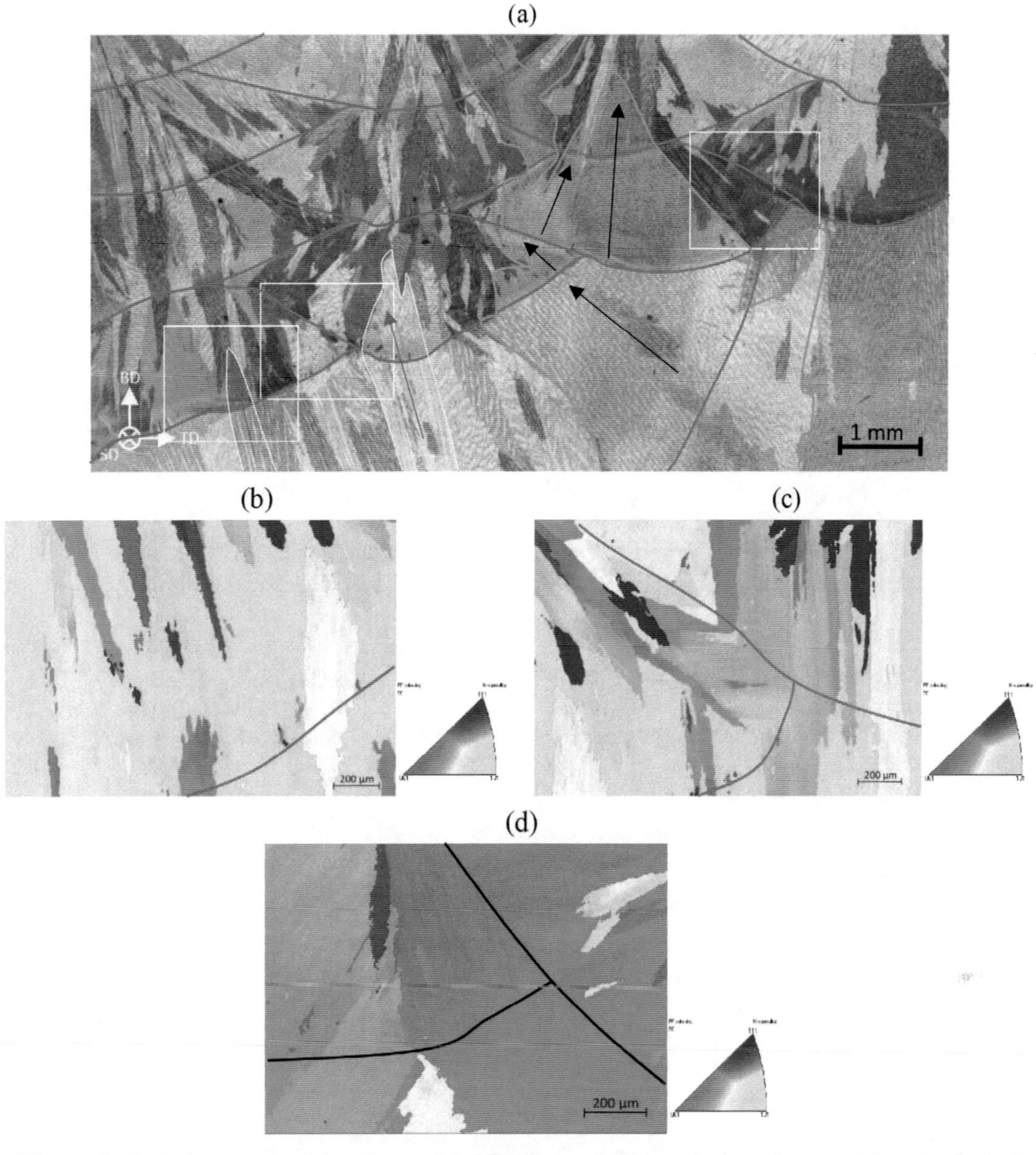

Figure 4. Optical macrostructure image (a); EBSD maps of the wire/powder transition: the central powder bead deposited on central wire bead (b), the intersection of powder beads deposited on the central wire bead (c), and the intersection of powder beads deposited on the wire bead intersection zone (d).

4. Discussion

First of all, during the first layer of powder deposition, the surface grain layer of wire should be fully remelted. After that, grain growth in the transition zone is epitaxial, so after remelting grain orientation of the previous wire layer is conserving and will be selected by the molten pool of the powder bead. Wire and powder unidirectional depositions have common microstructures features: continuous columnar dendritic grains in the central bead part and continuous columnar dendritic huge grains, altering their dendritic orientation (secondary dendritic branches transform to primary dendritic branches and so on [5]) in bead intersection zone from one layer to another. There is no columnar-to-equiaxed transition (CET) in the build. For the powder deposition with additional layers, we can see an almost total prevalence of huge columnar dendritic grains, having <111> Fibre parallel to SD orientation, of bead intersection zone over columnar dendritic grains of central bead zone, having Cubic <001> orientation, due to grain selection during powder deposition. On the contrary, the wire part has a solidification microstructure with an almost equal proportion between these two types of grains. The higher power density of powder deposition is given by the width of the bead: 1 wire bead equal to 3 powder beads. This difference could be a reason for this faster grain selection and prevalence of big grains of intersection zone in the microstructure of powder deposition. Another explanation of that could be an initial curvature of wire deposition before powder deposition. Moreover, wire deposition has a higher remelting capacity than powder deposition and different form of the molten pool. We can also see that molten pool dimensions are not constant for one powder layer. The first powder bead has a higher Depth/Width ratio than the last powder bead, due to thermal conditions, the curvature of the wire deposit, and support given by previous powder beads.

The microstructure of the transition zone is a mixture of these two zones. Three mechanisms are possible for powder grain selection and growth at wire/powder transition: the case of powder central zone deposited on wire central zone with fast selection columnar dendritic grains of the central zone; the case of powder intersection zone deposited on wire central zone with a slow selection of big columnar dendritic grains of intersection zone; and last one, the case of powder intersection zone deposited on wire intersection zone, that gives big grains, growing epitaxially, with almost no grain germination and fast grain selection: one powder layer could be sufficient. The wire/powder interface is not planar, it has a wavy morphology, defined by powder molten pools. The geometric form of wire/powder interface with previous wire deposition orientation, as well as the direction of thermal gradient during powder deposition, define further powder microstructure. From macrostructure and EBSD maps analysis, we can conclude that at least 3 powder layers are needed to complete wire/powder transition and delete wire microstructure.

Wire/powder transition could be prepared by the last wire layers or by the first powder layers, it could be also done by optimizing parameters of deposition, increasing power during the first layer of powder deposition, for example. Another option could be to deposit powder in another sequence, starting by right and left sides of the built, alternating the next beads to finish in the center of the powder deposition.

5. Conclusion and perspectives

- Wire and powder deposition microstructures have common microstructures: continuous columnar dendritic grains in the center of the bead and continuous columnar dendritic big grains in the intersection of beads, altering their dendritic orientation.
- Powder deposition has a more important proportion of big grains of intersection bead zone, due to the higher power density and initial curvature of wire build before powder deposition.
- Microstructure transition takes at least 3 layers of powder deposition to delete an influence of wire deposition.
- Wire/powder transition could be prepared by changing deposition conditions.

6. References

[1] DebRoy T, Wei H L, Zuback J S, Mukherjee T, Elmer J W, Milewski J O, Beese A M, Wilson-Heid A, De A and Zhang W 2018 Additive manufacturing of metallic components – Process, structure and properties *Progress in Materials Science* **92** 112–224

[2] Jafari D, Vaneker T H J and Gibson I 2021 Wire and arc additive manufacturing: Opportunities and challenges to control the quality and accuracy of manufactured parts *Materials & Design* **202** 109471

[3] Godec M, Malej S, Feizpour D, Donik Č, Balažic M, Klobčar D, Pambaguian L, Conradi M and Kocijan A 2021 Hybrid additive manufacturing of Inconel 718 for future space applications *Materials Characterization* **172** 110842

[4] Sui S, Tan H, Chen J, Zhong C, Li Z, Fan W, Gasser A and Huang W 2019 The influence of Laves phases on the room temperature tensile properties of Inconel 718 fabricated by powder feeding laser additive manufacturing *Acta Materialia* **164** 413–27

[5] Wei H L, Mazumder J and DebRoy T 2015 Evolution of solidification texture during additive manufacturing *Sci Rep* **5** 16446

Acknowledgments

This research work is a part of the French program PAM-PROD under the financial support of PSPC.

Understanding Ni-modified AlSi10Mg alloys from slow to rapid solidification

DA Moura[1,2], GL Gouveia[1,2], G Figueira[1,2], P Gargarella[1,2,3], JE Spinelli[1,2,4]

[1] Department of Materials Engineering, Federal University of São Carlos UFSCar, 13565-905 - São Carlos, SP, Brazil.
[2] Federal University of São Carlos, Graduate Program in Materials Science and Engineering, 13565-905 - São Carlos, SP, Brazil.
[3] Center of Characterization and Development of Materials (CCDM), Federal University of São Carlos, 13565-905 - São Carlos, SP, Brazil
[4] Author to whom any correspondence should be addressed.

Al-Si-Mg alloys are widely employed in a variety of industries, including aerospace, automotive, and microelectronics. This is because of its low density, acceptable mechanical properties, acceptable corrosion resistance, and inexpensive application cost. Because of its advantageous fluidity, limited solidification interval, and low volumetric contraction, Al-(9-11) wt.% Si-(0.2-0.5) wt.% Mg casting alloys have been employed in Laser-Powder Bed Fusion (LPBF) printing techniques. Despite being used as commercial alloys, their mechanical properties still need to be improved. In line with this, the current focus of aluminum-based alloys development for additive manufacturing is mostly on the modification of commercially available alloys. One of the possibilities is the reinforcement of cast alloys that may be processed using additive manufacturing technologies. Nickel is used as an alloying element in this study to generate the Al_3Ni intermetallic, which is distinguished by its improved strength. Furthermore, the thermal stability of the Al_3Ni may be a benefit, particularly for high-temperature applications. Under such context, the present study aims to investigate the solidification under low and high cooling rates of three alloys: AlSi10Mg-1Ni, AlSi10Mg-2Ni, and AlSi10Mg-3Ni (wt.%). Samples were obtained by directional solidification (DS) and laser surface remelting (LSR) and the cooling rates were calculated for both DS samples and with extrapolation for the more refined LSR samples as well as with the use of models from the literature. With the increase in Ni content, the Al_3Ni mass fraction increased from 3% to 5% and then to 7.5%, according to CALPHAD computations. The growth rate of the DS samples was lowered in the direction of Ni addition: 1 Ni > 2Ni > 3 Ni. SEM data will be presented, with a focus on the rapid solidified melt pool samples and their features.

1. Introduction

Due to its capacity to produce complicated geometry components in a near-net-shape fashion, Laser-Powder Bed Fusion (L-PBF) has recently arisen as an emerging Additive Manufacturing (AM) technology [1,2]. Even though this technology has several processing particularities, the components produced must solidify at high cooling rates ($\dot{T}$) of 10^4-10^7°C/s [3,4].
High cooling rates in the AM techniques have long been known to optimize the microstructure in terms of grain size and dendrite arm spacing (DAS). Owing to that, the Laser Surface Remelting (LSR)

technique can aid in the examination of preliminary data before the L-BPF processing. These previous investigations might be very valuable in rationalizing L-PBF assumptions. Moreover, transient directional solidification [5,6] is another key approach for rationalizing cooling rate-microstructure dependences in Al alloys.

It is well known that the cooling rates in the laser melt pool are not only variable at various points but also difficult to detect. As a result, the formation of a gradient microstructure is often reported [7]. The resulting features and benefits remain still unclear.

AM has extensively explored and manufactured commercial AlSi10Mg alloy parts [8,9]. This is one of the few Al alloys that have been successufully employed in AM. Under AM conditions, it evolved to a completely different solidification behaviour resulting in very refined microstructures. This is mainly because laser melting involves highly concentrated energy input, leading to large local temperature gradients during solidification. Nowadays, there is a current interest in developing alternative aluminum-based alloys for AM, which is mostly focused on the modification of commercially available alloys. Automotive piston alloys containing Ni, such as AlSi12CuMgNi, AlSi18CuMgNi and AlSi12Cu4Ni2Mg might be mentioned in the context of high-temperature resistant alloys. The usual piston alloy is a eutectic Al-12% Si alloy with around 1% each of Cu, Ni, and Mg. For increased strength at high temperatures, special eutectic alloys have been developed. Hypereutectic alloys containing 18 and 24 wt.% Si have lower thermal expansion and wear. In practice, Al piston suppliers utilize a variety of progressively improved alloy compositions, although they are all based on these fundamental alloy types [10].

Garmendia et al. [11] added 1wt.% Cu in the AlSi10Mg alloy during L-PBF processing. When compared to the AlSi10Mg, the tensile strength of the Cu-added specimens increased by 51 MPa. This was attributed to grain size reduction and subsequent Hall–Petch strengthening, as well as Orowan strengthening induced by Cu-rich particles. To address the paucity of commercial Al alloys for AM, the addition of Ni appears to be a promising possibility considering that Al3Ni particles may be generated, improving mechanical properties. AlSi10Mg-(Ni) alloys were chosen for this investigation to examine the influence of Ni content and cooling rate variations within the melt pool not only on the generation of gradient microstructures/morphologies but also on the formed phases. Moreover, either cooling rate extrapolation and simple model calculations were employed and compared to each other to estimate the melt pool cooling rates.

2. Experimental procedure

The AlSi10Mg-1Ni, AlSi10Mg-2Ni, and AlSi10Mg-3Ni alloys were directionally solidified (DS) in cylindrical bipartite AISI 310 stainless steel molds inner coated with a refractory material (to minimize radial heat losses). K-type thermocouples were distributed along with the height of the casting from the water-cooled bottom part, to obtain data on temperature as a function of time (acquisition frequency of 5Hz) and as a function of position (P) relative to the casting bottom surface. When the molten alloys reached the liquidus temperatures + 5% at the thermocouples closest to the ingot's bases, water flow was activated against the external bottom part, promoting upward directional solidification. Specific information about this experimental setup can be found in previous studies [5,6]. All DS castings have resulted in completely columnar structures.

Various samples obtained under different processing conditions (i.e., different P) were extracted from the DS castings. These samples were appropriately ground, polished, and etched to expose the resultant microstructures for further examination. After acquiring images by using an optical microscope (Olympus Corporation, GX41 model, Tokyo, Japan), the triangle technique was used to measure the primary dendritic spacing [12]. Furthermore, a SEM was used to get greater magnification images as well as to identify the produced phases (Philips XL 30 FEG, Eindhoven, The Netherlands).

Transverse samples associated with a cooling rate of approximately 0.7 °C/s were extracted to be used in the LSR experiments. The surface treatment experiments of the 3 alloys of interest were conducted with an Aurora Labs® S-Titanium Pro machine equipped with a CO_2 laser, which has a maximum power of 300W. The focused laser beam had a nominal spot size of 150 μm with a wavelength of 10.6 μm. The laser head is connected to a CNC-controlled XYZ table on which the sample is placed and moved

during the treatment. During the LSR trials, the following operating conditions were used for each alloy: laser beam speeds, v, of 5 and 20 mm/s; average laser power, Q, of 300W.
The sample surface was sandblasted before laser treatments to increase energy absorption from the laser beam. Conventional metallographic techniques were used to produce selected transverse (perpendicular to the laser trace direction) slices of the laser track. Optical and SEM microscopes were used to examine these sections. Finally, primary dendritic spacings were determined within the laser melt pool.

3. Results and discussion

The solidification precipitation sequences of the alloys may be described using the CALPHAD method through the Thermo-Calc software, as shown in Figure 1. For AlSi10Mg-1Ni and AlSi10Mg-2Ni alloys, the α-Al phase began to solidify first from the liquid. The Al_3Ni phase was the initial phase in the AlSi10Mg-3Ni alloy. The Si, Al_3Ni, and Mg_2Si phases were predicted to precipitate at lower temperatures. The impacts of Ni content change may be noticeable if the fractions of Al_3Ni at 100°C were examined, which were approximately 3%, 5%, and 7% with increasing Ni content. These diagrams in Figure 1 were also useful to confirm the experimentally determined *liquidus* and α-Al precipitation temperatures of 587.5°C, 583.5°C, and 580.0°C for the AlSi10Mg-1Ni, AlSi10Mg-2Ni, and AlSi10Mg-3Ni alloys, respectively.
Optical cross-section images are captured at key spots throughout the length of each casting, based on the evolution of the DS samples, as shown in the upper part of Figure 2 for the alloys containing 1% and 3%Ni. The generation of well-defined dendritic arrays with noticeable changes in length scale as a function of the cooling rate and α-Al growth rate encountered in different regions of the casting can be seen in both examples in Figure 2. The compositions of phases of the samples solidified at 0.5°C/s were determined by SEM/EDS. These results can be seen in the bottom part of Figure 2. All phases predicted by the CALPHAD computations were confirmed here.

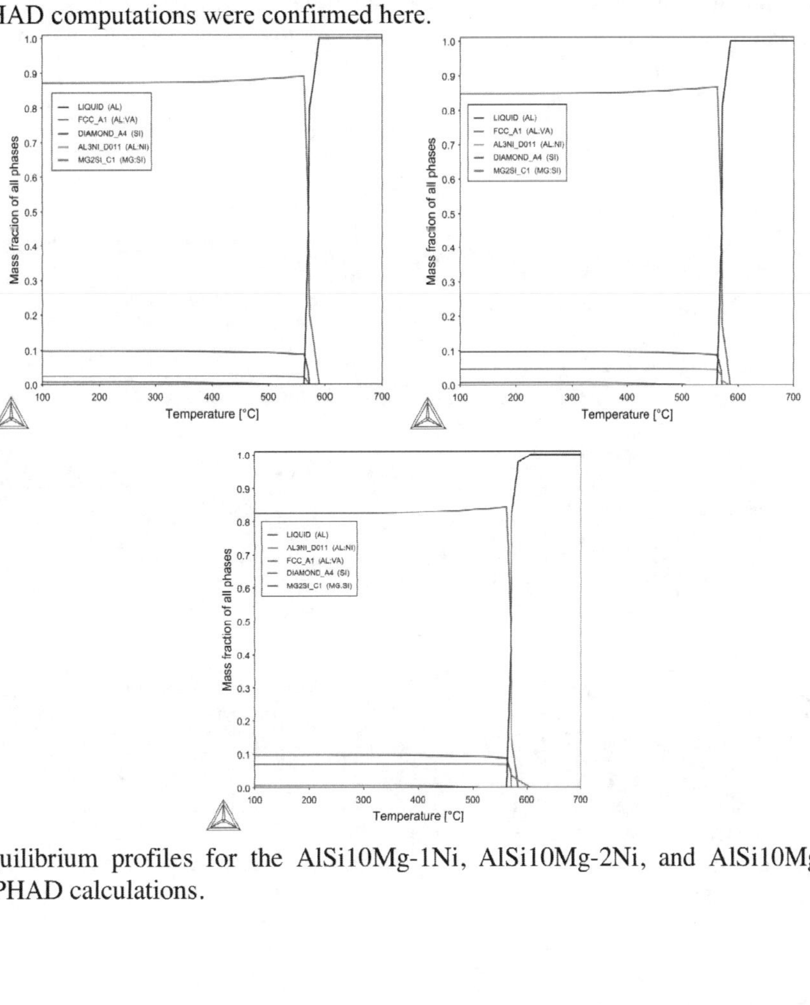

Figure 1. Equilibrium profiles for the AlSi10Mg-1Ni, AlSi10Mg-2Ni, and AlSi10Mg-3Ni alloys through CALPHAD calculations.

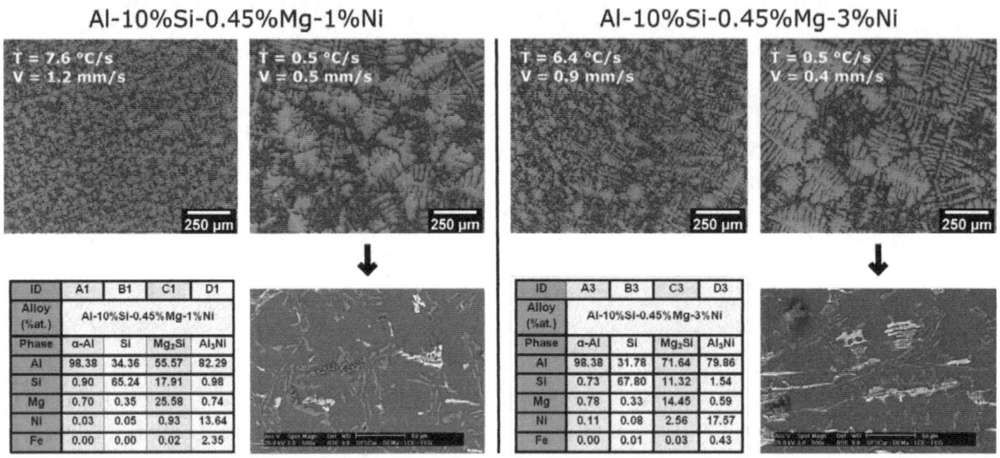

ID	A1	B1	C1	D1
Alloy (%at.)	Al-10%Si-0.45%Mg-1%Ni			
Phase	α-Al	Si	Mg_2Si	Al_3Ni
Al	98.38	34.36	55.57	82.29
Si	0.90	65.24	17.91	0.98
Mg	0.70	0.35	25.58	0.74
Ni	0.03	0.05	0.93	13.64
Fe	0.00	0.00	0.02	2.35

ID	A3	B3	C3	D3
Alloy (%at.)	Al-10%Si-0.45%Mg-3%Ni			
Phase	α-Al	Si	Mg_2Si	Al_3Ni
Al	98.38	31.78	71.64	79.86
Si	0.73	67.80	11.32	1.54
Mg	0.78	0.33	14.45	0.59
Ni	0.11	0.08	2.56	17.57
Fe	0.00	0.01	0.03	0.43

Figure 2. Optical and SEM typical microstructures and EDS results for the DS AlSi10Mg-(Ni) alloys.

The microstructural characteristics of the laser-treated zone and substrate, which were previously directionally solidified in a water-cooled mold, are depicted in Figure 3. The substrate had a coarse dendritic microstructure, and the resolidified zone (either low or high v) was characterized by a very fine dendritic arrangement at the melt pool bottom. Furthermore, Figure 3 depicts the shift in microstructure size at the top surface of the melt pool. In sum, a morphological transition was detected. Theoretical and experimental solidification investigations show that the increase in cooling rate resulted in a change of the micromorphology of the solid/liquid interface from planar to cellular and subsequently to dendritic, implying a direct effect of solidification kinetics on the morphology of the solidification interface. Some authors demonstrated that at higher cooling rates, the dendritic front may be transformed into a cellular and planar front, defining a reverse transition that may produced high-cooling rate cells towards the top [12–14]. In contrast, Wang et al. [15] affirmed that the dendrites may fragment due to Marangoni convection effects giving origin to equiaxed α-Al grains. These fragments may be trasnportated towards the top surface forming the final refined layer. For higher laser speeds, the layer of refined microstructure at the top surface was larger.

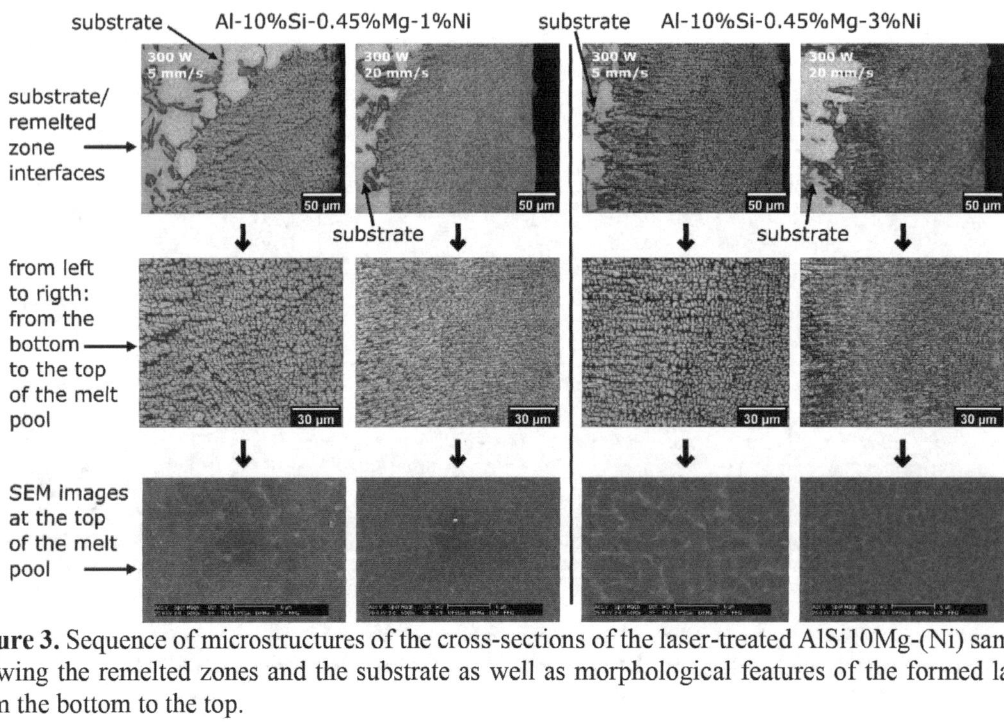

Figure 3. Sequence of microstructures of the cross-sections of the laser-treated AlSi10Mg-(Ni) samples showing the remelted zones and the substrate as well as morphological features of the formed layers from the bottom to the top.

The thermal profiles obtained through a number of thermocouples for all DS samples were analyzed to establish proper plots of position in the casting against time of α-Al isotherm passage during solidification: P = f(t). The α-Al growth rate (V_a) during directional solidification was calculated using the time derivatives of these functions, as shown in Figure 4. In addition, the time derivative of each cooling curve (dT/dt) was computed immediately after the α-Al isotherm passes through each thermocouple position to estimate the α-Al cooling rate ($\dot{T}$) as a function of casting position, as also shown in Figure 4. Figure 4(c) shows the λ_1 values as a function of cooling rate during DS. A single dendritic scaling relation may be typified by a power function, as can be seen in Figure 4(c). The -0.5 exponent described the dendritic development, being typical of those reported for multicomponent Al-based alloys [16]. The primary spacing varied from an average value of approximately 440 μm in the non-treated zone (i.e., the DS samples) to a spacing of about 1.38 μm in the top part of the resolidified zones of the investigated alloys, i.e., a remarkable decrease in length scale. The equation proposed by Ashby and Easterling [17] allowed for the cooling rate in laser melting traces to be estimated, based on the following expression:

$$\dot{T} = \frac{-2\pi k(T_\alpha - T_0)^2}{A(\frac{Q}{v})}, \tag{1}$$

where A is the absorptivity for the laser beam (0.18 as stated by Liu et al. [18]), v (m/s) is the laser scanning speed, Q (W) is the laser power, k (W/mK) is the liquid thermal conductivity determined through CALPHAD, T_0 is the room temperature and T_α is the transformation temperature of α-Al growth start.

Given that the thermal conductivity and T_α changed minimally as a function of Ni addition, the calculated melt pool cooling rates were similar for the alloys studied. $\dot{T}$ was about 12,600 °C/s for the 300W/5mm/s and 50,400 °C/s for the 300W/20mm/s. Extrapolations from the DS scaling law of λ_1 using average λ_1 values of 2.47 μm and 1.38 μm resulted in estimated melt pool cooling rates of 14,880 °C/s and 47,300 °C/s, which were adequately validated for the 300W/5mm/s and 300W/20mm/s LSR conditions, respectively. The extrapolation approach, which permits estimating point-to-point cooling rate inside the melt pool, was validated by the closeness of the cooling rate values determined by both methods.

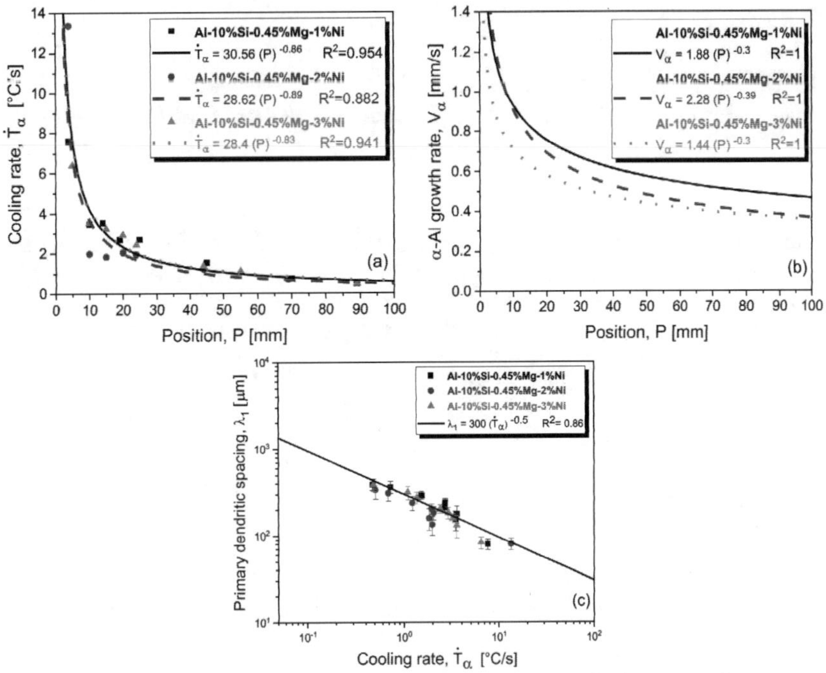

Figure 4. Variations of α-Al (a) cooling rate, (b) growth rate from the lower to the upper regions of the AlSi10Mg-(Ni) alloys castings, and (c) primary dendritic spacing as a function of cooling rate. R^2 is the coefficient of determination.

4. Conclusions

The addition of Ni (from 1 to 3 wt.%) had no effect on the spacing values, neither in the DS nor in the LSR AlSi10Mg-(Ni) samples considering the examined processing conditions. Furthermore, Ashby and Easterling's equation was used to validate the extrapolation approach for determining cooling rates. The melt pool was basically constituted of 2 layers: one (at the bottom) visibly dendritic and the other at the top surface, which will be further investigated.

References

[1] K. Kempen, L. Thijs, J. Van Humbeeck, J.-P. Kruth, *Phys. Procedia* 39 (2012) 439–446.
[2] N. Read, W. Wang, K. Essa, M.M. Attallah, *Mater. Des.* 65 (2015) 417–424.
[3] G. Yu, D. Gu, D. Dai, M. Xia, C. Ma, K. Chang, *Appl. Phys. A* 122 (2016) 891.
[4] P. Delroisse, O. Rigo, P. Jacques, A. Simar, in: TMS Annual Meeting and Exhibition, San Diego, California, US, March 2017.
[5] M. Gandolfi, M.G.C. Xavier, L.F. Gomes, R.A.V. Reyes, A. Garcia, J.E. Spinelli, *Metals (Basel)*. 11 (2021) 1019.
[6] R. Kakitani, R. V. Reyes, A. Garcia, N. Cheung, J.E. Spinelli, Metall. *Mater. Trans. A* 50 (2019) 1308–1322.
[7] K. Lu, *Science*. 345 (2014) 1455–1456.
[8] F. Trevisan, F. Calignano, M. Lorusso, J. Pakkanen, A. Aversa, E. Ambrosio, M. Lombardi, P. Fino, D. Manfredi, *Materials (Basel)*. 10 (2017) 76.
[9] M. Tang, P.C. Pistorius, S. Narra, J.L. Beuth, *JOM* 68 (2016) 960–966.
[10] Röhrle, M. D.: Pistons for Internal Combustion Engines, Verlag Moderne Industrie, 1995.
[11] X. Garmendia, S. Chalker, M. Bilton, C.J. Sutcliffe, P.R. Chalker, *Materialia* 9 (2020) 100590.
[12] R. Trivedi, J.A. Sekhar, V. Seetharaman, *Metall. Trans. A* 20 (1989) 769–777.
[13] W.W. Mullins, R.F. Sekerka, *J. Appl. Phys.* 35 (1964) 444–451.
[14] R.W. Cahn, P. Haasen, *Physical Metallurgy*, 4 ed., Amsterdam, 1996.
[15] D. Wang, C.H. Song, Y.Q. Yang, Y.C. Bai, *Mater. Des.* 100 (2016) 291–299.
[16] M.G.C. Xavier, B.J.M. Freitas, G.L. Gouveia, A. Garcia, J.E. Spinelli, *Adv. Eng. Mater.* (2021) 2100822.
[17] M.F. Ashby, K.E. Easterling, *Acta Metall.* 32 (1984) 1935–1948.
[18] B. Liu, B.-Q. Li, Z. Li, P. Bai, Y. Wang, Z. Kuai, *Results Phys.* 12 (2019) 454–459.

Acknowledgments

The authors acknowledge FAPESP (grant numbers 2019/23673-7 and 2017/27031-4) and CNPq. This study was financed in part by the Coordenação de Aperfeiçoamento de Pessoal de Nível Superior - Brasil (CAPES) – Finance Code 001.

IOP Conf. Series: Materials Science and Engineering 1274 (2023) 012017 doi:10.1088/1757-899X/1274/1/012017

Influence of nanoparticles on melting and solidification during a Directed Energy Deposition process analysed by simulation

G Zheng[1], R Laqua[1], P Rey[2], M Salgueiro[2], M Gipperich[3], J Riepe[3] and J Jakumeit[1]
[1] Access e.V. Intzestr. 5, 52072 Aachen, Germany
[2] Technology Centre AIMEN, Relva 27A - O Porriño, Pontevedra, 36410, Spain
[3] Fraunhofer Institute for Production Technology IPT, Steinbachstraße 1, 52074 Aachen, Germany

Abstract. Additive Manufacturing is a strategic tool for industrial applications. When large size structural parts are targeted, high deposition rates are important and Directed Energy Deposition (DED) is a preferred technique. A metal wire is melted by laser light and deposit on a substrate or already solidified material. Due to the small size of the melting zone, a detailed experimental analysis of the process is very difficult and simulation is an important tool to understand the manufacturing process and the influence of process and material parameter. Here the influence of the nanocomposite reinforcement on the deposition process are investigated by simulation single line tracks. A three-phase melting and solidification simulation methodology has been used to investigate the melting and solidification during DED printing of single line tracks of different wire materials. The approach uses the finite-volume method and arbitrary polyhedral control volumes to solve the governing equations. Heating of the wire by laser light is tackled using a volumetric heat source describing the specific absorption of the laser power by the metal wire. The influence of melting and solidification on the initially uniform nanocomposite distribution during the printing process is simulated using a Lagrangian approach. Firstly, simulations for steel ST 52-3 (1.0570) were performed for different process parameter settings and compared to experimental results of deposition width, height and form to validate the simulation approach. The method is then applied to Ti6Al4V wires with and without nanocomposites added to the wire material. Adding nanocomposites changes the melting and solidification behaviour of the wire materials. The influence of these changes of the material properties on the deposition process under different process conditions is analysed by simulation.

1. Introduction

Directed Energy Deposition (DED) becomes an increasingly interesting metal Additive Manufacturing (AM) technique, due to its flexibility in creating arbitrary shapes on even and uneven substrates. When large size structural parts are targeted, high deposition rates are important and DED is a preferred technique. However, due to the small size of the melting zone, a detailed experimental analysis of the process is very difficult and simulation is an important tool to understand the manufacturing process and the influence of process and material parameter.

Modelling and simulation of DED processes are challenging due to the multi-physics and multiscale nature of the process. Coupled thermal-fluid flow with phase transition [1], dependency of material properties on temperature [2], and interface topological changes controlled by strong surface tension effect [3] make the prediction of deposit structure in relation to the process parameters difficult.

Nanoparticles reinforced metal matrix composites (MMnCs) demonstrate their enhanced mechanical properties (sustainability, strength/stiffness, damping capacity, etc.) [4-6], improved wear [7-8], or corrosion resistance [9-10] in various industrial fields. The enhanced properties are strongly dependent on equal distribution of nanoparticles in the processes. The distribution and the structural states of nanoparticles or particle clusters in MMCs are of critically importance and determine the properties of the final products. The principles of material science reveal that a more uniform distribution of nanoparticles or particle clusters will generally lead to a significant enhancement in the material properties [11-14]. Therefore, to model and simulate the motion and the distribution of nanoparticles in MMCs becomes a key issue in the development of metal products with the desired properties.Addition of nanoparticles (NPs) affects not only the properties of the final products, but also on the manufacturing process itself.

Here, the influence of the NPs reinforcement on the deposition process are investigated by numerical simulation. Firstly, simulations for steel ST 52-3 (1.0570) were performed for different process parameter settings and compared to experimental results of deposition width, height and form to validate the simulation approach. The method is then applied to Ti6Al4V wires with and without NPs added to the wire material. Adding NPs changes the melting and solidification behavior of the wire materials. The influence of these changes of the material properties on the deposition process under different process conditions is analyzed by simulation.

2. Modelling and simulation of nanoparticle reinforced DED processes

2.1. Simulation of the DED process

A three-phase melting and solidification simulation methodology, implemented in the computational dynamics solver STAR-CCM+, has been used to investigate the melting and solidification during DED. Latent heat release during melting and solidification of the metal is handled using an enthalpy formulation. A sharp interface between gas and melt phase is ensured by a high-resolution interface capturing (HRIC) [15]. A detailed description of the simulation approach can be found in [16]. Gas phase is treated as ideal gas with physical properties of air. Heating of the wire by the laser is modelled as volumetric heat source with Gaussian distribution in radial direction for wire and substrate region. The volumetric heat source is active in the entire wire region and in the substrate region limited to a depth of 0.1 mm. The effective laser power, i.e. the power absorbed by the wire or substrate, is monitored separately for both regions.

2.2. Simulation of the nanoparticle effect

Effects of the material thermodynamic properties of nanoparticles (NPs) were considered for the simulation of DED-process. The following formulations were implemented into the software to determine the effective thermodynamic properties (density ρ, dynamic viscosity μ, thermal conductivity λ and specific heat c_p) of nanofluids [17].

$$\rho_{eff} = \varphi\rho_{pt} + (1-\varphi)\rho_f \tag{1}$$

$$\mu_{eff} = \frac{\mu_f}{1 - 34.87\left(d_{pt}/d_f\right)^{-0.3}\varphi^{1.03}} \tag{2}$$

$$\lambda_{eff} = \lambda_f\left(1 + 4.4Re^{0.4}Pr^{0.66}(T/T_r)^{10}\left(\lambda_{pt}/\lambda_f\right)^{0.03}\varphi^{0.66}\right) \tag{3}$$

$$c_{p,eff} = \frac{\varphi\left(\rho c_p\right)_{pt} + (1-\varphi)\left(\rho c_p\right)_f}{\rho_{eff}} \tag{4}$$

where φ is the volume fraction of NPs, d_{pt} the diameter of NPs, d_f the equivalent diameter of a base fluid molecule, T the temperature and T_{fr} the freezing point of the base liquid, Re the nanoparticle Reynolds number and Pr is the Prandtl number of the base liquid. The effect of NPs on the solidification of Ti6Al4V is described by the fraction solid curve calculated by the software Thermo-Calc depending on the volume fraction of nanoparticles.

Since NP involved simulations are computationally expensive due to huge amounts of particles, operability of such simulations under the present computational capability must be considered. By the integration of the thermodynamic properties of the nanoparticles into the melt properties, the influence of the NPs on the melt has already been considered. A so-called "one-way coupling" model is then supposed to be sufficient to solve nanoparticle involved casting processes, which can dominantly decrease the computing resources required for the simulations.

Particle motion results from various forces and torques acting on it, exerted by its surrounding medium. The Lagrangian approach computes the spatial position due to trajectory of each particle by integrating the forces and torques balance on each particle. To investigate particle motion and distribution during melt fluid simulations, interactions between melt and particles were defined by drag forces. The thermophoretic force and the Brownian force were newly implemented for NP tracking [18]. The direction and the amplitude of the Brownian force were determined by a so-called diffusion-adaptive method.

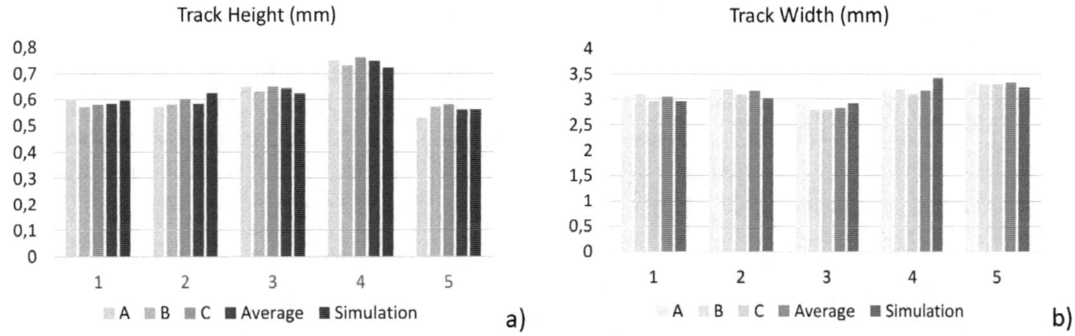

Figure 1 Three measured rack height (a) and width (b) in comparison with the simulation results

3. Validation of the DED track simulation

In collaboration with the Fraunhofer Institute for Production Technology IPT, the modeling of the DED process was developed at Access for different material pairings and validated using a tool steel wire (commercial name QuFe13 from Quada) and steel ST 52-3 (1.0570) for the substrate. The thermal-fluid data for both steels were taken from the ACCESS material database.

Table 1. Process parameter of steel DED-process

Set	Laser Power [W]	Travel Speed [mm/min]	Wire Feed Rate [mm/min]
1	2400	1000	1200
2	2600	1000	1200
3	2200	1000	1200
4	2400	1000	1500
5	2400	800	960

IPT conducted a series of single layer deposition experiments with lateral wire feed, each with different process parameters, and investigated the tracks produced in this process with regard to cross-sectional shape, track width and height. Table 1 shows a summary of the varied process parameters: laser power, wire feed rate and machine speed. Other relevant process parameters such as wire diameter, temperatures, shielding gas atmosphere and wire feeding angle remained unchanged.

Three tracks were evaluated for each parameter set. Figure 1 shows the three measured track height (a) and width (b) for each set, the average values and the results of the simulation. The simulated track heights are mostly within the range of variation of the experiments. The simulated track width tends to be slightly lower in the simulation. Figure 2

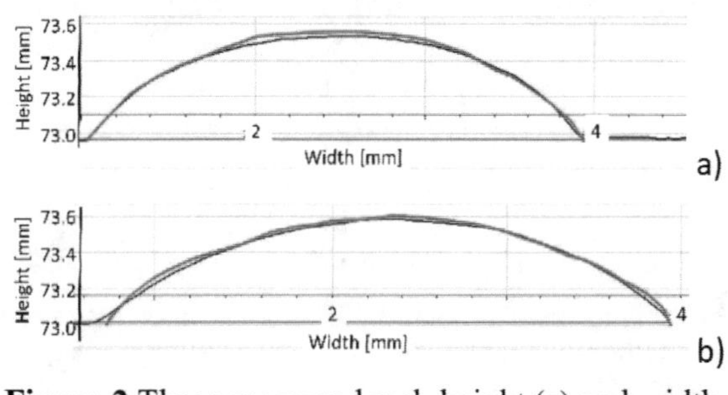

Figure 2 Three measured rack height (a) and width (b) in comparison with the simulation

shows exemplarily the good agreement between the measured and the simulation (pink line) track shape for set 1 (a) and set 5 (b).

4. Nanoparticle distribution in the molten track of a Ti64 DED process

Experiments of DED process using Ti6Al4V were carried out at AIMEN Technology Centre. The experimental setup is demonstrated in figure 2 of [19]. A coaxial laser wire configuration is utilized, which brings the most potential in the orientation of the process. A ring-shaped beam makes the coaxial feeding wire fused homogeneously from all directions and well connected with the local melt pool. So far, a Ti6Al4V wire with nanoparticles was not available and the effect of the nanoparticles on wire track shape and solidification behaviour could only be analysed by simulation. The operating parameters are given in table 2.

Table 2. Process parameters of a Ti6Al4V-TiC DED

D_{wire}	1.12	mm
V_{wire}	1.3	m/min
V_{laser}	12	mm/s
P_{laser}	1500	W
$W_{track,ref.}$	3.1	mm
$H_{track,ref.}$	0.9	mm

Table 3. Parameters of TiC nanoparticle

$T_{melting}$	3160	°C
$T_{boiling}$	4820	°C
ρ	4.94	g/m^3
h_{usion}	1.9E+5	J/kg
C_p	Table(T)	J/(kg K)
λ	Table(T)	W/(m K)
M_W	59.89	g/mol
$D_{particle}$	45	Nm

The numerical method, described in section 2, is applied to Ti6Al4V wires with and without NPs added to the wire material. The properties of nanoparticle TiC, given in the table 3, were used in equation 1-4 to calculate density, viscosity, thermal conductivity and specific heat of the wire material with NPs. The effect of the TiC NPs on the fraction solid curve are given in figure 3 for a Ti6Al4V wire with 1% and 4% TiC and without the addition NPs. The NP distribution and the influence of these changes of the material properties on the deposition process under different process conditions is analysed by simulation.

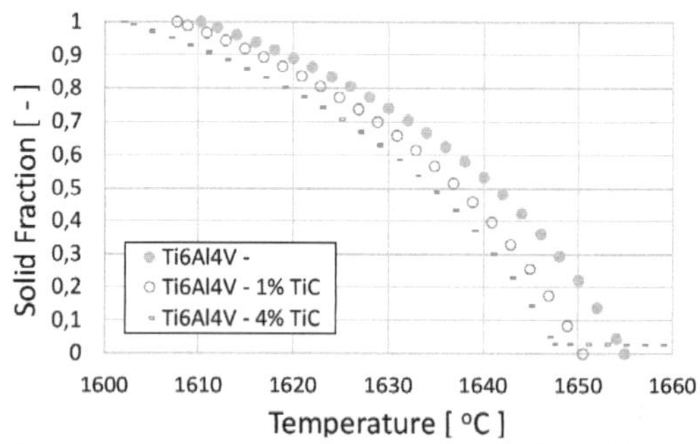

Figure 3 Fraction solid curve of Ti6Al4V with 1% and 4% TiC and without the addition nanoparticles.

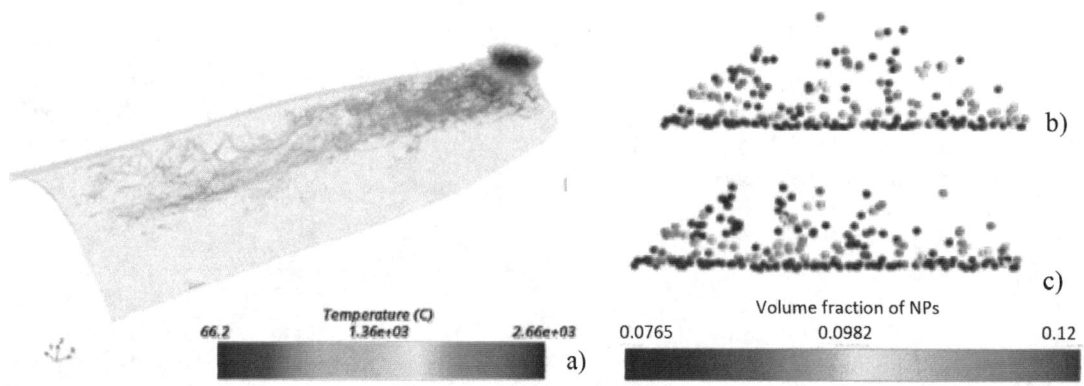

Figure 4 Distributions of nanoparticles during deposition after 0.6 s (a) and side view after 0.2 s (b) and 1.0 s (c).

It can be observed that NPs move with the melting flow, as demonstrated in figure 4 (a). During solidification, NPs were redistributed based on the local temperature gradients and the internal flow inside melt bead, which could be the resultant of the thermo-diffusion-oriented effect of the Brownian motion [18]. Figure 4 (b) and (c) show, that more NPs were observed at interface between the track and the substrate. Two mechanisms may lead to this observation: 1) Internal flow inside the melt track during deposition and starting solidification leads to NP recirculation and finally accumulation at the interface duo to the resistance of the substrate. 2) Temperature gradient forces NPs to migrate from the track centre to the interface and leads to the accumulation of NPs at the interface.

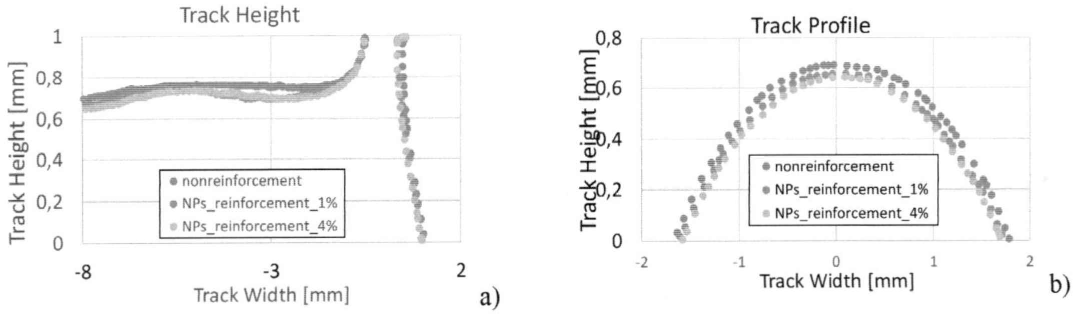

Figure 5 Influence of the nanoparticles on the height (a) and shape (b) of the single track.

Figure 5 shows the influence of the NPs on the single track height and shape for a NP volume fraction of 1 % and 4 %. Adding NPs reduces the track height and enlarges the track width slightly.
Surface tension and wetting angle plays a critical role in forming the track and both are influenced by the NPs. Here, the effect of wettability between the melt and substrate was studied by investigating the influence of contact angle on the dimensions of the single tracks.

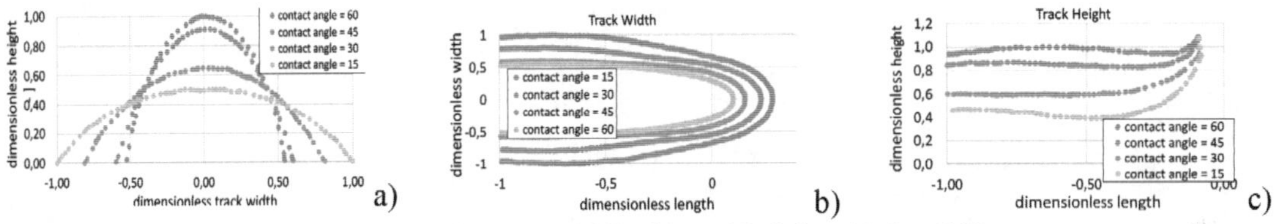

Figure 6 Comparison of track profiles (a), widths (b) and heights (c) for different contact angles.

Figure 6 shows normalized track dimensions (height, width and length) for different contact angles. As to expect, with a smaller contact angle, i.e., with a better wettability between melt and substrate, a flatter (larger width, smaller height) track can be formed. Addition of NPs will enhance the wettability [20]. However, for specific combination of wire and substrate materials applied to a DED process, a detailed determination of the contact angle is required. Furthermore, there is still a lack of reliable methods to quantify the effects of size and content of NPs on the dynamic contact angle. When experimental data from the DED-process at AIMEN are available, a more detailed investigation is planned.

5. Conclusions and outlook

The influence of TiC nanoparticles on melting and solidification during a Directed Energy Deposition process of a Ti6Al4V wire was analyzed using a continuum mechanic based simulation of single track experiments. The simulation was validated by comparing the simulated shape of the single line tracks with experiments of a steel DED-process. The methodology was then used to investigate the distribution of TiC nanoparticles inside Ti6Al4V single tracks printed by DED. An agglomeration of nanoparticles at the interface between melt track and substrate could be observed. Only a small influence of nanoparticles on shape of the single tracks could be found. The possible influence of an increased wettability due the addition of nanoparticles was investigated.
As soon as experimental data is available, a validation of the simulation results for the TiC-Ti6Al4V-DED-process is planned, followed by an investigation of the influence of the nanoparticles on surface tension and, consequently, the track shape.

6. Acknowledgements
The authors acknowledge the financial support of this work from EU in relation to the project 814552-LightMe-H2020-NMBP-TO-IND-2018-2020/H2020-NMBP-HUBS-2018. The authors acknowledge the support by Siemens PLM, agreement 60068580, for providing STAR-CCM+ licenses.

References
1. Zhao Z *et al.* 2021, Computer Methods in Applied Mechanics and Engineering, A thermal multi-phase flow model for directed energy deposition processes via a moving signed distance function, 373, 113518.
2. Yu J, *et al.* 2012, Physics Procedia, Material properties of Ti6Al4V parts produced by laser metal deposition, 39 pp 416 – 24.
3. Haley J C, *et al.* 2019, Materials Science and Engineering: A, Modelling Particle Impact on the Melt Pool and Wettability Effects in Laser Directed Energy Deposition Additive Manufacturing, 761, 138052.
4. Fudger S, *et al.* 2017, Material Science and Engineering A, Residual Stress Induced Mechanical Property Enhancement in Steel Encapsulated Light Metal Matrix Composites, pp 10-17.
5. Hemanth J, *et al.* 2009, Materials Science and Engineering A, Development and property evaluation of aluminum alloy reinforced with nano-ZrO2 metal matrix composites (NMMCs), pp 110-113.
6. Rajmohan T, *et al.* 2013, Trans Nonferrous Met Soc China, Evaluation of mechanical and wear properties of hybrid aluminum matrix composites, pp 2509-17.
7. Moorthy A, *et al.* 2012, Int J Mod Eng Res, Dry sliding wear and mechanical behavior of aluminum/fly ash/graphite hybrid metal matrix composite using taguchi method, pp 1224-1230.
8. Selvam J D R, et al. 2013, Energy Procedia, Synthesis and characterization of Al6061-Fly Ashp-SiCp composites by stir casting and compocasting methods, pp 637-646.
9. Alaneme K K, *et al.* 2013, Tribol Ind, Mechanical properties and corrosion behavior of aluminum hybrid composites reinforced with silicon carbide and bamboo leaf ash, pp 25-35.
10. Escalera-Lozano R, *et al.* 2007, Mater Charact, Corrosion characteristics of hybrid Al/SiCp/MgAl2O4 composites fabricated with fly ash and recycled aluminum, pp 953-960.
11. Ye N, *et al.* 2020, Journal of Materials Research and Technology, Microstructure and mechanical properties of Ni/Ti/Al/Cu composite produced by accumulative roll bonding (ARB) at room temperature, pp 5524–5532.
12. Zan Y N, *et al*, 2019, Materials and Design, Enhancing strength and ductility synergy through heterogeneous structure design in nanoscale Al2O3 particulate reinforced Al composites, pp 1-8.
13. Chen S H, *et al* 2022, Acta Mechanica, Size effects in the particle-reinforced metal-matrix composite, pp 113-127.
14. Pirouzi B, *et al.* 2018, Mechanics of Advanced Composite Structures, Effects of reinforcement distribution on the mechanical properties of Al–Fe3O4 nanocomposites fabricated via accumulative roll bonding, pp 131-139.
15. Teskeredžić A, Demirdžić I, & Muzaferija S, 2002, Numerical Heat Transfer: Part B: Fundamentals, Numerical method for heat transfer, fluid flow, and stress analysis in phase-change problems, 42(5), 437-459.
16. Jakumeit J, Zheng G, Laqua R, Clark S J, Zielinski J, Schleifenbaum J H & Lee P D, 2021, Additive Manufacturing, Modelling the complex evaporated gas flow and its impact on particle spattering during laser powder bed fusion, *47*, 102332.
17. Corcione M, *et al.* 2001, Energy Conversion and Management, Empirical correlating equations for predicting the effective thermal Conductivity and dynamic viscosity of nanofluids, pp 789-793.
18. Zheng G, Jakumeit J, *et al.* 2021, Light Metals, Computational Simulation of Nanoparticle Distributions in Metal Matrix Composite Casting Processes, pp 171-179.
19. Vaamondea E, Rey P, *et al.* 2021, Lasers in Manufacturing Conference, Effect of atmosphere conditions on additive manufacturing of Ti4Al6V by coaxial W-DED-LB process.
20. Munshi A M, Singh V N, Kumar M and Singh J P, 2008, Journal of Applied Physics, Effect of nanoparticle size on sessile droplet contact angle, 103, 084315.

IOP Conf. Series: Materials Science and Engineering 1274 (2023) 012018 doi:10.1088/1757-899X/1274/1/012018

Rapid solidification during welding of duplex stainless steels – in situ measurement of the chemical concentration by Laser-Induced Breakdown Spectroscopy (LIBS)

L Quackatz[1], A Griesche[1] and T Kannengiesser[1,2]

[1] Federal Institute for Materials Research and Testing, Unter den Eichen 87, Berlin 12205, Germany
[2] Otto-von-Guericke University Magdeburg, Faculty of Mechanical Engineering, Institute of Materials and Joining Technology, Universitätsplatz 2, 39106 Magdeburg, Germany

Lukas.quackatz@bam.de

Abstract. Duplex stainless steels (DSS) are frequently used, especially in applications requiring high strength combined with high corrosion resistance in aggressive media. Examples include power plant components and maritime structures. During welding of these steels, local variations in chemical composition can occur. This results in ferritization of the material and negatively affects the mechanical properties of the components. In this work, tungsten inert gas (TIG) welding experiments were performed with DSS. Chemical composition analysis was realized in situ by using Laser Induced Breakdown Spectroscopy (LIBS). The aim of the work is to quantitatively measure the chemical composition in the weld seam of various DSS and to identify possible influences of welding parameters on the microstructure of the material. The chemical concentrations of the main alloying elements Cr, Ni, Mn on the surface of the sample during the welding process and the cooling process were measured. Mn and Ni are austenite stabilizers and their content increases during welding by using certain high alloyed filler material. Spectra were recorded every 1.3 s at a spacing of approximately 2 mm. During the cooling process the location of the measurement was not changed. The LIBS method is proofed to be suitable for the quantitative representation of the chemical compositions during the welding process.

1. Introduction

The microstructure of duplex stainless steels (DSS) is characterized by a balanced ratio of austenite (γ) and ferrite (α). Austenite is embedded in the ferrite matrix. The suppression of the formation of martensite is due to the high proportions of alloying elements. With the high proportion of α-stabilizing elements such as Cr, Mo and additionally a low proportion of γ-stabilizing elements such as Ni, Mn, N, a ferritic-austenitic matrix is formed. The advantages of both phases are combined [1, 2]. Most DSS welds solidify completely ferritic. Dendrite growth follows the temperature gradient and a coarse-grained ferrite matrix results. The solid-state transformation of $\alpha \rightarrow \alpha+\gamma$ is diffusion-driven and thus time-dependent. Austenite precipitates as inter- and intragranular phase [1]. To achieve a balanced phase ratio in the microstructure of the DSS, workpieces are solution annealed (> 1,000°C) and quenched subsequently. In this way, the required microstructure originates and excellent mechanical properties of the DSS can be adjusted and utilized. Many industrial applications require a welding process and due to uncontrolled, rapid cooling and loss of alloying elements an unbalanced γ to α ratio can originate, resulting in low corrosion resistance, low ductility and solidification cracking susceptibility. Furthermore, too slow cooling can lead to precipitation of intermetallic phases. Particularly worthy to mention are chromium nitrides and carbides [3-5]. Recommendations for welding duplex steels are given in [6] and mainly relate to the heat input during welding, the chemical composition of the molten pool and the design of the weld.

Even the slightest element burnup during the welding process can lead to the critical change of the chemical composition and ferritization of the weld metal. Yang et al. [7] have shown that an unbalanced γ/α ratio with 80-90 % α in the weld metal (EN grade 1.4462) causes significant cracking problems in safety critical components. Martin et al. [8] have shown that the high temperature fracture strength of DSS increases dramatically with increasing γ-volume fraction. In order to predict the phase ratio of duplex steels after welding, some diagrams (Schaeffler, De Long and WRC1992) exist with which the microstructure in the weld metal, based on Cr_{eq} and Ni_{eq} can be calculated [9-11]. Recent research has shown that the calculation of Cr_{eq} and Ni_{eq} is inaccurate and needs adjustments [12, 13].

The aforementioned problems in welding duplex steels underscore the need for an in situ measurement method to predict the weld microstructure. Currently developed examination methods for welded components either inadequately predict the resulting solidification modes and ferrite contents of the welds or use post-mortem, destructive examination methods such as metallographic determination. Laser Induced Breakdown spectroscopy (LIBS) provides a highly accurate, effectively non-destructive, time- and spatially resolved quantitative measurement of chemical composition during welding, i.e., before, during, and after solidification of the weld metal in the fusion zone (FZ) and heat-affected zone (HAZ). In addition, the results from LIBS measurements can predict spatially resolved, localized buildup. LIBS is a spectroscopic technique based on the element-specific photon emissions of the plasma. A high-energy, pulsed laser is focused on the surface, of the sample under investigation and locally introduces heat into the material. The material begins to evaporate and a microplasma is formed. During the cooling of the plasma, element-specific radiation is produced, which can be evaluated with help of a spectrometer and suitable software. If the measurement method is calibrated, quantitative analyses can also be performed [14].

A common welding process is tungsten inert gas (TIG) welding. An arc is generated between a tungsten electrode and the workpiece by means of a sufficiently high current. A shielding gas (usually Ar) is used to prevent oxidation of the workpiece and electrode. The high heat flux of the welding plasma melts the workpiece and creates a weld pool. Vaporization of the molten material can release metal vapor from the surface of the weld pool. If the loss of material results in a critical change in chemical composition, the microstructure of the material can be altered and the mechanical properties of the weld can be affected. The complex processes of burning-off of alloying elements can be found in relevant literature and will not be described further here [15, 16].

LIBS could already be used as a suitable method for monitoring the weld pool chemistry as done by Taparli et al. [17]. Of particular interest was the detection of burnt manganese from the weld pool dejected on the HAZ of austenitic steels [18]. Previous research was able to demonstrate the feasibility of in situ measurement, as well as quantitative analysis of welded duplex steels and will now be further validated in this work. Furthermore, LIBS results could be validated with conventional EDS measurements [19]. With the help of an enhanced multivariate calibration, the quantitative analysis becomes more precise and reliable. The chemical composition of the weld pool can now be monitored using LIBS.

2. Materials and methods

2.1. Materials

Two different duplex steels (EN grade 1.4462 and 1.4162) were investigated in this research. Table 1 lists the chemical compositions of the sample materials and the filler metal. The chemical composition was measured with a conventional spark emission spectroscopy instrument from SPECTRO Analytical Instrument GmbH.

Table 1. Selected sample materials and their chemical compositions (measured with Spark-OES).

Grade	Chemical Composition (wt. %)						
(EN 10088-1)	C	Si	Mn	Cr	Ni	Mo	Fe
1.4462	0.020	0.320	1.680	22.650	5.790	2.970	Bal.
1.4162	0.018	0.600	4.950	21.940	1.540	0.310	Bal.
AWS ER 347[a]	0.020	0.960	1.170	18.81	9.770	0.060	Bal.

[a] filler material EN ISO 14343-A

Flat specimens, measuring 150 x 35 x 5 mm^3 (length x width x thickness) were fabricated for the experiments. Before welding experiments were carried out, the specimens were cleaned with acetone to remove contamination from production and processing.

2.2. Experimental

The welding experiments were carried out with the tungsten inert gas (TIG) welder CastoTIG 1611 DC power supply, from the company Castolin. A pure tungsten electrode, with a diameter of 1.6 mm, was used. All welding experiments were performed with an Ar shielding gas flow of 10 L/min. The electrode had a distance of 3 mm to the workpiece and the gas nozzle has a diameter of 6 mm. With the help of a linear motor, the specimens were welded at a constant speed.

The LIBS system includes a pulsed 1064 nm Nd:YAG laser, with a 15 Hz pulse rate and a maximum pulse energy of 200 mJ (Quantel CFR 200). The laser was synchronized with an Echelle spectrometer (Aryelle 200) from LTB Lasertechnik Berlin GmbH using a LIBS controller. To achieve the maximum time resolution, the spectrometer is connected to an Andor iStar ICCD detector from Oxford Instruments [20]. In figure 1 the experimental setup can be seen. The laser head is tilted by 20°, relative to the surface normal, so that the laser measurement spot can be focused 10 mm from the welding arc on the surface of the sample. Information is hardly lost due to this inclination. A light collecting optics (focal length: 300 mm) was used to focus the laser beam on the surface of the specimen. Subsequently, the electromagnetic radiation was collected with an optical fiber (same system) with a focal length of 10 mm and directed into the spectrometer. Due to the additional electromagnetic radiation from the welding plasma, additional optical filters are built into the LIBS system. For all experiments, the gate width was 40 µs and the delay time was 1 µs. These configurations were found to be promising in previously performed experiments [19]. For one spectrum 15 laser shots were accumulated.

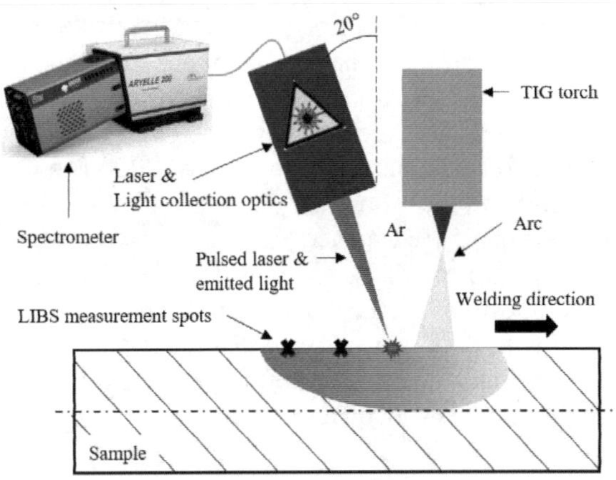

Figure 1. Sketch of the experimental set-up.

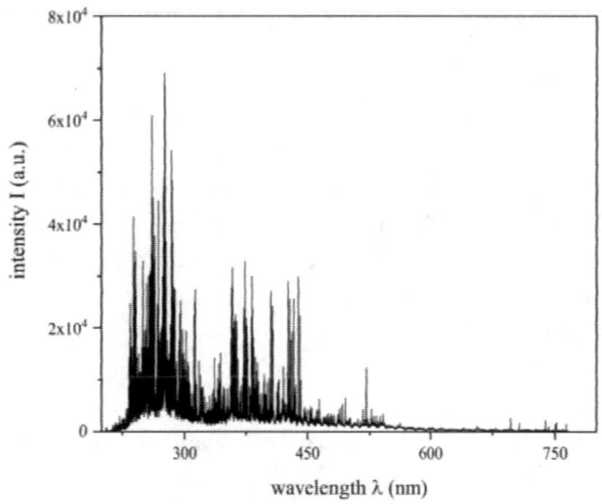

Figure 2. LIBS Spectrum of a Duplex stainless steel EN grade 1.4162.

The duplex steel specimens were welded at a constant welding speed of 10 cm/min and welding currents of 140 A and 160 A, respectively. Welding was performed both with and without filler metal. The total length of the welds was 50 mm. After the laser was started and spectra were recorded, the Ar shielding gas supply started after about 3 s. The welding arc then ignited and the linear motor started until the final length of 50 mm was reached. The welding arc was switched off and the shielding gas switched off after approx. 3 s. The LIBS laser was turned off after the specimen reached a surface temperature of 150°C. The surface temperature in the weld pool was measured using a non-contact laser pyrometer from Advanced Energy. The LIBS measuring spots were always recorded directly (10 mm) behind the welding torch. Accordingly, the measuring spots are initially in the base material, during welding in the molten material and at the end of the welding process in the solidified weld metal. The in situ LIBS spectra were recorded in the spectral range of 200 – 800 nm. The background noise in the recorded spectra was removed using the software AirPls [21].

3. Results and discussion

The solidified, cooling and molten material were recorded with the same configurations. Figure 2 shows a LIBS spectrum of a DSS EN grade 1.4162 with the corresponding spectral range.

In general, the overall goal of most chemical measurement methods is to produce a quantitative measurement with the highest possible precision. A considerable amount of literature has been published on calibration methods and procedures [22, 23]. Experimental calibrations are performed by measuring a series of calibration samples containing the analyte (or analytes) of interest in sensibly graded amounts. Materials whose contents are known with the greatest possible reliability are used as calibration samples. The contents of these calibration samples can be considered to be free of errors compared to the measurement results. The advantage of LIBS is that little preparation is required to measure the samples. For a quantitative measurement, calibration curves must be prepared in advance. A distinction is made between univariate and multivariate calibration. Multivariate models are usually associated with a higher computational and model building effort, such as the Partial Least Squares (PLS) method [22]. In previous work [19], simple univariate models were used for quantification. Now, the calibration using the PLS method, has been optimized and provides reliable results. For calibration, 30 certified reference materials were used. The calibration was validated in this work.

3.1. In situ measurements of element concentration during TIG welding

The welding tests were carried out both with and without filler material. Compared here are the experiments of two different grades of duplex steels (EN grade 1.4462 and 1.4162) with two different amperages (160 A and 140 A). The welding speed was kept constant at 10 cm/min. Figure 3 shows the results of the quantitative analysis of the in situ LIBS measurement during welding. The timing of the welding process can be read directly from the diagram. At t = 4 s, the welding torch started and there is a brief increase in concentration due to the flashing of the plasma and the additional electromagnetic radiation picked up by the spectrometer. This phenomen has been explained in previous research [19]. The concentrations of Mn increase sharply during welding. However, the concentration of Ni remains mostly constant. Mn can burn off during welding due to its low enthalpy of vaporization (220 kJ/mol). Ni has a much higher enthalpy of vaporization (372 kJ/mol) and therefore does not burn off [24]. The burned-off Mn is detected by the LIBS measurement and is therefore elevated [25]. After the welding process (t ~ 30 s), the concentrations drop. Then the cooling process of the samples down to 150°C starts. Table 2 shows the mean values of the chemical compositions in the weld metal during the cooling process of the main alloying elements (evaluated with PLS). It is striking that the Ni concentrations increase in all welds with filler metal. In contrast, the Mn concentrations increase only with a welding current of 140 A and the use of the filler metal. The Cr concentrations are inconspicuous and predominantly stable.

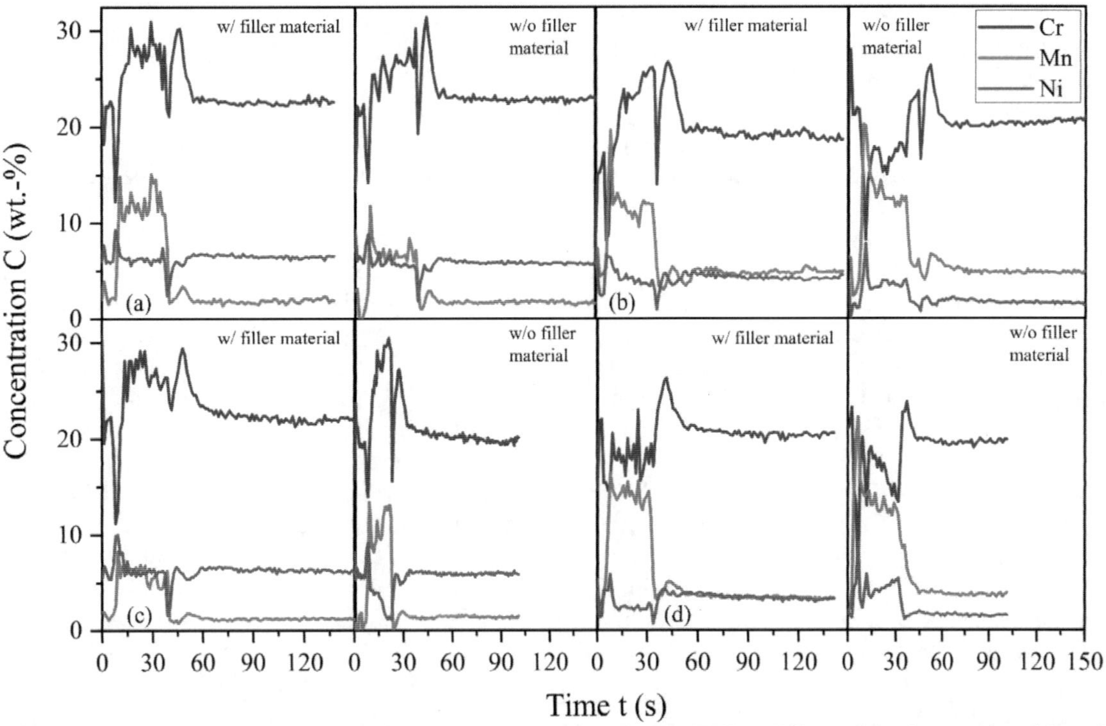

Figure 3. In situ measurement of chemical composition during TIG welding of duplex steels with and without filler material. Representation of the main alloying elements Cr, Ni und Mn. (a) EN grade 1.4462 140 A; (b) EN grade 1.4162 140 A; (c) EN grade 1.4462 160 A; (d) EN grade 1.4162 160 A.

Table 2: PLS quantification of the weld material after welding process with different current levels, both with and without filler metal. Mean values of the chemical concentrations of the main alloying elements.

Alloying element	Chemical Composition (wt.-%) EN grade 1.4162			
	140 A		160 A	
	With filler material	Without filler material	With filler material	Without filler material
Cr	19.3	20.0	20.4	19.7
Ni	4.2	1.7	3.5	1.6
Mn	5.0	4.9	3.5	3.8

	EN grade 1.4462			
	140 A		160 A	
	With filler material	Without filler material	With filler material	Without filler material
Cr	22.6	22.7	21.2	20.4
Ni	6.4	5.9	6.3	5.9
Mn	1.7	1.1	1.2	1.4

3.2. Analysis of the microstructure and phase ratio

After the welding tests, cross sections were made from the samples. These were taken from the locations where LIBS measurements were also taken. The LIBS measurements were only performed on the surface (i.e. at the top of the cross sections). Due to convection, it can be assumed that the surface measurements on the melt and in the solidified material is significant for the total melt volume.

Therefore, the comparison of the data with the microstructure is in order. Thus, the increased concentrations of Mn and Ni can be correlated with the phase ratio. The γ/α ratio was determined by image analysis. For this purpose, the samples were etched with the color etch Beraha II and captured with the light microscope at various locations. Figure 4 shows an example of the microstructure of an EN grade 1.4162 welded with 160 A and filler material. Overview images, detailed images of the base metal and the weld metal were taken. The microstructure shows a common appearance of a weld structure made of DSS. The base metal has a balanced phase ratio. In the weld metal, austenite precipitates at the grain boundaries, resulting in a Widmanstätten microstructure. Table 3 summarizes the ferrite contents (vol.-%) of the weld metal of the different weld configurations. The ferrite contents in the base metal were in the usual range between 50-60 vol.-% for all specimens. It is noticeable that the ferrite content decreases with filler materials. This is due to the additional Ni and Mn concentration in the filler material. Ni and Mn are austenite formers and promote the precipitation of austenite during cooling. The increased Ni and Mn concentration could be measured by the LIBS method and is also reflected in the microstructure, based on the increasing austenite contents. For EN grade 1.4462, the ferrite content is $\leq$ 70 vol.-%. In [26] it is required that welded joints of DSS must contain a ferrite content between 30-70 vol.-%. In contrast, the ferrite content without filler material in EN grade 1.4462 is over 90 vol.-%. The high ferrite contents lead to a degradation of the mechanical properties and a decrease in corrosion resistance. Even if the LIBS measurements with the chemical composition during welding do not correlate directly with the microstructure, correlations can be established based on the austenite formers (Ni, Mn) and the increasing austenite contents in the microstructure. The LIBS measurement can therefore be interpreted as successful and should be validated with further experiments. In further experiments, sections can also be taken from other locations of the weld seam for control purposes.

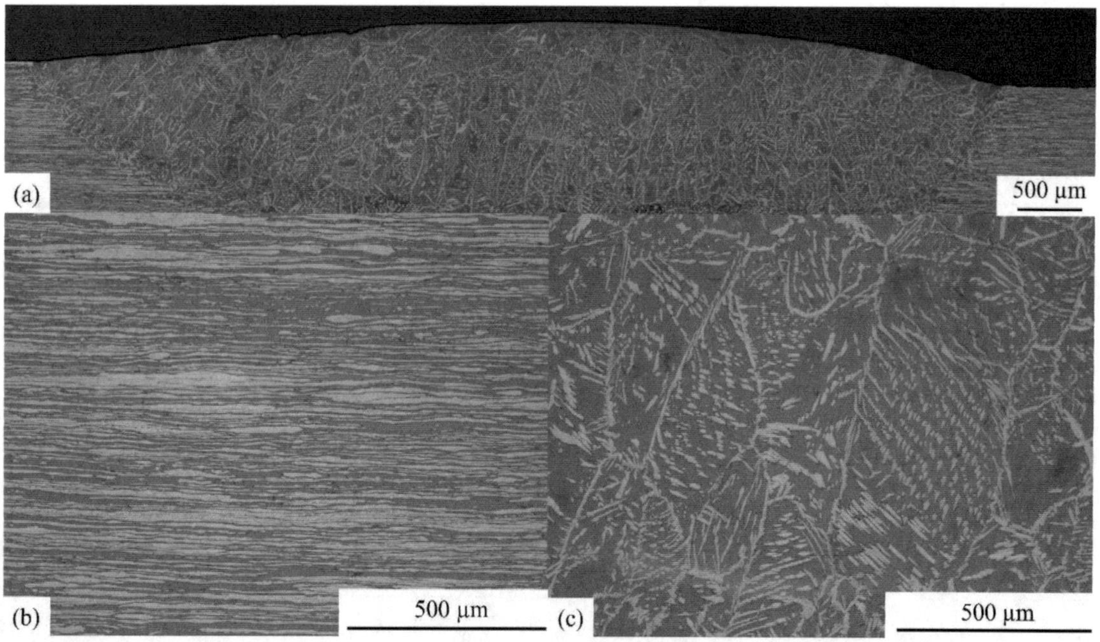

Figure 4. Exemplary image of the microstructure of a welded EN grade 1.4162, 160A with filler metal. (a) overview image of the weld; (b) image of the base metal with balanced phase ratio α/γ; (c) detail image of the weld metal. Visible ferritization of the material.

Table 3. Ferrite volume fraction in the weld microstructure of different DSS. Volume fraction of ferrite was determined by image analysis.

Grade (EN 10088-1)		Ferrite Content (vol.-%)	
		With filler material	Without filler material
1.4162	140 A	82	89
	160 A	75	86
1.4462	140 A	69	92
	160 A	70	93

4. Conclusion

In this work, it was shown that in situ LIBS measurement can be performed during TIG welding with and without filler material of duplex stainless steels. Main results of the work are:

- Quantitative LIBS measurements of the chemical composition of duplex steels during welding could be performed.
- Mn concentrations increase during the welding process. Both, with and without filler material.
- Ni and Cr concentrations remain constant during welding. Ni concentrations, however, are increased due to the filler material.
- The ferrite contents decrease due to the increased concentration of austenite-forming elements in the weld metal (Ni, Mn).

In this work, only the influence of one parameter (filler material) on the ferrite contents was investigated. In further work, other parameters such as heat dissipation, welding current and welding speed need to be investigated. Further developments should also record the temperature during the cooling process using LIBS.

Acknowledgements
Authors would like to thank Lasertechnik Berlin GmbH (LTB) for the analytical equipment. This work was funded by the Deutsche Forschungsgemeinschaft (DFG, German Research Foundation) – 442001176.

References

[1] Knyazeva M and Pohl M "Duplex Steels: Part I: Genesis, Formation, Structure,"2013 *Metallogr Microstruc.* **2** 113-121
[2] Solomon H D and Devine T "Duplex Stainless Steels--A Tale of Two Phases," 1982 *Duplex stainless steels* pp. 693-756.
[3] Knyazeva M and Pohl M "Duplex Steels. Part II: Carbides and Nitrides," 2013 *Metallography, Microstructure, and Analysis* **2** 343-351
[4] Verma J and Taiwade R V "Effect of welding processes and conditions on the microstructure, mechanical properties and corrosion resistance of duplex stainless steel weldments-A review," 2017 *J Manuf Process* **25** 134-152
[5] Karlsson L "Intermetallic phase precipitation in duplex stainless steels and weld metals: Metallurgy, influence on properties, welding and testing aspects," 1999 *Welding Research Council Bulletin* **438**
[6] Karlsson L "Welding Duplex Stainless Steels — A Review Of Current Recommendations," 2012 *Welding in the World* **56** 65-76
[7] Yang J, Wang Q, Wei Z and Guan K "Weld failure analysis of 2205 duplex stainless steel nozzle," 2014 *Case Studies in Engineering Failure Analysis* **2** 69-75
[8] Martin G *et al.* "A macro-and micromechanics investigation of hot cracking in duplex steels," 2012 *Acta Materialia* **60** 4646-4660
[9] Schaeffler A L "Constitution diagram for stainless steel weld metal," 1949 *Met Prog* **56** 680

[10] Delong W T "Constitution Diagram for Stainless-Steel Weld Metal .1. Delong Diagram," 1974 *Met Prog* **106** 226-226
[11] Kotecki D J and Siewert T A "Wrc-1992 Constitution Diagram for Stainless-Steel Weld Metals - a Modification of the Wrc-1988 Diagram," 1992 *Weld J* **71** 171-178
[12] Kotecki D J and Rajan V B "Submerged arc fillet welds between mild steel and stainless steel," 1997 *Weld J* **76** 57-66
[13] Wessman S "Evaluation of the WRC 1992 diagram using computational thermodynamics," 2013 *Welding in the World* **57** 305-313
[14] Cremers D A and Radziemski L J 2006 *Handbook of laser-induced breakdown spectroscopy* vol. 2 Chichester Wiley
[15] Collur M M, Paul A and Debroy T "Mechanism of Alloying Element Vaporization during Laser-Welding," 1987 *Metall Trans B* **18** 733-740
[16] Khan P A A and Debroy T "Alloying Element Vaporization and Weld Pool Temperature during Laser-Welding of Aisi 202 Stainless-Steel," 1984 *Metall Trans B* **15** 641-644
[17] Taparli U A, Jacobsen L, Griesche A, Michalik K, Mory D and Kannengiesser T "In situ laser-induced breakdown spectroscopy measurements of chemical compositions in stainless steels during tungsten inert gas welding," 2018 *Spectrochimica Acta Part B: Atomic Spectroscopy,* **139** 50-56
[18] Taparli U A, Kannengiesser T, Cieslik K, Mory D and Griesche A "In situ chemical composition analysis of a tungsten-inert-gas austenitic stainless steel weld measured by laser-induced breakdown spectroscopy," 2020 *Spectrochimica Acta Part B: Atomic Spectroscopy* 105826
[19] Quackatz L, Griesche A and Kannengiesser T "In situ investigation of chemical composition during TIG welding in duplex stainless steels using Laser-Induced Breakdown Spectroscopy (LIBS)," 2022 *Forces in Mechanics* **6** 100063
[20] Mueller M, Gornushkin I B, Florek S, Mory D and Panne U "Approach to Detection in Laser-Induced Breakdown Spectroscopy," 2007 *Analytical Chemistry* **79** 4419-4426
[21] Zhang Z M, Chen S and Liang Y Z "Baseline correction using adaptive iteratively reweighted penalized least squares," 2010 *Analyst* **135** 1138-1146
[22] Sharaf M A, Illman D L and Kowalski B R 1986 *Chemometrics* John Wiley & Sons
[23] R Noll 2012 *Laser-induced breakdown spectroscopy fundamentals and applications* Berlin: Springer
[24] Zhang Y, Evans J R G and Yang S "Corrected Values for Boiling Points and Enthalpies of Vaporization of Elements in Handbooks," 2011 *Journal of Chemical & Engineering Data* **56** 328-337
[25] Taparli U A, Kannengiesser T, Griesche A „Tungsten inert gas bead-on-plate weld chemical composition analysis by laser-induced breakdown spectroscopy," 2020 *IOP Conference Series: Materials Science and Engineering* **882** 012023
[26] DIN EN ISO 17781 *Erdöl-, petrochemische und Erdgasindustrie - Prüfverfahren für die Qualitätslenkung von Mikrostrukturen von ferritisch/austenitisch nichtrostenden Duplexstählen* 2017

Modelling of microstructures development in laser powder bed fusion process – Application to IN718 superalloy

T Camus[1,2,3], D Maisonnette[1], O Baulin[1], O Senninger[2], G Guillemot[2] and C A Gandin[2]

[1] Cetim, 52 Av. Félix Louat, 60300 Senlis, France
[2] CEMEF, Mines Paris, PSL University, UMR CNRS 7635, 1 Rue Claude Daunesse, 06904 Sophia Antipolis, France
[3] Author to whom any correspondence should be addressed.

Abstract. In laser powder bed fusion (L-PBF) process, a deposited powder layer is melted by a laser and solidifies when the laser moves away. During solidification, the microstructure of the part is formed due to both epitaxial grain growth and nucleation. This structure has a strong influence on the final mechanical properties of parts and is influenced by the choice of process parameters such as the laser power and velocity or the scanning strategy. As a consequence, the prediction and characterization of this microstructure is of prime interest considering size, crystallographic orientations and shapes of grains. Among the approaches reported in the literature to model microstructure development, the Cellular Automaton (CA) method is a relevant choice to describe grain structure evolution. This model has been adapted to investigate microstructure formation during L-PBF process applied on an IN718 nickel-base superalloy. The steady state thermal behaviour at the scale of the melt pool is analysed and used to compute the development of the structure. This structure is computed at the scale of the part also considering realistic scanning strategies.

1. Introduction

Among additive manufacturing processes, Laser Powder Bed Fusion (L-PBF) process builds parts layer by layer from a deposited metallic powder. This powder bed is spread on the surface and a laser is used to melt locally the metal. Once solidified, another powder layer is spread and the process is repeated one layer after another until part achievement. The power, velocity, radius and trajectory of the laser, and the thickness of powder layers have an influence on the generated microstructures in the build part. Mastering these process parameters is of prime interest in order to control the mechanical properties of the parts. It can be achieved by developing reliable models of the process, which is at a lower price than experimental tests. Moreover, numerical approaches bring a better understanding of the physical phenomena occurring during the process.

The Cellular Automaton method is used to predict the grain structure formation by modelling the nucleation and growth stages on a regular cellular grid. This method has indeed been proven to be reliable to model structures generated in additive manufacturing processes on millimetre scale domains, i.e. to representative scales of the grain structure [1–3]. The objective of this article is to model the grain structure generated in L-PBF at a representative volume and to characterize the influence of process parameters on the resulting microstructures.

2. Models description

A CA model previously proposed by Chen et al. and Xue et al. for welding process [4,5] has been adapted to predict microstructure development during L-PBF process. This model is coupled to a

mobile temperature field imposed onto a FE mesh separately computed before CA resolution. The growth rate of dendrites tips is taken as a power law of the undercooling encountered at these tips [6].
A Finite Element resolution of the heat flow and of the fluid dynamics in a level-set formulation has been investigated. The model has been developed by A. Saad [7], and then applied to L-PBF process by Q. Chen et A. Queva [8,9]. In this approach, the substrate, the powder, the liquid matter and the gas in the building chamber are considered in a single formulation. The powder is assumed as a homogeneous domain composed of 50% of solid and 50% of gas. The frontier between gas and metal is indicated by the isovalue $\phi = 0$ of a Level-Set function. During simulations, the Level-Set field moves as the track develops and the mesh is adapted in order to keep fine elements mesh at the frontier between the different domains. The heat, supplied by the laser to the melt pool, is modelled with a Gaussian distribution source of standard deviation $\sigma_L = d_L/4$, where d_L is the laser diameter. The computation of a single track of 1 mm was made with this "mesoscale" model in order to achieve a steady temperature field. On figure 1 (a), the temperature is represented in a part of the resolution domain.
The computation time is of 7 days to simulate the lasing of a 1 mm scan with this method on a 28 cores parallelized cluster. Regarding these time scales, the modelling of a multiscan/multilayer build-up cannot be done with this method. The proposed solution is to consider the FE field achieved in a steady state of this simulation and to apply the temperature to the CA model. At a steady state of the FE simulation, a box which includes the melt pool is consequently defined (figure 1 (a)). The temperature field of this box is saved for a single time step in a table. Then in a macroscopic model, an algorithm describes the laser trajectory, and the temperature is directly applied on the CA mesh using the tabulation also considering the current position of the laser (figure 1 (b)).

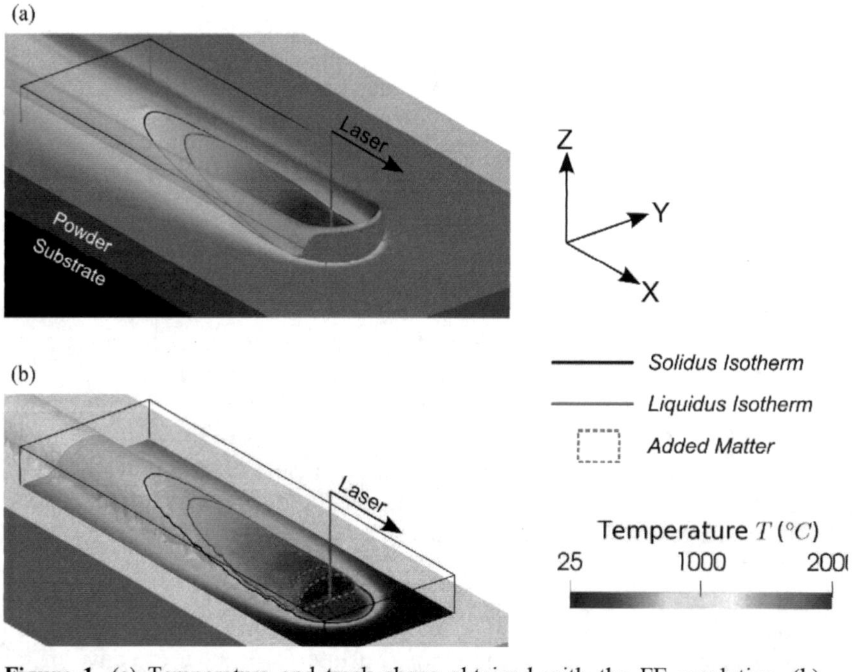

Figure 1. (a) Temperature and track shape obtained with the FE resolution. (b) Same temperature field applied in the CA model.

In the FE model, due to surface tensions between liquid and gas, the melted powder forms a rounded pool. At a steady state of the resolution, the profile of a transverse cross section of the track is extracted, close to a pseudo hemicircular shape. This 2D shape is saved and used in the macroscopic CA model to define the addition of material. An algorithm describes the laser displacement in the building domain. At each time step, this 2D shape is extruded between the current and the previous laser positions to model track development. On figure 1 (b), this volume is represented with dashed green lines.

With this method, the temperature field during the lasing is precisely computed for only one scan. Then it is applied as many times as needed to cover the multiscan build up. So the computation time is restricted only to the CA resolution (grains nucleation and growth). It is consequently possible to model grain structure genesis for large scale models without large computation resource.
For this study, four macroscale simulations for multitrack and multilayer constructions were run with different process parameters. They are described in table 1 also associated to two lasing strategies as presented in figure 2 (a) and (b) respectively. For cases 1 to 3, at each layer, the laser moves along X axis, but alternatively toward +X and −X directions while keeping same direction for a Y coordinate (see figure 2 (a)). In case four, the lasing direction is alternatively along X and along Y directions layer by layer (see figure 2 (b)). For even layers, the laser moves alternatively along +X, −X, +X... (Y increasing) such as for cases 1 to 3. For odd layers, the laser moves alternatively along +Y, −Y, +Y... (X increasing). Taking case 1 as the reference case, only one process parameter varies compared to case 1 in each other case.

Table 1. Process parameters for each simulation case.

Case	1	2	3	4
Laser power (W)	120	120	80	120
Laser velocity (mm s^{-1})		400		
Beam diameter (µm)		70		
Layer thickness (µm)		25		
Hatch spacing (µm)	76	50	76	76
Lasing direction	X	X	X	X/Y

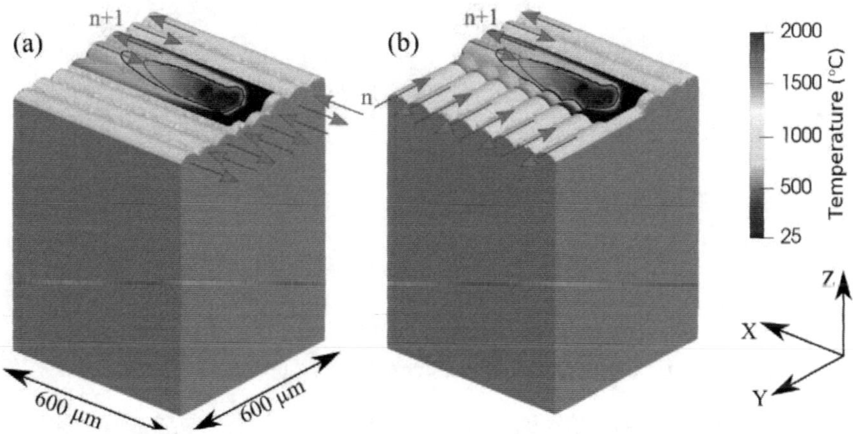

Figure 2. Global views of the CA domain during simulation, with (a) a unidirectional scanning strategy and (b) a bidirectional scanning strategy.

For the CA resolution, the cells size is set to 1.5 µm. The substrate has a grains density $d_s = 3 \times 10^4$ mm^{-3}. Nucleation can occur for undercooled liquid cells. The nucleation undercooling is $\Delta T = 0\ K$. The nucleated cells are randomly picked with a density $d_n = 1 \times 10^7$ mm^{-3}. When a new grain nucleates, a crystallographic orientation is also randomly picked.

3. Results

3D views of representative elementary volumes extracted from the simulated built parts are represented in figure 3. On each face of the cubes, the reference direction of the IPF map is the normal to the face. For each case, we can clearly see the signature of the laser trajectory on the microstructure. The grains in the centre of the scans are equiaxed and have not the same crystallographic orientation than the grains located between two consecutive scans (which are columnar). As can be seen on the top view of figure 3 (a), (b) and (c), these equiaxed grains have a crystallographic direction ⟨100⟩ aligned with the Z axis, which is not seen for the other grains. This phenomenon induces a periodic crystallographic structure along the Y direction for cases 1 to 3. The spatial period is the hatch spacing. We can clearly see the difference of periods between cases 1 and 2. The latter having a lower hatch spacing (table 1).

The bidirectional lasing strategy induces a double periodicity in the grain structure (along X and Y directions). So, we can see a checked pattern on the top surface of the (d) cube. It is observed that the structures obtained with the bidirectional lasing strategy have two equivalent directions: X and Y axis (both for grains shapes and for crystallographic orientations). On the contrary, a unidirectional lasing generates a more anisotropic grain structure.

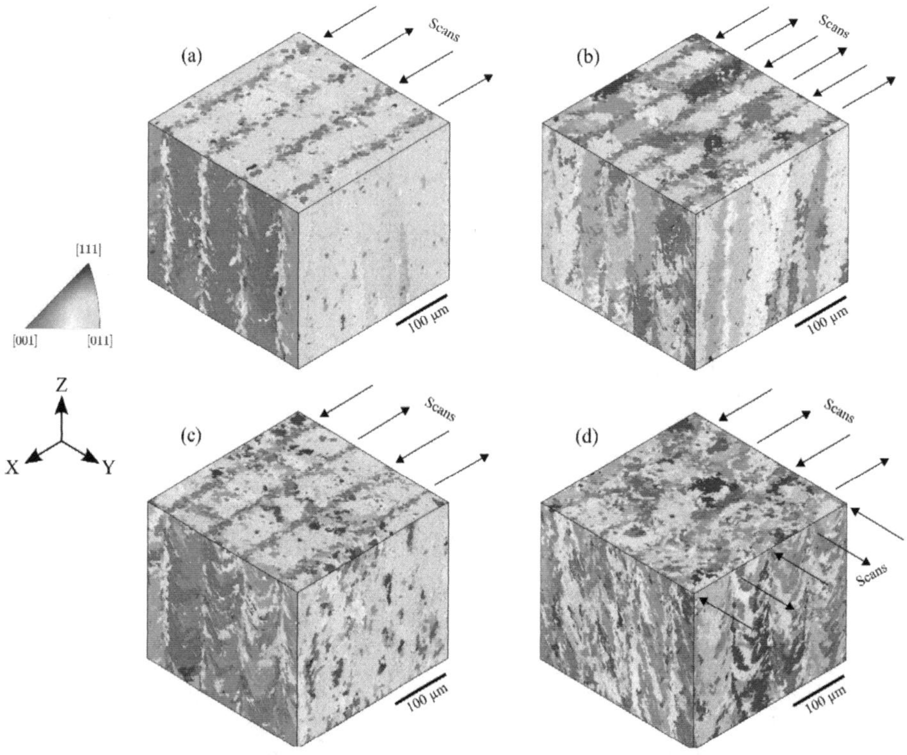

Figure 3. IPF maps on different cross-sections for (a) case 1, (b) case 2, (c) case 3 and (d) case 4 (table 1). On each face, the IPF orientation is the normal to the face.

This observation is confirmed by the pole density figures shown in figure 4. While for cases 1 to 3, some high intensity regions are observed, the case 4 has a far less textured crystallographic structure. In the reference case, the majority of the analysed surface has the same crystallographic orientation, with a ⟨100⟩ direction aligned with the X axis. The two other ones are normal to X and make a 45° angle with Y and with Z directions. These grains correspond to the ones in green on the top view of figure 3 (a). A few grains have a crystallographic direction ⟨100⟩ aligned with the Z axis.

In case 3, the laser power is lower than in reference case. As a result, the melt pool shape is smaller than in case 1. The IPF map of case 3 shows similarities with case 1 but it is more diffuse. The reason is that, with a smaller melt pool, less grains are remelted at each scan. Particularly, less nucleated grains, located at the top of the previous scans, are remelted. These grains have a completely random orientation, so they feed the structure with a very diverse texture.

For cases 1 and 3, the overlapping of melt pools is smaller than 50% from a given scan and the next one. But in case 2, due to a smaller hatch spacing, at each scan, half of the previous scan is remolten. So the resulting texture is not symmetrical (along X axis). This can be seen in figure 4 (b).

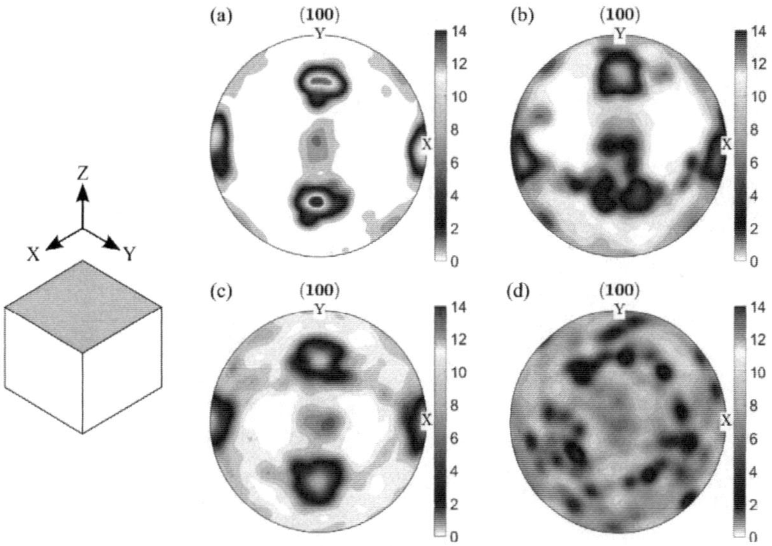

Figure 4. Pole density figures corresponding to top faces of figure 3.

4. Conclusion

A macroscopic model for simulation of the grain structure formation during L-PBF process is presented. Thanks to this model, the as-built grain structure in a part larger than 1 mm^3 can be simulated in reasonable computation time (7 days).

Different lasing strategies were investigated and their resulting grain structures were compared. For each studied case, the modification of the process parameters, from a reference case, induces visible differences on the simulated grain structures. The mean crystallographic orientations, the anisotropy and the periodicity of the grain structures are affected by the choices on process parameters. One of the main results is that the alternation of lasing directions generates more isotropic grain structures than unidirectionnal lasing strategies.

The limit of this macroscopic approach is that the prediction of structures in thermally unsteady areas (beginning and end of scans) can not been achieved. Moreover, the models do not consider the accumulation of heat in the part after several lasing scans. To go further, a finite element simulation of the heat accumulation could bring better results which may correspond to development of multiscale coupling strategies.

5. References

[1] Zinovieva O, Zinoviev A, Romanova V and Balokhonov R 2020 Three-dimensional analysis of grain structure and texture of additively manufactured 316L austenitic stainless steel *Addit. Manuf.* **36** 101521

[2] Mohebbi M S and Ploshikhin V 2020 Implementation of nucleation in cellular automaton simulation of microstructural evolution during additive manufacturing of Al alloys *Addit. Manuf.* **36** 101726

[3] Koepf J A, Gotterbarm M R, Markl M and Körner C 2018 3D multi-layer grain structure simulation of powder bed fusion additive manufacturing *Acta Mater.* **152** 119–26

[4] Chen S 2014 *Three dimensional Cellular Automaton-Finite Element (CAFE) modeling for the grain structures development in Gas Tungsten / Metal Arc Welding processes* PhD Thesis (Mines ParisTech)

[5] Xue C, Blanc N, Soulié F, Bordreuil C, Deschaux-Beaume F, Guillemot G, Bellet M and Gandin C-A 2022 Structure and texture simulations in fusion welding processes – comparison with experimental data *Materialia* **21** 101305

[6] PhysalurgY Librairy , tool KIND, https://physalurgy.cemef.mines-paristech.fr/

[7] Saad A, Gandin C-A and Bellet M 2015 Temperature-based energy solver coupled with tabulated thermodynamic properties – Application to the prediction of macrosegregation in multicomponent alloys *Comput. Mater. Sci.* **99** 221–31

[8] Chen Q, Guillemot G, Gandin C-A and Bellet M 2017 Three-dimensional finite element thermomechanical modeling of additive manufacturing by selective laser melting for ceramic materials *Addit. Manuf.* **16** 124–37

[9] Queva A, Guillemot G, Moriconi C, Metton C and Bellet M 2020 Numerical study of the impact of vaporisation on melt pool dynamics in Laser Powder Bed Fusion - Application to IN718 and Ti–6Al–4V *Addit. Manuf.* **35** 101249

IOP Conf. Series: Materials Science and Engineering 1274 (2023) 012020 doi:10.1088/1757-899X/1274/1/012020

Ultrasonic melt processing upon direct-chill casting of aluminium alloys

D G Eskin[1*], T Subroto[1], A Priyadarshi[2], C Beckwith[3], K Pericleous[3] and I Tzanakis[2]

[1] Brunel Centre for Advanced Solidification Technology, Brunel University London, Uxbridge UB8 3PH, United Kingdom
[2] Faculty of Technology, Design and Environment, Oxford Brookes University, Wheatley OX33 1HX, United Kingdom
[3] Computational Science and Engineering Group (CSEG), Department of Mathematics, University of Greenwich, Greenwich SE10 9LS, United Kingdom

Abstract. Ultrasonic melt processing (USP) has been known for decades for beneficial effects in as-cast aluminium alloys, including degassing, grain and structure refinement. In the last 10 years the authors performed a series of research projects dedicated to understanding the nature of these effects, by dedicated advanced experiments and by developing numerical models that adequately reflect the complicated physics involved, aiming at further optimising the technology so that it becomes feasible for scale-up applications and attractive for industrial use. Based on the main USP mechanisms previously studied by in-situ observations coupled with acoustic pressure measurements, i.e. fragmentation, deagglomeration and dispersion of the solid phases and inclusions, technological approaches are suggested and tested for grain refinement upon direct-chill casting of Al alloys. Results showed that USP in the melt flow in the launder significantly improves the as-cast structure of a billet, opening the way for upscaling.

1. Introduction

High-quality billets and ingots cast from aluminium alloys require uniform and fine grain structure with well controlled distribution of intermetallics, both primary and eutectic. This is typically achieved by adding specially designed grain refiners to the melt. These grain refiners are represented mostly by Al-Ti-B and less commonly Al-Ti-C master alloys where TiB_2 or TiC compounds act as activated substrates for nucleation of solid aluminium, while excess Ti acts as a growth-restricting solute element that controls grain growth. Even though the mechanisms behind grain refinement using these ceramic additions are extensively studied and the grain refining results are quite reliable, the efficiency of these grain refiners is not very high with only a fraction of added borides or carbides actually serving the purpose, while the majority ending up in filters, dross and, most undesirably, as agglomerates in the final cast products. This calls for two possible solutions: (a) increase the efficiency of grain refining additions via improved solubility and deagglomeration of particle clusters and (b) by

* Author to whom any correspondence should be addressed: dmitry.eskin@brunel.ac.uk

developing other means of grain refining that do not rely of additions of potentially harmful and expensive ceramic particles.

In the 1970-1980s a technology called ultrasonic melt processing (USP) emerged in the USSR and was trialled on the industrial scale for the purposes of melt degassing, filtration and grain refinement of aluminium- and magnesium-based alloys produced by direct-chill (DC) casting [1]. It was demonstrated that ultrasonic degassing can be successfully performed in a holding furnace or in a launder. This is quite understandable as degassing requires a relatively long time of melt processing, as the formed hydrogen-filled bubbles need to float to the surface of the melt. Grain refinement, on the other hand, was successfully performed upon ultrasonic processing in the DC casting mould, either in the sump of small billets or in the distribution bag of larger billets and ingots. The technological parameters of USP were thoroughly established in the 1980s, including the required ultrasonic power, processing time and the number of ultrasonic sources per treated volume [1, 2]. In this paper we will limit our consideration to the ultrasonic melt treatment for grain refinement. The mechanisms of USP have been suggested in the 1980s-1990s as the activation of insoluble non-metallic inclusions (e.g. oxides) in the melt by cavitation that induces wetting and surface activation of the particles [1]. These activated particles then act as substrates for the nucleation of primary phases, both intermetallics and matrix grains. It has been also noticed that an addition of Zr to aluminium wrought alloys significantly improves the grain refining effect upon USP. Further, it has been shown that the efficiency of introducing Al-Ti-B master alloys can be improved by USP [3].

The interest in USP has increased in the beginning of the 21st century, giving rise to in-depth research into the related cavitation-driven mechanisms, as well as to examining various schemes and regimes of USP, also applicable to DC casting. On the fundamental level the mechanisms of cavitation-assisted nucleation on oxides, fragmentation of solid phases, and de-agglomeration of solid particles have been well established [4-8]. Also the acoustic parameters of USP, including the distribution of acoustic pressure in the melt, the effects of ultrasonic power, acoustic flows and shock waves, have been studied in detail and incorporated in numerical models [9-11]. USP in a DC casting launder was confirmed to improve the efficiency of grain refining additions on the pilot scale by NIMR (Netherlands) and on the industrial scale by Southwire (USA) and Rusal (Russia). The mechanism of Zr-induced ultrasonic grain refinement was established in detail and demonstrated on the lab- and pilot-level giving hope for the efficient grain refinement without additions of borides or carbides [12]. Following the pioneering work in the USSR in the 1980s [1], in recent years, large-scale DC casting using multiple ultrasonic sources in the sump of a billet was reported by Central South university (China) [13], although this series of trials was compromised by the use of Ti-sonotrodes that were prone to rapid cavitation erosion.

The level of knowledge and expertise achieved to date allows us to suggest that there is a possibility of applying USP to the grain refinement of aluminium alloys without, or with significantly reduced, addition of borides or carbides. Two schemes of USP are known: application in the hot top or the sump of a billet and application in the launder. These two schemes have their advantages and disadvantages, employ different sets of mechanisms, and have different scalability factors. Up to now, processing in the launder is used mostly to improve the efficiency of added grain refiners (employing deagglomeration mechanisms) [2, 3] while the processing in the molten pool of the hot top or sump is aimed at grain refinement through enhanced heterogeneous nucleation, fragmentation of dendrites and efficient stirring by acoustic flows [1, 2, 13]. The processing time is limited by the melt flow velocity in the former approach, while the processing volume is limited by the cavitation zone and power attenuation in the latter approach. In this paper we will show that the ultrasonic processing in the melt flow in the launder can be successfully performed, making possible the upscaling of this technology without the need of multiple ultrasonic sources.

2. Experimental and numerical setup

In this work, an AA6XXX-series aluminium alloy was used as the base alloy with an addition of 0.25 wt% Zr, but without any AlTiB grain refiner. The chemical composition of the alloy was analysed using optical emission spectroscopy (OES) and the average composition of the alloy is shown in Table 1. DC casting was carried out in the Advanced Metal Casting Centre (AMCC) of the Brunel Centre of Advanced Solidification Technology (BCAST), Brunel University London. The diameter of the billet was 152 mm and it was cast with a hot top. Two billets were cast at a casting speed of 140 mm min^{-1}; one with ultrasonic processing and the other without ultrasonic processing as a reference. Each billet was about 1 m long.

The billets were cast at a pouring temperature of 730 ºC. A water-cooled 5-kW magnetostrictive transducer (Reltec) with a driving frequency of 17.3 kHz was used to power a cylindrical sonotrode 40 mm in diameter made of a Nb alloy. The input power of the transducer was 3.5 kW (corresponding to the 30 μm peak-to-peak amplitude at the tip of the sonotrode). The tip was immersed approximately 12 mm below the melt surface and the sonotrode was positioned in the middle between two partitions in the launder. The pair of refractory partitions were installed 266 mm apart (1λ – sound wavelength in liquid aluminium at the UST transducer frequency) in the launder. The position of the downstream partition was 200 mm from the hot top. The partitions were designed in such a way that the melt could flow over the top and bottom of the partitions to avoid melt overflow and entrapment in the launder.

For the structure observation, samples were cut from the billet along the entire radius. To reveal the grain structure, the samples were ground and then etched using Tucker's reagent. Meanwhile, for microscopic observation, the samples were ground, mirror polished, anodized in 5 wt% HBF_4 using 20 VDC, and subsequently examined under an optical microscope (Zeiss Axio Scope.A1) with polarized light. The grain size was measured using the linear intercept method taken randomly on optical microscope images and statistical analysis was carried out.

Table 1 Chemical composition (wt%) of the alloy obtained through optical emission spectroscopy.

Al	Si	Fe	Cu	Mn	Mg	Zr	Cr	Ti
Balance	0.75	0.24	0.76	0.49	0.63	0.25	0.10	0.05

Numerical simulations of ultrasonic treatment upon DC casting imitating the experimental conditions were carried out to provide guidelines on the optimisation of processing. A nonlinear Helmholtz-type model was especially developed to compute the acoustic pressure field, accounting for the effect of cavitation as described in [11]. To model the bubble activity, the Keller–Miksis (KM) approach [14] was used. In this work, individual bubble simulations were performed at a range of pressures, and the computed coefficients were referred to as a function of the local pressure in the Helmholtz equation.

For the fluid flow simulation, an OpenFOAM solver was developed using "*buoyantPimpleFoam*" (a combination of the PISO and SIMPLE algorithms) as a base. Solidification was modelled using a single continuum approach, with the solid region acting as a momentum sink. The simulation was run until the flow reached a steady state. Turbulence was accounted for using the k-Omega-SST turbulence model. The system was closed by the energy balance. In the slurry, the effective dynamic viscosity was calculated from the Stefanescu formula [15] with the dendrite coherency point chosen to be 0.3.

A summary of all the material properties used for liquid aluminium is given in [11]. The heat transfer across a solid boundary was approximated by the Fourier boundary condition $k\frac{\partial T}{\partial x} = q$, where k represents the material conductivity. For the water spray, the heat flux could be computed directly using a forced-convection heat transfer equation [9], but the resulting term could be also approximated as being a linear function depending on the average temperature $\bar{T}$ between the surface of the billet and the bulk fluid. The final heat transfer coefficients were taken from [11].

3. Results and discussion

The efficiency of ultrasonic melt processing depends to a great extent on the time that the melt stays within the active cavitation zone (residence time); that includes the cavitation zone itself and the area in the vicinity where the cavitation bubbles are transported by the acoustic or other flows and can still

IOP Conf. Series: Materials Science and Engineering 1274 (2023) 012020 doi:10.1088/1757-899X/1274/1/012020

oscillate and implode. This time can be extended either by using multiple sources and thereby increasing the melt volume subjected to the pressures above the cavitation threshold, or by recirculating the volumes of melt back into the active cavitation zone. Let us look at the flow patterns induced by the ultrasonic source in two typical settings: largely convective flow in the hot top/sump of a billet and directed melt flow in a launder. Based on the advanced modelling reported elsewhere [11], figure 1a-c shows these flow patterns. There are three interesting observations from these patterns: (i) in the hot top (figure 1a) the flow is largely directed downwards driven by acoustic streaming from the oscillating sonotrode with gentle recirculation at velocities below 0.1 m/s while (ii) in the launder the flow recirculates around the cavitation zone between partitions (figure 1c); and (iii) as a result of this efficient re-circulation in the launder, the distribution of seed particles in the sump after processing in the launder is much more uniform and the shape of the sump remains undisturbed (compare figure 1b and figure 1a). In contrast, processing in the hot top (figure 1a) leads to strong downward flow with velocities up to 1 m/s, which depresses and disturbs the liquidus isotherm as has been also shown in [9, 10]. This results in bringing some of the primary crystals (dendrites) into the active processing zone causing their fragmentation and multiplication, leading to significant grain refinement [2, 9, 10] that is, however, confined to the central and near-central parts of the billet, which creates a problem for large billets.

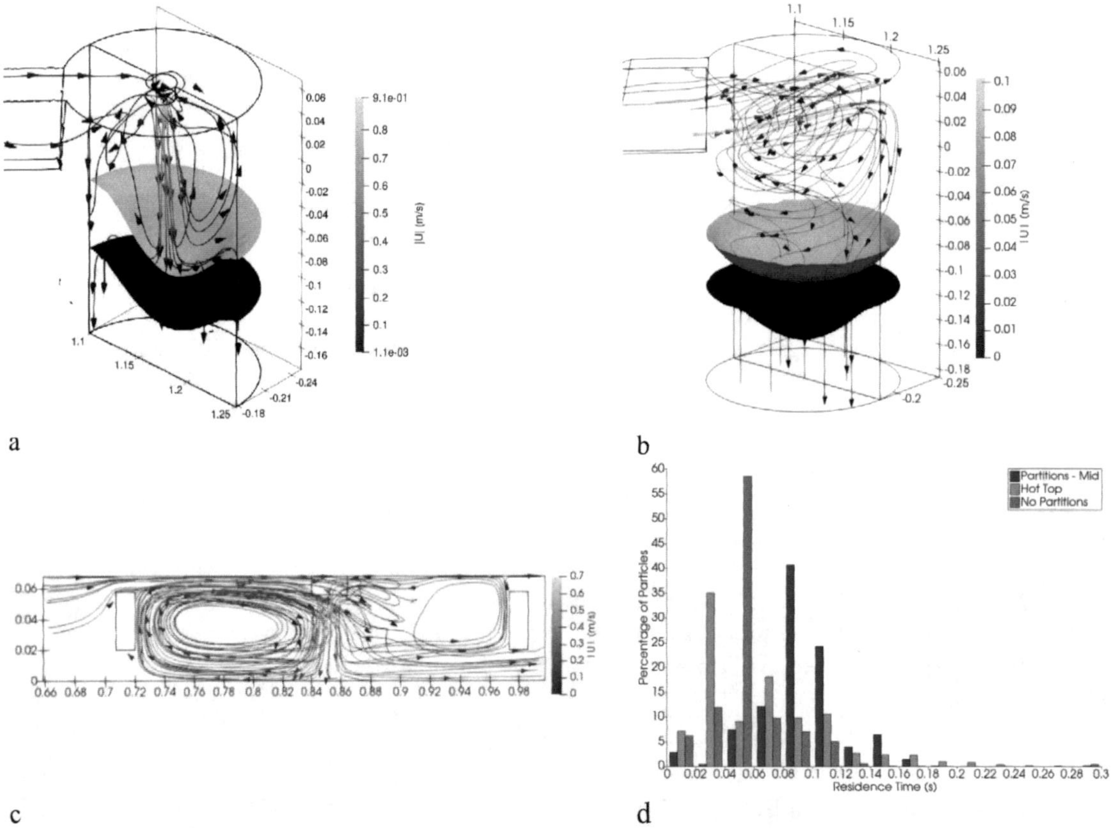

Figure 1. Computer simulation of flow patterns upon ultrasonic processing in a hot top (a) and in a launder (b, c): flow in the sump (a, b) and in the launder (c); and the residence time of tracer particles upon ultrasonic processing (quantified by tracking tracers) for the three cases of ultrasonic processing (d).

Multiple sonotrodes need, therefore, to be used in the sump, with their number depending on the processed volume [1, 2, 13]. It has been reported that a single sonotrode can be efficient in billets of up to 200 mm in diameter with 6 to 10 sonotrodes required for treating billets 500 to 1200 mm in

IOP Conf. Series: Materials Science and Engineering 1274 (2023) 012020 doi:10.1088/1757-899X/1274/1/012020

diameter [2]. This seems to suggest that a single source can be used for an extrusion billet. However, if one takes into account that extrusion billets are usually cast on a casting table with numerous moulds (12 to 48, and up to 160), it becomes obvious that using even a single sonotrode in each mould is not a technological way forward. On the other hand, if the ultrasonic processing is moved upstream into the launder, then the melt can be processed using a much smaller number of ultrasonic sources before it comes into the moulds. Sophisticated flow management is then required to ensure that the residence time is sufficient for processing. With having this in mind, we employ two principles for partitions design: (a) acoustic resonance that increases the acoustic pressure between two partitions and extends the active cavitation zone [16] and (b) flow recirculation that increases the residence time of the melt within the active cavitation zone. As a result, we managed to achieve a residence time (indicated by the mean time spend by a tracer particle in the cavitation-active zone) in the launder (0.094 s) that is 45% larger than the residence time upon ultrasonic processing in the hot top (0.064 s), and 70% larger than in the launder without partitions (0.055 s) as indicated in figure 1d.

In order to demonstrate the feasibility of this approach to DC casting, we have used an alloy without adding any standard grain refiner. In previous studies [3, 9, 10], USP was applied to melts already inoculated with TiB_2, and some additional grain refinement was demonstrated as a result of TiB_2 de-agglomeration and dispersion. However, de-agglomeration requires longer processing times due to the complexity of mechanisms recently revealed [8]. Instead, here we used additional alloying with Zr as an indicator of cavitation-induced grain refinement. It has been demonstrated on multiple occasions that cavitation facilitates heterogeneous nucleation of primary Al_3Zr particles on oxides [4, 5] and then additionally refines them by fragmentation using the power of shockwaves and microjets generated by cavitation bubbles implosion [6, 7]. These fine primary particles then act as substrates for nucleation of aluminium grains. Such an experiment demonstrates for the first time the pure effect of USP on grain refinement.

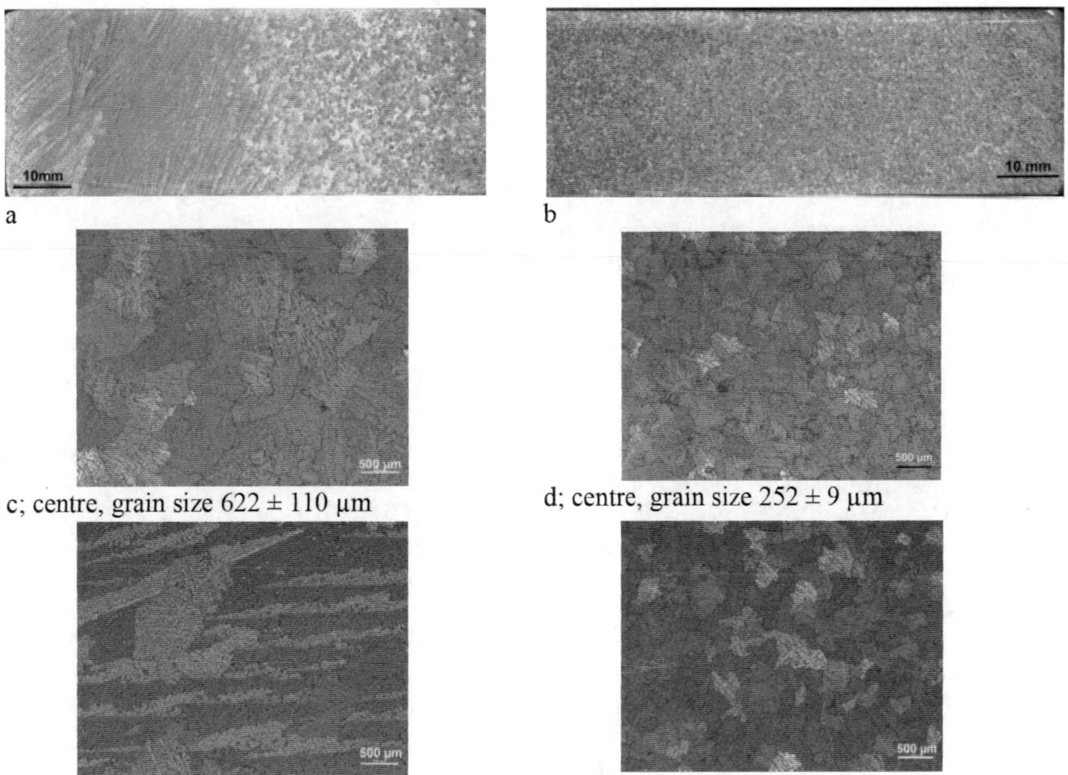

a b

c; centre, grain size 622 ± 110 µm d; centre, grain size 252 ± 9 µm

e; mid-radius, grain size 503 ± 95 µm f; mid-radius, grain size 272 ± 18 µm

Figure 2. Macro (a, b) and microstructures (c-f) of a 6XXX-series alloy cast without (a, c, e) and with (b, d, f) ultrasonic processing in the launder.

The results are demonstrated in figure 2. There is a remarkable change in the macrostructure where ultrasonic processing eliminates the feathery grain zone and shortens the subsurface zone of columnar grains, refining at the same time the equiaxed grains throughout the cross section. The reason behind these structural changes is the formation of numerous refined Al_3Zr primary crystals that act as substrates for the aluminium grains as shown in figure 3. After formation, fragmentation and refinement, these crystals are transported by the melt flow to the sump of the mould where they induce grain refinement throughout the billet cross section [12].

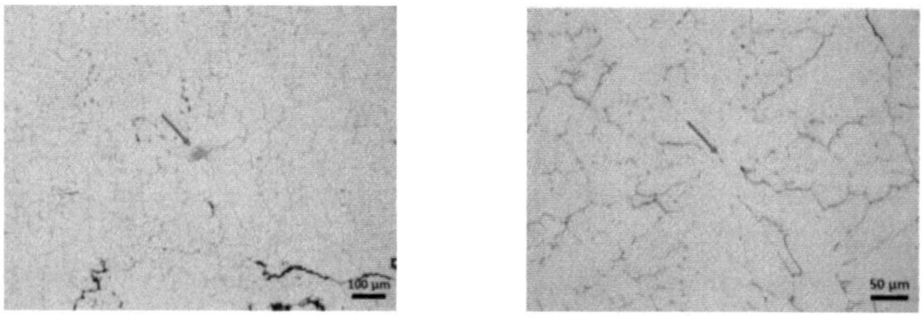

a b

Figure 3. Primary Al_3Zr intermetallics in the centre of grains in billets cast (a) without ultrasonic processing and (b) with ultrasonic processing in the launder. Note the different scale bars

4. Conclusion

A combination of numerical modelling and in-depth study of the mechanisms of ultrasonic melt processing allowed us to achieve significant grain refinement in a DC cast billet using an original scheme of ultrasonic processing in the launder, opening the way for upscaling of this melt processing technology.

Acknowledgements

The authors acknowledge financial support from the UK Engineering and Physical Science Research Council (EPSRC) for the Ultra-Cast (EP/L019884/1; EP/L019825/1; EP/L019965/1) and UltraMelt2 (EP/R011001/1, EP/R011044/1 and EP/R011095/1) projects as well as support from Constellium UK in performing casting trials.

References

[1] Eskin G I 1988 *Ultrasonic Treatment of Molten Aluminium* (Moscow: Metallurgiya).

[2] Eskin G I and Eskin D G 2015 *Ultrasonic Treatment of Light Alloys Melts*, 2nd Edition (Boca Raton: CRC Press).

[3] Eskin G I, Bochvar S G and Yelfimov V I 2010 *Tecknnol Legk. Spl.* no 1 38.

[4] Wang F, Eskin D, Connolley T and Mi J. 2017 *Trans. Nonferr. Met. Soc. China* **27** 977.

[5] Wang F, Eskin D, Mi J, Wang C, Koe B, King A, Reinhard C and Connolley T 2017 *Acta Mater.* **141** 142.

[6] Priyadarshi A, Khavari M, Subroto T, Conte M, Prentice P, Pericleous K, Eskin D and Tzanakis I 2021 Ultrason. Sonochem. **70** 105260.

[7] Priyadarshi A, Khavari M, Bin Shahrani S, Subroto T, Yusuf LA, Conte M, Prentice P, Pericleous K, Eskin D and Tzanakis I *Ultrason. Sonochem.* **80** 105820.

[8] Priyadarshi A, Khavari M, Subroto T, Prentice P, Pericleous K, Eskin D, Durodola J and Tzanakis I 2021 *Ultrason. Sonochem.* **79** 105792.

[9] Lebon G S B, Salloum-Abou-Jaoude G, Eskin D, Tzanakis I, Pericleous K and Jarry P 2019 *Ultrason. Sonochem.* **54** 171.

[10] Subroto T, Lebon G S B, Eskin D G, Skalicky I, Roberts D, Tzanakis and Pericleous K 2021 *J. Mater. Res. Technol.* **12** 1582.

[11] Beckwith C, Djambazov G, Pericleous K, Subroto T, Eskin D G, Roberts D, Skalicky I and Tzanakis I 2021 *Metals* **11** 674.

[12] Eskin D G 2020 *MATEC Web of Conferences* **326**, 06002.

[13] Zhang L, Li X, Li R, Jiang R and Zhang L 2019 *Mater. Sci. Eng. A* **763** 138154.
[14] Keller J B and Miksis M 1980 *The Journal of the Acoustical Society of America* **68** 628.
[15] Chang S. and Stefanescu D M 1996 *Metall Mater Trans A* **27** 2708.
[16] Khavari M, Priyadarshi A, Subroto T, Beckwith C, Pericleous K, Eskin D G and Tzanakis I 2021 *Ultrason. Sonochem.* **76** 105647.

Thermomechanical analysis of the solidification of fused cast alumina-zirconia-silica refractory blocks

T Houdard[1,2], I Cabodi[1], P Vespa[1], F Pigeonneau[2] and C-A Gandin[2]
[1] Saint-Gobain, SGR Provence, 84300 Cavaillon, France
[2] Mines Paris, PSL University, CEMEF UMR CNRS 7635, 06904 Sophia Antipolis, France

Abstract. Cooling history of fused cast Alumina-Zirconia-Silica (AZS) refractory blocks is recorded with thermocouples. Postmortem investigations are performed including hot tears. Discussion about solidification path is done with the purpose to better understand phenomena occurring during casting. Heat transfer simulations are compared with recorded temperature histories. Mechanical outputs from THERCAST® are analyzed, revealing the possibility to calibrate the risk of hot tears formation by considering a cumulated strain based criterion.

1. Introduction

Industrial oxide glasses are melted in large tanks at around 1 500 °C. The tank must be durable for up to twenty years. For this reason, alumina-zirconia-silica refractory is commonly used as soldier blocks in the infrastructure of glass furnaces. The fused cast foundry process is used to produce the soldier blocks. It works at extremely high temperatures, typically 1 850 °C. Usual casting defects can be found inside soldier blocks: hot tears along the block edges, macro-porosity in the riser and macro-segregation (or chemical heterogeneity) within the block. Several modeling approaches have been published in the literature to predict hot tear susceptibility in metallic alloys [1, 2]. Despite contributions that focused on thermal stress analyses [3], numerical predictions of hot tears and other defects applied to the fused cast products are still scare.

In the following, the fused cast process is presented in § 2 along with observations. Thenumerical model is briefly presented in § 3. Experimental and numerical predictions are compared in § 4, followed by a discussion in § 5.

2. Fused cast process

Fused-cast ceramic refractory blocks are composed of Alumina (Al_2O_3), Zirconia (ZrO_2), Silica (SiO_2) and Sodium oxide (Na_2O), hereafter referred to as AZS. The material is first melted in an electric-arc furnace before pouring into the sand mold. Block dimensions depend on customer's requirements. A typical block section will be around $0.4 \times 0.25\ m^2$.

Standard height is around 1.6 m. A block is composed of a rectangular cuboid and a riser above. The block mass is several hundreds of kilograms. Table 1 provides the composition of the AZS refractory material investigated in the current work.

Table 1: Composition of the AZS block.

wt%	Al_2O_3	ZrO_2	SiO_2	Na_2O
AZS41	45	42	12	1

The casting configuration is shown in Figure 1, limited to a quarter of the geometry. In addition to the AZS block (transparent), the casting is composed by the mold (blue) made of sand agglomerated with a resin, an insulating media (orange) controlling the cooling with the environment, a concrete base (brown) holding the mold and the media and a steel bin (red) that contains the entire system.

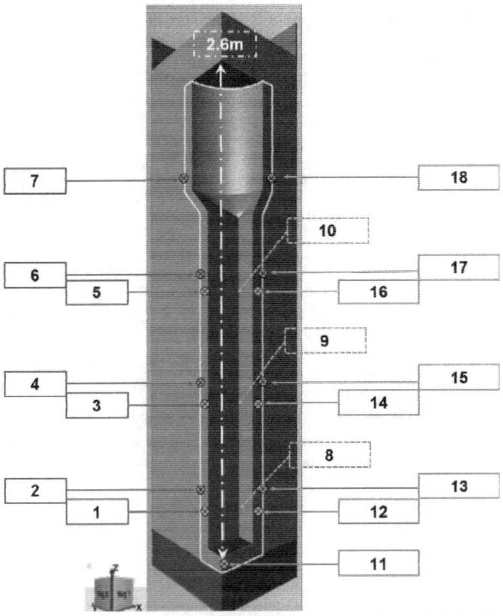

Figure 1: Casting configuration of AZS block (transparent) surrounded by a mold (blue), an insulating media (orange) and a concrete base (brown) contained in a steel bin (red). The 18 thermocouples, installed around the AZS block during an industrial casting, are displayed depending on their position: (orange circle with cross) inside the mold at mid-thickness and (blue circle with cross) outside of the mold. Locations of the thermocouples close to the edge of the block are shown by dashed arrows.

During the cooling period that lasts several days after casting, it is impossible to record temperature inside the AZS itself as liquid temperature is too high for available sensors. Nevertheless, to learn more about the cooling process, temperature inside the mold has been recorded using 18 thermocouples with locations reported in Figure 1. Thermocouples localized inside the mold at mid-thickness are shown with crossed orange symbols while those outside the mold are in crossed blue disks. Position of the thermocouples close to the edge of the block are shown by dashed arrows. The temperature recording is operated every minute and is used to assess the temperature history predicted by numerical simulations.

As already emphasized by Cockcroft *et al* [3], hot tears appear sometimes on the AZS blocks. Such defects are found at the edges of the block. They can extend over few centimeters within the block. An excessively long hot tear can lead to rejection of the block by glass manufacturers. SEM observation of a sample showing an inter-granular crack is presented in Figure 2, with an opening width reaching several hundreds of micrometers.

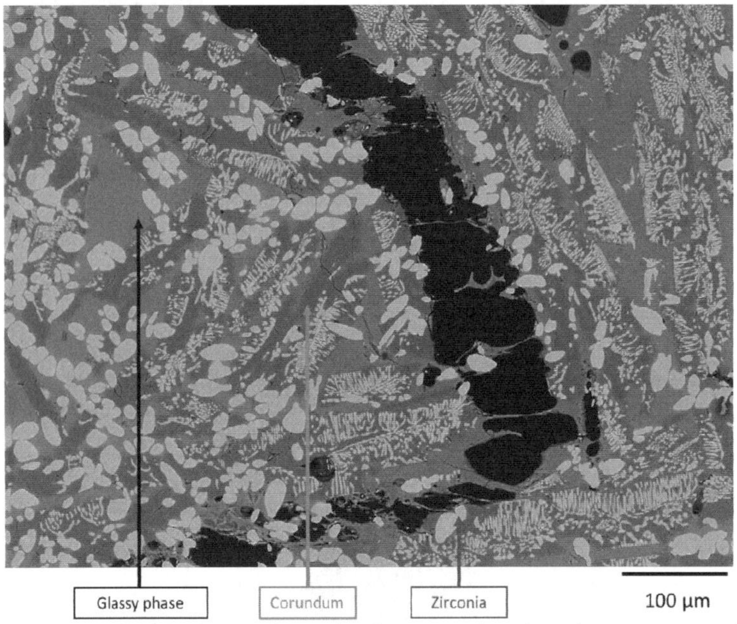

Figure 2: SEM observation of a hot tear showing the presence of glassy strings and the distribution of phases within the various microstructures.

Magnification in Figure 2 reveals the presence of glass filaments inside tears, confirmingthe opening of the crack while liquid phase was still present. No crystal is observed inside theglass filaments. This implies that those cracks occur when the solidification of zirconia andcorundum was over. Micro-probe measurements were achieved all along a hot tear. The glasscomposition was revealed to be similar to the glassy phase composition away from the crack. In complement to this SEM observation, the solidification path is investigated using the Thermo-Calc® software [4] with the TCOX10 database [5]. Roughly, the solidification path inthose conditions is relevant with the SEM observations depicted in Figure 2. Nevertheless, themullite phase is not observed inside the block while it is expected thermodynamically as thelast phase solidifying from the liquid. Despite the slow cooling process, the casting is too fastto reach thermodynamic equilibrium and the solidification of mullite. The thermodynamic solidification path can therefore be described as follows:

- solidification of 18 wt.% of zirconia from 1 895 °C to 1 759 °C in a tetragonal crystal structure with a dendritic microstructure;
- solidification of additional 29 wt.% zirconia plus 37 wt.% corundum from 1 759 °C to 1 011 °C by forming a two-phase eutectic made of the same tetragonal phase as before plus corundum;
- a martensitic reaction at 1 011 °C, transforming the tetragonal crystal structure into a monoclinic one;
- a glassy phase remains at low temperature, predicted as remaining 16 wt.% "liquid phase".

3. Thermomechanical simulation

Numerical simulations of the controlled cooling step have been performed with the THERCAST® software [6]. Based on the finite-element method, thermomechanical computation is done by considering the AZS block as a visco-plastic material above a critical temperature, Tc, and as an elasto-visco-plastic material below Tc. The critical temperature controlling the transition between the two behaviors is taken equal to 1 650 °C. Both thermal and mechanical fields are solved with temperature, velocity and pressure as primitive variables [7]. The thermal field is also determined outside the block but those other parts are assumed totally rigid. Properties of the fused-cast ceramic refractory material have been collected through literature [3, 8]. The thermodynamic properties of AZS (density, heat capacity and solidification path) are tabulated with temperature thanks to the TCOX10 thermodynamic database [5]. To agree with observations pointed out in the previous section, tabulations are done by rejecting the mullite phase.

4. Results

4.1. *Temperature history*

Numerical simulations are computed up to 28 h. Figure 3 displays the temperature as a function of time for the thermocouples #12 and #18 (Figure 1). After pouring, the temperature in the mold reaches its highest values in few hours. Temperature inside the mold is still around 500 °C after 1 day emphasizing that cooling is very slow. At most locations, numerical simulation results, red lines in Figure 3, are in good agreement with experimental data. Nevertheless, important differences between numerical and experimental results are observed in the riser for thermocouple #18 as will be discussed later.

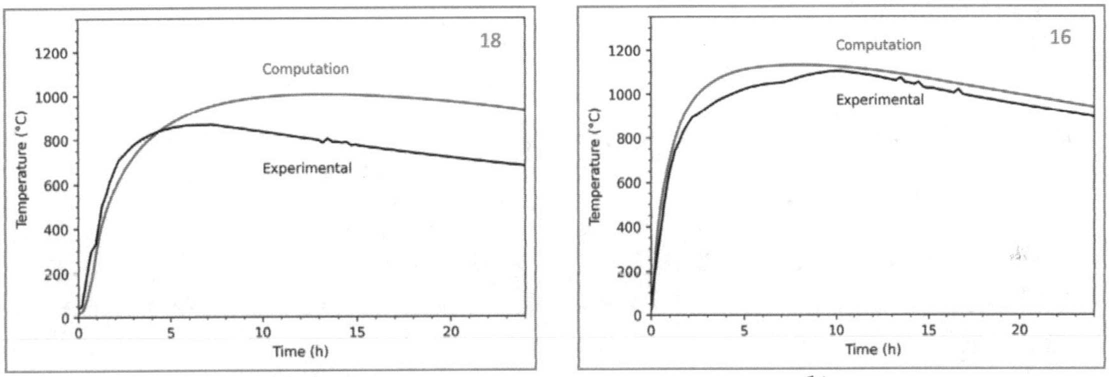

(a) (b)

Figure 3: Temperature history as a function of *t* time for two positions labelled in Figure 1, (a) #16 top of the block, (b) #18 riser. Experimental results (black) are compared to numerical results (red).

4.2. *Modeling of hot tear susceptibility*

Strain and stress distributions have been modeled in fused cast Alumina refractories [9] and AZS refractories [10]. It is then possible to switch from a strain distribution to a criterion about hot tear susceptibility. For metal alloys, several criteria have been defined to quantify hot tear susceptibility [1, 11]. A hot tear criterion takes into consideration accumulated strain during an interval of temperature that is called Brittle Temperature Range (BTR) $\int_{BTR} \dot{\varepsilon}\ dt$. The BTR is bordered with the zero-strength temperature (ZST) when the remaining liquid is trapped and the zero-ductility temperature (ZDT). The corresponding values hereafter chosen for the BTR are 1 650 °C and 1 300 °C as will be later discussed. Figure 4 shows the distribution of the accumulated strain in a horizontal cross section at height 0.8 m. Such deformation of the block, modeled in THERCAST®, generates strain at the middle of the faces, as shown in Figure 5, where the air layer between the block and the mold is wide. This contraction is anticipated industrially by casting the block with dimensions slightly over the expected standards. Mold combustion is not modeled here and an air gap is created between the mold and the block, that will impact heat exchanges. For the test with Consistency #1, the cumulated strain is maximal along the edge. Numerical sensors are set every 5cm along the edge of the block. Accumulated strain for each sensor is presented in Figure 4b. The distribution is quite similar all along the height of the block, after 24 h, when all sensors have cooled down below 1 300 °C.

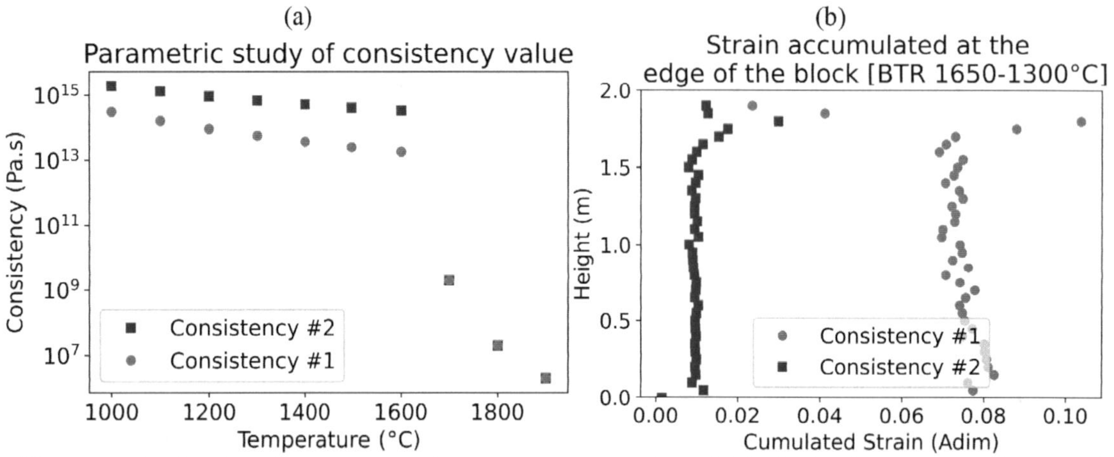

Figure 4: Study of the impact of consistency on sensors along AZS block edge at several heights. (a) tabulations of considered consistencies and (b) Strain cumulated during the range 1 300 °C - 1650 °C at the edge of the block for both consistencies' tabulations

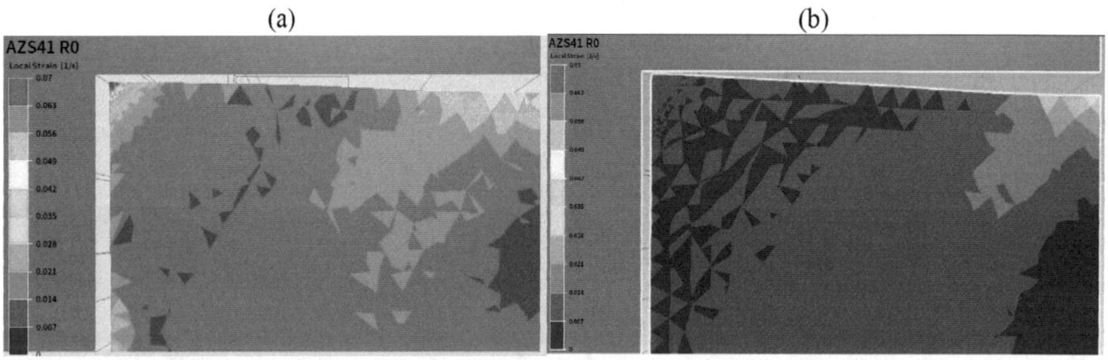

Figure 5: Prediction of Hot Tears Susceptibility inside an AZS block in BTR 1 650 °C to 1300 °C for a block cross section at 0.8 m height that shows also (white) the air gap layer created between the mold (grey) and the accumulated local strain in AZS block (color scale from 0 to 7%) depending on considered consistency: (a) Consistency 1 and (b) Consistency 2.

5. Discussion

5.1. Hot Tears Susceptibility predictions

The BTR is defined for metal alloys between the Zero Strength Temperature (ZST), set as a solid fraction g_s equal to 0.8 [11], and the zero-ductility temperature (ZDT) set at g_s =0.99. For considered refractory ceramics, the solidus is not reached as 15 wt.% of a glassy phase (considered as remaining liquid) is still present at the Glass Transition temperature. An adaptation is therefore required for such cumulated strain criterion. The upper boundary is defined as the temperature when almost all crystals have solidified and remaining liquid composition is very close to the glassy phase composition (i.e. 1 650 °C). The lower boundary is set as 1 300 °C when the liquid viscosity becomes high. The variations of hot tear susceptibility for different industrial cases are retrieved by simulations with this set of parameters.

This criterion is optimized for a given set of parameters. The constitutive equation considered in this article is an additive law considering the effect of strain and strain rate as independent. The considered consistency value is arbitrary lowered at high temperatures when solidification is not over. With such law, the deformation obtained at the edge of the block (in Fig.4b) is quite similar over the height of the block. The top of the block present variations in local accumulated strain that can be related to the time spent in the BTR. Elsewhere, even along the block, the averaged strain is extremely high as the deformation is close to 8 % whereas the critical strain in steel is around 2 % [2]. Validity for such large deformation values is still unclear and the thermo-mechanical properties could justify further investigations.

5.2. Mechanical predictions of shrinkage

After 24h of cooling, the shrinkage due to the solidification is over. Most of the solidificationshrinkage is localized in the riser. Current simulations do not retrieve the temperature history recorded inside the riser (Figure 3b). Also, the riser macro-porosity is predicted much thinnerthan the experimental one. By reducing artificially, the liquid density, solidification shrinkagegets a stronger effect on the riser macro-porosity. If macrosegregation is considered,the liquid inside the riser will be depleted in Zirconia and the density will indeed be lowered.A relevant riser macro-porosity prediction should impact the temperature exchanges with theenvironment.

6. Conclusion

Thermo-mechanical simulations have been run with THERCAST® to determine temperaturehistory and stress/strain distributions during the controlled cooling down stage of an AZS soldier block. Temperature and hot tear susceptibility predictions are in good agreement withmeasurement in the block, except in the riser. Explanations proposed involves the dependenceof thermophysical properties with macrosegregation.

Acknowledgments

This work has been supported by Saint-Gobain Research Provence and the French National Association Research Technology under ANRT CIFRE individual convention of education through research.

References

[1] Bellet M, Cerri O, Bobadilla M, and Chastel Y. 2009. *Metall. Mater. Trans. A*, **40**: pp2705–17.
[2] Won Y M, Yeo T J, Oh K H, and Seol D J. 2000. *Metall Mater Trans B*, **31**: pp779–94
[3] Cockcroft S L, Brimacombe J K, Walrod D G, and Myles T A. 1994. *J. Am. Ceram. Soc.*, **77**(6): pp1505–11.
[4] Andersson J. O., Helander T, Höglund L, Shi P and Sundman B. 2002. *Calphad*, **26**: pp273–312.
[5] Thermo-Calc Software. 2021a. Technical report: TCOX10. Metal Oxide Solutions Database.
[6] Transvalor. Thercast website. https://www.transvalor.com/en/thercast, 2022.
[7] Bellet M, Jaouen O, and Poitrault I. 2005. *Int. J. Numer. Methods Heat Fluid Flow*, **15**(2): pp120 42.
[8] Zhang Y, 2017, PhD thesis. Materials Sciences. PSL Research University.
[9] Au D, Cockcroft S L, and Maijer D. 2002. *Metall. Mater. Trans. A*, **33**: pp2053–65.
[10] Cockcroft S L, Brimacombe J K, Walrod D G, and Myles T A. 1994. *J. Am. Ceram. Soc.*, **77**(6): pp 1512–1521
[11] Yamanaka A, Nakajima K, and Okamura K. 1995. *Ironmaking Steelmaking* **22**(6): pp508–512

IOP Conf. Series: Materials Science and Engineering 1274 (2023) 012022 doi:10.1088/1757-899X/1274/1/012022

In-situ experimental investigations of the morphological characteristics of plumes and chimneys formed during the directional solidification of a binary alloy

Alok Kumar and Arvind Kumar*

Department of Mechanical Engineering, Indian Institute of Technology Kanpur, Kanpur 208016, India.

*Corresponding author: arvindkr@iitk.ac.in

Abstract. The defects such as freckles formed during the directional solidification of binary alloys severely affect the performance of cast products. The phenomenon of freckle formation is strongly influenced by the convective flows generated during solidification. The rejection of solute during the solidification at the solid/liquid interface induces the concentration gradients in the liquid. These concentration gradients in combination with existing temperature gradients, may induce buoyancy-driven convection in the mushy zone (plumes) and the parent liquid region (fingers). As the mushy zone evolves, small chimneys are formed, which leads to the formation of discrete plumes of interdendritic liquid. These plumes have a very complex flow structure which leads to the structural and compositional heterogeneities in the final cast product. The current study is an effort to investigate the morphological characteristics of plumes and chimneys formed during the directional alloy solidification. For this, controlled solidification experiments in a bottom-cooled rectangular cavity with an analogue transparent aqua ammonia system (NH_4Cl + 74 wt% H_2O) are performed. Particle Image Velocimetry (PIV), High Speed (HS) imaging and thermocouples temperature measurement techniques are used for real-time measurement of flow field, solidified and mushy zone dynamics and local temperature during the course of solidification. The dynamical nature of plumes is captured and the morphology of chimneys (shape, size, spacings, etc.) and plumes (velocity magnitude, shape, size, etc.) are quantified. The phase-locking with 180° phase difference due to the coupling between nearby plumes is observed. Also, the evolution of double-diffusive layers (DDLs) is captured and the critical value of thermal Rayleigh number for the onset of DDLs is estimated.

1. Introduction

The alloy solidification process has important applications across industries, geophysics and geology, such as metal castings, welding, metallurgy, crystal growth, sea ice growth, solidification in the Earth's interior, etc. [1]. During the solidification of hypereutectic alloys (in which solute is lighter for e.g. NH_4Cl-H_2O, Sn-Pb, Al-Cu etc.), the concentration gradients arise due to rejection of lighter solute in the mushy zone [2-3]. These composition gradients in combination with existing temperature gradients, may induce buoyancy-driven convection in the mushy zone and bulk liquid region [1-2]. In particular, during unidirectional solidification from the bottom, buoyant forces due to the rejection of lighter solute in the mushy zone results in the formation of fingers, chimneys, plumes, double-diffusive layers (DDLs), which in turn leads to the formation of channel-type defects called as 'freckles' [2-3]. In the castings, the freckles appear in the form of highly solute-concentrated, columnar regions spreading longitudinally in the direction of solidification and their presence causes severe material defects deteriorating the mechanical properties of final product [4]. Therefore, it becomes necessary to investigate the onset and dynamics of convective flows in mushy zone with chimneys [2]. In the past, several studies were carried out on the formation of plumes, chimneys, double-diffusive layers and freckles [2-10]. Aussillous et al. [5] investigated the transient growth of mushy zone from a fixed cold isothermal boundary and observed that chimneys form after the mushy zone exceeds a critical thickness. Experiments on steady-state mushy layer growth, in which the sample material is translated between two heat exchangers, were also performed in the past. These experiments show that several chimneys form when growth rates are slower than a critical threshold, depending on the initial liquid concentration [6-7]. Chen [8] performed shadowgraph experiments and qualitatively observed the formation of the double-diffusive layers (DDLs) during unidirectional solidification of 27 and 28 wt.% $NH_4Cl + H_2O$ in bottom cooled cavity. The critical thermal and solutal Rayleigh numbers for the onset of DDLs were reported to be of the order of 10^6 and 10^9, respectively. Shih et al. [9] suggested a strategy to suppress the emergence of double-diffusive convection by disrupting the flow at the top wall of test cell. Penney and Stastna [10] presented direct numerical simulations of three-dimensional double-diffusive flows in salt fingers, claiming that shear forces near the interface cause flux interleaving in the salt fingers. Even though several studies on the evolution of convection (plumes, bulk-fluid and double-diffusive layers) and formation of freckles were carried in the past, most of these studies are limited to qualitative observations. Furthermore, several key details about the complete evolution cycle of a double-diffusive layer are unexplored and the morphology of chimneys (shape, size, spacings etc.) and plumes (velocity magnitude, shape, size etc.) has not been quantified.

To address the above issues, controlled unidirectional solidification experiments are performed in a bottom-cooled rectangular cavity with an analogue transparent aqua ammonia system (NH_4Cl + 74 wt% H_2O). Particle Image Velocimetry (PIV), High Speed (HS) imaging and thermocouples temperature measurement techniques are used for real-time measurement of flow field, solidified and mushy zone dynamics and local temperature during the course of solidification to obtain the insights of plumes, DDL, chimneys and freckles formation. The dynamical nature of plumes is described, and the morphology of chimneys (shape, size, spacings, etc.) and plumes (velocity magnitude, shape, size etc.) are quantified. Also, the evolution of double-diffusive layers (DDLs) is delineated and the critical value of thermal Rayleigh number for the onset of DDLs is estimated.

2. Experimental setup

The complete experimental facility consists of a particle image velocimetry (PIV) system, test apparatus (test cavity), cooling system and data acquisition system (see Fig. 1a). The test apparatus consists of a double wall rectangular cavity separated by a gap of about 15 mm (see Fig. 1b). This gap is filled with N_2 gas to prevent condensation on the cavity walls. It also provides effective insulation resulting in minimization of end effects. The inner cavity is 80 mm in length (L), 110 mm in height (H) and 45 mm in depth (W). The cavity walls were made of 10 mm thick transparent plexiglass except for the bottom wall. A heat exchanger made from 10 mm thick oxygen-free electrolytic copper forms the bottom surface of the cavity. It is used to maintain a constant temperature at the bottom surface of test cavity by using a programmable microprocessor-controlled constant temperature bath (SISKIN PROFICHILL-AC200). To take advantage of the counter flow heat exchanger, the coolant's path in the heat exchanger is planned so that two adjacent paths carry incoming and return fluid streams flowing in opposite directions. The PIV system used to measure the fluid flow consists a Nd:YAG (200 mJ (×2) at 15 Hz, 532 nm wavelength) laser with an 8MP CCD camera (having resolution of 3320×2496 pixel at 8.5 fps frame rate) synchronized via LaserPulse synchronizer using Insight 4G® software. The data acquisition system consists an Intel (R) Xeon (R) CPU-2620V3 @2.40GHz (2 processors, RAM: 128GB DDR3) based computer.

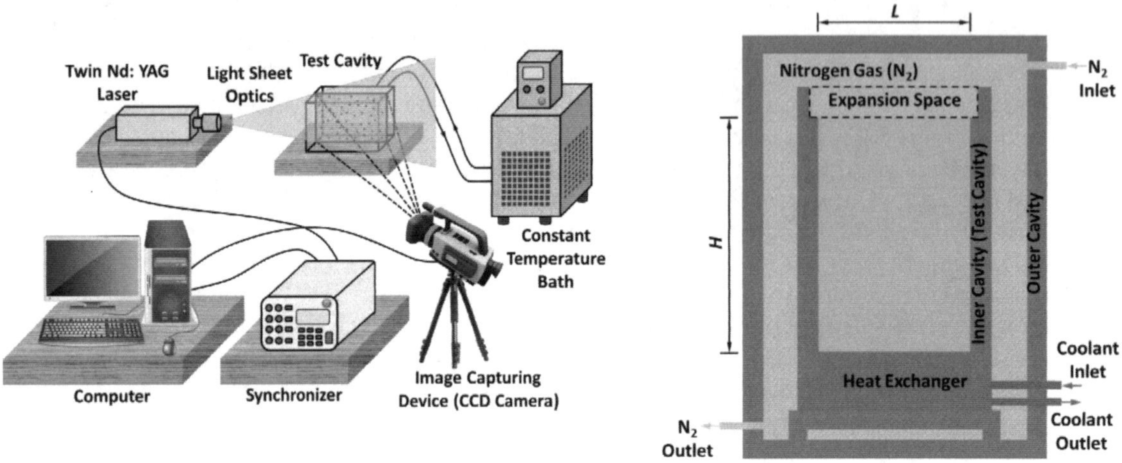

(a) (b)

Figure 1. Schematic of the (a) experimental facility and (b) test apparatus.

3. Methodology

To investigate the formation of freckles and morphological characteristics of plumes and chimneys, unidirectional solidification of NH_4Cl+74 wt% H_2O solution in the bottom cooled cavity is analysed. The NH_4Cl-H_2O solution is chosen as it has the characteristic of solidifying like metallic alloys and it is transparent. Therefore, it enables the direct observation of the transport phenomena during solidification using optical techniques. Before starting experiments, the test apparatus is levelled correctly with the help of spirit level to avoid the gravity-induced effects. Precautions are taken to ensure that no air bubbles become trapped inside the cavity, as they can act as nucleation sites, resulting in the loss of consistency. The flow field in the liquid region during the solidification has been characterized using the particle image velocimetry (PIV) technique. For this, an aqueous ammonium chloride solution having 74 wt% H_2O is prepared and seeded with silver coated hollow glass seed particles (density: 1.7 g/cc, diameter: 14 μm). The solution is stirred well to form a homogeneous dispersion. The size of silver particles for PIV measurement is chosen so that the Stokes number is in the order of 0.01. The concentration of seeding particles is optimized through several trial experiments and kept at 10 to 15 particles in an interrogation window. The temperature at the bottom surface of cavity is reduced and maintained at −25 °C. A green-colored Nd-YAG pulsed laser light sheet is generated using a concave cylindrical lens and passed through the centre plane of cavity (W/2 = 22.5 mm) filled with NH_4Cl-H_2O solution seeded with silver-coated hollow glass particles. The light is scattered through seeding particles in the flow and the images are captured periodically using a CCD camera. The captured images are processed using Insight4G® software (provided by TSI Incorporated), which calculates instantaneous velocity field by cross-correlation of the successive images. The transient evolution of plumes from the mushy zone and solid/liquid interface is recorded using 4MP PHANTOM-V641 high-speed (HS) camera (resolution 2560×1600 pixel at 1450 fps, RAM:16 GB). The transient temperature data at strategic locations is recorded with the help of 'T' type Teflon coated copper-constantan thermocouples (range: -270ºC to 370ºC) through NI cDAQ 9171 USB Chassis and NI-9211 4-channel thermocouple card. The thermocouples were calibrated against SISKIN PROFICHILL-AC200 with an accuracy ± 0.01 °C. The uncertainty of the thermocouples is found to be ± 0.5 °C. The PIV and thermocouple data is processed in Intel (R) Xeon (R) CPU-2620V3 @2.40GHz (2 processors, RAM: 128GB DDR3) based computer.

4. Results and discussion

During solidification of the aqueous ammonium chloride solution, three phases, namely a solid phase (S), a liquid phase (L) and mushy zone (S+L), are formed (see Fig. 2a). In the present case, the concentration gradients arise due to the rejection of lighter solute (water) in the mushy zone. We observed three different kinds of fluid flow during the solidification. They include an ascending flow of fingers and plumes, a descending flow of bulk fluid and horizontal convection in double-diffusive layers (DDLs) (see Fig. 2). The finger regime, which consists of ascending solute-rich fluid fingers, develops immediately after the start of the solidification process (see Fig. 2b). When significant amount of solute is rejected at the growth front during solidification, the fine-scale finger regime is overtaken by the appearance of solute plumes (see Fig. 2c and 2d). The maximum velocity of the plumes measured by the PIV is found to be ~2.5 mm/s. To balance the mass in mushy zone, the bulk fluid gradually moves downward (see Fig. 2c). In the present case, the concentration field is nearly stable (except plumes) in the horizontal direction and therefore, the concentration gradients mainly exist in a vertical direction. During the solidification, horizontal temperature gradients are developed as a result of the stochastic existence of a greater number of plumes on one side of the test cavity (see Fig. 3a). In combination with the vertical concentration gradients, these horizontal temperature gradients result in the buoyancy driven double-diffusive convection cells across the width of the test cavity and the formation of double-diffusive layers (DDLs) (see Fig. 2e and 2f). These layers are diffusive interfaces that start to form into the process at about $t = 100$ min. Before the onset of DDLs, the horizontal difference in the temperature at the left and right wall increases (0.5°C < ΔT_H < 2°C) as the solidification progresses. However, it nearly vanishes after the onset of DDLs as convection in DDLs homogenises the temperature horizontally (see Fig. 3a). We also observed that double diffusive layers are slightly inclined from the horizontal due to the vertical motion of the plume. It denotes that the existence of plumes is required for the formation of DDLs, but presence of that plumes does not necessarily result in the initiation of DDLs. The thermal Rayleigh number ($Ra_T = \frac{g\beta_T \Delta T_H h_l^3}{\alpha \nu}$) is calculated using the temperature difference at the left and right wall (ΔT_H) and height of liquid region (h_l). It is observed that for the onset of DDLs, a critical thermal Rayleigh number of ~20×10⁶ is needed to be achieved (see Fig. 3b).

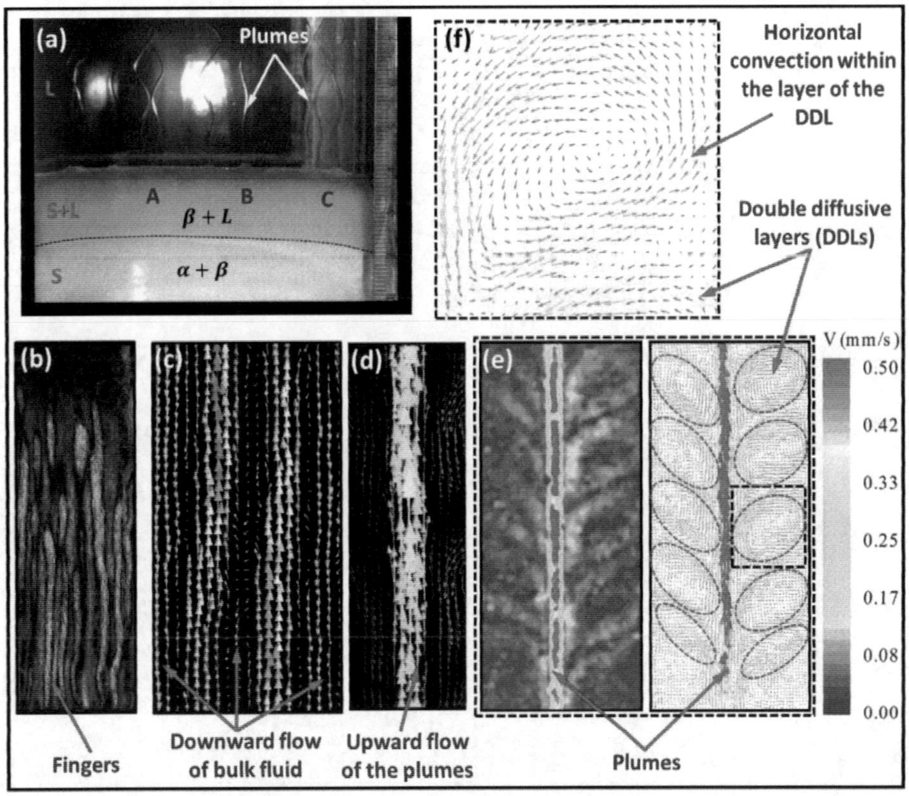

Figure 2. Figure shows (a) solid, liquid and mushy zones at $t = 250$ min (b) finger regime at $t = 15$ min (c) downward flow of the bulk fluid at $t = 50$ min (d) upward flow of the plumes at $t = 150$ min (e) velocity magnitude in the plumes and DDLs at $t = 150$ min (f) flow in the DDLs (zoomed view of the rectangular region marked in Fig. 2e).

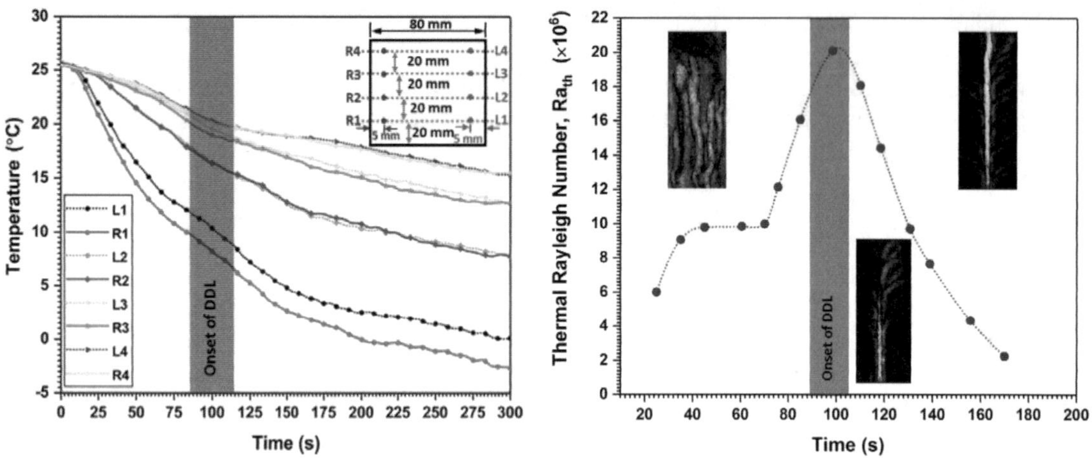

(a) (b)

Figure 3. Transient variations of the (a) temperature at the left and right walls of cavity, (b) thermal Rayleigh number (Ra_T).

Figure 4 shows morphological characteristics of the plumes and chimneys formed during alloy solidification. We observed 180° phase locked pairs of oscillating plumes (see Fig. 4a). The reason for phase locking with 180° phase difference is believed to be due to coupling between neighbouring jets. The phase synchronisation of neighbouring plumes occurs due to the dynamical instabilities caused by the entrainment of fluid and counter-propagating shear flow. The wavelength of the oscillating plumes is measured and found to be ~17 mm. As salt-rich fluid near the plume cools, it moves downward as a sheath along the plume. Due to shearing force, these sheaths are unable to traverse the plume bends, forcing them to detach from the plume (wisps) and descend freely (see Fig. 4b). Due to the rejection of solute, buoyancy driven convective solutions are focused into the distinct solid free areas, which punctures mushy region at approximately periodic intervals. These defects are known as the chimneys (see Fig. 4c). It appears as narrow dendrite-free, cylindrical shapes and rejects a part or all the lighter components of the mixture (water). As the chimneys get solidified, it results in the formation of freckles. The diameter and spacings of the chimneys are measured and found to be ~1.6 mm and ~15 mm, respectively (see Fig. 4d and 4e). These values are also shown in Table 1.

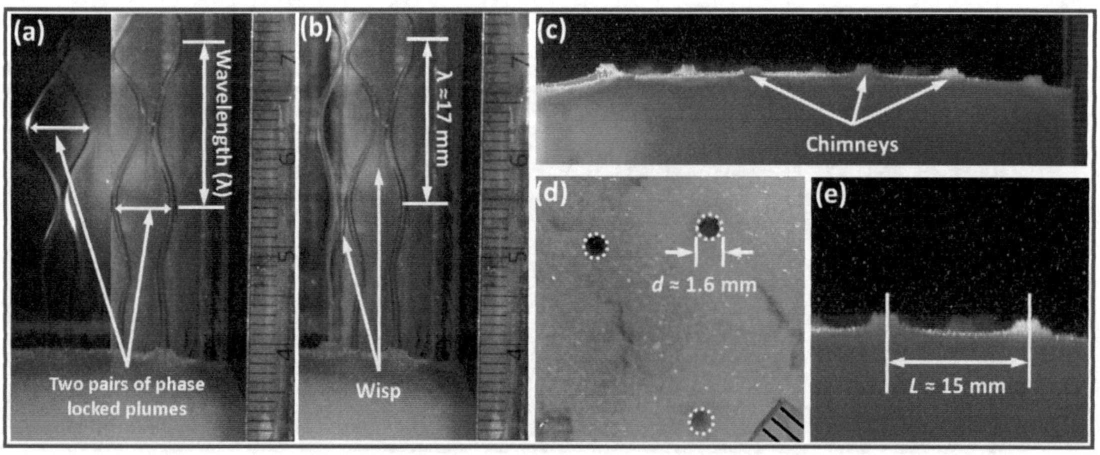

Figure 4. Figure shows (a) the pairs of phase locked plumes, (b) the wisp formation (c) chimneys, (d) average diameter of chimneys (top view) (e) average spacing between chimneys.

Table 1. Morphological parameters for plumes, chimneys and double-diffusive layers (DDLs)

Parameters	Values
Maximum velocity of the plumes (V_P)	2.5 mm/s
Average diameter of the chimneys (d)	1.6 mm
Average spacing between chimneys (L)	15 mm
Average wavelength of the oscillating plumes (λ)	17 mm
Critical thermal Rayleigh number for onset of DDLs (Ra_{Tcr})	20×10^6

5. Conclusion

Solidification experiments were performed in a bottom-cooled rectangular cavity with a metal analogue transparent aqua ammonia system (NH_4Cl + 74 wt% H_2O) for in-situ investigation of the freckles formation and morphological characteristics of the plumes and chimneys. Particle Image Velocimetry (PIV), High Speed (HS) imaging and thermocouples temperature measurement techniques were used for real-time measurement of flow field, solidified and mushy zone dynamics and local temperature during solidification. The following conclusions are drawn from this study:

1. Three different kinds of fluid flow, namely upward flow of fingers and plumes, descending flow of bulk fluid and horizontal convection in double-diffusive layers (DDLs), were observed during solidification.
2. The existence of plumes is required for the formation of DDLs, but presence of that plumes does not necessarily result in the initiation of DDLs. For the onset of DDLs, a critical thermal Rayleigh number of $\sim20\times10^6$ need to be achieved.
3. Maximum velocity in the plumes obtained by PIV measurements is ~2.5 mm/s.
4. Few pairs of the phase locked plumes with 180° phase differences were observed. The wavelength of these oscillating plumes is found to be ~15 mm.
5. Average diameter and average spacing between the chimneys are found to be ~1.6 mm and ~15 mm, respectively.

Acknowledgments

The financial support of MHRD, India (Grant No. MHRD/ME/2016408H) is gratefully acknowledged.

References

[1] Kumar A and Kumar A 2021 *Materials Today: Proceedings* **46** (17), 7764–7769.
[2] Copley S M, Giamei A F, Johnson S M and Hornbecker M F 1970 *Metallurgical and Materials Transactions B* **1**, 2193–2204.
[3] Tan F L 2009 *Heat and Mass Transfer* **45**, 255–265.
[4] Chen C F and Chen F 1990 *Journal of Fluid Mechanics* **227**, 567–586.
[5] Aussillous P., Sederman A J, Gladden L F, Huppert H E and Worster M G 2006 *Journal of Fluid Mechanics* **552**, 99–125.
[6] Peppin S S L, Huppert H E and Worster M G 2008 *Journal of Fluid Mechanics* **559**, 465–476.
[7] Whiteoak S H, Huppert H E and Worster M G 2008 *Journal of Crystal Growth* **310** (15), 3545–3551.
[8] Chen F, 1997 *Journal of Crystal Growth* **179** (1-2), 277–286.
[9] Shih Y C, Tu S M, and Chiu C C 2013 *Applied Thermal Engineering* **50** (1), 1055.
[10] Penney J and Stastna M, 2016 *Physics of Fluids* **28** (8), 086602.

On modelling conjugated heat transfer in the thin slab CC mold and solid shell formation under the applied EMBr

A Vakhrushev[1], E Karimi-Sibaki[1], M Wu[2], A Ludwig[2], G Nitzl[3], Y Tang[4], G Hackl[4], J Watzinger[5], J Bohacek[6], A Kharicha[1],

[1]Christian-Doppler Lab for Metallurgical Applications of Magnetohydrodynamics, University of Leoben, Franz-Josef-Str. 18, 8700 Leoben, Austria
[2]Chair of Simulation and Modeling of Metallurgical Processes, University of Leoben, Franz-Josef-Str. 18, 8700 Leoben, Austria
[3]RHI Magnesita GmbH, Kranichberggasse 6, 1120, Vienna, Austria
[4]RHI Magnesita GmbH, Technolgy Centre, Magnesitstrasse 2, 8700, Leoben, Austria
[5]Primetals Technologies, Turmstrasse 44, 4031, Linz, Austria
[6]Heat Transfer and Fluid Flow Laboratory, Faculty of Mechanical Engineering, Brno University of Technology, 61669 Brno, Czech Republic

E-mail: abdellah.kharicha@unileoben.ac.at

Abstract. Continuous casting (CC) became one of the dominant steel production technologies throughout last decades. Better quality, energy savings and high production rates are the main aims of the research especially in the field of the thin slab casting (TSC). The electromagnetic brake (EMBr) is applied to control the highly turbulent flow after the fresh melt is fed through the ports of a submerged entry nozzle (SEN). The numerical modelling is a perfect tool to investigate the multiphase phenomena of the turbulent flow in the CC mold, heat transfer and solidification coupled with the effects of the magnetohydrodynamics (MHD). Traditionally the heat transfer in the CC mold during the numerical simulations is predefined by the heat flux profile which could be taken from the plant measurements, published data, or is described by the semi-empirical formulas. In all these cases the heat extraction in the CC mold cavity is strictly predefined and is not significantly influenced by the transient flow behavior. Moreover, the heat flux, used in a simulation, is frequently measured for the different flow pattern inside the mold. That is especially important when the EMBr effects on the solid shell formation are investigated. Thereby, the presented study considers the coupled heat transfer in the water-cooled copper mold, including the averaged thermal resistance between the slab and mold, implemented using OpenFOAM® open-source CFD software. The melt flow, the temperature field, and the induced electric current density are compared between the traditional approach (the applied heat flux) and the modelled heat transfer in the TSC mold. Different scenarios are studied without and with the applied magnetic field.

1. Introduction

During a thin slab casting (TSC) due to the narrow profile of the continuous casting (CC) mold a strong turbulence is generated after the fresh melt is fed through the ports of the submerged entry nozzle (SEN). As it was shown previously, the turbulent flow has a strong influence on the solidification by interaction with the mushy zone [1–3]. Due to the partial remelting of the solid shell the risks of a breakout dramatically grow [4–6]. The application of the magnetic field in the form of the electromagnetic braking (EMBr) became an efficient flow control technology through last decades [7]. A very recent study has enlightened that the complex topology of the Lorenz force is not limited (as expected) to the damping effects only, but results in the entrainment of the quiescent melt into the motion and results in the formation of the reverse flow zones [8–10].

The melt flow and the distribution of the superheat in the TSC process are significantly altered under the EMBr action. Thereby an accurate model of the heat transfer in the CC mold is desired to predict those changes. Typically, such models consider the melt/shell, skin/shell and mold/skin contact and formation of the solid and liquid layers of the slag film as well as the air gap growth [11]. However, they become computationally demanding and, to overcome such a bottle neck, are reduced to the 2D-slice or even 1D models where the turbulent melt flow is over-simplified.

In the presented study, a single mesh approach is applied to simulate the turbulent flow together with the solidification, thermal resistance of a slag skin and the air gap, and with the heat transfer in the water cooled copper mold [12]. The electric potential method was used to tackle the magnetohydrodynamics effects [13]. The proposed method appeared to be robust and showed a good agreement with the corresponding experimental measurements [1,14].

2. Numerical model

2.1. *Mold heat transfer model*

Figure 1 display the simulation domain. A thin gap region was introduced between the melt bulk and the mold. The numerical mesh was significantly refined to resolve the gap. The geometry of the simulated mold corresponds to the effective thickness of the copper plates along the wide face (25 mm) and at the narrow side (18 mm) published elsewhere [14]. The cooling water temperature was assumed to be uniform (323.15 K) and will be adjusted in the future studies according to the recent publications [15]. The casting speed of 5.5 m/min was used in the simulation.

Traditionally, to model the solidification during CC process, a heat flux along the casting direction at the mold side of the solid shell is defined as boundary condition. It is commonly taken from literature for the similar mold designs, estimated from the on-site temperature measurements, defined by the semi-empirical formulas or modeled directly [1,14–16]. Accordingly, the temperature normal gradient is determined at the shell surface from the imposed heat flux distribution $Q(z)$ as following:

$$-\lambda \cdot \nabla_{\mathbf{n}} T = -Q(z), \tag{1}$$

where z is the distance from the meniscus, λ corresponds to the thermal conductivity and $\mathbf{n}$ is the wall normal vector oriented inward. It should be emphasized that the boundary condition (1) defines a fixed rate of the extracted heat. In such conditions, the model cannot respond accurately to the strong and fast changes in the temperature field caused by an EMBr or due to a SEN flow asymmetry [3].

In the presented study an alternative to the Neumann-type boundary condition (1) was chosen. As schematically shown in figure 2(a), it includes the heat transfer across TSC mold plates considering a thermal resistance between the shell surface and the mold hot face. The Robin-type boundary condition at the copper mold / cooling water interface is applied in the form:

$$-\lambda_{\text{mold}} \nabla_{\mathbf{n}} T|_{\text{mold-water}} = -\text{HTC}(z) \cdot (T_{\text{cold}} - T_{\text{water}}), \tag{2}$$

where $\text{HTC}(z)$ corresponds to the heat transfer coefficient between the mold cold face and the cooling water. The condition (2) and the complex heat transfer inside the mold plates and along the gap region give more degrees of freedom in comparison to the expression (1). Since the heat transfer coefficient is not provided, the procedure described below is applied to estimate the $\text{HTC}(z)$ distribution.

Figure 2(b) displays the mold heat flux (blue dashed line) along the casting direction z. Based on the solidification modelling using the traditional approach a time-average slab surface temperature T_{shell} was obtained, red line in figure 2(b). The temperature curve slightly recovers towards the mold's exit, which is typically observed by thermocouples measurement for the continuous casting [11,17]. The HTC is estimated based on the applied heat flux Q and predicted slab surface temperatures T. The effective gap resistance R_{gap} is introduced representing the slag skin layer and formed air gap; it linearly increases from the meniscus level down to the mold exit. The slag rim is not considered.

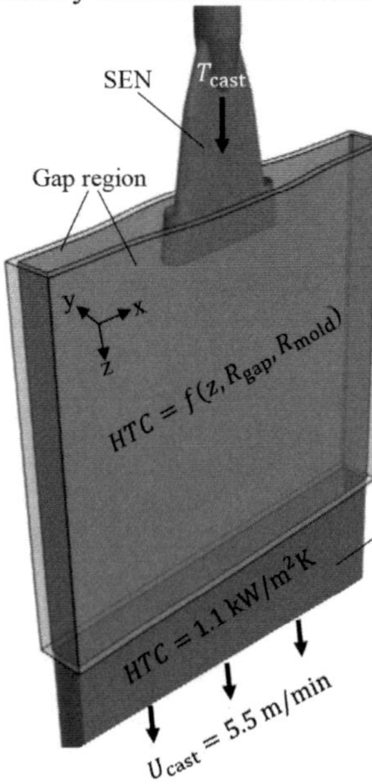

Figure 1. Simulation domain including SEN, TSC mold, gap region, strand part and corresponding boundary conditions.

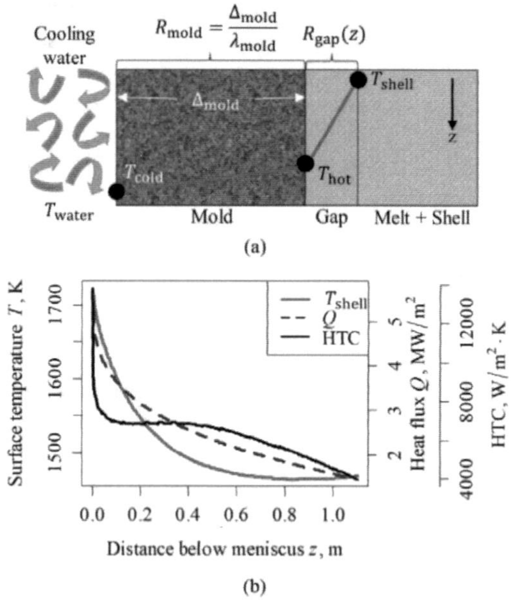

Figure 2. Heat transfer model: (a) shell-to-mold region; (b) imposed heat flux Q, modelled shell surface temperature T_{shell}, and estimated HTC(z).

By adding the mold thermal resistance R_{mold}, see figure 2(a), the HTC is obtained from:

$$\text{HTC}(z) = \left((T_{\text{shell}} - T_{\text{water}})/Q(z) - (R_{\text{gap}} + R_{\text{mold}})\right)^{-1}, \tag{3}$$

where the reference temperature of coolant T_{water} is 323.15 K. The only unknown in equation (3) is the effective resistance R_{gap}. Assuming its linear variation along the casting direction, it was adjusted to fit the experimental measurements of the mold's cold face temperatures and the modelled predictions of the liquid slag layer [11,18]. The calculated distribution of HTC(z), shown in figure 2(b) with a black line, was used in further modelling. The cooling of the strand is described by an effective HTC of 1100 W/m^{2}K, obtained from the experimental measurements by solving the inverse conduction problem [19].

2.2. Turbulent flow with solidification and MHD effects

The momentum equation was used in the mixture volume average approach as

$$\rho\left[\frac{\partial \vec{u}}{\partial t} + \nabla \bullet (\vec{u} \otimes \vec{u})\right] = \nabla \bullet [\text{dev}(\boldsymbol{\sigma}_{\text{tot}}) - p\mathbf{I}] + \vec{F}_{\text{mush}} + \vec{F}_{\text{MHD}} + \vec{F}_{\text{damp}}, \tag{4}$$

where the total stress $\boldsymbol{\sigma}_{\text{tot}}$ incorporate molecular viscous forces and the turbulence contribution by Reynolds stress modelling using large eddy WALE sub-grid scale model [20]. The Darcy force $\vec{F}_{\text{mush}}$ represents the drag between the mushy zone and the liquid melt [1]. The Lorentz force $\vec{F}_{\text{MHD}}$ acting due to the applied magnetic field is calculated using the electric potential method [13]. The details of the MHD model implementation considering the shell conductivity can be found elsewhere [8,9]. The damping source $\vec{F}_{\text{damp}}$ refers to a single mesh approach restricting motion to the melt region only [12].

The energy equation considering solidification is derived with respect to temperature T [3]:

$$\rho C_{\mathrm{p}}\left[\frac{\partial T}{\partial t}+\nabla\bullet(\vec{u}T)\right]=\nabla\bullet\lambda\nabla T+\rho L\left[\frac{\partial f_{\mathrm{s}}}{\partial t}+\nabla\bullet(f_{\mathrm{s}}\cdot\vec{u}_{\mathrm{s}})\right], \quad (5)$$

with the specific heat C_{p}, latent heat of fusion L, and liquid and solid fractions satisfying $f_{\ell}+f_{\mathrm{s}}=1$. The important details of an implementation, considering the TSC funnel curvature, distribution of the solid velocities $\vec{u}_{\mathrm{s}}$ and the effects of the latent heat advection were published earlier [1,2,21,22]. All models were developed as an in-house code using OpenFOAM® finite volume libraries [23].

3. Simulation results

Next, the described heat extraction model was used to simulate the solidification during the TSC process without and with the magnetic field of the electromagnetic brake. The schematics of the applied EMBr is shown in figure 3(a). The modeled induced electric current lines are detailed in figure 3(b) indicating the insulating slag skin layer preventing the e-current going through the highly conductive copper mold. Simultaneously, the closure of the e-current lines through the solid shell is observed. The predicted hot and cold face temperatures of the TSC mold are resented in figure 3(c).

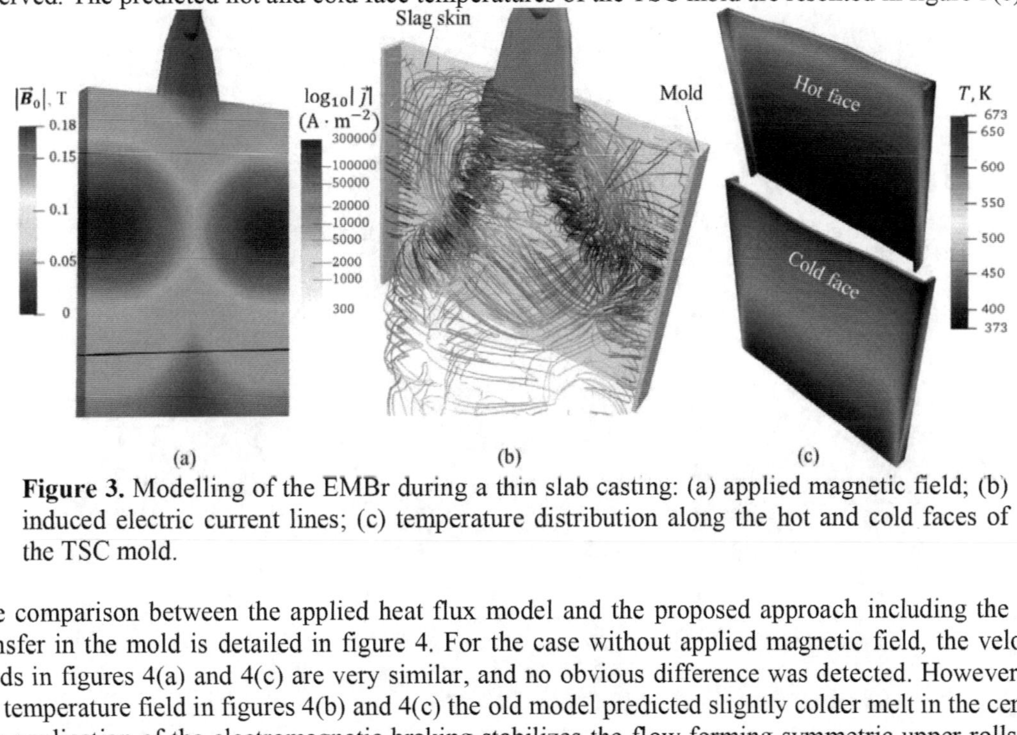

Figure 3. Modelling of the EMBr during a thin slab casting: (a) applied magnetic field; (b) induced electric current lines; (c) temperature distribution along the hot and cold faces of the TSC mold.

The comparison between the applied heat flux model and the proposed approach including the heat transfer in the mold is detailed in figure 4. For the case without applied magnetic field, the velocity fields in figures 4(a) and 4(c) are very similar, and no obvious difference was detected. However, for the temperature field in figures 4(b) and 4(c) the old model predicted slightly colder melt in the center. The application of the electromagnetic braking stabilizes the flow forming symmetric upper rolls and uniform flow closer to the molds exit, as observed from figure 5. More superheat is kept at the upper part of the mold at the meniscus level, which is highly desired in the CC process due to higher flux melting rate, uniform solidification, reduction of a frozen meniscus risk, etc.

The MHD flow including solidification is compared for two models in figure 5. With the EMBr the new model showed faster melt flow in the upper rolls by comparing figures 5(a) and 5(d). Again, colder central part was detected near molds exit for the old model, figure 5(b). Slightly more induced electric current is presented at the top part of the mold for the case of the new model in figure 5(f).

Thus, it can be seen, that with the dramatically changing flow pattern the difference between two models grows. Next study is to verify and detail of this difference using experimental data.

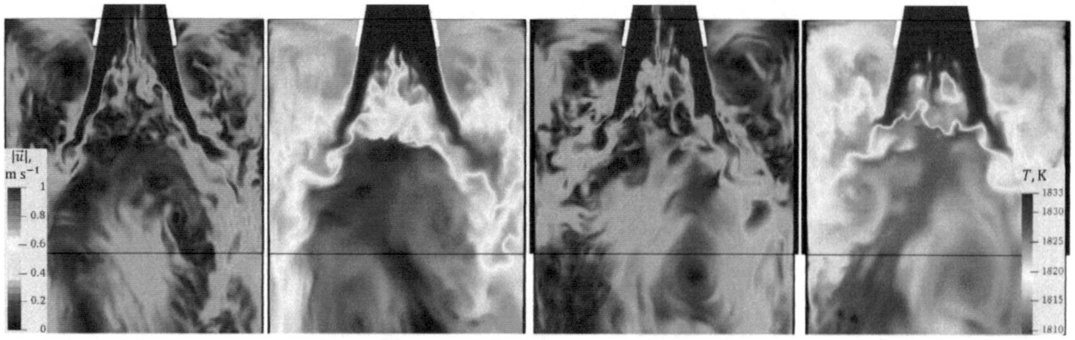

Figure 4. Simulation results for the flow velocity $\vec{u}$ and the temperature field T for the cases without EMBr: (a)-(b) imposed heat flux model; (c)-(d) model with the heat transfer across the mold.

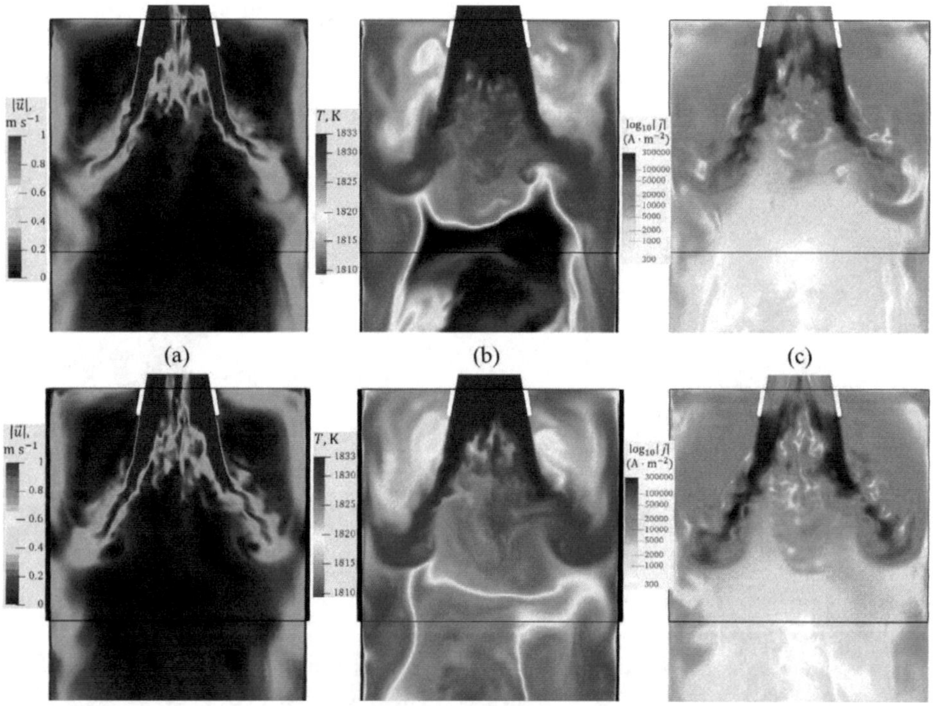

Figure 5. Simulation results for the flow velocity $\vec{u}$, temperature field T and the induced electric current $\vec{j}$ considering applied EMBr of 180 mT: (a)-(c) imposed heat flux model; (d)-(f) model with the heat transfer across the mold.

4. Conclusions

In the presented study a heat extraction model was introduced including a thermal resistance between the shell and the TSC mold hot wall. The simplified water cooling of the mold's cold face was modelled by estimating the heat transfer coefficient distribution along the casting direction. The magnetohydrodynamics forces were introduced to reflect the applied electromagnetic braking during TSC process. The simulated electrically insulating slag skin layer kept the induced e-current inside the liquid melt bulk, and the closure of the e-current lines through the solid shell was observed. The model showed good computational performance due to the employed single mesh approach. The obtained simulation results agreed with the old model. However, under drastically changing flow conditions, for example with the applied EMBr, the difference between two approaches grows. In the future study the model will be verified in detail using experimental data. The method will be extended by introducing the shell to skin and skin to mold contact models and considering temperature dependent properties.

5. Acknowledgments

The authors acknowledge the financial support by the Austrian Federal Ministry of Economy, Family and Youth and the National Foundation for Research, Technology and Development within the framework of the Christian Doppler Laboratory for Metallurgical Applications of Magnetohydrodynamics.

References

[1] Vakhrushev A, Wu M, Ludwig A, Tang Y, Hackl G and Nitzl G 2014 *Metall. Mater. Trans. B* **45** 1024–37
[2] Wu M, Vakhrushev A, Ludwig A and Kharicha A 2016 *IOP Conf. Ser.: Mater. Sci. Eng.* **117** 012045
[3] Vakhrushev A, Kharicha A, Karimi-Sibaki E, Wu M, Ludwig A, Nitzl G, Tang Y, Hackl G and Watzinger J 2022 *steel research int.* srin.202200088
[4] Iwasaki J and Thomas B G 2012 *Supplemental Proceedings* ed Tms (Hoboken, NJ, USA: John Wiley & Sons, Inc.) pp 355–62
[5] Luk'yanov S I, Suspitsyn E S, Krasilnikov S S and Shvidchenko D V 2015 *Int J Adv Manuf Technol* **79** 1861–8
[6] Zappulla M L S, Cho S and Thomas B G 2019 *Steel Res. Int.* **90** 1800540
[7] Thomas B G and Cho S M 2018 *IOP Conf. Ser.: Mater. Sci. Eng.* **424** 012027
[8] Vakhrushev A, Kharicha A, Liu Z, Wu M, Ludwig A, Nitzl G, Tang Y, Hackl G and Watzinger J 2020 *Metall. Mater. Trans. B* **51** 2811–28
[9] Vakhrushev A, Kharicha A, Karimi-Sibaki E, Wu M, Ludwig A, Nitzl G, Tang Y, Hackl G, Watzinger J and Eckert S 2021 *Metall. Mater. Trans. B* **52** 3193–207
[10] Kharicha A, Vakhrushev A, Karimi-Sibaki E, Wu M and Ludwig A 2021 *Phys. Rev. Fluids* **6** 123701
[11] Meng Y and Thomas B G 2003 *Metall. Mater. Trans. B* **34** 685–705
[12] Vakhrushev A, Kharicha A, Wu M, Ludwig A, Tang Y, Hackl G, Nitzl G, Watzinger J and Bohacek J 2021 *Metals* **11** 1375
[13] Davidson P A 2001 *An Introduction to Magnetohydrodynamics* (Cambridge University Press)
[14] Camporredondo S. J E, Castillejos E. A H, Acosta G. F A, Gutiérrez M. E P and Herrera G. M A 2004 *Metall. Mater. Trans. B* **35** 541–60
[15] Niu Z, Cai Z and Zhu M 2020 *Ironmaking & Steelmaking* **47** 1135–47
[16] Zhang H and Wang W 2017 *Metall. Mater. Trans. B* **48** 779–93
[17] Santillana B, Hibbeler L C, Thomas B G, Hamoen A, Kamperman A and Knoop W V D 2008 *ISIJ Int.* **48** 1380–8
[18] Meng Y and Thomas B G 2003 *Metall and Materi Trans B* **34** 707–25
[19] Bohacek J, Kominek J, Vakhrushev A, Karimi-Sibaki E and Lee T-W 2021 *OpenFOAM Journal* **1** 27–46
[20] Nicoud F and Ducros F 1999 *Flow, Turbulence and Combustion* **62** 183–200
[21] Vakhrushev A, Kharicha A, Wu M, Ludwig A, Nitzl G, Tang Y, Hackl G, Watziger J and Rodrigues C M G 2019 *IOP Conf. Ser.: Mater. Sci. Eng.* **529** 012081
[22] Vakhrushev A, Kharicha A, Wu M, Ludwig A, Nitzl G, Tang Y, Hackl G, Watzinger J and Rodrigues C M G 2022 *J. Iron Steel Res. Int.* **29** 88–102
[23] Weller H G, Tabor G, Jasak H and Fureby C 1998 *Comput. Phys.* **12** 620

Influence of stirring process during slurry formation on the casting defects and high-temperature fatigue of rheocast AlSi7Mg alloys

P Zaffarani[1], Q Zhang[2], A E W Jarfors[2] and G Timelli[1]

[1] Department of Management and Engineering, University of Padova, Stradella S. Nicola, 3, 36100, Vicenza, Italy

[2] Department of Materials and Manufacturing, School of Engineering, Jönköping University, Jönköping, Sweden

Email: anders.jarfors@ju.se

Abstract. The effects of the stirring phase during slurry formation on the defect's content and the influence on the high-temperature fatigue behaviour of a rheocast AlSi7Mg alloy were investigated. Single or double-stir step was used during RheoMetal processing. Metallographic and image analysis techniques were performed to quantitatively examine the defect's changes occurring with different stirring methods. Samples were cross-sectioned along the gauge length, and the pores content was measured. Uniaxial fatigue tests were performed at high temperature (150 °C) with the staircase method, applying a stress ratio of $R = -1$. Fracture surfaces were examined using a scanning electron microscope to determine the fatigue crack initiation sites, type and size of the defects. Shrinkage porosity has more probability of acting as a crack initiator than gas porosity or oxides. The double-stir step effectively reduces the size of pores compared to the conventional technique; it reduces porosity to less than half if compared to the single-stir process. Shrinkage pores distribution show a higher population of smaller pores for the double-stir process. According to the Dixon-Mood formula, the mean fatigue strengths are comparable. The capability of the double-stir to reduce porosity is evident, even having no significant effect on the high-temperature fatigue life.

1. Introduction

AlSi7Mg casting is used in the automotive and aerospace industries and their application is also developing to components subjected to high cycle fatigue [1]. The predominant effect on controlling fatigue performance is addressed to porosity [2], in both semisolid and die casting [3], considering that pores act as preferential crack initiation places [4]. Porosity is mainly divided into gas porosity and shrinkage porosity [5]. In some applications, the effect of the temperature above 150-200 °C could not be neglected, the resistance of the sample is reduced, leading to a softer and less resistant material [6].

Thixoforming increases the fatigue limit for the same cycles compared to standard casting methods [7]. Semisolid casting can potentially minimise porosity and oxide films, depending on the shape complexity and the casting parameters [8]. The RheoMetal® process and its applications in the automotive industry are widely explained by Jarfors et al. [9,10]. During this process the Enthalpy

Exchange Material (EEM), a solid cylinder with the same composition of the alloy being cast, is dipped under rotation into the ladle to create the slurry. The stirring is essential for creating a uniform slurry, breaking dendrites and avoiding the entraining of the air into the melt/slurry. A recent study [11] evidenced how a double-stir increases the slurry homogeneity, improving the slurry quality and developing a low porosity level. Inhomogeneity refers to solid clusters in the slurry due to the incomplete EEM dissolution during the slurry preparation [12]. The intensive stirring disperses α-Al particles increasing the liquid feeding during the solidification, and reducing porosity [13]. The present work aimed to investigate the effect of the double-stir in the RheoMetal® process, during the semisolid slurry formation of an AlSi7Mg alloy, on casting defects and on high-temperature fatigue life.

2. Materials and methods

2.1. Materials

The component consists of a baseplate for a thermal engine, with a wall-thickness of 30-40 mm, prepared from aluminium EN AC 42000 alloy (Table 1) cast via two different processes and tested as-cast. One batch consists of the standard RheoMetal® process (single-stir), in which the EEM with a mass equal to 5 % of the mass of the liquid is dipped into the melt (660-680 °C) and the stir takes place at 1000 rpm for 45s, during which the solid volume is partially melted and dispersed, the semisolid slurry is then injected into the die cavity at temperatures ranging from 600 to 615 °C, ensuring semisolid conditions for the given alloy composition. The EEM provides the solid fraction within the semisolid slurry. The second batch (double-stir) consists of a secondary stir step with a speed of 600 rpm for 10 s, dipping a rod provided with two SiC coated wings immediately after the first stir, before the injection into the mould cavity, as described in a previous publication [11]. Fifteen samples for each casting condition were extracted and shaped according to ASTM E466, with a gauge diameter of 6 mm and length of 14 mm. The two batches shared similar microstructure in terms of diameter, shape factor [14] and numerical density of the primary α-Al particles (Table 2).

2.2. High-temperature fatigue test

All axial fatigue tests were carried out in a servohydraulic test system at a temperature of 150 °C in a high-temperature furnace controlled by three thermocouples along the longitudinal direction. Constant-amplitude uniaxial stress-controlled test were performed at a frequency of 20 Hz at stress ratio R = -1. The staircase method was used to determine the fatigue strength at 2 x 10^6 cycles with a stress amplitude step of 5 MPa evaluated by the Dixon-Mood formula. Mean fatigue strengths and the standard deviations [15] were calculated adopting the testing conditions represented in Table 3.

After testing, the fracture surfaces of the failed specimens were examined in a JEOL JSM-7001F scanning electron microscope. The crack initiation site was determined, and its visible area was measured using image analysis software.

2.3. Porosity content

The gauge length of the samples which failed at 2 x 10^6 cycles was cut on the transversal direction in seven cross-sections to evidence pores, divided into gas or shrinkage pores as a function of their morphology, shape and development along the length of the cross-sections, as described in [16]. Ninety-eight cross-sections were metallographically prepared, the number of the pores was calculated, and their area was measured with the same procedure above. The overall porosity reveals the percentage of area related to shrinkage pores. The shrinkage pores distribution was obtained.

Table 1. Average chemical composition in wt % of the Al-7Si-Mg alloy used in this study.

	Chemical analysis (wt %) Elements											
Alloy	Si	Fe	Cu	Mn	Mg	Cr	Ni	Zn	Pb	Sn	Ti	Al
EN AC 42000	6,37 to 7,42	0,36 to 0,43	0,087 to 0,12	0,21 to 0,23	0,33 to 0,39	0,013 to 0,016	0,004 to 0,006	0,1 to 0,12	0,003 to 0,009	<0,0010	0,051 to 0,064	Balance

3. Results and discussion

3.1. High-temperature fatigue

The predominant defect observed on the fracture surfaces was shrinkage porosity, these appear as thin holes surrounded by Al particles, evidenced in the central region of the thick-walled baseplate, the last region to become solid and for this reason, the most subjected to volume shrinkage. The staircase curves (Figure 1) show that approximately half of the samples failed using the staircase method. The fatigue strength at 2 x 10^6 cycles and the pore diameter on the fracture surfaces were measured. The single-stir process has a fatigue strength of 70 ± 7 MPa and a pore diameter of 1080 ± 380 μm; whereas the double-stir 68 ± 4 MPa and 1143 ± 440 μm respectively. Shrinkage pores on the fracture surfaces were much larger than the shrinkage pores observed along the gauge length. The results suggest the stirring method did not influence the reduction of large shrinkage pores in this particular thick-walled component; suggesting why the high-temperature fatigue life was not affected (Figure 2).

Table 2. Microstructural characterisation of the two different batches in terms of diameter, shape factor and numerical density of the primary α-Al particles, semisolid casting, EN AC 42000.

Process	Diameter *μm*	Shape factor	Numerical density *particles/mm²*
Single-stir	52 ± 28	1,55	26
Double-stir	48 ± 22	1,78	22

Table 3. High-temperature fatigue test setup adopted for the staircase method.

Test conditions	Value
Fatigue test	Load controlled
Waveform	Sine tapered
Frequency	20 Hz
Stress ratio, R	-1
Mean stress	0
Step amplitude	5 MPa
Starting stress	60 MPa
Max stress amplitude	75 MPa
Temperature	150 °C

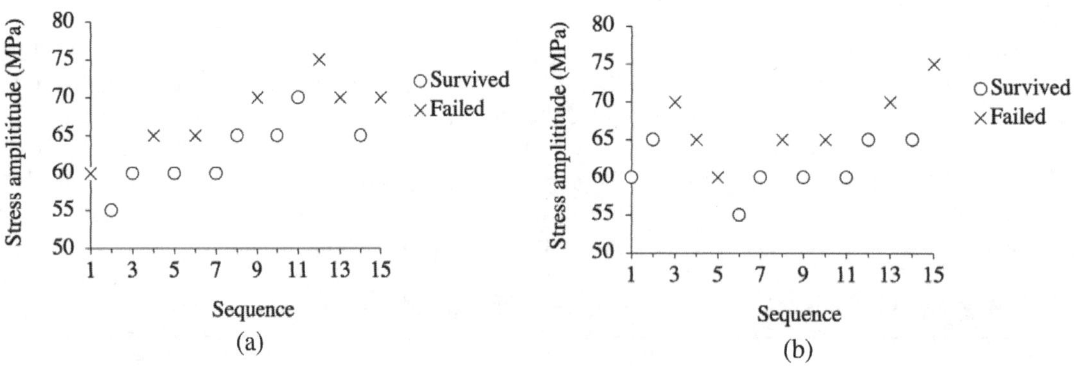

Figure 1. Staircase curves for single-stir semisolid casting (a); and double-stir semisolid casting (b). High-temperature fatigue test at 150 °C, EN AC 42000 alloy.

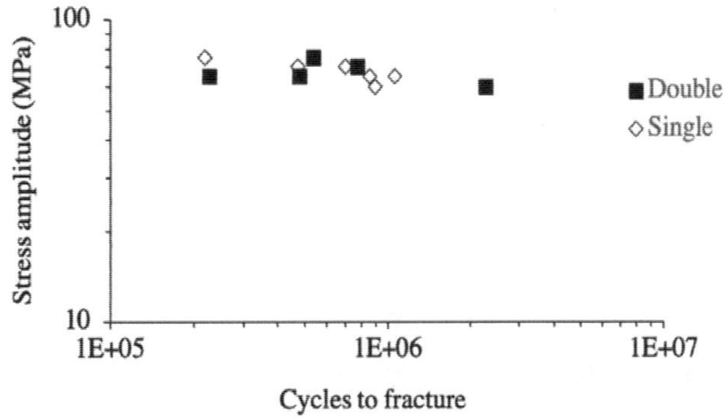

Figure 2. Stress amplitude vs cycles to failure, fatigue test at 150 °C, semisolid casting, EN AC 42000.

3.2. Porosity content

The results revealed how the double-stir influences the reduction of the shrinkage pores during the solidification. The number of pores is comparable between the two processes, equal to 145; whereas the double-stir is characterised by having 42% of the pores of the single-stir (Figure 3). Two representative cross-sections of the different casting conditions are shown in Figure 4. Shrinkage pores distribution evidences a higher population of smaller pores for samples made by the double-stir process (Figure 5). The reduction of the porosity content can only be attributed to the secondary stirring step, with certain reliability due to a large number of cross-sections analysed. The ability of the double stir to reduce the amount of porosity could be referred to as the uniform dispersion of the solid particles from the Enthalpy Exchange Material into the molten alloy during the semisolid slurry preparation. The second-stir can break the solid clusters which did not break during the single-stir, this condition favours the feeding of the liquid within the primary α-Al particles during the filling of the die cavity to compensate for the volume shrinkage. In that sense, when the liquid fraction becomes solid the liquid is still able to flow within the primary α-Al particles, which are more likely separated and distributed, to compensate for the voids created by the solidification.

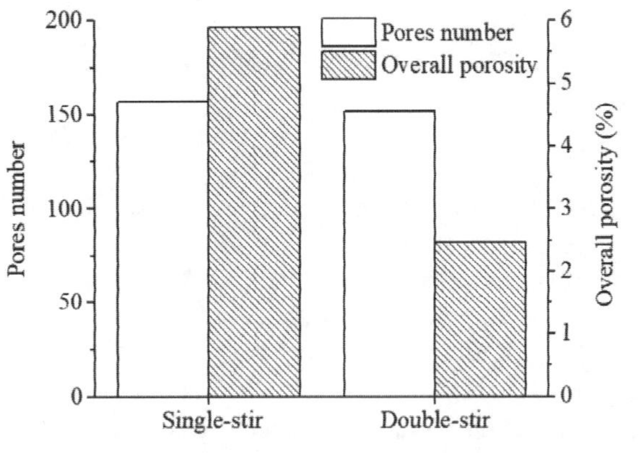

Figure 3. Porosity content along the gauge length. The pores number is comparable between the two processes. The overall porosity shows the beneficial effect of the double-stir.

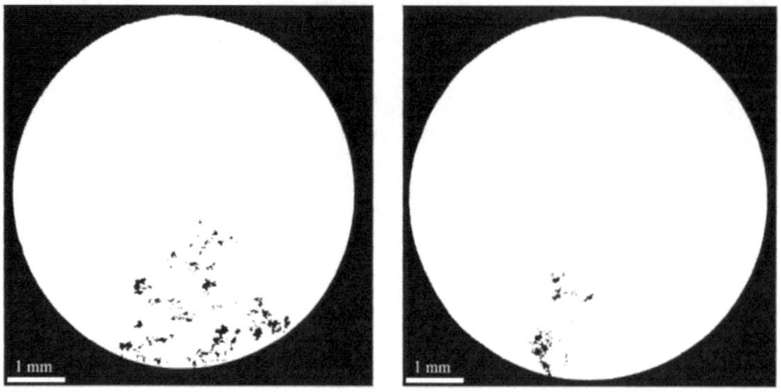

(a) (b)

Figure 4. Light optical microscope micrographs show cross-sections along the gauge length of fatigue samples obtained from a thick-walled component. Shrinkage pores are black within the bright area, single-stir (a) and double-stir (b), semisolid casting, EN AC 42000 alloy.

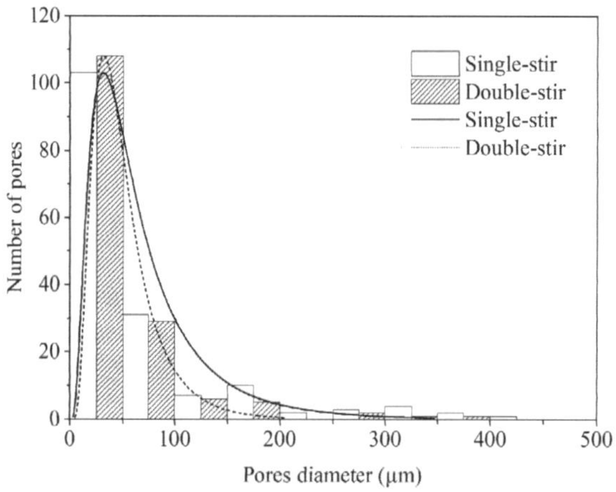

Figure 5. Distribution of shrinkage pores for the single-stir and double-stir processes. Log-normal fitting curves are also drawn. semisolid casting, EN AC 42000 alloy.

4. Conclusions

This experimental study demonstrated the influence of the secondary-stir step during the slurry formation on the development of casting defects of EN AC 42000 alloy, focusing on porosity. Specimens were extracted from a component made by the standard RheoMetal® process and a variation, then subjected to an axial high-temperature fatigue test at 150 °C. Even though defects found in specimens showed some dispersion, the results revealed that:

- Fatigue strength of approximately 70 MPa, is comparable between the two processes.
- The diameter of the defects that initiate the crack is similar, around 1100 μm.
- The stirring process does not influence fatigue life in this study.
- Shrinkage porosity is reduced to 42 % in samples made by the double-stir, compared to the single-stir.
- The shrinkage pores distribution shows the higher population of smaller pores for the double-stir.

The double-stir process can reduce the size of shrinkage pores compared to the single-stir. The main reason could be referred to as the ability of the double-stir to break the solid EEM when it is stirred into the molten alloy. This process reduces the solid clusters, during the feeding of the die cavity the liquid can easily flow within the solid particles to compensate for the volume shrinkage, which is the maximum in the middle region, the latter to become solid. The additional stir step has no evident influence on fatigue life at high temperature since the diameter of the pores associated with crack initiation site are comparable between the two processes; and much larger than the pores observed along the gauge length. There should be an investigation on the samples which survived after 2 x 10^6 cycles. In the present work, the double-stir process reduces porosity to less than half of the single-stir process. Larger pores are reduced, favouring a higher number of smaller pores, less harmful for the failure. The results are promising, and the double-stir step should require more investigation in the future, avoiding the influence of a single large pore which might substantially increase the measured porosity content.

Acknowledgements
This research work is supported by Vinnova contract 2018-02831.

References
[1] Dezecot S and Brochu M 2015 Microstructural characterization and high cycle fatigue behavior of investment cast A357 aluminum alloy *Int. J. Fatigue* **77** 154–9
[2] Rodrigo P D D, Murray M, Mao H, Chandrasekar V, Kisioglu Y, Deshpande A, Brevick J, Mobley C and Esdaile R 2000 Fatigue Properties of Die Cast Magnesium Alloys SAE 2000 World Congress pp 2000-01-1122
[3] Davidson C J, Griffiths J R, Badiali M and Zanada A 2000 Fatigue properties of a semi-solid cast Al-7Si-0.3Mg-T6 alloy **18** 5
[4] Donlon W T, Paige C, Morris C J and Allison J E 1996 The effects of casting defects and microstructure on the mechanical properties of die cast AM50 magnesium and 356 aluminum. *Proc Aluminum and Magnesium for Automotive Applications* (Cleveland, OH: Minerals, Metals and Materials Society/AIME (USA)) pp 17–27
[5] Akhtar S, Arnberg L, Di Sabatino M, Dispinar D and Syvertsen M 2009 A Comparative Study of Porosity and Pore Morphology in a Directionally Solidified A356 Alloy *Int. J. Met.* **3** 39–52
[6] D. Zhao, Johnson Controls Inc. and Lampman S 2000 Hot Tension and Compression Testing *Mechanical testing and evaluation* vol 8 (ASM International)
[7] Fan Z 2002 Semisolid metal processing *Int. Mater. Rev.* **47** 49–85
[8] Chen Z W, Peck S R and Davidson C J 1999 Semisolid casting using a vertical-injection squeeze casting machine: material flow and tensile properties of Al-7Si-0.3Mg alloy *Int. J. Cast Met. Res.* **12** 127–35
[9] Jarfors A E W 2020 A Comparison Between Semisolid Casting Methods for Aluminium Alloys *Metals* **10** 1368
[10] Jarfors A E W, Zheng J C, Chen L and Yang J 2019 Recent Advances in Commercial Application of the Rheometal Process in China and Europe *Solid State Phenom.* **285** 405–10
[11] Jarfors A E W, Zhang Q and Jonsson S 2020 Feeding and Pore Formation in Semisolid Metal Casting *Metals* **10** 1560
[12] Santos J, Jarfors A E W and Dahle A K 2016 Filling, Feeding and Defect Formation of Thick-Walled AlSi7Mg0.3 Semi-Solid Castings *Solid State Phenom.* **256** 222–7
[13] Zhang Q, Jonsson S and Jarfors A E W 2021 On the Role of Stirring on Microstructure and Ductility of Rheocast Al Alloys *JOM* **73** 3848–57
[14] Xing B, Li Y, Ma Y, Chen T J and Hao Y 2013 Evolution of rheocast microstructure of AZ31 alloy in semisolid state *China Foundry* **10** 221–6
[15] Ekaputra I M W, Dewa R T, Haryadi G D and Kim S J 2020 Fatigue Strength Analysis of S34MnV Steel by Accelerated Staircase Test *Open Eng.* **10** 394–400
[16] Anson J P and Gruzleski J E 1999 The Quantitative Discrimination between Shrinkage and Gas Microporosity in Cast Aluminum Alloys Using Spatial Data Analysis *Mater. Charact.* **43** 319–35

Modeling the Fluid Flow for Recent Solidification Experiments in Microgravity Electromagnetic Levitation

G P Bracker[1] and R W Hyers[1]

[1]Mechanical Engineering, University of Massachusetts, Amherst, MA 01003, USA

Correspondence to gbracker@umass.edu

Abstract Microgravity electromagnetic levitation provides a unique processing environment for solidification experiments. The contactless processing allows for greater access to the undercooled region of the melt by isolating the sample from its environment and reducing the available heterogeneous nucleation to consistently achieve deep undercoolings. In the ISS-EML, a variety of solidification studies have investigated the relationship between nucleation temperature and flow conditions in metallic melts. These studies include recent work by Kelton *et al.* in which the coupled flux nucleation model was used to explore solidification in supercooled liquids using glass forming alloys: Vit106, $Cu_{50}Zr_{50}$, and $Ti_{39.5}Zr_{39.5}Ni_{21}$. The relationship between fluid flow and nucleation is also a critical factor in recent experiments investigating dynamic nucleation. Unexplained solidification events occurred during both Spacelab Missions IML-2, MSL-1R, and ISS-EML. These experiments encompassed both steady and transient, accelerating flows. Repeatable, anomalous solidification was observed in both flow conditions. The conditions are consistent with dynamic nucleation, which relies heavily on the flow conditions within the sample for nucleation to occur. During these experiments, flow is not directly observable; however, magnetohydrodynamic models allow the flow to be calculated using the properties of the melt and the experimental conditions. These models provide key insights into the flow conditions and the impacts of flow on nucleation in these experiments.

1. Introduction

Containerless processing provides a fully isolated environment in which a sample may be investigated. This allows for highly reactive melts to be processed at high temperatures. Metallic melts are one such class of materials that are both highly reactive and require very high temperatures to study. Further, containerless processing eliminates the interface between a melt and its container. This significantly reduces the available heterogeneous nucleation sites. With fewer heterogeneous nucleation sites, deeper undercoolings can be achieved and sustained for various investigations. Containerless processing allow for the study of metallic melts to investigate the thermophysical properties, nucleation, growth, and phase selection.

1.1. Electromagnetic Levitation

Electromagnetic levitation (EML) is a containerless processing technique by which a sample is supported by an electromagnetic field. In the electromagnetic levitator on board the International Space Station (ISS-EML), the electromagnetic field is generated using two superposed currents operating at different frequencies. The superposition levitation method produces two independently tuneable force fields: a quadraploe positioner field and a dipole heater field. During an EML experiment, the heater field induces current within the sample which then heats the sample by Joule heating. By tuning the heater field, the temperature of the sample is controlled throughout the experiment.

1.2. Magnetohydrodynamics for Electromagnetic Levitation

The induced currents interact with the applied field to produce levitation forces which position the sample. The same forces also drive flow in molten samples; however, the properties of the flow are very difficult to directly measure. Metallic melts are opaque and featureless, preventing the use of Doppler velocimetry. Additionally, sensors cannot be inserted into the flow as the samples are less than 8mm in diameter and the probe would have a large impact on the flow pattern. Further any probe used would need to remain inert and solid when in contact with the high temperatures at which the highly reactive melt is processed. Any electromagnetic sensor would be dominated by the levitation field. Particle image velocimetry has also been considered; however, solid particles are swept into the stagnation lines of the flow pattern. While solid particles observed in PdSi allowed the laminar to turbulent transition to be quantified as a function of the calculated Reynolds number [1], solid particles are not useful for particle image velocimetry.

Absent techniques to directly measure the flow, computational fluid dynamics (CFD) provides the only insight into the behaviour of the internal flow of the sample. To model the flow conditions during EML experiments, the electromagnetic force field and its distribution must be coupled with CFD to accurately calculate and quantify the flow pattern for a given set of conditions. The magnetic model must incorporate the geometry of the levitation coils, frequencies and amplitudes of the oscillating currents, conductivity of the sample, and geometry of the sample. The geometry of the levitation coils is described by Lohöfer and Piller in [2] for the ISS-EML facility. Using the given parameters, Maxwell's equations are solved using the method of mutual inductances [3] to determine the distribution of induced currents in the sample. These currents are then used with the Biot-Savart law to calculate the magnetic field and the Lorenz forces in the sample.

The electromagnetic forces act on the sample as body forces to drive flow. The model is evaluated using commercially available CFD software to solve the flow pattern by evaluating the distribution of the force field with the properties of the melt. In the model, axial symmetry is applied to evaluate the model in 2D space. The boundaries of the model are such that flow does not cross the free surface of the drop; however, the free surface is free of traction. Additionally, the axis of symmetry is a symmetry boundary at which flow may not cross and derivatives must be zero.

The models presented here have been validated against a rare experimental case in which measurement of the flow by particle tracking was possible. A CoCu sample separated into a two phase liquid in which the smaller, cobalt rich phase was more emissive. These unique conditions allowed particle image velocimetry to be used to measure the flow velocity [4]. Lee *et al.* found that the model agreed with the experimental results to within the accuracy of the experimental measurement of velocity.

2. Coupled Nucleation and Shear Strain Rate

In levitation experiments, the available heterogeneous nucleation sites are reduced which allows undercooling. In the liquid, the formation of nuclei is a stochastic process in which the nuclei may grow or shrink. Those nuclei above a critical size are thermodynamically favoured to increase in size. The coupled-flux model is used to describe the interactions between a nucleating phase whose composition is distinct from that of the surrounding melt. As each nuclei grows, the change in the local composition interacts with the long-range diffusion field which results in the stochastic processes of interfacial kinetics coupling with diffusion. When the interfacial attachment/detachment rates are faster than the diffusion rate in the melt, the kinetics of nucleation are reduced. However, in moving liquids, the fluid flow transports of solute throughout the melt and is expected in to increase the nucleation rate. [5], [6].

The internal flow and resulting shear rates were calculated for the conditions present during the experiments using the Vit106, $Ti_{39.5}Zr_{39.5}Ni_{21}$, and $Cu_{50}Zr_{50}$ samples. The flow rates are calculated using a magnetic model, described in [1], [3], in conjunction with a 2D axisymmetric model of the fluid flow. The model is evaluated using computational fluid dynamics in ANSYS Fluent. A sample of the results of the model are given in Figure 1 for a cycle in which the Vit106 sample solidified at about 240°C

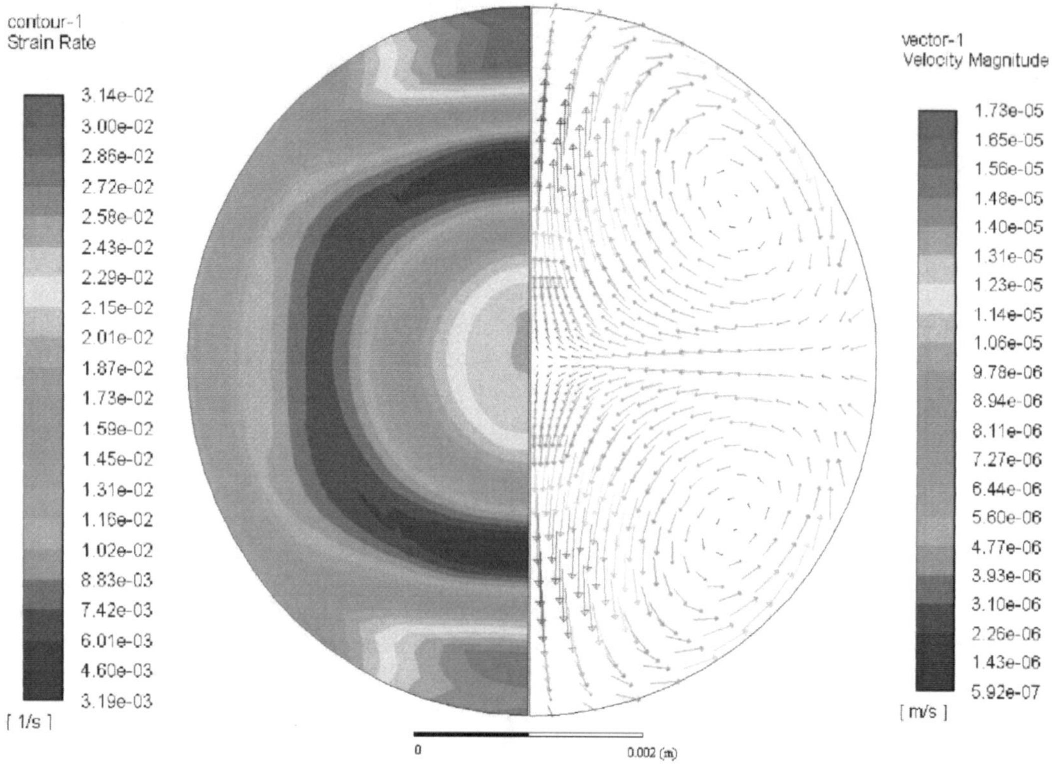

Figure 1: Shear strain rate contours are plotted on the left and the velocity vectors are plotted on the right for Vit106 modeled at 577°C using 15.097 Pa·s from the KKZNT viscosity model [6]. The oscillating heater current is 17.5A and the oscillating positioner current is 98.5A. The maximum shear-strain rate is 0.031 1/s with the majority of the sample below 0.0187 1/s.

below the solidus temperature. At the time of solidification, the maximum velocity in this sample 1.73×10^{-5} m/s and gives a Reynolds number of 4.49×10^{-5}. The low Reynolds number indicates that the flow is below the laminar-turbulent transition. Additionally, the maximum shear-strain rate in the sample is calculated to be 0.031 1/s while the majority of the sample is below 0.0187 1/s. While this indicate small amounts of stirring in the sample, this processing cycles had a cooling rate of approximately 1.4 °/s.

The fluid flow conditions were used to analyze the changes that stirring has on the nucleation rate, *I(T)*, through changes in the prefactor, A^*, which contains kinetic parameters, such as the diffusion coefficient, atomic jump distance, and the number of possible sites for nucleation, used in calculating the homogeneous nucleation with classical nucleation theory. The prefactor was found to correlate with both the maximum velocity and shear-strain rate in the sample in [5]. By comparing the maximum shear-strain rate in the flow to the prefactor, it was calculated that an order of magnitude increase in the fluid flow corresponds to an increase in the prefactor of approximately 5 order s of magnitude in Vit106 [5].

$$I(T) = A^* \exp\left(-W^*/k_B T\right) \quad (1)$$

Similar calculations were done for experiments on $Cu_{50}Zr_{50}$ and $Ti_{39.5}Zr_{39.5}Ni_{21}$ to quantify the fluid flow at the time of nucleation. In $Cu_{50}Zr_{50}$ the prefactor increased by approximately 4 times over a 40% increase in the fluid flow. [5] Meanwhile in the $Ti_{39.5}Zr_{39.5}Ni_{21}$, the prefactor was calculated to decrease by a factor of 33 with a factor of 4 increase in the fluid flow [5].

In the Vit106 alloy, the nucleation rate was significantly enhanced with increasing fluid flow. These results are consistent with the coupled flux nucleation model, in that a melt which experiences compositional changes during solidification will experience an enhanced nucleation rate during

IOP Conf. Series: Materials Science and Engineering 1274 (2023) 012025 doi:10.1088/1757-899X/1274/1/012025

increased stirring. The results from this work suggest that the fluid flow effects are much smaller in alloys like $Cu_{50}Zr_{50}$ in which the compositional changes during solidification are smaller. [5]

3. Dynamic Nucleation and Pressure within the Sample

During electromagnetic levitation experiments, subcritically undercooled melts are expected to be very stable and to remain liquid for extended periods of time. [7]–[9] However, this has not always been observed in practice. During Spacelab Mission MSL-1R in 1997, pure zirconium was free cooled in over 110 free cooling cycles and observe to undercool by 334°±4°C [10]. During other experiments on this same sample, the sample was held between 30°C and 80°C below the melting temperature. However, during 5 of these experiments at subcritical undercoolings, the sample solidified. These solidification events were not consistent with the deep undercoolings achieved for these compositions. These solidification events were attributed to dynamic nucleation induced by the collapse of cavities in the fluid [10], [11] similar to the theories presented by Hunt and Jackson [12] and Childs and Frawley [13]. Work presented in [14] replicated the solidification events reported by Hofmeister [10] and further explored hypotheses explaining the anomalous solidification events.

A variety of hypotheses have been proposed to explain these solidification events including chemical contamination and solid particles contacting the sample. It was proposed that the anomalous solidification events may be explained by chemical contamination of the sample. If this were to be the case, it is expected that the sample would be unable to reach the very deep undercoolings expected for pure zirconium. However, the zirconium sample used in [14] was allowed to free cool and achieved undercoolings of 321.7±5.7°C below the melting temperature before, during and after the experiments in which the sample solidified at subcritical undercoolings. A subset of this hypothesis, is that the sample may have been contaminated with oxygen, providing for interstitial atoms to increase subcritical nuclei beyond the critical size. While any oxygen in the system would remain in solution, oxygen is known to stabilize the crystalline phase of zirconium. This would serve to increase the melting temperature of the sample by a few degrees, and therefore increase the minimum achievable temperature. However, the experiments are processed at shallow undercoolings and unlikely to be affected by such a shift.

Another mechanism that has been suggested to account of the anomalous solidification events is that solid material contacted the sample and provided heterogeneous nucleation sites. There are two forms of solid material in the sample chamber that could possibly accomplish this: "dust" particles and "flakes" of material. While larger "flakes" of material could, in theory, trigger heterogeneous nucleation, the flakes of material are rare in EML and are unlikely to account for the large number of solidification events observed in [14]. Solid "dust" particles have also been suggested to provide heterogeneous nucleation sites for solidification in the sample. However, dust requires the cycle to be processed in gas for the dense clouds of particles to form. During cycles processed in vacuum, evaporated material travels radially, away from the sample, and condenses on contact with a solid surface, usually the wall of the sample chamber.

Dynamic nucleation is a series of stochastic events in the liquid leading to solidification. In a liquid, the free volume is not evenly distributed but rather in voids ranging from atomic-scale to micro-scale, analogous to the subcritical nuclei in an undercooled liquid. The sizes of the voids within the melt are distributed over a wide range, in water for example Shima [15] measured voids over a range from a few cubic microns to over 2000 cubic microns. These voids are carried along the streamlines of the flow through the liquid. When a void interacts with a low pressure region, such as at the center of a circulation loop, the void expands and contracts in response to the changes in the local pressure. The excitation can cause to void to collapse and induce a high pressure wave to travel through the sample. The local high pressure shifts the local melting temperature according to the Claussius-Claperon equation. In a system in which the melting temperature is isothermally raised, the undercooling experiences a corresponding increase. In the high pressure region, the deeper undercooling may provide sufficient excess free energy to drive nucleation. [12], [13]

3.1. Dynamic Nucleation during Rapid Stirring

In 2016 experiments in the ISS-EML observed a pure zirconium sample to solidify when it was expected to remain liquid [16]. Further experiments were done in 2018 on the same zirconium sample in the ISS-EML to confirm the anomalous solidification event. These experiments are described in detail in [14]. The 2018 experiment series, held the molten zirconium sample at subcritial undercoolings from 45° to 290° below the melting temperature. During free cooling, this sample was observed to reach a critical undercooling of 321.7±5.7°C. The isothermal holds were maintained for up to 600s or until the sample solidified. During these experiments, 18 of 19 processing cycles observed the sample to solidify on the hold in less than 600s.

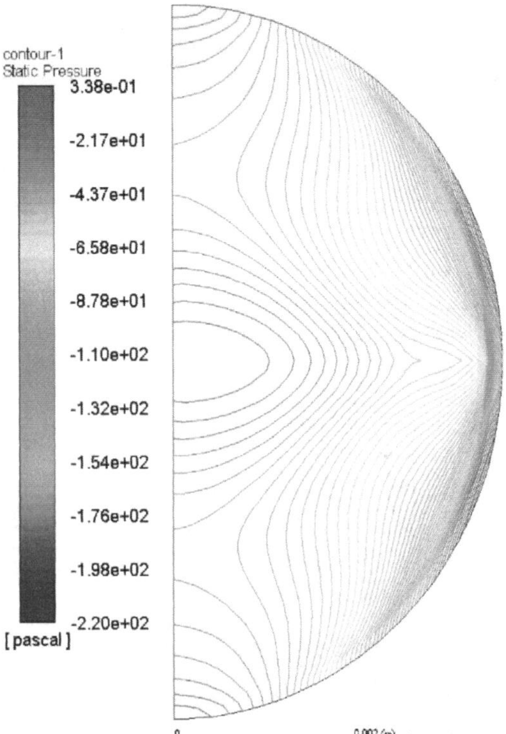

Figure 2: The pressure contours in the drop show that there is a sharp gradient from the deepest hydrostatic tension along the surface of the drop. Shown here, the static pressure contours are plotted for the zirconium drop at 110° below the melting temperature.

The magnetohydrodynamic model was used to calculate the flow during the isothermal holds. During these holds, the maximum velocity in the drops was calculated to be between 0.137 m/s and 0.209 m/s which corresponds to Reynolds numbers from 987 up to 1906. The Reynolds numbers for the flow are above the laminar to turbulent transition, indicating that the flow was turbulent during the isothermal hold. Additionally, pressure within the drop was calculated, for the different isothermal holds. It was found that the maximum hydrostatic tension in the drop ranges from approximately 140 Pa to 263 Pa depending on the flow conditions. The distribution of static pressure in the drop is shown in Figure 2 which shows hydrostatic tension within the drop at 110°C below the melting temperature. The deepest hydrostatic tension, or negative pressure, in the drop is approximately 220 Pa and occurs near the surface of the drop where the flow is at high velocities. Additionally, there is a sharp gradient from the surface of the drop toward the interior. While the deep hydrostatic tension is greatest along the surface between 45° above and below the equator of the sample, the highest pressure regions are toward the center of the drop.

3.2. Other Conditions in which Dynamic Nucleation has been observed

Another instance of a sample solidifying occurred during IML-2 experiments on $Zr_{64}Ni_{36}$. In these experiments, oscillations were induced in the sample at shallow undercoolings. At the onset of the excitation pulse, the sample solidified. [11] Recent work has explored the effects of the excitation pulse on the flow conditions of the drop.

This work aims to further understand the different ways that flow can affect nucleation and solidification in EML samples. To better understand the mechanism, the sample was cooled to a targeted subcritical undercooling and an excitation pulse, similar to those used to excite surface oscillations in thermophysical property measurements, was applied to the sample. The excitation pulse is a brief period of elevated EML forces driving rapidly accelerating flow in the sample. To model the flow conditions during the excitation pulse, a transient model was developed.

4. Conclusions

During EML experiments, the environment reduces the available heterogeneous nucleation sites allowing molten samples to be stable deep into the undercooled region. This provides an environment in which the relationship between fluid flow and solidification can be better explored. Most recently, work has been done to explore the coupled-flux nucleation model using glass forming alloys, such as Vit106, $Cu_{50}Zr_{50}$, and $Ti_{39.5}Zr_{39.5}Ni_{21}$. Other experiments are exploring anomalous nucleation events during past processing. These past events have been attributed to dynamic nucleation. While other hypotheses have been proposed to explain these events, they are not consistent with the observations of the more recent experiments. In the experiments presented here, understanding and predicting the conditions and mechanisms for solidification relies on an understanding of the internal flow within the sample, which, in turn, requires the support of magnetohydrodynamic models for this key insight.

5. References

[1] R. W. Hyers, G. Trapaga, and B. Abedian, "Laminar-turbulent transition in an electromagnetically levitated droplet," *Metallurgical and Materials Transactions B*, vol. 34, no. 1, pp. 29–36, Feb. 2003, doi: 10.1007/s11663-003-0052-7.

[2] G. Lohoefer and J. Piller, "The new ISS Electromagnetic Levitation Facility - 'MSL-EML,'" in *40th AIAA Aerospace Sciences Meeting & Exhibit*, 0 vols., American Institute of Aeronautics and Astronautics, 2002. doi: 10.2514/6.2002-764.

[3] J. C. Zong, J. Szekely, and E. Schwartz, "An Improved Computational Technique for Calculating Electromagnetic Forces and Power Absorptions Generated in Spherical and Deformed Body in Levitation Melting Devices." *IEEE Transactions on Magnetics*, vol. 28, no. 3, pp. 1833-1842, May 1992, dio: 10.1109/20.141293.

[4] J. Lee, D. M. Matson, S. Binder, M. Kolbe, D. Herlach, and R. W. Hyers, "Magnetohydrodynamic Modeling and Experimental Validation of Convection Inside Electromagnetically Levitated Co-Cu Droplets," *Metallurgical and Materials Transactions B*, vol. 45, no. 3, pp. 1018–1023, Jun. 2014, doi: 10.1007/s11663-013-9995-5.

[5] A. K. Gangopadhyay *et al.*, "Demonstration of the effect of stirring on nucleation from experiments on the International Space Station using the ISS-EML facility," *npj Microgravity*, vol. 7, no. 1, pp. 1–6, Aug. 2021, doi: 10.1038/s41526-021-00161-9.

[6] K. Kelton, "Time-dependent nucleation in partitioning transformations," *Acta Materialia*, vol. 48, pp. 1967–1980, May 2000, doi: 10.1016/S1359-6454(99)00455-3.

[7] D. Holland-Moritz, T. Schenk, P. Convert, T. Hansen, and D. M. Herlach, "Electromagnetic levitation apparatus for diffraction investigations on the short-range order of undercooled metallic melts," *Meas. Sci. Technol.*, vol. 16, no. 2, pp. 372–380, Jan. 2005, doi: 10.1088/0957-0233/16/2/007.

[8] G. W. Greenwood *et al.*, "Magnetic ordering and crystal nucleation in undercooled Co–based melts," *Philosophical Transactions of the Royal Society of London. Series A: Mathematical, Physical and Engineering Sciences*, vol. 361, no. 1804, pp. 497–515, Mar. 2003, doi: 10.1098/rsta.2002.1140.

[9] C. Bührer, U. Holzwarth, K. Maier, D. Platzek, and J. Reske, "Positron annihilation in solid, liquid and undercooled melts of Co80Pd20," 1996, doi: 10.1007/BF01567649.

[10] W. H. Hofmeister, R. J. Bayuzick, R. Hyers, and G. Trapaga, "Cavitation-induced nucleation of zirconium in low earth orbit," *Applied Physics Letters*, vol. 74, no. 18, pp. 2711–2713, May 1999, doi: 10.1063/1.123945.

[11] Team Tempus, "Containerless processing in space: Recent results," in *Materials and Fluids Under low Gravity*, Berlin, Heidelberg, 1996, pp. 233–252. doi: 10.1007/BFb0102523.

[12] J. D. Hunt and K. A. Jackson, "Nucleation of Solid in an Undercooled Liquid by Cavitation," *Journal of Applied Physics*, vol. 37, no. 1, pp. 254–257, Jan. 1966, doi: 10.1063/1.1707821.

[13] W. J. Childs and J. J. Frawley, "Dynamic nucleation of supercooled metals.," Feb. 1968, Accessed: Jan. 04, 2022. [Online]. Available: https://ntrs.nasa.gov/citations/19680039839

[14] G. Bracker, S. Schneider, R. Wunderlich, H.-J. Fecht, J. Zhao, and R. Hyers, "Confirmation of Anomalous Nucleation in Zirconium," *JOM*, vol. 72, Jul. 2020, doi: 10.1007/s11837-020-04257-7.

[15] A. Shima and I. Sakai, "On the equation for the size distribution of bubble nuclei in liquids (Report 2)," *Reports of the Institute of High Speed Mechanics, Tohoku University*, vol. 54, pp. 51–59, 1987.

[16] R. Wunderlich, presented at the EML International Working Group Meeting, Dec. 2016.

Acknowledgments
Support for this projected was provided through NASA grants NNX16AB54G and 80NSSC21K0103. The experiments discussed were processed in the ISS-EML facility, formerly MSL-EML.

Potential eutectic accumulation in single crystal turbine blade due to geometry effect: a numerical study

H J Zhang[1], X S Liu[2], D X Ma[2], A Ludwig[1], A Kharicha[1], M H Wu[1,*]

[1] Chair for Simulation and Modelling of Metallurgy Processes, Department of Metallurgy, University of Leoben, A-8700 Leoben, Austria

[2] Central South University, Powder Metallurgy Research Institute, Changsha, 410083, China

* Corresponding author: menghuai.wu@unileoben.ac.at

Abstract. Inhomogeneous distribution of the eutectic phase in single crystal (SC) turbine blades is detrimental to its mechanical properties. In a recent publication, it was reported that an accumulation of eutectics was observed on the upper surface of solidification units (e.g. shroud of a turbine blade), whereas fewer eutectics were detected at the bottom. This kind of casting defect brings a huge challenge to the production of turbine blades since the accumulated eutectics cannot be dissolved completely by heat treatment. A sample with three pairs of platforms/shrouds was designed to study the geometry effect on eutectic accumulation. In the current study, the simulation was performed by coupling ProCAST with our previously developed multiphase volume-average solidification model. In the first step, the thermal field including radiation in the furnace and heat conductivity in the casing and mould was calculated via ProCAST. Then, the obtained temperature profiles were set as thermal boundary conditions of the casting to predict the eutectic accumulation and formation of freckles. Based on the simulation results, convective plumes arising from the thermosolutal buoyancy transport solute-enriched liquid upward, which causes the solute pile-up and the final eutectic accumulation at the top surface in each platform. Freckles accompanied with eutectic accumulation was also observed at the corner of the sample. The calculated inhomogeneous distribution of the eutectics agrees well with experimental observations. Knowledge about the geometrical effect on eutectic accumulation is extended.

1. Introduction

The single-crystal (SC) turbine blades fabricated from the Ni-based superalloys have been widely used for aero-engines and industrial gas turbines, because of the high strength, microstructural stability, and corrosion resistance of Ni-based superalloys at elevated temperatures [1, 2]. Typically, the SC blades are directionally solidified in the Bridgman furnace, through which the temperature gradient and withdrawal rate can be well controlled. The solidification of the SC blades is starting with the development of the columnar dendrite matrix phase, i.e. γ phase, and the residual liquid is mainly solidified as the γ/γ' eutectic phase and a minor of intermetallic precipitates [3]. A study by Brewster et al. [4] found that a thin layer of γ/γ' eutectics accumulated on the external surface of the casting along the solidification length. This was explained by the contraction of the solid dendritic network, which squeezed the residual solute-enriched liquid into the sample surface area. The addition of refractory element Re was found to enhance the formation of surface γ/γ' eutectics [4, 5]. In a recent publication [6], it was reported that an accumulation of eutectics was observed on the upper surface of solidification units (e.g. shroud/platform of a turbine blade), whereas fewer eutectics were detected at the bottom. This kind of casting defect poses great challenge to the production of turbine blades since the accumulated eutectics cannot be dissolved completely by heat treatment, and an increased solutioning temperature may cause incipient melting [7]. The aggressive mechanical operation prior to heat treatment will pose additional concerns, e.g. inducing recrystallisation and altering the product dimensions [5].

In this paper, the previously developed multiphase volume-average solidification model [8] is employed to simulate the solidification of a sample with three pairs of platforms. To investigate the geometrical effect on the eutectic accumulation, the liquid flow as raised from the thermo-solutal buoyancy and the as-solidified structure was calculated. The calculated distribution of the γ/γ' eutectics is consistent with the experimental observations.

2. Model description and calculation method

The directional solidification (DS) of the SC casting is carried out in a vacuum Bridgman furnace. As shown in Figure 1(a), the configuration of the furnace includes an inducing hot zone on the top and a cold zone with water cooling at the bottom connected by a graphite baffle in the middle, Al_2O_3 shell mould, superalloy casting, chill base, and grain selector. The shell mould is mounted on the chill base, which drives the mould to move during the casting process. To study the geometrical effect, i.e. the abrupt transition of the cross-section, as shown in Figure 1(b), a specifical geometry with three pairs of platforms (H1–H3, S1–S3), which protruded 15.0 mm on each side of the casting, was designed to simplify the complex turbine blades. One narrow side of the casting is facing the heating wall (heating side), while the other side is back to the heating wall (shadow side). Only 1/12 of the furnace, which is occupied by one casting is calculated. Before the casting process, the mould and the superalloy were heated to 1520 °C, and then the metal melt was poured into the shell mould. The DS of the SC casting was triggered by withdrawing the shell mould from the hot zone to the cold zone at a velocity of 3.0 mm/s. The third generation of the superalloy was used for this study. Its main compositions are presented in Table I.

Table I. The main composition of the superalloy.

Elements	Ni	Ta	Cr	Al	Co	W	Re	Ti	Mo
Content (wt. %)	base	8.07	3.39	5.69	5.97	6.52	4.89	0.15	0.41

To simplify the simulation, the current multi-element superalloy is treated as eight Ni-X binary alloys with constant liquidus slope (m) and solute partition coefficient (k). The equivalent solute concentration was calculated via $\bar{c}_0 = \sum_{i=1}^{N} c_{0,i}$ [9], in which N denoted the number of solute elements. By means of the same method, the equivalent liquidus slope $\bar{m}$, the equivalent solute partition coefficient $\bar{k}$, and the equivalent solute expansion coefficient $\bar{\beta}_c$ can be evaluated via $\bar{c}_0 \cdot \bar{m} = \sum_{i=1}^{N} m_i c_{0,i}$, $\bar{c}_0 \cdot \bar{m} / \bar{k} = \sum_{i=1}^{N} m_i c_{0,i} / k_i$, and $\bar{c}_0 \cdot (1-\bar{k}) \bar{\beta}_c = \sum_{i=1}^{N} c_{0,i} (1-k_i) / \beta_{ci}$, respectively. The material properties and processing parameters are listed in Table II.

The simulations were performed by coupling the commercial software ProCAST and ANSYS Fluent. As shown in Figure 1(c), in the first step, the thermal field was calculated via ProCAST with the consideration of the radiation in the furnace, the heat conductivity in the casing and the shell mould, and the release of the latent heat during the solidification. The geometry used for the thermal calculation in ProCAST is shown in Figure 1(a). It should be mentioned that the effect of the liquid flow on the thermal field was not considered due to its low intensity. Then, the obtained temperature profiles on the casting surface were extracted and set as thermal boundary conditions of the casting for calculation of flow-solidification interaction. Only the casting domain, Figure 1(b), was solved using our previously developed multiphase volume-average solidification model [8] to predict the eutectic accumulation and formation of freckles. Since the nature of the single crystal growth, only the liquid melt and the solid columnar dendrites were considered here with their volume fractions equal to unit, i.e. $f_l + f_c = 1$. The columnar dendrites can only start to grow from the bottom of the domain, which is connected to the top of the grain selector. The no-slip boundary condition was applied to all walls, and the alloy was assumed to be incompressible with constant density and viscosity.

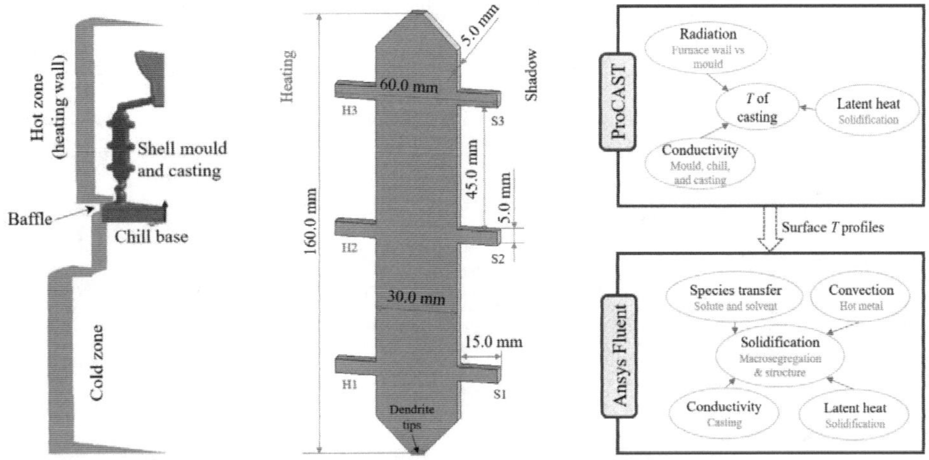

(a) (b) (c)

Figure 1. (a) Layout of the Bridgman furnace and the assemble of the casting; (b) Dimension of the casting domain; (c) Sketch of the calculation strategy.

Table II. Material properties and processing parameters [9–12].

Properties/parameters	**Symbol**	**Units**	**Values**
Thermophysical			
Specific heat of the alloy	$c_{p,l}$ $c_{p,s}$	$J \cdot kg^{-1} \cdot K^{-1}$	500.0
Latent heat	Δh_f	$J \cdot kg^{-1}$	2.4×10^5
Liquid diffusion coefficient	D_l	$m^2 \cdot s^{-1}$	3.6×10^{-9}
Liquid thermal conductivity	k_l	$W \cdot m^{-1} \cdot K^{-1}$	33.5
Solid thermal conductivity	k_s	$W \cdot m^{-1} \cdot K^{-1}$	24.6
Thermal expansion coefficient	β_T	K^{-1}	-1.16×10^{-4}
Solutal expansion coefficient	$\bar{\beta}_c$	$wt.\%^{-1}$	−0.228
Density	ρ	$kg \cdot m^{-3}$	7646.0
Viscosity	μ_l	$kg \cdot m^{-1} \cdot s^{-1}$	4.9×10^{-3}
Thermodynamic			
Eutectic temperature	T_{eut}	K	1627.0
Liquidus slope	$\bar{m}$	$K\ (wt.\ \%)^{-1}$	−1.145
Equilibrium partition coefficient	$\bar{k}$	-	0.57
Primary dendritic arm spacing	λ_1	μm	500.0
Melting point of the solvent	T_f	K	1728.0
Others			
Initial concentration	$\bar{c}_0$	wt.%	35.09
Initial temperature	T_0	K	1773.0
Withdrawal velocity	v	mm/min	3.0

3. Simulation results and discussions

3.1, Comparison of the temperature field

The calculated temperature field in the Bridgman furnace is shown in Figure 2(a). A positive temperature gradient is formed along the casting direction. Since the heating and cooling by the furnace wall above and below the baffle, respectively, the temperature gradient of the shell mould on the left side (heating side) is higher, resulting in a narrower mushy zone on this side, Figure 2(b). The temperature profiles along two vertical lines, as marked in Figure 2(b) are plotted in Figure 2 (c). Above the baffle, the temperature on the heating side of the casting can be 13 K higher than that on the right side, while below the baffle, it is about 17K lower on the heating side. From Figure 2(c), it can also be seen that the calculated temperature profiles by these two different solvers are identical, which ensure the credibility of the thermal boundary conditions that are used for the flow calculation.

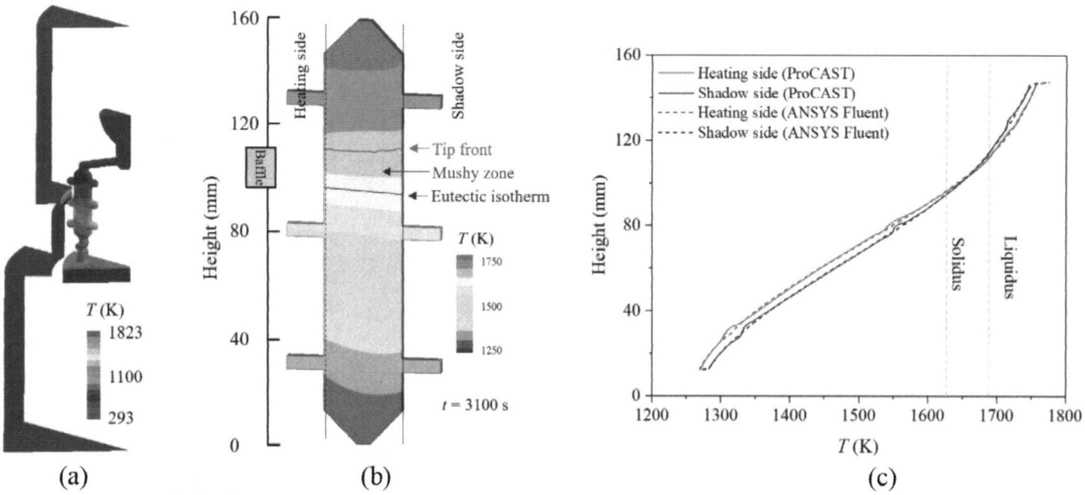

Figure 2. Calculated temperature results: (a) *T* filed in the Bridgman furnace; (b) *T* filed on the casting surface with the identification of the solidification front and the eutectic isotherm; (c) comparison of *T* profiles along the two vertical lines (as marked in Figure 2(b)) calculated by ProCAST and ANSYS Fluent.

3.2, Flow pattern and solidification sequence

During the solidification of the current SC casting, the liquid flow is driven by the thermal-solutal buoyancy. In the bulk liquid of the casting, due to the nearly homogeneous solute concentration, the liquid convection is dominated by the thermal buoyancy. As shown in Figure 3(a), a clockwise flow up to the strength of 3 mm/s is generated in the top part of the casting domain. Some isolated vortices, as indicated by the light blue vectors in Figure 3(a), can be found in the platforms. The flow in these platforms is more than one order of magnitude lower (~ 0.1 mm/s) than the flow in the main casting. In the mushy zone, the solute rejection due to the solidification causes the density inversion, i.e. the solutal buoyancy exceeds the thermal effect. The lighter solute-enriched liquid enclosed in those convective plumes floats upward, as indicated by the green vector on those plumes in Figure 3(a). Due to the blowing effect of the main flow (black vectors) in the bulk liquid, the plumes tilt to the left.

The solidification sequence when the solidification front passes the first pair of platforms (S1 and H1) is presented in Figs. 3(b)–(c). Before the solidification front reaches S1, although the melt in S1 has been sufficiently undercooled, it does not solidify until the solidification front advances there. Once the dendrites develop to S1, the side branches extend quickly to the right side and the alloy solidifies rapidly, which causes the convex solidification front in S1. The small vortex in S1 persists its previous flow pattern that is presented in Figure 3(a). With the solidification of the casting, as shown in the zoom-in view of Figure 3(c), local density inversion due to the solute pile-up in the interdendritic region drives the upward flow (~ 0.6mm/s) in S1. Some short plumes, as marked by the green iso-surfaces, can also be seen above the solidification front. Although S1 is connected to the casting, since the fact that the narrow gate at the connection has been partially solidified, the liquid is difficult to flow out of S1. The local circulation in S1 transports more and more solute-enriched liquid to the top surface, which causes the final eutectic accumulation there. The solute diffusion due to the solute concentration gradient is also one possible mechanism for the eutectic accumulation on the top, but plays a small role compared to the convective effect. The solidification in the heating side, H1, is a bit different from S1. As shown in Figure 3(d), a flat solidification front was observed in H1. This can be understood with the assist of Figure 2. At the solidification front, the liquid on the heating side is warmer than the liquid on the shadow side. There is no intensive undercooling like S1. When the dendrites develop to H1, the side branches can extend gradually to the left side. The flow and solidification in a later moment (Figure 3(e)) are similar to that of Figure 3(c). The local circulation leads to the enrichment of the residual solute-enriched liquid at the top surface of H1, and hence the final eutectic accumulation. Freckles are also observed on this casting. As displayed in Figure 3(f), all freckles are formed at the four vertical corners of the casting, and the shadow side is prone to form freckles. Accumulation of eutectic was also found in these freckle areas, which can be seen through the zoom-in view in Figure 3(f). Formation mechanism and quantitative comparison with experiments of the freckles will be made in our subsequent journal paper.

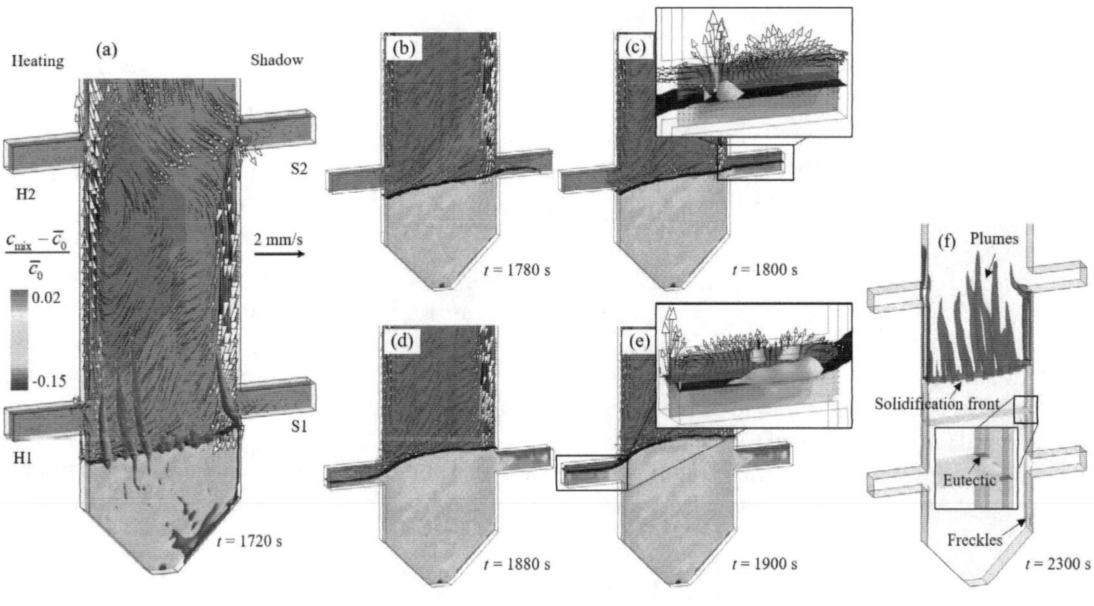

Figure 3. The calculated flow pattern and solidification sequence of the sample. (a) Contour of segregation index $(c_{mix}-\bar{c}_0)/\bar{c}_0$ in the central section overlaid by the black vectors of liquid velocity $\vec{u}_l$; the solidification front is indicated by a horizontal iso-surface, and the convective plumes are denoted by the vertical columns overlaid by the green vectors of $\vec{u}_l$. (b)-(e) Contour of $(c_{mix}-\bar{c}_0)/\bar{c}_0$ in the central section overlaid by the black vectors of $\vec{u}_l$; the horizontal iso-surface indicates the solidification front. (f) Distribution of plumes over the solidification front and freckles below the solidification front.

3.3, Comparison with experiments

The calculated final distribution of the eutectics (f_{eut}) on the central section of the casting is presented in Figure 4(a). A quantitative analysis to show the evolution of f_{eut} from the bottom to the top of these six platforms is made in Figure 4 (b). Specifically, a thin eutectic accumulated layer (~1.5 mm) can be observed near the top surface of each platform, while a relatively low f_{eut} at the bottom of them. Based on the zoom-in view of Figure 4(a), the current simulation results agree well with the experimental observation [6]. Additionally, more eutectics form in the platforms at a higher position. The same tendency was also observed in the main casting. From point A, B to C, f_{eut} is equal 9.5%, 11.6%, 20.1%, respectively. A eutectic concentrated region can also be seen at the top of the casting. As shown in Figure 3(a), the convective plumes transport solute-enriched liquid that is rejected from the interdendritic region into the up-bulk liquid and raise the solute concentration of the unsolidified liquid, which is responsible for increasing f_{eut} along the casting direction.

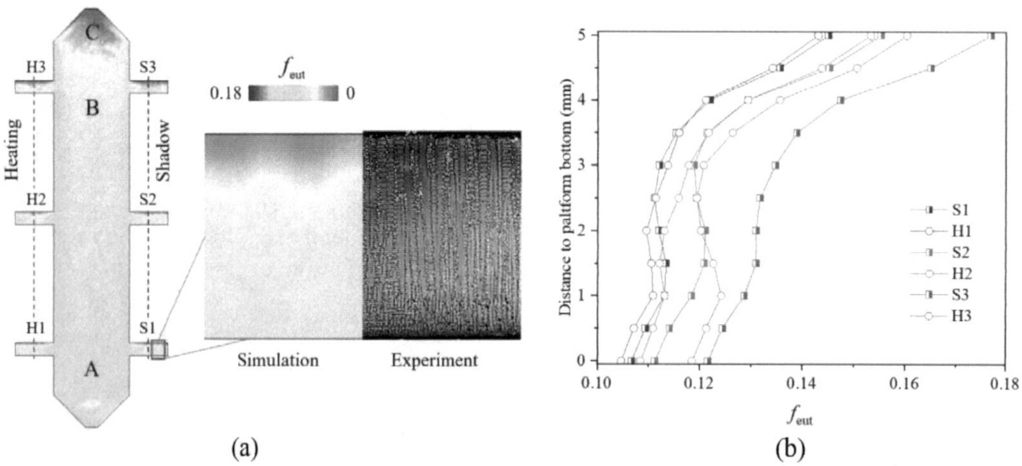

Figure 4. Validation of the calculated f_{eut}. (a) Distribution of f_{eut} on the central section of the casting, and a zoom-in view to show the local solidification structure. (b) Evolution of f_{eut} from the bottom to the top of these platforms along the dash lines in **Figure 4(a)**.

4. Conclusions

The Potential eutectic accumulation in SC turbine blade due to geometry effect was numerically studied based on a special casting with three pairs of platforms. The simulation was performed by coupling ProCAST with a volume-average multiphase solidification model. Based on the current simulation results, the convective plumes arising from the thermosolutal buoyancy transport the solute-enriched liquid upward. During the solidification, the enriched liquid can hardly flow out of the partially solidified platform through the narrow gate at the connection. The solute-enriched liquid finally solidified as a thin layer (~1.5 mm) of eutectic phase at the top surface. The convective plumes also cause the increased f_{eut} along the casting direction in the main casting, and the eutectic concentrated area at the top of the casting. Freckles accompanied with eutectic accumulation were observed at the four corners of the sample.

5. Acknowledgement

This work was financially supported by the Austrian Science Fund in the framework of the FWF-NKFIN joint project (FWF, I4278-N36).

6. References

[1] D. Ma and A. Bührig-Polaczek: *Metall. Mater. Trans. A*, 2014, vol. 45, pp. 1435–44.
[2] G.P. Sabol and R. Stickler: *Phys. status solidi*, 1969, vol. 35, pp. 11–52.
[3] K. Cao, W. Yang, J. Zhang, C. Liu, P. Qu, H. Su, J. Zhang, and L. Liu: *J. Mater. Res. Technol.*, 2021, vol. 11, pp. 474–86.
[4] L. Cao, L. Yao, Y. Zhou, T. Jin, and X. Sun: *J. Mater. Sci. Technol.*, 2017, vol. 33, pp. 347–51.
[5] L. Cao, Y. Zhou, T. Jin, and X. Sun: *J. Mater. Sci. Technol.*, 2017, vol. 33, pp. 1308–13.
[6] D. Ma, Y. Zhao, W. Xu, L. Pi, and Z. Li: *Acta Met. Sin*, 2021, vol. 57, pp. 1539–48.
[7] G. Brewster, H.B. Dong, N.R. Green, and N. D'Souza: *Metall. Mater. Trans. B*, 2008, vol. 39, pp. 87–93.
[8] M. Wu, A. Ludwig, and A. Kharicha: *Metals (Basel).*, 2019, vol. 9, p. 229.
[9] N. Ren, J. Li, C. Panwisawas, M. Xia, H. Dong, and J. Li: *Acta Mater.*, 2021, vol. 206, p. 116620.
[10] 11 S.D. Felicelli, D.R. Poirier, and J.C. Heinrich: Metall. *Mater. Trans. B*, 1998, vol. 29, pp. 847–55.
[11] P.N. Quested, R.F. Brooks, L. Chapman, R. Morrell, Y. Youssef, and K.C. Mills: *Mater. Sci. Technol.*, 2009, vol. 25, pp. 154–62.
[12] D. Szeliga: Metall. *Mater. Trans. B*, 2018, vol. 49, pp. 2550–70.

IOP Conf. Series: Materials Science and Engineering 1274 (2023) 012027 doi:10.1088/1757-899X/1274/1/012027

Forecasting the steelmaking and casting of a medium manganese steel

B. Santillana*, **R. Rana and A. Burghardt**

Tata Steel, 1970 CA IJmuiden, The Netherlands
* corresponding author: begona.santillana@tatasteeleurope.com

Abstract. The steelmaking and casting of medium Mn steels brings some challenges in the production route because of their high alloying content. In this paper, the evaluation of the casting processes for these steel chemistries is assessed with a set of models and laboratory-scale tests, in order to find possible casting conditions. Results evaluation from this study pointed out to the regular conventional slab casting and hot rolling route, so that it can be produced through a conventional production line.

1. Introduction

Medium manganese transformation induced plasticity (TRIP) steels, or simply "medium Mn steels" containing Mn typically in the range between 3 and 10 wt%, are being investigated as a 3rd generation advanced high strength steel worldwide [1, 2], including in Tata Steel. It has been reported that these steels can also be subjected to hot press forming fetching a host of advantages in the product characteristics such as ultrahigh strength (tensile strength > 1000 MPa), high elongation (>15%) and bendability (VDA bending angle > 90° at 1.0 mm gauge) [3, 4].

However, some challenges in the production route of such high alloyed steels are expected. The steelmaking route for alloying with large quantities of Mn poses challenges that have to be met with process developments. In this paper the evaluation of the casting processes for a medium Mn steel nominally containing 0.13C-7.2Mn-1.5Al-0.25Si (wt.%) is assessed with a set of models and laboratory-scale tests. All the process studies focused on the regular conventional slab casting and hot rolling route.

2. Steelmaking

As the considered medium Mn steel contains significantly high amounts of Mn (7.2 wt.%), it raises some concerns for the very first step in the manufacturing process, i.e. steel melting. The steelmaking consist of two major steps – converter processing followed by ladle alloying and refining. The foreseen issues are whether:

1) the high amount of Mn can be added to the melt without compromising the target carbon content
2) the high amount of Mn can be dissolved into the melt and it will have any adverse effects on the melt temperature and homogeneity.

Manganese is added to the melt in one of three following forms, higher carbon Fe-Mn ferroalloy, decarburised Fe-Mn ferroalloy and pure manganese flakes. For standard low Mn-alloyed steel, relatively cheap high carbon Fe-Mn ferroalloy can be used since due to the lower target Mn contents, the ferro-manganese addition does not increase the C content of the alloy much. But in a 7.2 wt.% Mn steel, the high carbon ferro-manganese (~70 wt.% Mn, ~7 wt.% C) cannot be used since it will cause an overshoot of the carbon content that cannot be further corrected. On the other hand, using all manganese flakes will be too costly. Therefore, calculations were made to balance the different options in order to achieve a maximum C content of ~0.14-0.15 wt.% in the steel. This level of C is perfectly acceptable for the designed steel, and therefore poses no problem.

Once the relatively high amount of manganese is added to the steel melt, it needs to be homogeneously mixed. Furthermore, large amount of ferro-manganese or Mn-flakes may cool down the melt. Therefore, an alloying strategy has been proposed for the converter tapping and ladle treatment.

It is worth mentioning that addition of 7.2 wt.% Mn will also lower the liquidus of the steel from 1536 °C for pure iron to about 1493 °C for this alloy. To be aware of such a lower liquidus temperature it helps on further processing of this grade, in order to adjust the superheat and secondary cooling strategy to minimise macrosegregation.

3. Slab Casting

From the compositional point of view, the steel grade does not show any unexpected elements in the composition, except that the Mn concentration is somewhat higher than other standard steels that have been cast until now as a regular carbon steel grade.

3.1. Phase evolution during solidification

The thermodynamic calculations (phase fractions as a function of temperature during solidification) performed at equilibrium using ThermoCalc (database TCFE9) are shown in Figure 1.

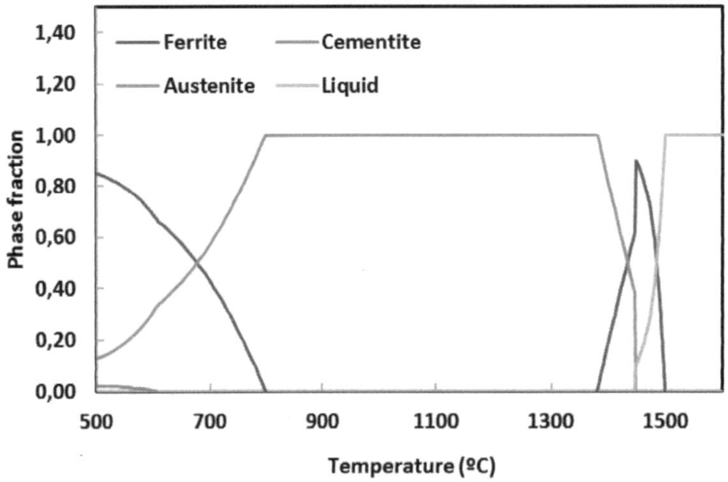

Figure 1. Phase fractions versus temperature for the solidification of the medium Mn grade as calculated by ThermoCalc.

Since it is a peritectic grade, thin-slab casting via direct strip plant (DSP) is not possible due to the high casting speed. Therefore, all the casting studies focused on the regular thick-slab (~225 mm thickness) casting which was also considered for downstream processing of the steel as well. It is important to note from Figure 1 that this grade has no microalloying elements or elements that may have a precipitate formation during cooling or casting. The phases which would appear are regular iron lattice structures of ferrite, austenite and cementite. Thus, considering that there are no other phases (i.e. precipitates) appearing during solidification and cooling, there do not seem to be any complications during casting. If precipitation would appear, then extra considerations should have to be taken at the secondary cooling practise to minimise the susceptibility for transverse cracking [5].

However, it seems that the casting of this steel will need a specialised mould powder due to the relatively high Al (1.5 wt.%) and Mn (7.2 wt.%) contents of this steel because both elements most probably will react with the mould slag due to their high affinity to oxygen, raising issues with friction, sticking and uneven heat extraction.

IOP Conf. Series: Materials Science and Engineering 1274 (2023) 012027 doi:10.1088/1757-899X/1274/1/012027

Furthermore, since the level of Mn is high and it has to be considered in combination with sulphur levels to avoid MnS formation as it may lead to internal cracking. However, because there is no requirement of S addition to the medium Mn steel from a metallurgical point of view, S level of the steel can be kept low (e.g. below 20 ppm) which should not cause any problems. The high level of Mn may cause some segregation, particularly at the centre of the slab. The metallurgical consequences of this need to be evaluated in future studies.

3.2. Steel strength evaluation

It is to be noted that the above discussion is based on only thermodynamic calculations which informs nothing about the strength of the steel. As it is a steel with high austenite fractions during solidification, it may lead to very strong forces in the caster. Therefore, dilatometer tests were performed under conditions that are normally encountered during continuous casting.

The stress-strain data of the steel at high temperatures was measured from compression tests in a dilatometer on specimens of 5 mm diameter and 10 mm height extracted from laboratory cast rough-rolled material. These tests were performed in the temperature range of 800-1100°C using strain rates in the range of 0.01-0.1 s^{-1}. The true stress-true strain plots of the samples are given in Figure 2.

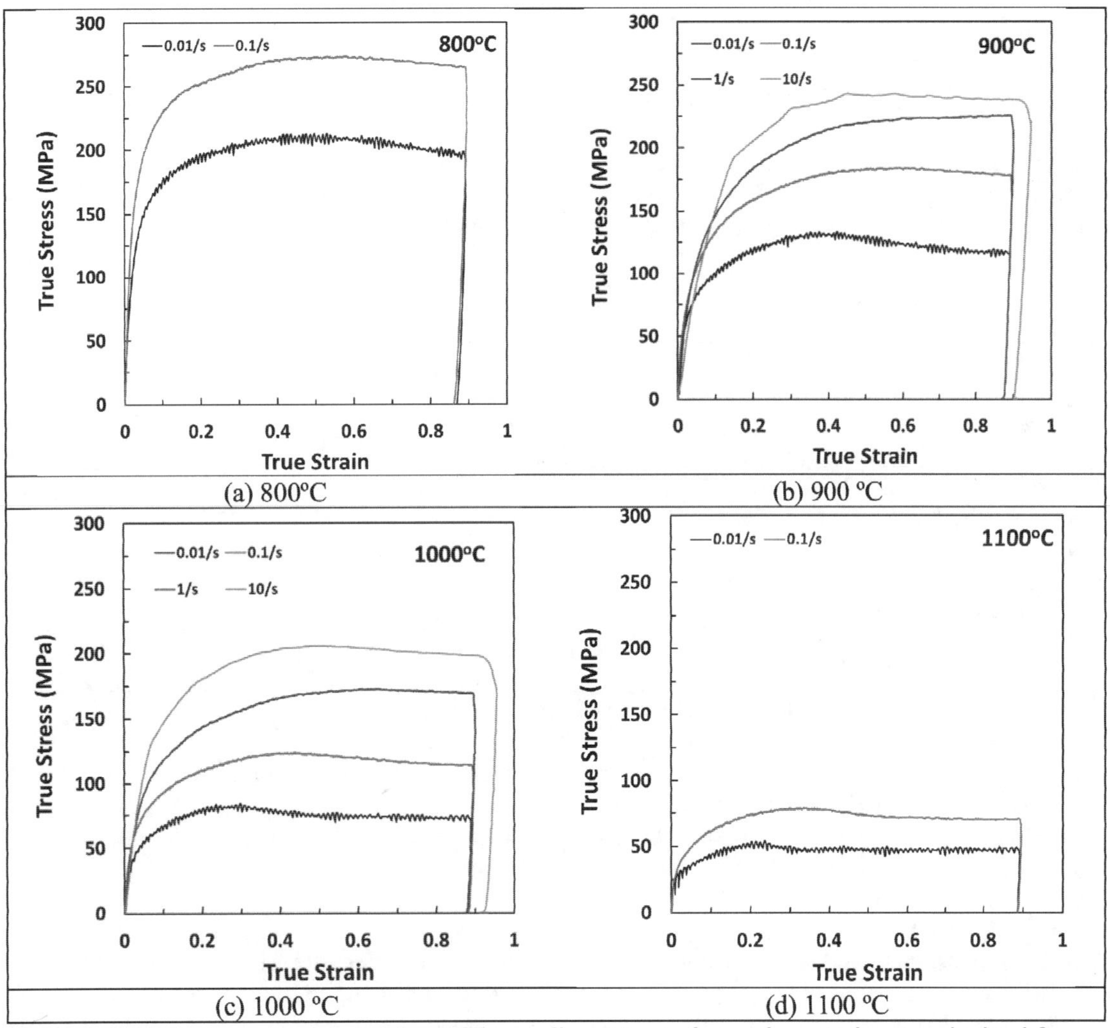

(a) 800°C (b) 900 °C

(c) 1000 °C (d) 1100 °C

Figure 2: True stress- true strain plots of the medium Mn steel at various strain rates obtained from the dilatometer hot compression tests at various temperatures.

The yield strength was used as a measure for the hot strength of the material as it passes through the caster (shown as "distance from meniscus" in abscissa in Figures 3 and 4). The yield strength calculations were done for two of the slab casters at Tata Steel in IJmuiden, the Netherlands (continuous caster 21 and 22: named as CC21 and CC22) for two casting speeds, 1.25 and 1.45 m/min, and with three cooling patterns (cp2, cp3, cp5 in the plots) associated with some reference steel grades which have actually been cast in the past [6].

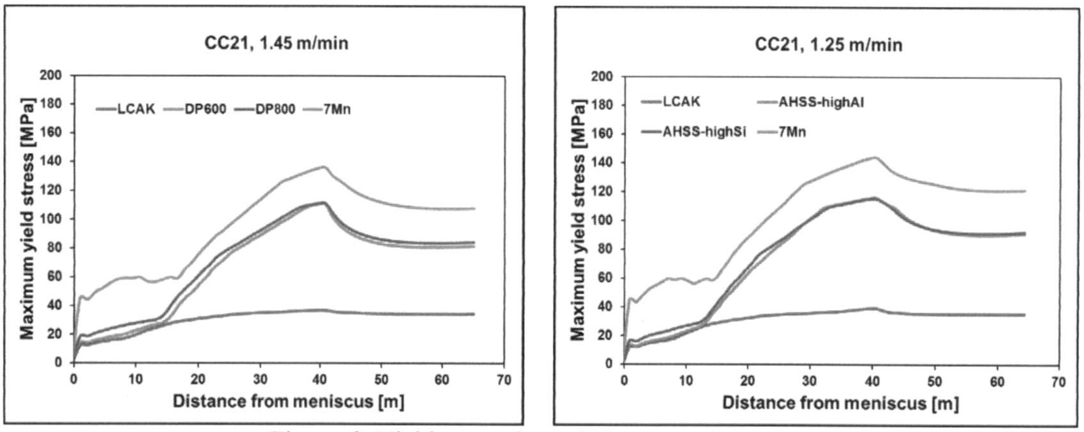

Figure 3. Yield stress through the caster for CC21.

One of the strongest grades actually cast at Tata Steel in IJmuiden is the high C grade containing ~0.7 wt.% C, therefore this grade has been used as a comparison for the medium Mn grade. Additionally other grades have been included in the graphs just for a comparison.

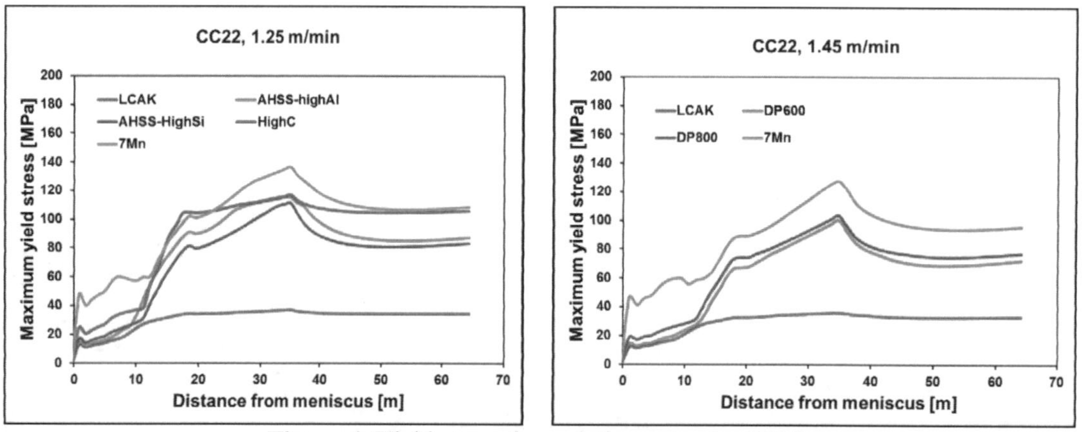

Figure 4. Yield stress through the caster for CC22.

From Figures 3 and 4, it is obvious that the medium Mn steel is stronger than the other grades in both the casters CC21 and CC22, but the maximum yield strength profile of the medium Mn steel and the high C grade look similar and the difference is not significant. The difference between both casters is that the CC21 is a vertical bending machine and the CC22 is a curved caster.

3.3. Slab Reheating

Considering that the medium Mn steel will be prone to centreline segregation, it is necessary that a high reheating temperature and longer time are allowed during slab reheating in order to reduce segregation as much as possible. No laboratory data were produced for reheating, but a temperature of 1250 °C and a time of 2 hours were used for reheating of laboratory ingots of 100 mm thickness. Although segregation was not measured, it is recommended that these relatively high reheating temperature and long time should be used for reheating of works slabs (225 mm thick) to homogenise the composition minimising Mn segregation. 1250 °C is a quite high temperature for the plant, but it is achievable. Furthermore, walking beam furnace (instead of pusher-type) should be used to avoid the skid marks.

4. Conclusions

The following conclusions can be drawn from this paper:

- Even when the medium Mn grade is a considerably higher alloyed grade than the grades that have actually been cast, the composition of this steel is simple and it does not show any complex phases to be taken into account during solidification and cooling under continuous casting conditions.
- Moreover, the strength of this grade in the caster is a bit higher than that of some grades already cast, but the difference in strength in comparison to a strong high carbon steel grade is not large.
- Taking into account some general considerations regarding a specific mould powder and a cooling pattern to avoid segregation, medium Mn steel can be produced in a conventional production line.

References:

[1] R. Rana, "Special issue on 'Medium Mn steels', Mater. Sci. Technol. 2019, 35(17) 2039-2044.
[2] J. Speer, R. Rana, D. Matlock, A. Glover, G. Thomas, E. De Moor, "Processing variants in medium Mn steels", Metals 2019, 9(7), 771
[3] R. Rana, C.H. Carson, J.G. Speer, "Hot forming response of medium manganese transformation induced plasticity steels", Proc. 5th CHS2 Conference, Toronto, ON, Canada, 31 May-3 June 2015; AIST: Warrendale, PA, USA, pp. 391-400.
[4] R. Rana, T. Kop, "Innovative hot forming of a medium Mn steel", Proc. 4th High Manganese Conference (HMnS), Aachen, Germany, 1-3 April 2019, pp. 299-302.
[5] B. Santillana et al, "Temperature control in secondary cooling to reduce cracking susceptibility in hsla steels at the Tata Steel continuous caster in Kalinganagar, India", 10th ECCC, Bari, Italy, 2021, Manuscript nr 042
[6] A. Burghardt et al, "Modelling of strand deformation to assess the castability of high alloyed steels", LMPC conference, Leoben, Austria, 2015, pp 426-432.

IOP Conf. Series: Materials Science and Engineering 1274 (2023) 012028 doi:10.1088/1757-899X/1274/1/012028

Numerical modelling of porosity with combined gas and shrinkage effects in HPDC

U Godwal[1], S Bhagavath[1], B Ghaffari[2], M Li[2], P D Lee[3, 4] and S Karagadde[1*]

[1]Department of Mechanical Engineering, Indian Institute of Technology Bombay, Mumbai, 400076, India

[2]Ford Research and Advanced Engineering, Dearborn, USA

[3]Univeristy College London, WC1E 6BT, UK

[4]Research Complex at Harwell, Harwell Campus, OX11 0FA, UK

[*]Corresponding author: s.karagadde@iitb.ac.in

Abstract. High-pressure die casting is a manufacturing process in which near-net-shape components are produced rapidly under a pressurized environment. However, due to the relatively higher cooling rate prevailing during the process, isolated liquid pockets form at certain locations, leading to increased porosity formation. A one-dimensional deformable grid numerical model has been developed for predicting the evolution of a single pore in an elementary volume, which combines the diffusion model with the shrinkage affected growth. The model accounts for the change in pore size due to shrinkage and inter-granular growth. This model can provide predictions in representative volumes and be used for component level predictions by combining with a macroscopic model.

1. Introduction

High-pressure die casting (HPDC) is a popular casting process used to manufacture near-net shape components at a high production rate with a good surface finish, making it a desirable choice for manufacturing automotive parts [1]. However, HPDC components are also prone to defects, such as porosity and shear bands [2]. Due to the high pressures employed in HPDC, the interdendritic liquid metal feeding during solidification is better than in low-pressure or gravity casting. However, due to high cooling rates, the thin regions solidify early and might block the applied pressure (termed intensification when performed after filling) from reaching certain locations [3]. These isolated liquid pockets also suffer from the lack of liquid metal compensation during the later stages of solidification, leading to shrinkage porosity formation. Thus, isolated pockets can be subjected to a large range of pressure values, especially if an intensification stage is utilized, and it is important to estimate the resulting porosity level. Further, the shrinkage effects augment the diffusion-driven pore growth, leading to the unexpected presence of larger porosity compared to other regions [4]. Although several studies have modelled porosity in the past [5,6], the role of pressure and above combined effects remain largely unexplored.

In this study, we present a one-dimensional deforming grid numerical model to predict combined gas and shrinkage pore growth with different controlling parameters such as initial hydrogen concentration, pressure, and cooling rate. The growth of porosity into the interdendritic regions is also incorporated in the model.

2. Model Theory

For simulating the porosity formation in isolated liquid pockets undergoing different pressures, the combined gas and shrinkage porosity with high-pressure effects were considered. Initially, pure gas pore growth was modelled and verified with the literature. A 1D diffusion-controlled model was used to simulate the growth of an isolated pore [7], with the domain boundary representing an advancing solidification front. For pure gas pore growth, the diffusion equation was solved with the source term for hydrogen representing hydrogen partitioning during solidification. As shown in figure 1(a), the hydrogen influx, due to concentration gradient across a representative elementary volume (REV), drives the pore growth. A nucleated pore with an initial radius of 10 μm within an REV with a radius of 100 μm was considered with an initial hydrogen concentration uniformly distributed in the liquid metal.

The diffusion of hydrogen in the REV, is modelled using the species conservation equation in the liquid phase. The second term in the right-hand side of equation 1 is the source term, which accounts for the hydrogen rejection from the solid due to solubility difference during solidification [6].

$$\frac{\partial C_l}{\partial t} + \nabla.(uC_l) = \nabla.(D_l \nabla C_l) + (1-k)\frac{d(C_l f_s)}{dt} \quad (1)$$

where, k is the hydrogen partition coefficient, C_l is the hydrogen concentration in the liquid, D_l is the diffusion coefficient, f_s is the solid fraction and u is the convective velocity.

2.1 Pore growth

The pore was considered to be surrounded by the liquid metal supersaturated by hydrogen [6,8,9]. A concentration gradient is developed due to low hydrogen content at the pore-liquid interface and higher amount of hydrogen at the solid-liquid interface (solid rejection). This eventually initiated the hydrogen influx into the pore. The pore pressure was given by the Young-Laplace equation [6,10,11]. The equation to determine the concentration at the pore-liquid interface was taken from Lee & Hunt [12] and Pequet et al. [13], which was governed by Sievert's law given as follows.

$$C_{lp} = S_l \sqrt{\frac{P_P}{P_l}} \quad (2)$$

where, C_{lp} is the hydrogen concentration at the pore-liquid interface, S_l is Sievert's constant, P_P is the pore pressure and P_l is the liquid metal pressure. The concentration inside the pore was calculated using the ideal gas law.

The pore growth is assumed to be dominated by hydrogen diffusion, and the pore growth velocity can be obtained by calculating the rate of increase of mass in the pore [6]. The rate of increase of mass is balanced by two phenomena:

(i) The rate of concentration influx by virtue of concentration gradient and diffusion of hydrogen into the pore; and

(ii) The rate of mass flow due to pore-liquid interface movement. This is represented in a global coordinate frame of reference taken from the centre of the pore. The mass at the interface is given by $(4\pi R^2)\ C_{lp}$. With the movement of pore-liquid interface, the mass flow would be given by the second term on RHS in equation 3.

$$4\,\pi\, R^2 C_P \frac{dR}{dt} + \frac{4}{3}\pi R^3\ \frac{dC_P}{dR}\frac{dR}{dt} = 4\pi\, R^2 D_l \frac{\partial C_l}{\partial r} + 4\pi\, R^2 C_{lp}\frac{dR}{dt} \quad (3)$$

where, C_p is the concentration in the pore and R is the instantaneous pore radius. The left-hand side of equation 3 represents the increase in mass of the pore, and the right-hand side is composed of the rate of concentration influx and the rate of mass flow due to pore-liquid interface movement. Since the concentration at the pore-liquid interface is assumed to be at equilibrium, we can explicitly add the local pressure term in equation 3. Also, it is appropriate to carry with the C_{lp} as it includes the local pressure governed by Sievert's law (equation 2). Simplifying the above equation provides the instantaneous pore interface velocity:

$$V_{inst}^{P} = \frac{1}{C_P - C_{lp} - C_d} D_l \left.\frac{\partial C_l}{\partial r}\right|_{r=R} \quad (4)$$

where C_d is given as:

$$C_d = -\frac{R}{3}\frac{dC_P}{dR} = \frac{2\sigma C_C}{(3R_G T)R} \quad (5)$$

where R_G is the universal gas constant and C_C is the conversion factor for relating number of moles per unit volume of hydrogen in terms of ml/100 gm of aluminium [6]. The equations are only solved in the liquid region, which reduces in volume as the solidification proceeds. It is assumed that the solid is permeable, and the liquid is pushed away from the domain during pore growth. The pore-liquid interface will move as the pore growth is incorporated due to hydrogen influx into the pore. A deformable grid approach ensures that at each time step, a node is present at the moving boundary [14]. A fixed number of nodes (100 nodes) are considered with the shrinking grid.

Pore growth is captured by solving the species equation, calculating the hydrogen influx and tracking the pore-liquid interface. The code was also extended to the solid and pore growth simultaneously. The boundary movement of the solid is also captured, and the liquid region shrinks between the pore and solid. A constant velocity is assumed with linear movement in the radial direction for solid boundary movement.

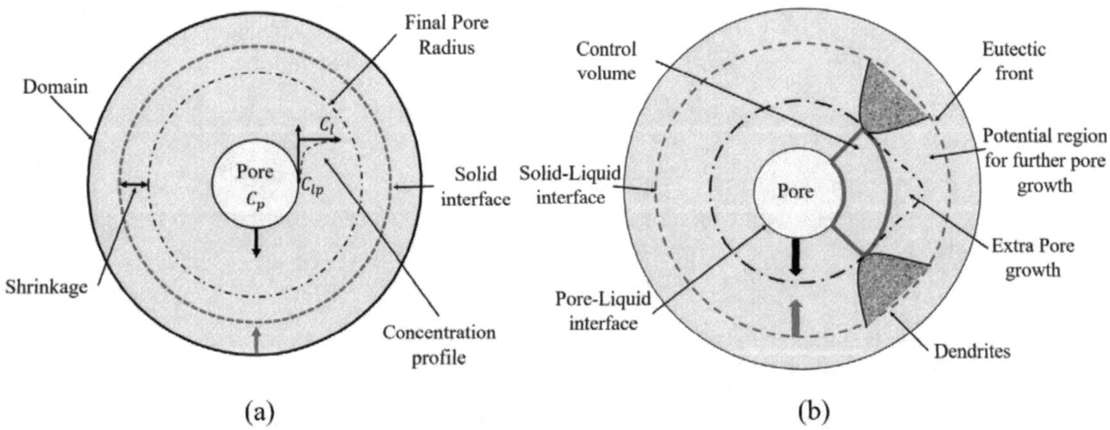

Figure 1. (a) Schematic considered for 1D diffusion model and addition of shrinkage effect, (b) Schematic of restricted pore growth

2.2 Restricted gas pore growth

A gas pore grows as a sphere until it encounters a restriction, as shown in figure 1(b). Once the pore impinges on a growing dendrite, the spherical growth of a pore is terminated, and the further growth as a perturbation out of a spherical surface is initiated [10–12]. The solid-liquid interface was tracked simultaneously with the pore-liquid interface to capture the pore-solid interaction.

The perturbation after the pore-solid interaction is modelled using the approximation as a spherical cap, and the radius for the spherical cap is adopted from Atwood et al. [10]:

$$R_c = \sqrt{R_G^2 - (z - R_G)^2} - R_G \tag{6}$$

where, R_G is the radius of the solid grain impinging the pore, R_C is the radius of the spherical cap, and z is the growth length. After the pore-solid interaction starts, the pore usually adopts the 3-D negative shape of the solid [15]. The curvature effects also enhance the pore growth, but these effects are not included in this study.

2.3 Combined gas-shrinkage effects

During solidification, after the dendritic coherency is reached [16,17], the feeding is restricted due to the poor permeability of the mush. In this work, the shrinkage effects are combined with the gas effects to obtain an updated pore evolution when shrinkage effects exist. An instantaneous shrinkage volume is calculated, which is added once the gas pore is fully grown. The total radius is calculated after adding the shrinkage effects.

It is assumed that there is no liquid feeding to compensate for the shrinkage effects after the dendritic coherency is reached [1], and the control volume is treated as an isolated pocket which forms during HPDC. After entrapment of gas pore due to coherency, the metal flow for volumetric shrinkage compensation is restricted, and hence shrinkage effects enhance the pore radius. The gas pore becomes the nucleation point for the micro-shrinkage formation. The above points can be well understood by the pressure balance and the effect of shrinkage pressure for enhancing the pore size [17,18].

3. Results and discussion

3.1 Hydrogen concentration

Pore growth and final pore radius are captured for different initial hydrogen concentrations, as shown in figure 2. The results are quantified, and the trend matches well with the literature [10–12] (see figure 2(a)). For validating with the Lee & Hunt model [12], all the boundary conditions were kept the same with a grid spacing of 200 μm and a time step of 25 ms. The error is found to be less than 10% at all hydrogen content. The 10% error is due to difference in capturing the pore growth in both models in spite of considering same hydrogen content. The present model was extended for higher hydrogen content. The concentration gradient was high for higher hydrogen concentration, resulting in faster growth. With increasing the initial hydrogen concentration from 1 to 5 ml/100 g of Al, an increment from 30 μm to 63 μm (i.e. twice) of final pore radius is reported (see figure 2(b)).

3.2 Pressure

Radius-pressure trends were observed from the 1D diffusion code for high pressure values. All other parameters were kept constant, and only pressure values were changed. The initial hydrogen content was taken as 2 ml/100 gm of Al for all cases. A reduction of 50% was observed for the final pore radius, for increasing the pressure from 1 to 50 atm (see figure 2(b)). Specific experimental data for isolated gas pores at higher pressures are not available in literature, to the best of the author's knowledge. In-situ experiments capturing the isolated gas pores at higher pressures will be carried out to validate the numerical model. However, a similar decreasing trend with pressure was observed by C. Wang et al. [19]. According to the governing physics, the external pressure restricts the pore growth and the radius variation is expected to flatten at higher pressure, which can be observed in figure 2(b). Due to the higher cooling rate and faster solidification in HPDC, the time for hydrogen diffusion is less, and hence the pore is smaller in size than other casting processes. Although solidification rate was used in this case, it should be noted that many other effects, such as dilatancy effects, can also affect pore growth during high pressure die casting.

3.3 Restricted gas pore growth

Predictions of the model for pore fraction and equivalent pore radius are plotted in figure 3(a) for different initial hydrogen concentrations, showing a near-linear variation. The pore-liquid and solid-liquid interface movements were tracked until their interaction. The same trend for radius variation with hydrogen content has been observed in literature [10,20]. The pressure was kept constant at 1 atm. The equivalent pore radius increased by 50%, for the initial hydrogen content varying from 1 to 5 ml/100g of Al.

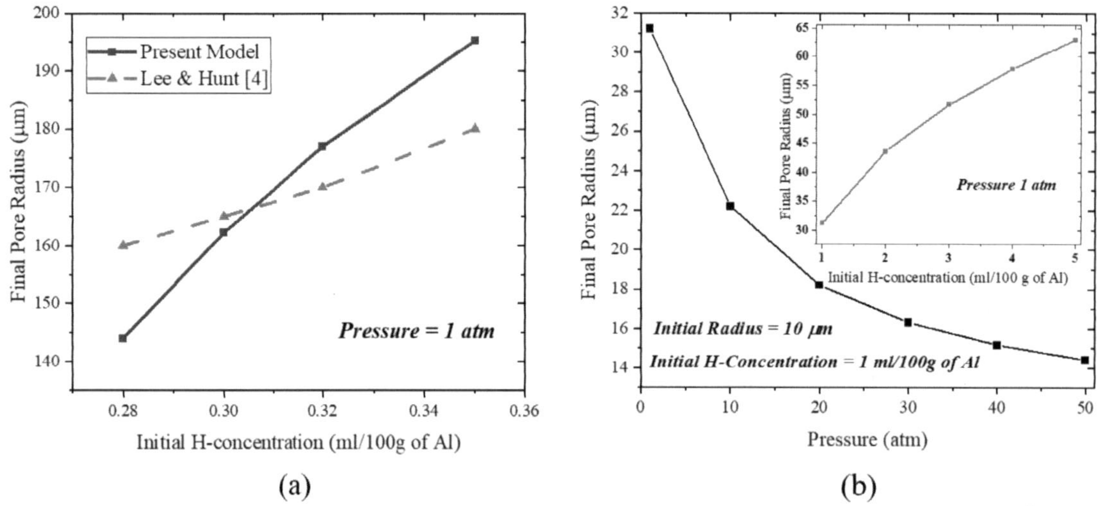

Figure 2. (a) Final pore radius with initial hydrogen content (validation case grid spacing 200 μm and time step as 25 ms with initial hydrogen content as 0.28, 0.30, 0.32 and 0.35 ml/100 gm Al), and (b) Final pore radius variation with pressure and initial hydrogen content for initial pore radius as 10 μm and domain size of 100 μm (present model)

The final pore radius and pore fraction variation with the pressure is also plotted. Figure 3(b) shows the decreasing trend for final pore radius and pore fraction with increased pressure. As the pressure is increased from 1 to 50 atm, the final pore radius showed a decrement from 33 μm to 14 μm.

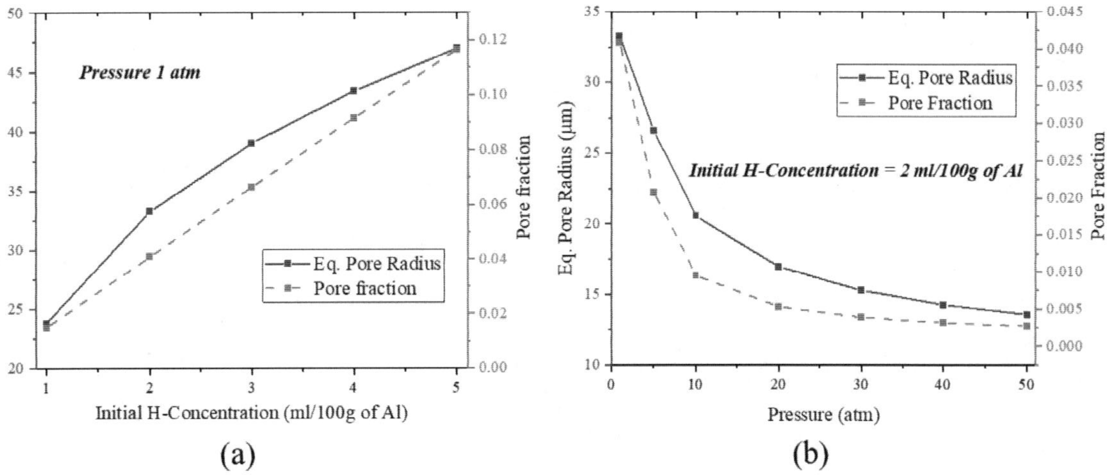

Figure 3. Restricted pore growth (a) Eq. Pore radius and pore fraction variation with Hydrogen concentration and (b) Eq. pore radius and pore fraction variation with pressure

3.4 Combined gas and shrinkage effects

The porosity formed in isolated liquid pockets in HPDC is affected by both gas concentration and shrinkage. The shrinkage effects enhance the pore growth, and hence the combined gas and shrinkage pores are bigger in size than the pure gas pores. Initially, the pore grows purely due to hydrogen diffusion, but at later stages of solidification shrinkage-driven effects dominate the pore growth (see figure 4(a)).

The final pore radius with varying initial hydrogen content for pure gas and combined gas and shrinkage effects, displaying an increase with higher hydrogen concentration is shown in figure 4(a). For higher hydrogen content, the difference in the pure gas and combined gas and shrinkage pore is reduced due to the smaller amount of liquid metal present in REV, eventually having less shrinkage. An increment of 33% of pore radius was observed at 1 ml/100g of Al, which decreased to 11% increment at 5 ml/100g of Al, depicting less contribution of shrinkage effect at higher H-content.

Figure 4(b) shows the final pore radius variation with pressure for pure gas and combined effects. At high pressure values, the contribution of shrinkage effect is larger than the contribution from the gas content, compared to lower pressure values [21]. At low pressure values, the increment in the combined gas and shrinkage pore radius is 25%. At a higher pressure value, the increment is 61%, which depicts the more significant contribution of shrinkage porosity at higher pressures.

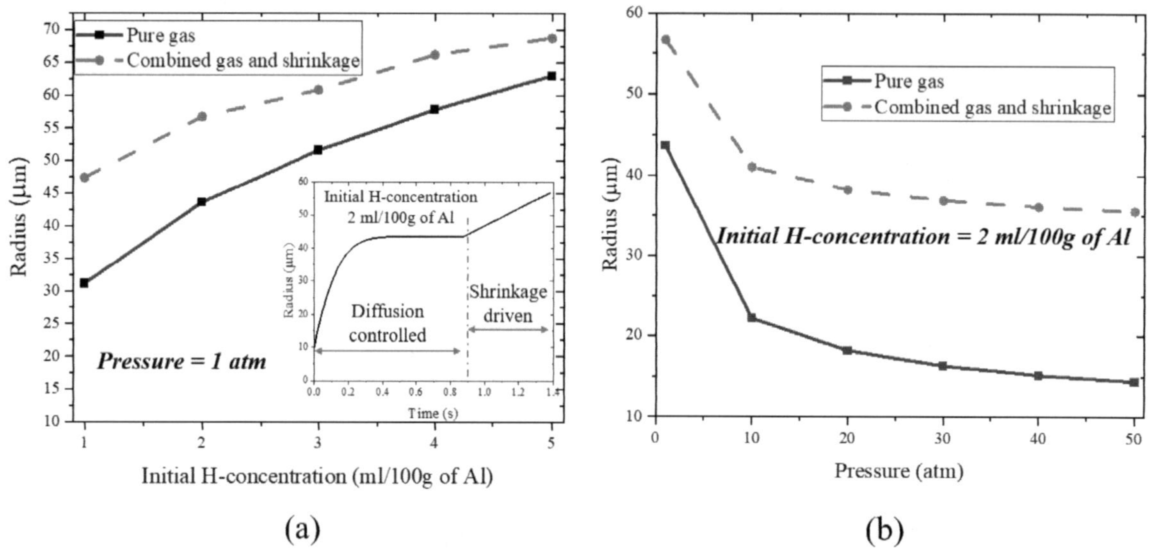

Figure 4. Combined gas and shrinkage pore growth (a) Final pore radius variation with initial hydrogen concentration and (b) Final pore radius variation with pressure

4. Conclusions

The growth of porosity in HPDC due to combined gas and shrinkage effects was studied with respect to different process parameters and its growth with different varying parameters was studied. A one-dimensional diffusion-driven numerical model was used for estimating the gas pore growth. The variation of pore radius with pressure and hydrogen concentration levels was studied. The model established the expected reduction in pore radius with pressure. More importantly, the quantitative extent of reduction with the applied pressure was successfully captured. The shrinkage was found to have a considerable influence on the final porosity (gas and shrinkage) at higher pressure values. The methodology provides porosity estimates for all ranges of pressure and hydrogen level that represent the local variation in a typical HPDC casting. The model is expected to help in making macroscopic predictions of porosity in HPDC components, aiding in locating probable defect locations.

Acknowledgments

The authors thank the University Research Program at Ford Motor Company, USA for partial financial support. S. Karagadde and P.D Lee acknowledge the sanction of project P1299 under the SPARC (Scheme for Promotion of Academic and Research Collaboration) initiative. This work was supported by the Royal Academy of Engineering (CIET1819/10) and Material Made Smarter Centre (EP/V061798/1). U. Godwal would like to acknowledge Supriyo Roy (IIT Bombay) for the technical discussions during the work.

5. References

[1] Otarawanna S, Laukli H I, Gourlay C M and Dahle A K 2010 Feeding mechanisms in high-pressure die castings *Metall. Mater. Trans. A Phys. Metall. Mater. Sci.* **41** 1836–46

[2] Dahle A K and StJohn D H 1998 Rheological behaviour of the mushy zone and its effect on the formation of casting defects during solidification *Acta Mater.* **47** 31–41

[3] Bhagavath S, Gong Z, Wigger T, Shah S, Ghaffari B, Li M, Marathe S, Lee P D, Karagadde S, Gong Z, Wigger T, Shah S, Ghaffari B, Li M, Marathe S, Lee P D and Karagadde S 2021 Role of the local stress systems on microstructural inhomogeneity during semisolid injection

[4] Lee S G and Gokhale A M 2006 Formation of gas induced shrinkage porosity in Mg-alloy high-pressure die-castings *Scr. Mater.* **55** 387–90

[5] Lee P D, Chirazi A and See D 2001 Modeling microporosity in aluminum-silicon alloys: A review *J. Light Met.* **1** 15–30

[6] Karagadde S and Dutta P 2016 A comparison of time-scales governing the interaction and growth of hydrogen bubbles with a solidifying front *Int. Commun. Heat Mass Transf.* **79** 16–20

[7] Yu W, Ma C, Ma Y and Xiong S 2021 Correlation of 3D defect-band morphologies and mechanical properties in high pressure die casting magnesium alloy *J. Mater. Process. Technol.* **288** 116853

[8] Bhagavath S, Cai B, Atwood R, Li M, Ghaffari B, Lee P D and Karagadde S 2019 Combined Deformation and Solidification-Driven Porosity Formation in Aluminum Alloys *Metall. Mater. Trans. A Phys. Metall. Mater. Sci.* **50** 4891–9

[9] Karagadde S, Sundarraj S and Dutta P 2012 A model for growth and engulfment of gas microporosity during aluminum alloy solidification process *Comput. Mater. Sci.* **65** 383–94

[10] Atwood R C, Sridhar S, Zhang W and Lee P D 2000 Diffusion-controlled growth of hydrogen pores in aluminium-silicon castings: In situ observation and modelling *Acta Mater.* **48** 405–17

[11] Atwood R C and Lee P D 2002 A three-phase model of hydrogen pore formation during the equiaxed dendritic solidification of aluminum-silicon alloys *Metall. Mater. Trans. B Process Metall. Mater. Process. Sci.* **33** 209–21

[12] Lee P D and Hunt J D 2001 Hydrogen porosity in directionally solidified aluminium-copper alloys: A mathematical model *Acta Mater.* **49** 1383–98

[13] Pequet C, Rappaz M and Gremaud M 2002 Modeling of microporosity, macroporosity, and pipe-shrinkage formation during the solidification of alloys using a mushy-zone refinement method: Applications to aluminum alloys *Metall. Mater. Trans. A Phys. Metall. Mater. Sci.* **33** 2095–106

[14] Voller V R, Boundary M and Notes P Chapter 6 Deforming Grid Solutions of the Stefan Problem 1–35

[15] Felberbaum M and Rappaz M 2011 Curvature of micropores in Al-Cu alloys: An X-ray tomography study *Acta Mater.* **59** 6849–60

[16] Kubo K and Pehlke R D 1985 Mathematical modeling of porosity formation in solidification *Metall. Trans. B* **16** 359–66

[17] Stefanescu D M 2005 Computer simulation of shrinkage related defects in metal castings - A review *Int. J. Cast Met. Res.* **18** 129–43

[18] Gu C, Ridgeway C D, Cinkilic E, Lu Y and Luo A A 2020 Predicting gas and shrinkage porosity in solidification microstructure: A coupled three-dimensional cellular automaton model *J. Mater. Sci. Technol.* **49** 91–105

[19] Wang C tao, Yao J, Zhao H dong and Yang R sheng 2019 Influence of intensification pressures on pores in die-cast ADC12 alloys *China Foundry* **16** 184–9

[20] Atwood R C and Lee P D 2003 Simulation of the three-dimensional morphology of solidification porosity in an aluminium-silicon alloy *Acta Mater.* **51** 5447–66

[21] Zhang Y, Lordan E, Dou K, Wang S and Fan Z 2020 Influence of porosity characteristics on the variability in mechanical properties of high pressure die casting (HPDC) AlSi7MgMn alloys *J. Manuf. Process.* **56** 500–9

IOP Conf. Series: Materials Science and Engineering 1274 (2023) 012029 doi:10.1088/1757-899X/1274/1/012029

High-precision solidification simulation by estimating heat transfer coefficients through data assimilation

Y Natsume[1], T Oikawa[2] and M Ohno[3]

[1] Department of Materials Science, Faculty of Engineering Science, Akita University, 1-1 Tegata-Gakuenmachi, Akita 010-8502, Japan
[2] Graduate School of Engineering Science, Akita University, 1-1 Tegata-Gakuenmachi, Akita 010-8502, Japan
[3] Division of Materials Science and Engineering, Faculty of Engineering, Hokkaido University, Kita 13 Nishi 8, Kita-ku, Sapporo, Hokkaido 060-8628, Japan

Abstract. Solidification analysis is important for accurately predicting casting processes. A high-precision casting simulation requires accurate data of the physical properties of materials and heat transfer parameters. In particular, the heat transfer coefficient is a major parameter influencing the casting simulation results. Because the heat transfer coefficient is not uniquely determined by the material, it is typically estimated by trial and error, which requires considerable time for accurate estimations. Therefore, we developed a method to quickly estimate the heat transfer coefficient in a casting process by combining data assimilation and solidification simulations. With this method, the time-dependent heat transfer coefficient can be estimated in an appropriate and easy manner. To reproduce the 3D temperature distribution in a casting with high accuracy, we developed a method to estimate multiple heat transfer coefficients simultaneously by combining data assimilation and 3D solidification simulation. Al–5 mass% Si alloy was cast into a sand mold with a casting size of 30 mm × 30 mm × 100 mm, and the temperature was measured at five points in the casting to obtain a cooling curve. Two cooling curves corresponding to locations of 5 mm from the casting wall were used to estimate the two heat transfer coefficients on the side and bottom surfaces, which were simultaneously estimated by data assimilation. The results confirmed that all the five cooling curves could be reproduced with a high accuracy and that high-precision casting simulation was possible.

1. Introduction

An appropriate casting process design is required to improve the quality of casting products. In recent years, numerical simulations have become indispensable in the design of casting processes. In a solidification simulation, several parameters significantly influence the simulation results. Although the heat transfer coefficient is a key parameter, it is difficult to obtain reference values from literature data because it is not a material-specific property. Therefore, the heat transfer coefficient that can reproduce the measured cooling curves is estimated by trial and error; that is, the evaluation of the heat transfer coefficient corresponds to an inverse analysis. To this end, various inverse analysis methods have been recently proposed using a neural network [1], Bayesian approach [2], and genetic algorithm [3]; however, in most existing inverse analysis methods, the application conditions are limited, such as the heat transfer coefficients in the steady state and lack of consideration of the time dependency. Although a trial-and-error approach is required, the essential limitations of this approach cannot be neglected. Data assimilation [4-6] is an interesting technique to overcome these limitations, and it can be used in a more generalized manner under various casting conditions than other methods. Data assimilation methods include Kalman filter (KF) [7], particle filter (PF) [8], and four-dimensional variational (4D-Ver) methods [9]. The PF is easy to implement in simulation models and is often used for parameter estimation. Recently, Oka *et al.* [4, 5] proposed a method to simultaneously estimate the heat transfer coefficient and thermal conductivity in unidirectional solidification using PF. This study was conducted by twin experiments using only numerical simulations, and the possibility of parameter estimation in the solidification analysis by data assimilation was clarified. Based on these methods, we developed a method to estimate the time-dependent heat transfer coefficient in unidirectional casting experiments [6] and confirmed that the heat transfer coefficient estimated by this method can be used to accurately simulate the cooling curves obtained from unidirectional casting experiments on Al–1 mass% Si alloy. Generally, there are multiple heat removal directions and boundaries during a casting process. In other words, to perform a high-precision solidification simulation of general casting products, multiple heat transfer coefficients corresponding to individual heat removal conditions are required. However, it is difficult to appropriately evaluate the multiple heat transfer coefficients, particularly, because of their time-dependent characteristic. Applying our previous study [6], although it is desirable to simultaneously estimate multiple time-dependent heat transfer coefficients by combining 3D solidification simulation and data assimilation, the effectiveness of such an approach has not been fully investigated. In this study, we developed a method to estimate multiple heat transfer coefficients simultaneously by combining data assimilation and 3D solidification simulation and investigated the effectiveness of this method in reproducing the 3D temperature distribution in a casting with a high accuracy.

2. Experiment and model

2.1. Experimental procedure

Al–Si alloy was cast into a sand mold with a casting size of 30 mm × 30 mm × 100 mm, and the temperature was measured at five points in the casting to obtain a cooling curve. **Figure 1** shows the schematic of the apparatus. Al–5 mass% Si alloy prepared with 99.99% pure Al and 99.999% pure Si was used. The sample alloy was melted in an electric furnace and poured into a sand mold at a casting temperature of 705 °C, with the cap, which was preheated at 700 °C before pouring the molten alloy and was immediately placed on the mold. The temperatures at the five points in the alloy (T_1, T_2, T_3, T_4, and T_5) and two points in the mold (T_6 and T_7), shown in Figure 1, were measured, and the cooling curves were obtained under a time interval of 1 s.

2.2. Simulation model

A 3D solidification analysis was performed to simulate the experimental cooling curves. The 3D calculation scale was 30 mm × 30 mm × 100 mm, which corresponds to the alloy part that did not include the mold, and the mesh size was 2.5 mm. The boundary conditions at the bottom and side were given by the heat flux $q = h(T\text{-}T_{\mathrm{mold}})$ and that at the top was the zero-Neumann condition; here, h is the time-dependent heat transfer coefficient, and T_{mold} is the measured temperature in the mold (T_6 and T_7). To calculate the heat flow, including the release of latent heat during solidification, the enthalpy method was adopted. In this explicit method, the variation in the enthalpy within each time step, ΔH, was calculated using Equation (1), and the new enthalpy was calculated by $H^{t+\Delta t} = H^{t} + \Delta H$. The new temperature $T^{t+\Delta t}$ was determined using the relationship between the temperature and enthalpy. **Figure 2** shows the relationship between the temperature and enthalpy for the Al–5 mass% Si alloy used in this study. This relationship was obtained using the PANDAT software with the PanAl database.

$$\frac{\partial H}{\partial t} = \frac{\lambda}{\rho}\frac{\partial^2 T}{\partial z^2} \qquad (1)$$

where H is the enthalpy per unit mass, T is the temperature, t is time, λ is the thermal conductivity, and ρ is the density. The values of λ and ρ used in the simulations were 95 $\mathrm{WK^{-1}m^{-1}}$ and 2.39×10^3 kg m^{-3}, respectively [10]. In reality, the physical properties, such as thermal conductivity and density, should be temperature dependent. However, such dependences are neglected and constants are used for simplicity because we focus on the estimation of the heat transfer coefficient in this study.

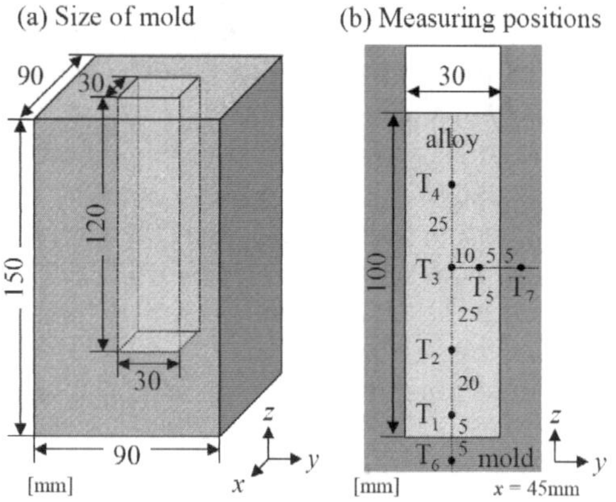

Figure 1. Sand mold size and temperature measurement position.

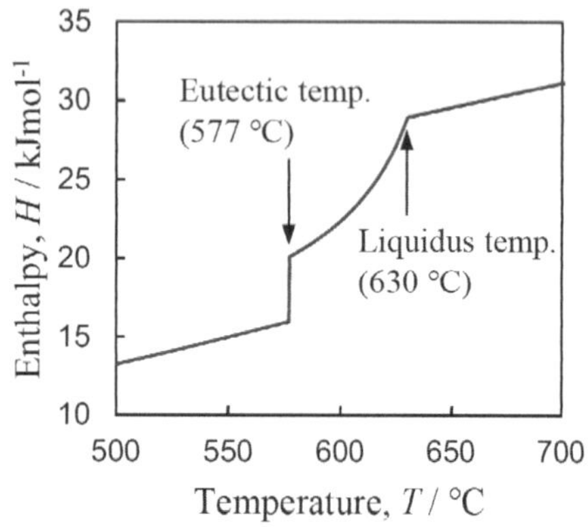

Figure 2. Relationship between temperature and enthalpy in Al–5 mass% Si alloy.

2.3. Data assimilation (particle filter)

The PF is a data assimilation method, and the algorithm used is simple and can be easily implemented. Thus, we adopted the PF to estimate the time-dependent heat transfer coefficient. The procedure for estimating the time-dependent heat transfer coefficient is briefly described below. The method used in this study was extended to simultaneously estimate two h values by applying the single h estimation method used in a previous study [6]. To explain the essence of the method, a single h estimation method is described herein. **Figure 3** shows a schematic of the estimation process of the time-dependent heat transfer coefficient (h) using the PF. At time t, arbitrary values of h were randomly assigned to each particle. Subsequently, a heat flow calculation was conducted for each particle. The value of the likelihood for each particle was calculated, and "resampling" was done. This operation is called "filtering." The likelihood of particle i, given by Eq. (2), was estimated by comparing the measured and calculated cooling curves.

$$\lambda_t^i = \frac{1}{\sqrt{(2\pi)^m |\Sigma|}} \exp\left[-\frac{1}{2} \left(\mathbf{T}_{meas} - \mathbf{T}_{cal}^i \right)^T \Sigma^{-1} \left(\mathbf{T}_{meas} - \mathbf{T}_{cal}^i \right) \right] \quad (2)$$

where Σ is a variance-covariance matrix ($m \times m$ matrix), m indicates the number of positions of the thermocouples used for the temperature measurement, $\mathbf{T}_{\text{meas}} = (T_{\text{meas,T1}}, T_{\text{meas,T2}}, \ldots, T_{\text{meas,T5}})^{\text{T}}$ is the column vector of the measured temperatures for all the thermocouple positions, and $\mathbf{T}^i_{\text{cal}} = (T^i_{\text{cal,T1}}, T^i_{\text{cal,T2}}, \ldots, T^i_{\text{cal,T5}})^{\text{T}}$ is the column vector of the calculated temperatures of particle i. This study used $\Sigma = \sigma_{\text{T}}^2 I$, assuming that the errors at all measurement points follow the same variance (σ_{T}^2), where I is an $m \times m$ unit matrix. For resampling after calculating the likelihood, the normalized likelihood of the ith particle (β_t^i) is given by Eq. (3).

$$\beta_t^i = \lambda_t^i \Big/ \sum_{j=1}^{N} \lambda_t^j \quad (3)$$

Here, N is the total number of particles. In the subsequent resampling, particles with lower normalized likelihoods, which correspond to the smaller circles in Figure 3, were randomly chosen and removed, whereas those with higher normalized likelihoods (larger circles in Figure 3) were duplicated. Finally, the value of h was determined for particles with the highest likelihood as the estimated heat transfer coefficient at t. The values of h of each particle after the resampling were used as the initial h for the heat flow calculations in the next step. However, the produced particles have the same heat transfer coefficient as the original parent particles. Thus, a system noise (v_t) was added to the heat transfer coefficients of all the particles, and $h + v_t$ was the heat transfer coefficient of each particle after updating. The system noise was taken as the product of the value of h for particles having the highest likelihood and a random number given by the Gaussian distribution, where the mean and standard deviation are 0 and σ_h, respectively. The temperature profile calculated for the particles with the highest likelihood was updated as the initial temperature profile of the next step. By iterating these steps, we can estimate the time-dependent heat transfer coefficient. In the case of simultaneous estimation, the combination of h_1 and h_2 was replaced by h, as shown in Figure 3. For example, if there are three h_1 and h_2, respectively, it is equivalent to estimating nine combinations of h_1 and h_2. The values of σ_T and σ_h used in this study were 2.0 K and 0.3, respectively. The temperature measuring points ($m = 2$) used for the estimation were T_1 and T_5, as shown in Figure 1. Therefore, the column vectors of temperatures used in the estimation were $\mathbf{T}_{meas} = (T_{meas,T1}, T_{meas,T5})^T$ and $\mathbf{T}^i_{cal} = (T^i_{cal,T1}, T^i_{cal,T5})^T$. The number of particles was set under four conditions of 32^2, 64^2, 128^2, and 256^2, and the effect of the difference in the number of particles on the prediction accuracy of the cooling curves was investigated.

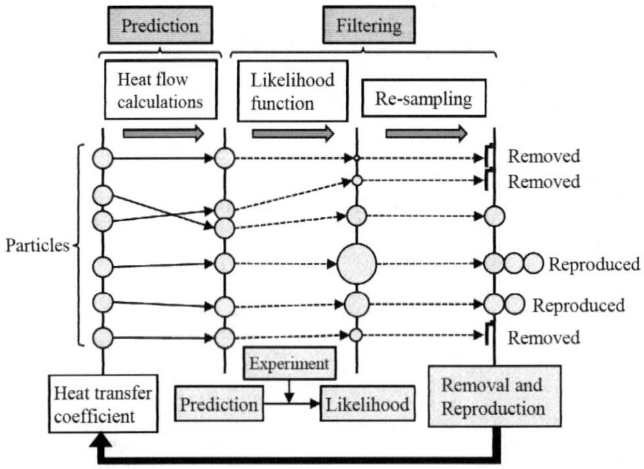

Figure 3. Schematic of the method for estimating a single time-dependent heat transfer coefficient using the particle filter.

3. Results and discussion

Figure 4 shows the cooling curves obtained by the experiment (solid line) and simulations (dashed line) and the heat transfer coefficient estimated using this method. The figures on the left (Figures 4 (a), (c), and (e)) show the results of the estimation by limiting the heat transfer coefficient to positive values. This is a condition in which the heat transfer coefficient is limited to a positive value because the heat transfer during solidification is from the alloy to the mold. However, those on the right (Figures 4 (b), (d), and (f)) show the results of the estimation without limiting the value of the heat transfer coefficient. This condition allows for free estimation by data assimilation. Figures 4 (a) and (b) show the cooling curves at the temperature measurement positions (T_1, T_5) used to estimate h, and (c) and (d) are those at the positions (T_2, T_3, T_4) not used to estimate h. The measured cooling curves shown in Figures 4 (a) and (b) were used to estimate h_1 and h_2. Therefore, h_1 and h_2 are estimated to minimize the error between the measured and calculated cooling curves. Consequently, the calculated cooling curves were in good agreement with the measured ones, and the measured cooling curves not related to the estimation of h (Figures 4 (c) and (d)) were also accurately simulated. However, in Figures 4 (a) and (c), a deviation from the measured cooling curves can be observed around the liquidus temperature at which the primary phase nucleates. **Figures 5 (a) and (b)** show enlarged versions of Figures 4 (a) and (b), respectively, showing the cooling curves from 0 s to 50 s. As shown, when the estimated value range of h is limited, the cooling curves in the undercooled region before and after the nucleation are not accurately simulated. This is because the relationship between the temperature and enthalpy in the equilibrium state (Figure 2) was used in the solidification simulation in this study, and the estimated value of h was limited to a positive value. Therefore, to accurately simulate the cooling curves while limiting the estimated value of h to a positive value, it is necessary to introduce a model that can handle the undercooling phenomenon in general. In comparison, when the estimated value of h is not limited, the cooling curves in the undercooled region can be accurately simulated. In particular, the cooling curves after recalescence (after 15 s at T_1 and after 23 s at T_5) are in good agreement with those obtained experimentally. This is because a high positive value of h was estimated to simulate the temperature drop due to undercooling, and then a negative value of h was estimated because of the temperature rise due to recalescence (Figure 4(f)). The same can be confirmed before and after the start of the eutectic solidification. Note that the negative value of h here means that the heat transfer is from the mold to the alloy, and not the result of latent heat release by nucleation. However, from the viewpoint of model simplification, it may be useful to accurately simulate the cooling curves using the relationship between the temperature and enthalpy in the equilibrium state and without introducing a model of undercooling and recalescence. The PF enables the accurate reproduction of the experimental results, including the undercooling and recalescence.

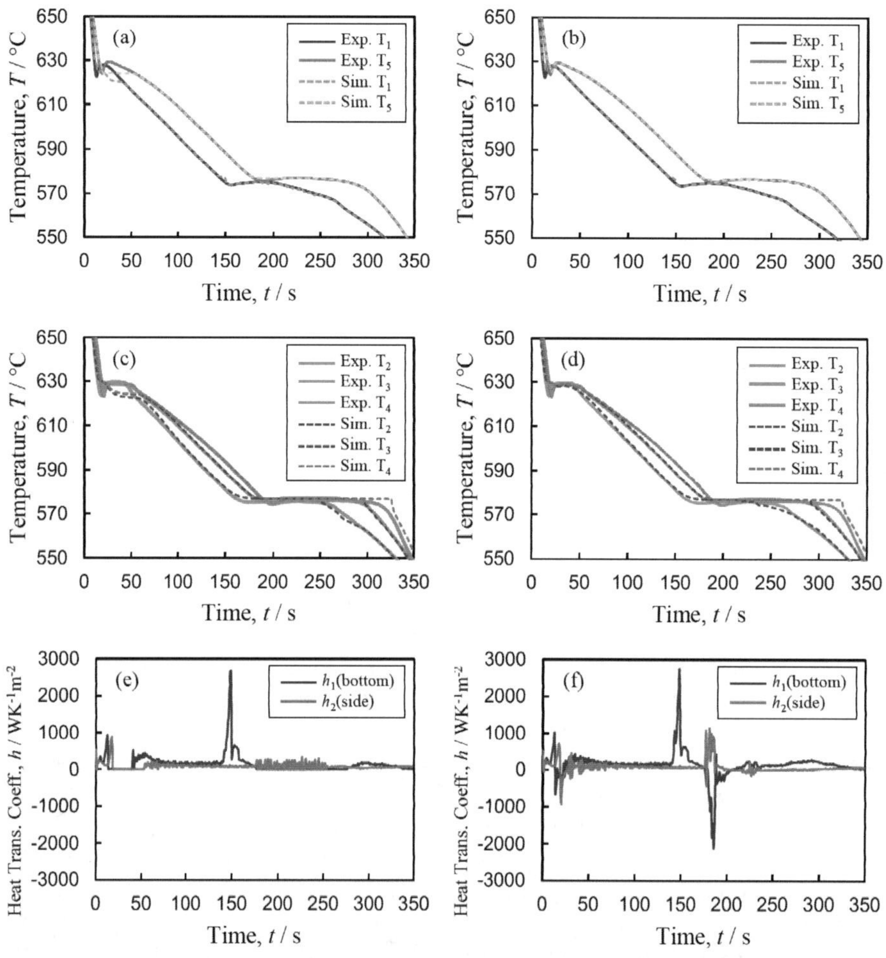

Figure 4. Cooling curves obtained by experiments and simulations, and the relationship between the heat transfer coefficient and time. The number of particles used for the estimation is 128^2. (a) and (b) Cooling curves used for estimating h with and without limiting the estimated value range of h, respectively. (c) and (d) Cooling curves not used for estimating h with and without limiting the estimated value range of h, respectively. (e) and (f) Relationship between the heat transfer coefficient and time with and without limiting the estimated value range of h, respectively.

Figure 6 shows the cooling curves simulated with the number of particles as 32^2 and 256^2 under the condition that the estimated value range of h is not limited. The cooling curve can be accurately simulated even with 32^2, which is a small number. However, there are cases where the simulation results deviate significantly, as in the cooling curve at position T_1. This is because, in principle of the PF, the greater the number of particles, the more appropriate the estimation. In this study, a highly accurate simulation was possible when the number of particles was approximately 100^2.

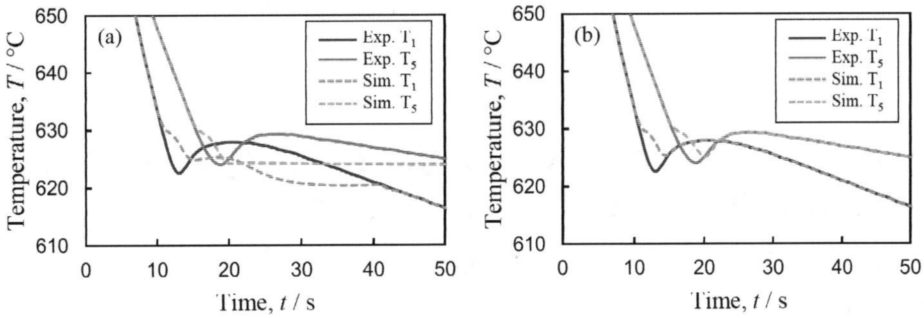

Figure 5. Enlarged view of the cooling curves before and after the start of solidification at positions T_1 and T_5. (a) and (b) Cooling curves obtained with and without limiting the estimated value range of *h*, respectively.

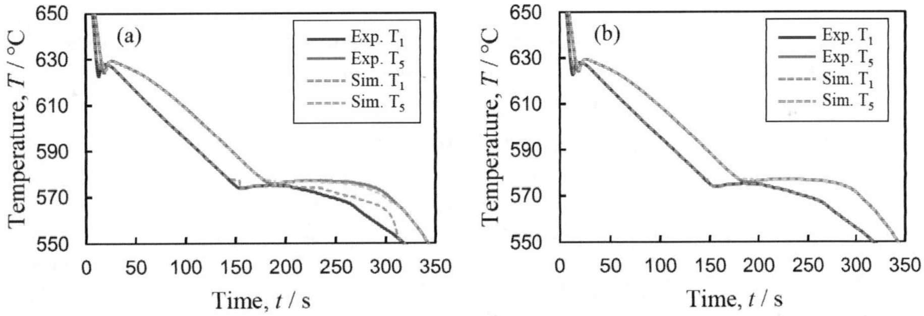

Figure 6. Cooling curves obtained by simulation when the numbers of particles are (a) 32^2 and (b) 256^2 without limiting the estimated value range of *h*.

4. Conclusion

We developed a method to estimate multiple heat transfer coefficients simultaneously by combining data assimilation and 3D solidification simulations. To verify that the 3D temperature distribution during solidification could be reproduced with a high accuracy, sand casting using Al–5 mass% Si alloy was performed, and the cooling curves were compared. The experimental cooling curves could be accurately simulated. Moreover, although it is not a correct physical model, the temperature change in the undercooled region can be simulated by allowing a negative value for the estimated heat transfer coefficient.

References

[1] Azizi S and Ahmadloo E 2016 *Appl. Therm. Eng.* **106** 203
[2] Gnanasekaran N and Balaji C 2011 *Int. J. Heat Mass Transf.* **54** 3060
[3] Czél B and Gróf G 2012 *Int. J. Heat Mass Transf.* **55** 4254
[4] Oka Y, Ohno M and Matsuura K 2017 *Tetsu-To-Hagane* **103** 755
[5] Oka Y and Ohno M 2020 *Int. J. Heat Mass Transf.* **149** 119232
[6] Natsume Y, Oka Y, Ogawa J and Ohno M 2020 *Int. J. Heat Mass Transf.* **150** 119222
[7] Kalman R E 1960 *Trans. ASME.* **82D** 35
[8] Kitagawa G 1996 *J. Comp. Graph. Stat.* **5** 1
[9] Courtier P, Thépaut J -N and Hollingsworth A 1994 *q QJ Royal Met. Soc.* **120** 1367
[10] Kurz W and Fisher D J 1998 *Fundamentals of Solidification* p.293

Anisotropic Mechanical Behaviour of Nickel Based Single Crystal Superalloy CMSX-4 for Prediction of Grain Recrystallization during Homogenization Heat Treatment Following Solidification

E Hazemann[1,2], Y Zhang[2], K Inal[2], Ch A Gandin[2], M Long[1], N T Niane[3], A Pinto Mora[2], V Maguin[1], M Leduc[1], V Jaquet[1] and M Bellet[2]

[1] Safran Advanced Turbine Airfoils, 171 Boulevard Valmy, 92700 Colombes, France
[2] Mines Paris, PSL Research University, Centre de Mise en Forme des Maté riaux, UMR CNRS 7635, Sophia Antipolis 06904, France
[3] Safran Aircraft Engines, 171 Boulevard Valmy, 92700 Colombes, France

Abstract. In the present work, a methodology to identify parameters of an elastic viscoplastic behaviour law is presented. Orthotropic elasticity and Hill48 plastic anisotropy for as-cast CMSX-4 are considered. In order to obtain the corresponding parameters by inverse analysis, an experimental database is built by tensile/relaxation tests at constant temperatures between 800°C and 1200°C on specimens oriented along <001>, <110> and <111> directions at different strain rates. Tests are performed with a resistive heating machine, using infrared thermography and digital image correlation. The obtained mechanical behaviour is very dependent on specimen orientation at all tested temperatures. Recrystallization behaviour is also very relevant to the specimens' orientation.

1. Introduction

In jet engines, high pressure turbine blades undergo the most severe operating conditions. To face both extreme temperature and stress, single crystal nickel base superalloys are extensively used as they show higher resistance to creep and thermal fatigue compared to conventional microstructures. Composition of these alloys is characterized by the absence of grain boundaries strengthening elements. As a result, the presence of unexpected grain boundaries, caused by defects such as recrystallized grains (RGs), is seriously detrimental to the mechanical performances of the final part. Recrystallization is related to the accumulated plastic strain introduced during solidification and cooling of the molten alloy when casting a part. Being a thermally activated phenomenon, recrystallization may occur during the homogenization heat treatment that follows solidification.

In order to predict accurately RGs occurrence in a turbine blade, a representative description – through finite element calculation – of stresses and strains formed during the casting process is required. Therefore, an anisotropic elastic-viscoplastic behaviour law of the single crystal nickel-based superalloy in the as-cast state is necessary.

Different works have studied the conditions of recrystallization occurrence with different thermomechanical process such as indentation, compression or traction at different temperatures [Zam2007, Cox2003, Pan2013, Li2015]. The effects of deformation temperature, level of plastic strain, annealing temperature have been established in these studies. However, stresses were studied at a constant temperature, which is not representative of the mechanical load undergone by a turbine blade during solidification.

The present paper gives a methodology to identify parameters of a behaviour law for as-cast CMSX-4 with elastic and plastic anisotropy considerations. Thermomechanical tensile tests with constant cooling rate and loading ramp are presented for three different orientations of specimens to reproduce solidification and cooling paths of a CMSX-4 turbine blade. These specimens are subjected to subsequent standard homogenization heat treatment to reveal recrystallization occurrence.

2. Materials and Methods

2.1. Tensile samples oriented along <001>, <110> and <111> directions

Second generation nickel-based superalloy CMSX-4 in the as-cast state was used for this investigation. Its composition is given in table 1. Several plates of 50 x10 x145 mm^3 and 80 x10 x 165 mm^3 were cast in Bridgman type furnace employing grain selectors and single crystal seeds, respectively.

All specimens were machined by Electro Discharge Machining (EDM). Samples with tensile direction aligned with <100> crystallographic orientation were obtained within plates of 50 x10 x145 mm^3. Whereas samples with tensile direction aligned with <110> or <111> crystallographic orientation were obtained within plates of 80 x10 x 165 mm^3. Specimens were cut off by 1 mm from each surface to avoid surface defects.

The exact orientation of tensile specimens was measured by X-Ray Diffraction (XRD). Only specimens with less than 10° of misorientation were used in this study. Standard electrochemical etching technique was performed in Safran Advanced Turbine Airfoils to ensure that neither stray grains, multi-crystals nor freckles were present in gauge part of tensile specimen. Additional etching using 5 % HF, 27.6 % HNO_3, 67.4 % H_2O for 60 minutes removed residual surface affected by EDM. As-cast microstructure cartographies of one specimen oriented along the <001> direction were obtained by Scanning Electron Microscopy (SEM) with Zeiss Supra 40 apparatus [Zei2022]. Local composition was measured on two specimens oriented along <001> by Energy Dispersive Spectroscopy (EDS) at different heights to account for macrosegregation with Brücker system and Maya3 Tescan apparatus [Bru2022, Pol2022]. For SEM imaging and EDS analyses, samples were cut on a XY plane (figure 1) and mechanically polished to a final step of colloidal silica. A high voltage of 20 kV was used for the imaging.

Table 1. Nominal composition of CMSX-4.

Element	Al	Co	Cr	Hf	Mo	Ni	Re	Ta	Ti	W
wt%	5.6	9.7	6.4	0.1	0.6	Balance	3.0	6.5	1.0	6.4

2.2. Tensile tests at high temperature for behaviour law identification

Tests are conducted in tensile machine DEDIMET developed at CEMEF [Pig2018, Gao2022]. A schematic view of the sample in the machine is presented in figure 1, where the sample is heated by Joule effect. All the tests are carried out under primary vacuum conditions to limit oxide formation on sample surface. Tests are performed at a constant nominal temperature between 1000 and 1200 °C. Setpoint temperature is monitored by a type K thermocouple welded at mid hight of the gauge length. A bichromatic pyrometer measure maximum temperature within the gauge length. The upper jaw is fixed, setpoint velocity is imposed on the lower jaw. The machine is monitored by a Labview code. During heating and cooling, a zero-set point load is fixed to prevent stresses caused by thermal dilatation/contraction. With Joule effect heating, strain is localized due to high vertical temperature gradient. Strain and displacement are tracked locally within gauge length with Digital Image Correlation (DIC). Temperature resistant speckles made of Aremco Pyro-Paint 634-ALP are applied on sample surface [Are2022]. VIC-Snap is used for images acquisition [Sut2009]. Two cameras Point Grey Grasshopper2 are used for the tests in the same vertical axis with an angle inferior to 30° between them [Fli2022]. They are combined with Schneider-Kreuznach Xenoplan 2.0 optics [Sch2022]. Acquisition frequency varies between 5 and 10 Hz. VIC-3D is used to analyse images to obtain displacement and deformation of the speckles in the gauge length with respect to a reference image [Sut2009].

For each sample, 3 cycles of tensile/relaxation steps are performed with 0.01, 0.003 and 0.008 mm. s^{-1} set point velocities for the three tensile steps, respectively. The duration of each tensile step depends on set point temperature, which is adapted to introduce sufficient plastic deformation (from 2 to 20%) within the gauge length at the end of the test. Durations were calculated using preliminary tensile tests with one tensile/relaxation sequence. Heating rate and cooling rate were 6 °C. s^{-1} and -7.5 °C. s^{-1}, respectively. A temperature stabilisation step of 100 s precedes the first tensile/relaxation sequence.

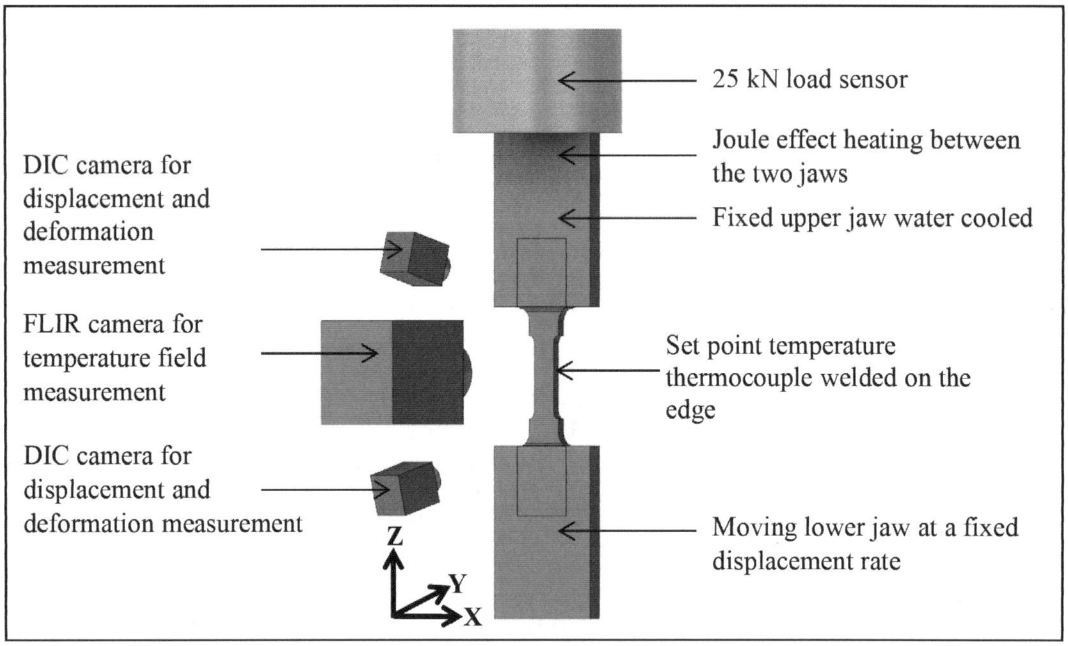

Figure 1. Schematic view of DEDIMET machine.

2.3. Tensile tests at high temperature, constant cooling rate and constant load rate for recrystallization criterion identification

Additional thermomechanical tests with constant cooling rate and load ramp are performed in DEDIMET to study recrystallization occurrence. This type of test is presented in figure 2. The test parameters are gathered in table 2. Three different tests with various final loading F_{end} were carried out on specimens oriented along the <100>, <110> and <111> directions. Set point temperature was monitored with a bichromatic pyrometer.

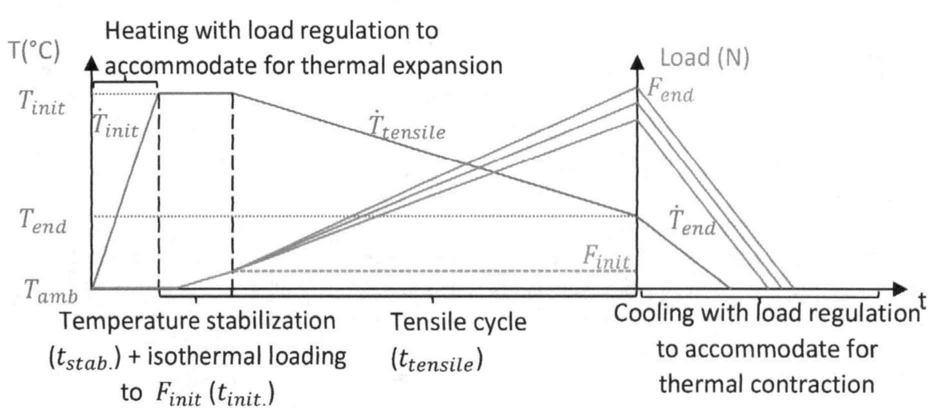

Figure 2. Schematic representation of non-isothermal tests.

Table 2. Testing parameters for non-isothermal tests.

T_{init} (°C)	$\dot{T}_{init}$ (°C.s^{-1})	$\dot{T}_{tensile}$ (°C.min^{-1})	$\dot{T}_{end}$ (°C.s^{-1})	F_{init} (N)	F_{end} (N)	$t_{tensile}$ (s)	$t_{tinit.}$ (s)	$t_{stab.}$ (s)
1200	6	-10.6	-7.5	1000	5000	2300	500	60
					7000			
					9600			

2.4. Homogenization heat treatment and subsequent etching to highlight recrystallization occurrence
Specimens used for non-isothermal tests were subjected to the standard homogenization heat treatment in a primary vacuum heat treatment furnace Carbolite STF16-180 [Dat2022]. Specimens were attacked for 2 hours with Kalling etchant to highlight RGs when recrystallization occurred [Zam2007].

3. Mathematical model

3.1. Behaviour law
Only main equations are presented in this work. More details can be found elsewhere [Gao2022]. The mathematical model is implemented in a library developed at CEMEF (Cimlib).
For small strains, one can assume an additive decomposition for the strain rate:

$$\dot{\boldsymbol{\varepsilon}} = \dot{\boldsymbol{\varepsilon}}^{el} + \dot{\boldsymbol{\varepsilon}}^{vp} + \dot{\boldsymbol{\varepsilon}}^{th} \tag{1}$$

With $\dot{\boldsymbol{\varepsilon}}^{el}$, $\dot{\boldsymbol{\varepsilon}}^{vp}$ and $\dot{\boldsymbol{\varepsilon}}^{th}$ respectively the elastic, viscoplastic and thermal contributions to the strain rate tensor. The elastic strain rate is adapted for orthotropic properties. To consider anisotropic viscoplasticity, the Hill48 criterion, adapted for orthotropic material, is chosen [Hill1948].
The developed expression for the behaviour law is as follows:

$$\bar{\sigma} = \left(\sigma_{Y0,T_{ref}} + Q_{T_{ref}} * \left(1 - \exp(-b\bar{\varepsilon})\right)\right) * \exp\left[\frac{Q_1}{R}\left(\frac{1}{T} - \frac{1}{T_{ref}}\right)\right] + k_{T_{ref}}\dot{\bar{\varepsilon}}^m * \exp\left[\frac{Q_2}{R}\left(\frac{1}{T} - \frac{1}{T_{ref}}\right)\right] \tag{2}$$

With Q_1 and Q_2 activation energies and T_{ref} the reference temperature.

3.2. Boundary conditions of the model
Only the gauge length is considered in the model. Temperature profile is directly imposed and velocity along tensile direction Z is considered as the boundary conditions during the simulation. They both are adapted from experimental measurements. More detailed methodology is presented in [Gao2022].

3.3. Identification strategy
An inverse analysis software developed at CEMEF is used to identify the different parameters of the behaviour law. Optimization strategy is similar to the one presented in [Gao2022]. Cost function expression is as follows:

$$F_c = \sum_i \omega_i \sqrt{\frac{\sum_{i,j}\left(F_{i,j}{}^{exp} - F_{i,j}{}^{sim}\right)^2}{\sum_{i,j}\left(F_{i,j}{}^{exp}\right)^2}} \tag{3}$$

where i is the test index, j the sampling instant index, ω_i the sampling instant for test i, $F_{i,j}{}^{exp}$ and $F_{i,j}{}^{sim}$ are measured and calculated load values for test i at sampling instant j. Considering the high total number of parameters used in the model, elastic properties were chosen as inputs, which are tabulated with temperature and are adapted from [Her1996].
A first step of identification strategy consisted in identifying all the parameters for the orientation <001>. After identifying all the parameters for <001> orientation, grouped optimizations are performed for one temperature and three different orientations. All parameters are identified in this step, including those governing plastic anisotropy.

4. Results and Discussion

4.1. Model development

The results of grouped optimization for specimens oriented along <001> direction at 1021°C, 1116°C and 1201°C are presented in figure 3, with a reference temperature of 1116°C. The 3 tensile/relaxation cycles are indicated by the vertical dashed lines for the specimen deformed at 1021°C. One can see that a very good agreement between simulated and experimental loads is obtained. Even though elastic properties are given as inputs of the model, simulated elastic part agrees well with experimental data. One can conclude that the model can predict accurately the behaviour of CMSX-4 between 1021°C and 1201°C for the orientation <001>. For the test at 1021°C, simulated load is slightly overestimated compared to experimental results for the third tensile/relaxation cycle. This result could depend on the choice of Arrhenius law and T_{ref} value. The effect of reference temperature on grouped optimization was studied with three different values (1021°C, 1116°C and 1201°C), to estimate the relevance and influence of an Arrhenius-type law choice for temperature dependence in eq. (2). The cost function values for reference temperatures of 1021, 1116°C and 1201°C are 0.044, 0.045 and 0.044, respectively. Although the global coherency is not influenced by reference temperature, results are affected locally. An overestimation of simulated load for the third tensile/relaxation cycle of test realised at 1021°C is observed for reference temperatures of 1021°C and 1116°C. For reference temperature of 1201°C, the simulated load agrees well with experimental data for the third tensile/relaxation cycle of test conducted at 1021°C. However, with reference temperature of 1201°C, simulated load for tests conducted at 1116°C and 1201°C are slightly different with experimental data. Hence, choice of reference temperature strongly affects the local fit quality between experimental and simulated loads. Additionally, the Arrhenius temperature dependence of the different parameters is questionable. In [Zha2013], it is shown that for heat-treated CMSX-4, yield strength divided by Young's modulus exhibits a temperature dependence according to a two-regime Arrhenius law. Thus, the single-regime Arrhenius law chosen to describe temperature dependence of all parameters presented in equation (2) could explain the poorer fit between simulated and experimental data for the third tensile/relaxation cycle at 1021°C. To verify this interpretation, an additional tensile test was realized at 907°C. A grouped optimization including the three previous test and the additional one (not presented in this work) reveals an increased gap between simulated and experimental data, with the highest difference for tests realised at 907°C and 1021°C.. This confirms the need to include a two-regime Arrhenius law with two different activations energy values for both Q_1 and Q_2 whether temperature is higher or lower compared to a critical temperature.

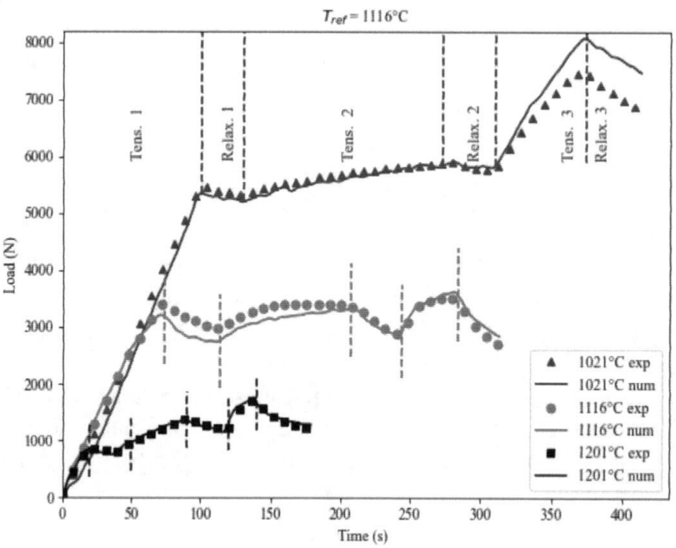

Figure 3. Grouped optimization for <001> oriented specimens at 1021, 1116 and 1201 °C with $T_{ref} = 1116°C$.

4.2. Recrystallization occurrence

Optical micrographs of specimens oriented along <111> direction after non-isothermal load ramps to 7000 N and 9600 N, homogenization heat treatment and attacked by Kalling etchant are presented in figures 4 and 5, respectively. No recrystallization is visible on gauge length for the specimen loaded up to 7000 N, while numerous RGs are visible on almost all gauge length for specimen loaded up to 9600 N. Additionally, one can observe strong necking for the latter, indicating large plastic strain compared to the sample loaded to 7000 N. Interestingly, most of the recrystallized grain boundaries appear to follow the direction of the primary dendritic arms along [100] direction, indicated by arrows in figure 4 and 5. Recrystallized grain size is quite large, indicating that only few orientations were favoured to grow.

These observations could be explained by the as-cast microstructure. Dendritic arms, interdendritic areas and eutectic areas differs by their compositions. Consequently, the solvus temperature of γ' precipitates is different in the three zones, with the lowest solvus temperature being in the dendritic arms. It has been shown that second-phase particles, such as γ' precipitates, may exert a pinning effect on recrystallized grain boundaries [Hum2004]. Hence, during homogenization heat treatment, γ' precipitates firstly dissolve in dendritic arms, deleting pinning effect in these areas. Grain boundary mobility is favoured in primary dendritic direction [100]. This interpretation is in accordance with observations in [Wan2018].

The results for recrystallization occurrence of three different orientations and three different load ramps are gathered in table 3. The absence of recrystallization is indicated with a - mark. Recrystallization occurrence is evaluated qualitatively between + and +++ marks, with + the smallest recrystallized zone and +++ the largest. For ramp to 9600 N, one can conclude that specimen orientation has significant effect on recrystallization occurrence: recrystallized area is larger for specimen oriented along <111> direction than <100> direction, while no recrystallization is visible for specimen oriented along <110> direction. A recrystallization criterion can be identified in term of load ramp at constant cooling rate between 7000 and 9600 N for <111> oriented specimens. For <110> orientation, a load ramp up to a higher value than 9600 N is necessary. For <100> orientation, the recrystallization criterion can be identified between 5000 and 7000N.

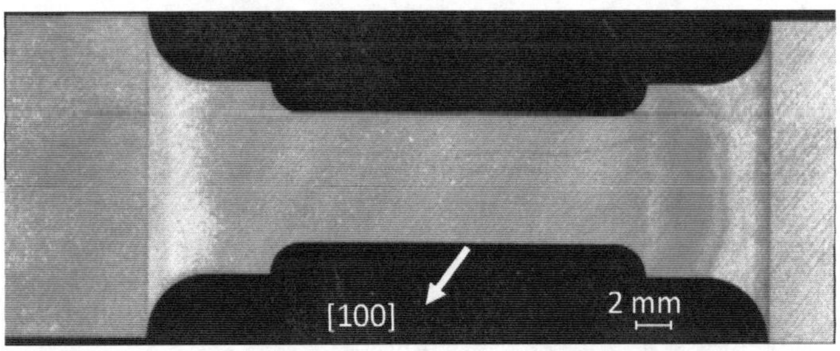

Figure 4. <111> specimen after non-isothermal loading up to 7000 N, standard homogenization heat treatment and subsequent etching.

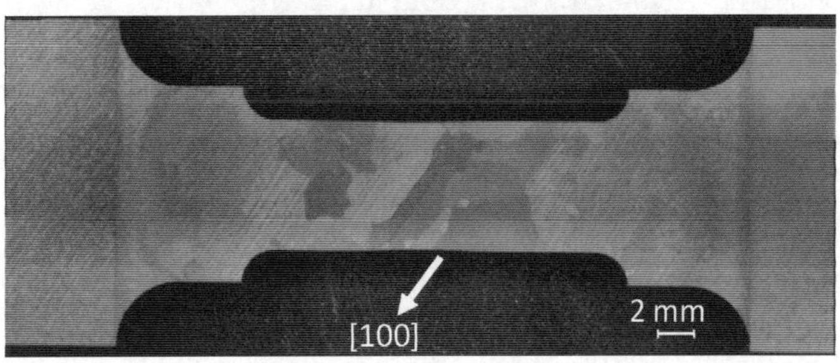

Figure 5. <111> specimen after non-isothermal loading up to 9600 N, standard homogenization heat treatment and subsequent etching.

Table 3. Recrystallization results for non-isothermal tests of three different orientations.

F_{end} (N)	5000	7000	9600
<001>	-	+	++
<110>	-	-	-
<111>	-	-	+++

5. Conclusions

A methodology to identify parameters of anisotropic elastic viscoplastic behaviour law for as-cast CMSX-4 has been presented. Results show that parameters identified for <100> orientation and temperature between 1021°C and 1201°C agrees well with experimental results. Plastic and elastic anisotropies are considered in the model, identification of plastic anisotropy parameters for more temperatures is ongoing. Elastic anisotropy has been validated with analytic calculations.

Orientation appears to have a strong effect on recrystallization behaviour for specimens tested on non-isothermal load ramps with subsequent homogenization heat treatment. Ongoing works include direct simulation on thermomechanical paths to identify a recrystallization criterion. Additionally, several experimental tests are still required to identify a restrained window of recrystallization occurrence for <110> oriented specimens. The final perspectives of current work are to implement the anisotropic constitutive law and recrystallization criterion in the process simulation software, to predict recrystallization occurrence in cast geometries of new turbine blades.

6. References

[Are2022] https://www.aremco.com/high-temp-refractory-coatings/
Last consulted: 2022/01/25

[Bru2022] https://www.bruker.com/en/products-and-solutions/elemental-analyzers/eds-wds-ebsd-SEM-Micro-XRF/software-esprit-family.html
Last consulted: 2022/01/25

[Cox2003] Cox, D C, Roebuck, B, Rae, C M F, & Reed, R C (2003). Recrystallisation of single crystal superalloy CMSX–4. *Materials Science and Technology*, **19**(4), 440–446. https://doi.org/10.1179/026708303225010731

[Dat2022] https://datasheets.globalspec.com/ds/3961/CARBOLITEGERO/C66B7056-7922-48C0-9461-60924048CDD1
Last consulted: 2022/01/25

[Fli2022] https://flir.app.boxcn.net/s/btwotjdov5nh89u0qewxyj5ro6hh2qyj/file/418659124558
Last consulted 2022/01/25

[Gao2022] Gao, F, Macquaire, B, Zhang, Y, Bellet, M (2022). A new localized inverse identification method for high temperature testing under resistive heating: Application to the elastic-viscoplastic behaviour of L-PBF processed In718. Strain 12409 ital. doi.org/10.1111/str.12409

[Hill1948] Hill, R (1948) A theory of the yielding and plastic flow of anisotropic metals. *Proc. Roy. Soc. London A*, **193**, pp. 281-297

[Her1996] Hermann, W, Sockel, HG, Han, J, Bertram, A (1996). Elastic properties and determination of elastic constants of nickel-based superalloys by a free-free beam technique. Superalloys 1996, *The Minerals, Metals & Materials Society.*

[Hum2004] Humphreys F.J and Hatherly M. (2004). *Recrystallization and Related Annealing Phenomena*. Elsevier. ISBN: 0080441645

[Li2015] Li, Z, Xiong, J, Xu, Q, Li, J, & Liu, B (2015). Deformation and recrystallization of single crystal nickel-based superalloys during investment casting. *Journal of Materials*

Processing Technology, **217**(March 2015), 1–12. https://doi.org/10.1016/j.jmatprotec.2014.10.019

[Pan2013] Panwisawas, C, Mathur, H, Gebelin, J, Putman, D, Rae, C M F, & Reed, R C (2013). Prediction of recrystallization in investment cast single-crystal superalloys. *Acta Materialia*, **61**(1), 51–66. doi.org/10.1016/j.actamat.2012.09.013

[Pig2018] Pignolet, A, Combeaud, C, Fournier, F, Fiorucci, G, Pradille, C, Zhang, Y, Pinto-Mora, A, Gao, F, Bellet, M (2018). Experimental approach for metals mechanical behaviour characterization at high temperature: development of a complex tensile test machine, ICEM18, *18th Int. Conf. on Experimental Mechanics*, Brussels, July 1-5, 2018, in Proceedings, an open access journal published by the Multidisciplinary Digital Publishing Institute, vol. 2 355-360

[Pol2022] https://www.polyu.edu.hk/umf/facility/mrc/110-field-emission-scanning-electron-microscope-tescan-maia3/
Last consulted: 2022/01/25

[Rou2020] Roux, E, Tillier, Y, Kraria, S, Bouchard, PO (2020). An efficient parallel global optimization strategy based on Krieging properties suitable for material parameters identification. *Archive of mechanical engineering, 67, 169*
DOI: 10.24425/ame.2020.131689

[Sch2022] https://www.vision-dimension.com/de/objektive/c-mount-objektive-und-cs-mount-objektive/festbrennweite-manuell/23-mm-c-mount-schneider-kreuznach-objektiv-xenoplan-1-4/23-0902/1151?number=303885
Last consulted 2022/02/12

[Sut2009] Sutton, M A, Orteu, J J, Schreier, H W (2009). Image Correlation for Shape, Motion and Deformation Measurements. Basic Concepts, Theory and Applications. (Springer, Boston, MA). XX, 322 DOI: 10.1007/978-0-387-78747-3

[Wan2018] Wang, R, Xu, Q, Gong, X, Su, X, & Liu, B. (2018). Experimental and numerical studies on recrystallization behavior of single-crystal Ni-base superalloy. *Materials*, *10*(7). https://doi.org/10.3390/ma11071242

[Zam2007] Zambaldi, C, Roters, F, Raabe, D, & Glatzel, U (2007). Modelling and experiments on the indentation deformation and recrystallization of a single-crystal nickel-base superalloy, Mat. Sci. and Eng. A 454–455, 433–440. https://doi.org/10.1016/j.msea.2006.11.068

[Zei2022] https://wiki.smfi.unipr.it/dokuwiki/lib/exe/fetch.php?media=lmn:brochure.pdf
Last consulted 2022/01/25

[Zha2013] Zhang, L, Yan, P, Zhao, M, Li, J, Zhao, J, Zeng, Q, Han, F. (2013). Tensile anisotropy of a single-crystal superalloy, The 8th Pacific Rim International Congress on Advanced Materials and Processing, The Minerals, Metals & Materials Society. Edited by Fernand Marquis

Acknowledgments
The authors are thankful to all the CEMEF staff that helped for preparation of specimens, microscopic observations and mechanical discussion: Cyrille Collin, Gilbert Fiorucci, Suzanne Jacomet, Christophe Pradille, Salim Kraria, Guillaume Corvec, Arnaud Pignolet and Francis Fournier.

IOP Conf. Series: Materials Science and Engineering 1274 (2023) 012031 doi:10.1088/1757-899X/1274/1/012031

Gabriel Faivre: his legacy on eutectic solidification

S Akamatsu[1]

[1]Sorbonne Université, Institut des Nanosciences de Paris, CNRS UMR 7588, 4 place Jussieu, 75005 Paris, France

Abstract. In the honor of Gabriel Faivre (1944-2020), I will present a review of major scientific contributions to the understanding of the dynamics of eutectic growth patterns. From the end of the 1980s, Gabriel Faivre undertook a systematic research in solidification guided by the new concepts of the nonlinear physics of out-of-equilibrium pattern formation. Drawing on his outstanding capabilities as an experimentalist, he refined the method of in situ directional solidification of model alloys. With constant reference to physics and metallurgy, he succeeded in carrying out a high-level research, keen to reach strong qualitative impact and quantitative accuracy. Gabriel Faivre made key discoveries, together with coworkers and young researchers in Paris, and in collaboration with materials scientists and physicists in France and abroad. From symmetry breaking instabilities to eutectic cells and dendrites, over rod-like and labyrinth patterns, full light has been shed onto new phenomena, fascinating to the eye and the mind. During the last decade, Gabriel Faivre mentored an in-depth analysis of interfacial-anisotropy effects on coupled-growth patterns, thus reconciliating the theories of regular eutectics and crystal-orientation dependent eutectic-grain growth. Being both a rigorous scientist and a generous colleague, he left us a vast legacy of prospective research topics in solidification and crystal-growth science. Sharing his knowledge of fine arts and humanities, Gabriel Faivre also instilled the best of intellectual thinking in those who were fortunate enough to work with him.

1. Introduction

Gabriel Faivre was born on December 16, 1944 in Nancy, France. After graduating at École des Mines de Paris (1966), he spent a few years as a metallurgical engineer at IRSID in Saint-Germain-en-Laye. He graduated a master degree (DEA) in Solid-state physics (Orsay, 1967), keeping unforgettable memories of the teaching by French physicists and metallurgists. Gabriel Faivre pursued his academic curriculum with a PhD in Physical Sciences, which he defended in 1974, under the direction of Georges Saada (University Paris XIII). He was hired by the CNRS as an associate researcher in 1971. He carried out a precursory work on the coherency constraints of precipitates [1], abnormal absorption in transmission electron microscopy of noncentrosymmetric crystals [2], and self-diffusion involving bivacancies at high temperatures [3]. Gabriel Faivre also studied the structures of dislocations in low-symmetry crystals [4,5]. He then turned to the equilibrium-polymer properties of liquid selenium [6,7], and the crystallization of spherulites [8,9]. In the late 1980s, Faivre moved to the Groupe de Physique des Solides (GPS) of University Paris VII (now Institut des Nanosciences de Paris, Sorbonne University) in the Jussieu campus, close to river Seine in the Quartier Latin of Paris, at the invitation of Christiane and Bernard Caroli [10,11]. From the beginning of the 1990's as a Directeur de Recherche at CNRS, Gabriel Faivre led an impressive, nearly 30-year journey through the study of the fascinating structures of solidification fronts. He was joined by myself (SA) as a young CNRS researcher in 1992, and then by Sabine Bottin-Rousseau in 2000. During his career, Gabriel Faivre supervised several doctorate students –Jacques Mergy (1992), Marie Ginibre (1997), Ghislain Boutet (1999) and Mikael Perrut (2007) as concerned solidification– and post-doc researchers –among them,

Thomas Ihle, Tamás Börzsönyi and Melis Şerefoğlu. Emeritus since 2012, he actively pursued his research until 2019.

Gabriel Faivre played an important role in the refoundation of the field of solidification that took place in the 1980s and 1990s. His research work on solidification put strong focus on the non-linear physics of self-organized structures in extended systems (nonequilibrium pattern formation), with constant reference to solid-state physics and metallurgy. His studies cast new light on the phenomenological equivalence between the dynamics of crystal melt-growth shapes –cellular, dendritic and eutectic growth– in metals and that of other well-studied nonequilibrium systems such as Rayleigh-Bénard convection [12,13]. He also provided renewed evidence of the overwhelming power of demonstration of in situ experimentation by real-time observation of the solidification front in model systems, and was keen to make it quantitative. This was achieved thanks to instrumental innovations developed with his collaborators in Paris. Faivre was one of the initiators of a microgravity project coordinated by the European space agency (ESA), thanks to long-term support by the French space agency (CNES) [14,15]. Fortunately, he had the opportunity to see the images of the first campaign in 2018. Gabriel Faivre also elaborated new theories, and inspired collaborations with physicists performing numerical simulations.

Faivre liked to quote Jacques Friedel, Raimond Castaing and André Guinier among his admired masters. His work also owes much to sustained discussions he had with non-linear physicists. He greatly appreciated meetings organized in the 1990's by the European research network *Pattern Formation, Noise and Spatio-Temporal Chaos in Complex Systems* (L. Kramer). Gabriel Faivre enjoyed fruitful collaborations, noticeably those with Alain Karma (Northeastern University, Massachusetts) and Mathis Plapp (Ecole Polytechnique, Palaiseau), but also Laurent Limat (CNRS, Paris) and Efim Brener (FZ Jülich, Germany), as well as with Stephan Rex, Ulrike Hecht and their colleagues (Access e.V., Aachen, Germany). He also established links with research groups in Marseille (Bernard Billia; Alain Pocheau) and Grenoble (Denis Camel) in France, and Zürich (Jörg Bilgram) and Lausanne in Switzerland –he frequently remembered the Zermatt workshops organized by Wilfried Kurz and Michel Rappaz (EPFL).

Gabriel Faivre's contributions to spherulite growth [8,9], dendritic/cellular solidification [16-18], subboundaries [19] and faceted growth [20,21] are all of great value. This makes too many topics for the present paper. I shall limit myself to a brief account of Gabriel Faivre's research activity in the field of eutectic growth [22] during, say, the latest 20 years of his career, and focus on (i) the spatio-temporal dynamics of thin and bulk eutectics, and (ii) the effects of the free-energy anisotropy of interphase boundaries on the microstructures of eutectic grains.

2. Eutectic solidification patterns

2.1. Morphology diagram

Binary eutectic alloys present a three-phase equilibrium between the liquid and two different solids at a given eutectic temperature T_E. This makes possible a two-phase growth from the melt upon cooling. In directional solidification at imposed velocity V in a fixed temperature gradient G, eutectic growth occurs in the form of two-phase patterns. In nonfaceted alloys, the dynamics is primarily determined by solute diffusion in the liquid and by capillary effects. The resulting composite microstructure is a frozen-in trace left behind in the bulk solid by the coupled-growth patterns. A reference in the field of eutectic growth is the theoretical and experimental work by Jackson and Hunt (JH) in the 1960's [23]. These authors established the existence of a scaling length, the so-called minimum-undercooling spacing λ_m, which is typically of order 10 µm for V in a 1-µms^{-1} range, and varies with V as $V^{-1/2}$. Experimentally, Jackson and Hunt showed the general-demonstration power of thin-sample directional solidification using model transparent nonfaceted alloys for in situ observation. It should be mentioned that Gabriel Faivre, who suspected the extraordinary potential of the model eutectic CBr_4-C_2Cl_6 alloy, was confronted to the detrimental fact that the CBr_4 molecule reacts with foreign impurities, and degrades in the liquid state. In the late 1980's, he initiated a collaboration with Robert Mellet (CNET), who was an expert on high-level purification of organic systems by zone-refining. Purified CBr_4 and

C_2Cl_6 with a good chemical stability could be mixed, and used for both precise thermodynamic assessment [24] and unprecedented solidification studies –including that of ternary eutectics (see, e.g., [25] and [26]). During the doctorate work by Jacques Mergy, and then by Marie Ginibre, major aspects of the dynamics of lamellar-eutectic patterns in two dimensions (2D), in thin samples, were explained. Basic theories (the JH analysis of regular eutectics, but also the Mullins-Sekerka theory of cellular patterns) predict the existence of continuous branches of steady periodic structures with the spatial periodicity λ varying over a finite range at fixed control parameters. The problem is to determine the interval of accessible λ values, and to identify the mechanisms that limit that interval. In thin lamellar-eutectic samples, secondary instabilities occur, on the large-λ side, via spontaneous breaking of basic symmetries. The term spontaneous means, as a general feature of pattern formation in spatially extended systems [12], that the morphological instabilities are not imposed from outside, but are made possible by the nonlinear coupling between the shape of the interface and the diffusion field. Faivre and Mergy discovered the tilt bifurcation, corresponding to a change from steady-state, mirror-symmetry patterns to travelling, tilted ones. It can occur uniformly [27,28] or in the form of finite domains [29]. Interestingly, the drifting motion of tilt-domain walls entails a dynamic lamellar-spacing selection [29,30]. The tilt bifurcation is independent of any crystallographic effect. During Marie Ginibre's thesis, a comprehensive description of 2D symmetry-broken shapes could be achieved [31] –including some oscillatory patterns that had been observed previously in Al-Cu [32]. In addition, a comparison of experimental observations and time-resolved numerical simulations by Karma and Sarkissian was achieved, with quantitative agreement [33]. A few years later, the question of how incipient spacing modulations in a lamellar eutectic pattern smooth out and evolve towards a fully periodic state was addressed by Gabriel Faivre and myself, in collaboration with Alain Karma and Mathis Plapp [34]. This led to revisit the so-called "phase-diffusion" phenomenon in reference to the previous work by Langer [35].

In 2000, Bernard Zappoli at CNES brought a strong support to our project dedicated to in situ observation of bulk eutectics, with the prospect of a science-in-microgravity campaign aiming at studying eutectic growth in purely diffusion-controlled conditions [15]. That was a challenging task, both theoretically and experimentally, as regards the number of degrees of freedom, the possibility of forming lamellar, rod-like and more complex shapes, and the presence of thermo-solutal convection. The problem of how to observe eutectic growth patterns in situ over the whole solidification front was not obvious as well. With Sabine Bottin-Rousseau, it was decided (i) to use flat-wall samples of the CBr_4-C_2Cl_6 alloy with a useful thickness in the millimeter range, and (ii) to observe the solidification front from the exterior, at an oblique view. With some improvements –mainly a diaphragm for dark-field viewing– the method (DIRSOL) worked fine [36]. With the new instrument, the zig-zag instability of lamellar eutectics was discovered [37]. It was also observed that, in fully isotropic conditions, a lamellar eutectic initially forms "labyrinth" patterns that evolve slowly, and appear almost "frozen" [38]. These observations also prepared us to address an even more complex question: that of the so-called lamellar-to-rod transition –a typically 3D morphological change that obviously cannot occur in thin samples. For this purpose, Mikael Perrut studied a rod-like eutectic, namely the succinonitrile-d,camphor (SCN-DC) alloy [39,40], with the help of Victor Witusiewicz and Laszlo Sturz at Access e.V. in Aachen (Germany) [41]. Thanks to Mikael Perrut's remarkable thesis work, the decision was made at ESA to engage a collaborative project (Paris, Aachen, Leoben) and build a new apparatus, starting from the DIRSOL method, for the study of solidification processes in transparent alloys by real-time optical monitoring. After a long maturation process –let us also mention the work by Melis Şerefoğlu during that period of time [42]– the apparatus (TRANSPARENT ALLOYS) was built by ESA contractors, sent to the ISS (Dec. 2017), and installed in the Microgravity Science Glovebox (MSG) facility. The first scientific campaign onboard the ISS in January-March 2018 brought full evidence of a lamellar-to-rod transition in the SCN-DC alloy [43].

2.2. Anisotropy effects

Faivre and Mergy made several observations during thin-sample directional solidification showing that, in a given sample, the lamellar-growth dynamics was varying from one eutectic grain to another

[11]. This indicated that the orientation of the crystals could influence the coupled-growth dynamics. In particular, in certain eutectic grains, the lamellae were growing tilted with respect to the main solidification axis, and, simply speaking, followed an inclination, which did not change whatever the value of V and λ. This dynamics was called "locked-lamellar" growth by Christiane Caroli, which illustrates well that, in the anisotropic case, the tilt dynamics is imposed by the crystallography, as opposed to a spontaneous breaking of the symmetry. In Ref. [11] the term "floating" dynamics was also introduced for designating regular lamellar eutectic grains within which the interphase boundaries can freely rotate (Fig. 1a). The question of the formation, growth and competition of eutectic grains has been formulated and extensively documented in the 1970's in the metallurgy literature. The main background at that stage was that of the identification (in particular by electron microscopy) of (heterophase) coincidence lattice planes, thus defining crystal orientation relationships (ORs) [44]. As a common wisdom, it was stated that lamellae were driven to grow along a coincidence plane corresponding to a low-energy inclination of the interphase boundary. The observations were convincing, and the intuition inspiring, but no rigorous argument was advanced. Gabriel Faivre, with Sabine Bottin-Rousseau and myself, decided to reconsider the question on the following bases.

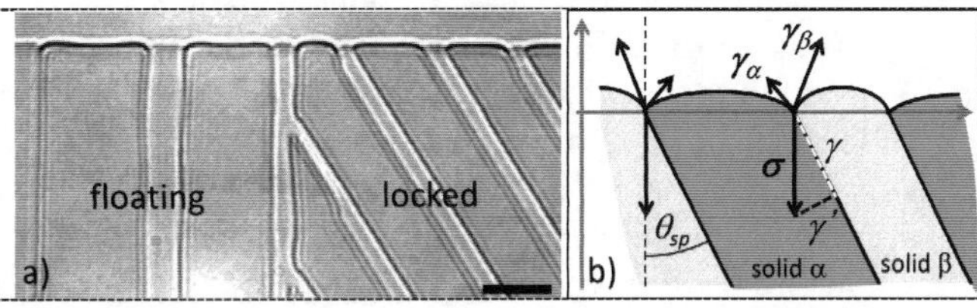

Figure 1. a) In situ optical image of a floating and a locked eutectic grains. Directional solidification of a hypo-eutectic CBr_4-C_2Cl_6 alloy in a thin sample. The liquid is on top. Bar: 20 µm. b) Schematic representation of the symmetric-pattern approximation.

The locked-lamellar dynamics, and, more generally, the eutectic-grain dependence of coupled-growth patterns, is not described by the JH theory of regular eutectics, which assumes fully isotropic interfaces. Generally speaking, it has been shown that, with some anisotropy of solid–liquid or solid-solid interfaces, lamellar eutectic patterns must travel laterally in steady-state, and deliver tilted-lamellar microstructures in the solid. Gabriel Faivre carried out an analysis "à la JH" of tilted-lamellar patterns, and brought a demonstration that a weak solid-liquid anisotropy, to be expected in nonfaceted eutectics, can entail the formation of tilted-lamellar patterns with tilt angles not larger than a few degrees [45]. In contrast, the free energy γ of the interphase boundaries in the solid can be substantially anisotropic, in particular in eutectic alloys that present special ORs with dense coincidence planes. Lamellar tilt angles can then be large. Two questions were open: 1- how decisive progress could be made on a theoretical level –that is, how a realistically large anisotropy of γ could be incorporated in the equations? and 2-what could be done experimentally to inspire and/or support that theory?

The anisotropic quantity γ enters the condition of local equilibrium at the trijunctions. It can be expressed by introducing the Hoffmann-Cahn surface-tension vector $\boldsymbol{\sigma}$ in the Young-Herring equation [46]. The vector $\boldsymbol{\sigma}$ is the sum of a component parallel to the boundary, of value γ, and a component normal to the boundary, of value γ' (the angular derivative of γ). In two dimensions, an anisotropic surface tension of the interphase boundary means that, at fixed relative orientation of the two solids, the value of $\gamma \equiv \gamma(\theta)$ depends on the inclination angle θ of the interphase boundary measured from a fixed axis –e.g., the axis of the temperature gradient **z** in directional solidification. The $\gamma(\theta)$ polar plot (γ-plot) is a circle for an isotropic system, or a slightly deformed circle (with a centrosymmetric shape) for a mildly anisotropic system. In contrast, the γ-plot exhibits deep, singular minima in strongly anisotropic cases –for example in a eutectic grain with an OR with dense lattice planes in coincidence. At the trijunction, except for special configurations, the $\boldsymbol{\sigma}$ vector is no longer parallel to the interphase boundary, and the lamellae can be expected to grow tilted. The question then is to determine the value of the lamellar tilt angle θ_t in steady-state. As a key point, in situ observations during thin-sample directional solidification of the CBr_4-C_2Cl_6 alloy were made, evidencing that in locked-lamellar patterns, the shape of the solid-liquid interfaces was keeping (nearly) the same mirror symmetry as in a regular, isotropic case. This indicates a weak coupling between the lateral drift motion of the pattern and the diffusion field. It can then be realized that a strict symmetric-pattern condition imposes that the $\boldsymbol{\sigma}$ vector is parallel to the axis of the temperature gradient (Fig. 1b) [47]. In the symmetric-pattern (sp) approximation, the steady-state tilt angle $\theta_t \equiv \theta_{sp}$ of the lamellae is fully determined. It can be calculated from a nonlinear equation for the inclination θ of the interphase boundary for an arbitrary anisotropy function $\gamma(\theta)$, knowing the orientation of the eutectic grain. The contribution of Gabriel Faivre to this analysis was determining. The qualitative relevance of the sp-approximation was confirmed numerically [48] (also see [49] and refs. therein). A similar approach was used to analyse the growth of anisotropic subboundaries [50].

Aiming at a better understanding of the mechanisms via which eutectic grains form during early solidification stages, a careful experimental study of the so-called eutectic invasion was carried out by Gabriel Faivre and myself with Sébastien Moulinet [51]. This process occurs without nucleation of new crystals, and can deliver floating, isotropic eutectic grains with no special OR.

The rotating directional solidification (RDS) method was then developed (see [52] for a previous attempt, and [21] for the first use of the RDS apparatus in Paris). During RDS, a standard directional-solidification configuration is used, and a thin sample, instead of being translated, is rotated about an axis perpendicular to the sample plane. On one side of the sample, the solid grows (it melts on the other side) while the orientation of the crystals is continuously varied. In an isotropic eutectic grain, the interphase boundaries remain locally perpendicular to the isotherm, and describe mere circles in the sample reference frame. In an anisotropic grain, it can be intuited that the shape of lamellae in the solid reflects the γ-plot. Indeed, within the symmetric-pattern approximation, the RDS microstructure is homothetic to the shape of the interphase boundaries predicted by the Wulff construction [53]. A more recent RDS setup has been built by Melis Şerefoğlu in her laboratory in Turkey, and used, in particular, for the study of interfacial-anisotropy effects in ternary, three-phased eutectics [54].

The next step was to establish a clear link between the locked-lamellar growth dynamics, the presence of deep minima in the γ-plot, and coincidence planes in anisotropic eutectic grains. For this purpose, in situ optical observations were coupled with ex situ crystallographic measurements. The thin-sample technique was adapted to metallic eutectic alloys. Using first the low-melting In-In_2Bi system, Gabriel Faivre helped us –with Oriane Senninger and Laurent Carroz– to define a new experimental protocol including the analysis of x-ray diffraction pole figures, and to establish a method for identifying dense coincidence planes, and ORs [55]. More recently, this methodology was applied to thin Al-Al_2Cu samples [56].

3. Conclusion

Gabriel Faivre was a gifted physicist, a man of rare intelligence. From the reading of the above text, it should be clear that he was viewing laboratory experimentation and theoretical analysis as intimately linked and interdependent activities. He accomplished outstanding achievements on both levels.

Gabriel Faivre was a discreet person, sometimes even a little secretive. He was uncompromising and rigorous in his scientific arguments, but very attentive in his relations with his peers – at the same time demanding and benevolent. Being indefatigable at passing his knowledge to his collaborators, Faivre also knew how to leave the initiative to his close collaborators and thus bring out new ideas.

Gabriel Faivre was also a man of great culture. During his whole life, he has been immensely fond of books, history, music and fine arts. He also learned a lot, according to his own words, from rock climbing, which he practiced intensely in his youth, mainly in the French Alps.

Gabriel Faivre passed away on April 17, 2020 in Paris, in the protective presence of his beloved family. His colleagues are deeply, and forever indebted to him, being conscious how fortunate they are to have known him, and worked with him.

4. References

[1] Faivre G 1969 *Phys Status Solidi* **35** 249
[2] Faivre G and Legoff J J 1979 *Acta Cryst. A* **35** 604
[3] Chiron R and Faivre G 1985 *Phil. Mag. A* **51** 865
[4] Faivre G and Saada G 1972 *Phys Status Solidi B* **52** 127
[5] Faivre G 1974 *Phil Mag* **29** 1289
[6] Faivre G and Gardissat G 1986 *Macromolecules* **19** 1988
[7] Simon C, Faivre G, Zorn R, Batallan F and Legrand J F 1992 *J. Phys I France* **2** 307
[8] Ryschenkow G and Faivre G 1988 *J. Cryst. Growth* **87** 221
[9] Bisault J, Ryschenkow G and Faivre G 1991 *J. Cryst. Growth* **110** 889
[10] Caroli B, Caroli C, Roulet B and Faivre G 1989 *J. Cryst. Growth* **94** 25
[11] Caroli B, Caroli C, Faivre G and Mergy J 1992 *J. Cryst. Growth* **118** 135
[12] Cross M and Hohenberg P 1993 *Rev Mod Phys* **65** 851
[13] Faivre G, in *Branching in nature* Fleury V, Gouyet J F, Leonetti M eds., EDP Sciences, vol. 13, Springer, Les Houches, 2001
[14] Plapp M, Bottin-Rousseau S, Faivre G and Akamatsu S 2017 *C. R. Mecanique* **345** 56
[15] Ludwig A et al 2012 *JOM* **64** 109
[16] Akamatsu S, Faivre G and Ihle T 1995 *Phys. Rev. E* **51** 4751
[17] Akamatsu S and Faivre G 1998 *Phys. Rev. E* **58** 3302
[18] Garandet J P et al 2005 *J. Cryst. Growth* **279** 195
[19] Bottin-Rousseau S, Akamatsu S and Faivre G 2002 *Phys. Rev. B* **66** 054102
[20] Börzsönyi T, Akamatsu S and Faivre G 2002 *Phys. Rev. E* **65**, 011702
[21] Börzsönyi T, Akamatsu S and Faivre G 2009 *Phys. Rev. E* **80**, 051601
[22] Hecht U et al 2004 *Mat Sci Eng R* **46** 1
[23] Jackson K A and Hunt J D 1966 *T. Metall. Soc. AIME* **236** 1129
[24] Mergy J, Faivre G, Guthmann C, Mellet R 1993 *J. Cryst. Growth* **134** 353
[25] Akamatsu S, Perrut M, Bottin-Rousseau S and Faivre G 2010 *Phys. Rev. Lett.* **104** 056101
[26] Bottin-Rousseau S, Şerefoğlu M, Yücetürk S, Faivre G and Akamatsu S 2016 *Acta Mater.* **109** 259
[27] Faivre G and Mergy J 1992 *Phys. Rev. A* **45** 7320
[28] Kassner K and Misbah C 1991 *Phys. Rev. A* **44** 6533
[29] Faivre G and Mergy J 1992 *Phys. Rev. A* **46** 963
[30] Coullet P, Goldstein R E and Gunaratne G H 1989 *Phys. Rev. Lett.* **63** 1954
[31] Ginibre M, Akamatsu S and Faivre G 1997 *Phys Rev E* **56** 780
[32] Zimmermann M, Karma A and Carrard M 1990 *Phys. Rev. B* **42** 833
[33] Karma A and Sarkissian A 1996 *Met. Trans. A* **27** 635
[34] Akamatsu S, Plapp M, Faivre G and Karma A 2004 *Metal. Mater. Trans. A* **35** 1815
[35] Langer J S 1980 *Phys. Rev. Lett.* **44** 1023
[36] Bottin-Rousseau S, Perrut M, Picard C, Akamatsu C and Faivre G 2007 *J. Cryst. Growth* **306** 465
[37] Akamatsu S, Bottin-Rousseau S and Faivre G 2004 *Phys. Rev. Lett.* **93** 175701
[38] Perrut M, Parisi A, Akamatsu S, Bottin-Rousseau S, Faivre G and Plapp M 2010 *Acta Mater.* **58** 1761
[39] Perrut M, Akamatsu S, Bottin-Rousseau S and Faivre G 2009 *Phys. Rev. E* **79** 032602
[40] Perrut M, Bottin-Rousseau S, Faivre G and Akamatsu S 2013 *Acta Mater.* **61** 6802
[41] Akamatsu S, Bottin-Rousseau S, Perrut M, Faivre G, Witusiewicz V T and Sturz L 2007 *J. Cryst. Growth* **299** 418
[42] Şerefoğlu M, Bottin-Rousseau S, Akamatsu S and Faivre G 2011 *IOP Conference Series: Materials Science and Engineering* **27** 012030
[43] Bottin-Rousseau S, Witusiewicz V T, Hecht U, Fernandez J, Laveron-Simavilla A and Akamatsu S 2022 *Scripta Mater.* **207** 114314
[44] Hogan L M, Kraft R W and Lemkey F D 1971 *Adv. Mater. Res.* **5** 83

[45] Faivre G 2004, in: Rappaz M, Beckermann C, Trivedi R, editors, *Solidification processes and microstructures: a symposium in honor of Prof. W. Kurz*, Warrendale (PA): The Minerals, Metal and Materials Society, p. 239
[46] Hoffmann D W and Cahn J W 1972 *Surf Sci* **31** 368
[47] Akamatsu S, Bottin-Rousseau S, Şerefoğlu M and Faivre G 2012 *Acta Materialia* **60** 3199
[48] Ghosh S, Choudhury A, Plapp M, Bottin-Rousseau S, Faivre G and Akamatsu S 2015 *Phys. Rev. E* **91** 022407
[49] Akamatsu S and Bottin-Rousseau S 2021 *Metal. Mater. Trans. A* **52** 4533
[50] Ghosh S, Karma A, Plapp M, Akamatsu S, Bottin-Rousseau S and Faivre G 2019 *Acta Mater.* **175** 2144
[51] Akamatsu S, Moulinet S and Faivre G 2001 *Met. Mater. Trans A* **32A** 2039
[52] Oswald P, Moulin M, Metz P, Géminard J C, Sotta P and Sallen L 1993 *J Phys France III* **3** 1891
[53] Akamatsu S, Bottin-Rousseau S, Şerefoğlu M and Faivre G 2012 *Acta Mater.* **60** 3206
[54] Mohagheghi S and Şerefoğlu M 2019 *Met. Mater. Trans A* **50A** 5221
[55] Bottin-Rousseau S, Senninger O, Faivre G and Akamatsu S 2018 *Acta Mater.* **150** 16
[56] Bottin-Rousseau S, Medjkoune M, Senninger O, Carroz L, Soucek R, Hecht U and Akamatsu S 2021 *J. Cryst. Growth* **570** 126203

Acknowledgments
I gratefully acknowledge the organisers of the ICASP 6 conference for giving me the opportunity to honor the memory of Gabriel Faivre, along with other colleagues and friends whom I warmly thank as well. I thank Sabine Bottin-Rousseau and Mathis Plapp for critical reading of the manuscript.

IOP Conf. Series: Materials Science and Engineering 1274 (2023) 012032 doi:10.1088/1757-899X/1274/1/012032

Observation of peritectic couple growth for a hyper-peritectic alloy under microgravity conditions

A Ludwig, J P Mogeritsch

Department Metallurgy, Montanuniversitaet Leoben, Leoben, Austria

email: johann.mogeritsch@unileoben.ac.at

Abstract: Under specific conditions, peritectic alloys can form microstructures that behave similar to regular eutectic alloys, with two solid phases growing in a coupled manner directly from the melt. This so-called peritectic couple growth (PCG) is significantly affected by convection. Thus, Bridgman-type experiments were performed onboard the International Space Station using the transparent peritectic alloy, tris(hydroxymethyl)aminomethane–neopentyl glycol (TRIS–NPG). Under these purely diffusive conditions, the formation of PCG, its development over time, and its dependence on the applied process conditions were studied. In this paper, we provide novel insights into the appearance of PCG and the challenges associated with corresponding experiments.

1. Introduction

Peritectic reactions are characterised by the transformation of a solid pro-peritectic α-phase and a liquid phase into a new solid peritectic β-phase upon cooling. At the peritectic temperature T_p, the three phases exist in equilibrium with each other. Trivedi [1,2] reported that for an initial composition within the peritectic plateau and for conditions wherein solidification morphologies of both phases are planar and solute transport is only diffusive, two different peritectic layered structures occur. The first structure represents an oscillation of the solute concentration in the liquid ahead of the solid/liquid (s/l) interface, leading to the formation of an alternating nucleation and a growth sequence such that subsequent bands of both the phases are formed parallel to the solidification front [3], and the second structure represents simultaneous and coupled growth of both phases similar to a regular eutectic growth front, called peritectic coupled growth (PCG), which occurs slightly below T_p. Peritectic layered structures have been reported to be extremely sensitive to convection ahead of the solidification front [4]. Consequently, in directional solidification experiments, peritectic alloys exhibit a variety of complex microstructures, such as isothermal PCG, cellular PCG, island bands, and oscillatory tree-like structures [3-5].

Over the past few years, the authors of this study have employed the organic peritectic metal-like solidifying tris(hydroxymethyl)aminomethane–neopentyl glycol (TRIS–NPG) system [6] to investigate the dynamics of the formation of peritectic layered structures [7-10]. The advantage of such a transparent model alloy is that it enables in-situ observations of the occurring phenomena. However, it has also been demonstrated that on Earth, natural convection affects the formation of peritectic layered structures, thereby complicating unambiguous interpretations of observations. Therefore, experiments under purely diffusive conditions were performed aboard the International Space Station (ISS). In this paper, the first observations are presented and discussed.

2. Experimental Set-up and Method

Investigations were performed using alloys of the following organic compounds: TRIS and NPG as model systems for metal-like peritectic solidification. The experiments were conducted between 17 March and 24 April 2021 aboard the ISS using the Bridgman technique. For this purpose, the so-called Transparent Alloy (TA) instrument, specifically designed by the European Space Agency (ESA) for investigations with organic model substances, was installed in the Microgravity Science Glovebox (MSG) onboard the ISS.

TRIS and NPG are organic substances that exhibit faceted phases at low temperatures and transparent plastic crystals at high temperatures. The plastic TRIS-rich pro-peritectic α-phase forms a body-centred cubic lattice, and the plastic NPG-rich peritectic β-phase forms a face-centred cubic lattice. Barrio et al. [6] investigated the TRIS–NPG phase diagram. Figure 1 presents the peritectic plateau and alloy composition used in the present study. The compounds were supplied as powders by Sigma-Aldrich [12], with a reported purity of 99.9+% for TRIS and 99% for NPG. Both these compounds are extremely hygroscopic [13], which is why all manipulations with organic compounds were performed under a protective atmosphere. The water content of the organic substance NPG was further reduced by drying at 375 K for 24 h. TRIS was used without further processing because it is sensitive to long-term annealing at elevated temperatures. Alloying and filling of special syringes were performed by the authors. The material was then shipped to QinetiQ Space (Antwerp, Belgium) [14], where the cartridges for space experiments (TAC) were filled. The cartridges possessed a solidification volume of 100 mm (length) × 6 mm (width) × 1 mm (depth). Owing to specific requirements for the contact area between the cartridge and hot and cold clamps, only an effective solidification length of 66 mm was available. This length was conceptually separated into six segments, with lengths of 11 mm each. The solidification experiments were conducted only with 'fresh' segments that were never melted before.

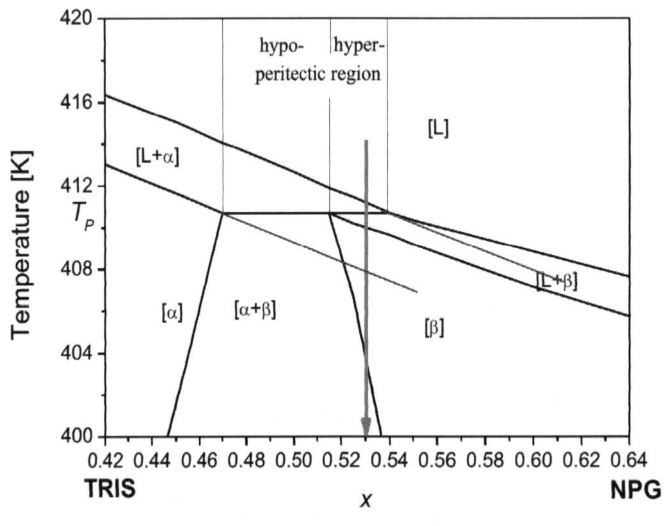

Figure 1. Peritectic region of the TRIS–NPG system. The dashed blue lines represent the metastable extension of the liquidus and solidus lines for the pro-peritectic α-phase. The red arrow indicates the nominal sample composition, x = 0.53 mol fraction of NPG. The peritectic temperature, T_p, is 410.7 ± 0.5 K.

The TA instrument was developed by QinetiQ Space on behalf of the ESA. It was specifically designed for experiments with organic transparent alloys under microgravity (μg) conditions. The instrument is a Bridgman furnace with a control unit and an optical assembly. The furnace consisted of two parallel clamps as a hot zone (rear and front clamp) and two parallel clamps as a cold zone (rear and front clamp), and these zones were separated by an adiabatic zone of 7 mm. This represents the area for optical observation. In the adiabatic zone, a temperature gradient governed by the temperature of the clamps was established.

The optical recording system of the TA apparatus included a light-emitting diode source for illumination and a charge-coupled device camera centred at the adiabatic zone with a field of view (FOV) of 6.1 x 5.1 mm^2. During the experiments, a set of three images was acquired with a time step of 3 s at the following three different focus points: ff_1 = 0 mm, directly on the inside of the front glass wall; ff_2

= 0.5 mm, at the centre of the cartridge; and ff_3 = 0.8 mm, near the rear glass wall. A set of new images was acquired every 30 s. However, only once per hour a set of three images with different focus points was broadcast directly to Earth. The present findings are primarily based on these images.

The experiment we report here will be referred to as TAC4S2. The experiment was performed between 19 April 2021 07:32 GMT and 20 April 2021 06:33 GMT. The nominal alloy composition was 0.53 mol fraction of NPG. The experiment consisted of the following steps. First, the temperature of all clamps was set to 403 K, and the sample was annealed for 2 h. Subsequently, the temperature in the cold zone was set to 379 K, and the temperature in the hot zone was set to 439 K. The sample was maintained at rest for an additional 1 h to achieve thermal equilibrium within the device. For the selected clamp temperatures, we expected a temperature gradient within the adiabatic zone of 3.0 ± 1 K/mm. Thereafter, the cartridge was moved with a constant pulling velocity of V_p = 0.10 µm/s from the hot zone towards the cold zone. Under these conditions, the s/l interface was well-observable within the adiabatic gap. After 22 h of growth, the motion was stopped, and the sample was maintained at rest for another 4 h. The microgravity level during the processing time of TAC4S2 was measured using the 'es09'-sensor at the ceiling right side of the MSG. The mean µgRMS[1] was approximately 2.3 µg between 08:00 and 22:00 GMT and less than 0.5 µg during night's rest for the astronauts[2]. The median frequency measured by the sensor was constant at approximately 0.182 Hz.

3. Results

Immediately after activation of the gradient, liquid appeared at the FOV's 'hot' side. The solid in direct contact with the liquid consisted of several small grains, which immediately began to coarsen and elongate with time. However, as the two involved solid phases are optically indistinguishable, determining whether these grains consisted of the pro-peritectic α- or peritectic β-phase or both was difficult. In addition, several small bubbles were observed. These bubbles formed a substructure in the colder part of the solid, which gradually disappeared at higher temperatures. During pulling, this bubble substructure moved without any visible alteration towards the FOV's 'cold' side until it finally disappeared. Occasionally, larger bubbles were observed in the solid/mush close to the melt.

A few smaller bubbles and tracers in the liquid were tracked during the course of the experiment. The origin of the tracers was unclear. They could be contaminants already present in the purchased components. However, they were extremely small and difficult to observe. The bubbles and tracers tracked with a focus point ff_2 = 0.5 mm were located in the middle of the sample and thus did not stick to the glass walls. Based on the trajectory of the small bubbles, it was confirmed that the solid moved at the pulling speed. The tracers in the liquid, however, moved faster by approximately 4%, indicating the density difference between the solid and liquid phases and, thus, the need for a feeding flow to compensate for solidification shrinkage.

The growth dynamics of the s/l interface morphology are illustrated in figure 2. Note that the images were captured with a focus point at ff_2 = 0.5 mm. Hence, all structures with a sharp appearance are located around the middle of the sample, whereas all blurry patterns appear either in the back or towards the front glass wall. The bright blurry structures lie in the back, and the grey blurry structures appear at the front. Therefore, the image presented is not a top view of the inclined s/l interface as it might seem but a view from beneath. As outlined in another paper written by the authors [10], this inclination of approximately 6–7° against the sample normal was caused by an unplanned thermal bias between the rear and front clamps. In retrospect, this turned out to be useful because the details at the s/l interface were easier to observe through this inclined view.

After 3 h of pulling, solidification occurred with a planar growth of larger grains, which most probably consisted of a single solid phase, that is, either α or β (figure 2a). After 4 h, the growth morphology clearly changed (figure 2b). Patches of a second solid phase could now be identified. Several irregularly surrounded and unconnected patches growing in wavy troughs on the first growing primary solid phase can be observed in figure 2b. After 1 h (figure 2c), the number of patches increased, but their size decreased. In addition, the patches appeared more spherical in shape. After another hour, most of the patches became elongated, and some demonstrated curved shapes (figure 2d). For the next couple of hours, this was the dominant growth morphology, whereby some of the curved lamellae resembled the zigzag lamellae that formed during eutectic growth (figures 2d and e). Furthermore, as

[1] RMS: root-mean-square acceleration
[2] Information taken from „https://gipoc.grc.nasa.gov/pims/roadmap/"

the experiment progressed, the lamellae changed into fibres (figure 2f). It is worth mentioning that as shown in figures 2a–f, the phase that had formed always grew in depressions of the primary phase, independent of its morphology. Only the tip of the growing fibres (cell-like growth, figure 2f) was at the same level (or even higher) as the primary phase. During the entire 22 h of growth, the s/l interface recoiled by approximately 3 mm, which corresponds to an interface temperature reduction of approximately 9 K (assuming a gradient of $G = 3$ K/mm).

After a period of constant pulling, the motion stopped, but the thermal gradient was maintained. Lateral growth of the phase that formed the fibres could now be observed (figure 2g). The fibres turned into patches that increased in size by growing along the s/l interface (figure 2h). Simultaneously, it appeared that the opposite occurred in the background. However, this observation requires more attention and will be considered in detail in the future. Along with the overgrowth, the s/l interface moved toward higher temperatures so that the interface gradually reached a final position at a somewhat higher temperature level.

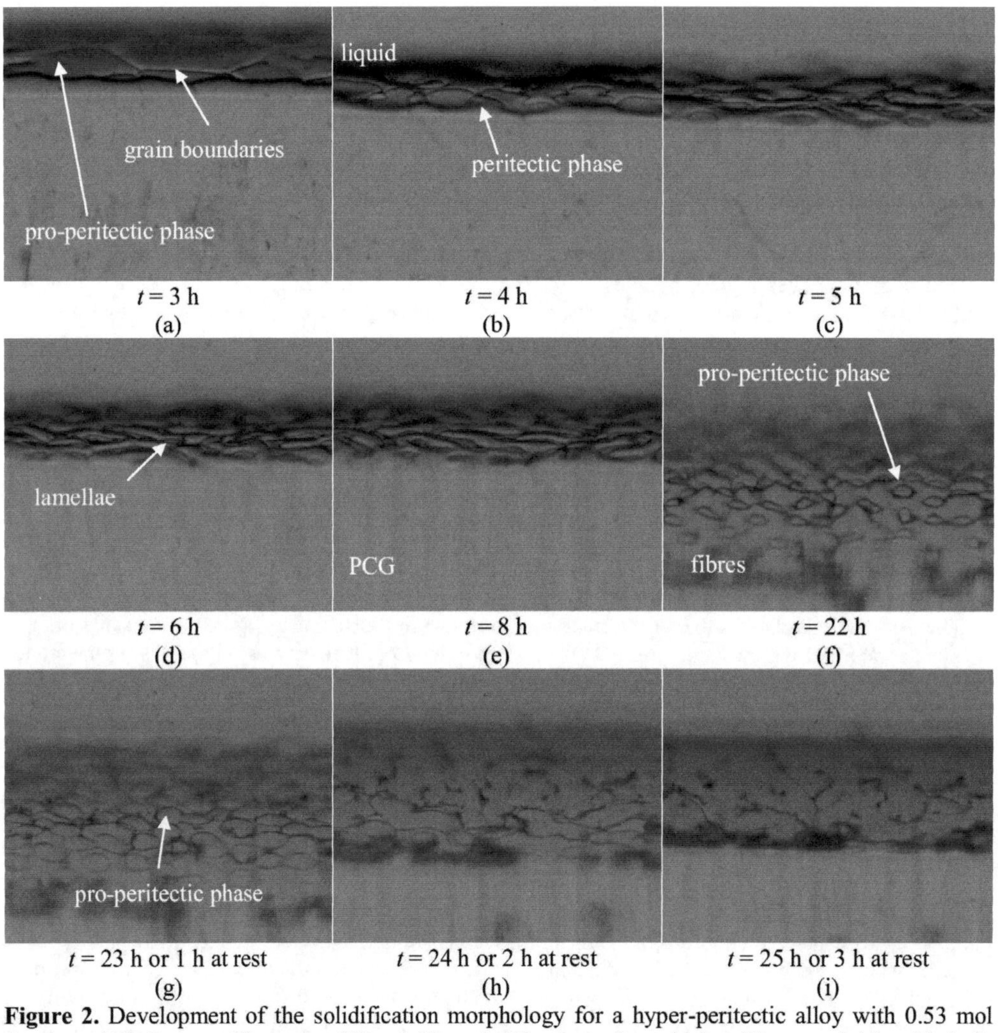

Figure 2. Development of the solidification morphology for a hyper-peritectic alloy with 0.53 mol fraction of NPG at a pulling rate of $V_p = 0.10$ µm/s. The time after which pulling was initiated is indicated below each image. All images have a width of 1 mm.

4. Discussion

Figures 2b–f present the changes occurring in PCG over a duration of more than 18 h. However, the observed phases cannot be clearly classified into α or β phases. Particularly, the occurrence of patches of the second solid phase after approximately 4 h of growth raised doubts regarding their formation mechanism. Thankfully, in the meantime we have got excess to the full picture set, with pictures being taken every 30 s. It turned out that the patches were formed by unstable finger-like overgrowth of the primary s/l interface with a second solid phase. Nucleation was not involved in this transition. Therefore, it was clear that the patches, which later turned into lamellae and fibres, consisted of the primary solid phase, and the matrix consisted of the second solid phase.

Based on this finding, we determined the identities of the different phases. According to the TRIS–NPG phase diagram (figure 1), solidification should begin with the pro-peritectic α-phase. Thus, it is natural to assume that the primary solid phase, as well as patches, lamellae, and fibres, consisted of the α-phase, and the matrix phase consisted of the β-phase. We have labelled figure 2 accordingly. As the interface continued to recoil to lower temperatures, the amount of the α-phase decreased; hence, the patches transformed into lamellae and later into fibres. This is in consistence with the phase diagram. However, that the α-phase fibres finally cover, at least partly, the s/l interface at the at the end of the experiment is still curious.

With the suggested phase identification in mind, we can now discuss the solidification experiment from the beginning. Clearly, the α-grains grew with a planar but tilted s/l interface (figure 2a). However, the corresponding solutal recoil was far from being completed, as the interface displacement still varied approximately linearly with time. Three hours after the pulling was initiated, the s/l interface moved by 0.4 mm, which corresponds to a reduction of the interface temperature of around 1.2 K. Based on the phase diagram information, this fact leads to an interface temperature of approximately 0.2 K below the peritectic temperature, T_p. Four hours after the initiation of pulling (figure 2b), when the peritectic β-phase first appeared, the interface temperature was approximately 0.5 K below T_p. However, herein, we assumed that the s/l interface temperature after the 1-h thermal gradient stage was just at the liquidus of the nominal alloy composition (1.0 K above T_p). Note that this interface temperature can be lower owing to solutal balancing in solid and liquid phases prior to pulling or due to the squeezing of intergranular segregated liquid from the mush, as recently reported by the authors [10]. More importantly, than the actual temperature at which the β-phase first appeared is the fact that the interface continued to move towards lower temperatures. This indicates that the growth temperature for the subsequent PCG is definitely below T_p.

Figure 2b indicates that the pro-peritectic α-phase patches are located at the interface depressions of the peritectic β-phase. From figures 2b–e, it appears that the pro-peritectic α-phase is unwilling to leave the field for the peritectic β-phase, although the overall interface position continues to move towards lower temperatures. The temperature at the triple line (α-phase, β-phase, liquid) is always at a reduced level, which is understandable from the viewpoint of energy. However, from figures 2b–e, it appears that the mean temperature of the peritectic β-phase is slightly higher than the mean temperature of the pro-peritectic α-phase, and it increases with time. Figures 2b–f also indicate that the typical size of the pro-peritectic α-phase lamellae decreases, until a transition from lamellar growth to fibre growth occurs, attended by a gradual decrease in the interface temperature by approximately 9 K. Evidently, we can conclude that PCG led to an increase in solute accumulation ahead of the front, similar to the initial transient for single-phase growth. It also appears that the amount of α-phase decreased. However, this is not easy to determine quantitatively from the available images. We also observed that the coupled-growth pattern experienced a global drift along the inclined isotherm from the front towards the back. This was also observed in recent studies on growing eutectics with tilted s/l interfaces [15].

5. Conclusions and Summary

The experimental findings based on directional solidification under purely diffusive conditions with a rather small pulling velocity using a hyper-peritectic TRIS–NPG alloy can be summarised as follows:

- Solidification commenced via a planar growth of the pro-peritectic α-phase.
- Below the peritectic temperature, the peritectic β-phase grew over the α/l interface in an unstable manner, such that patches of the pro-peritectic α-phase remained in contact with the liquid.
- These α-phase patches grew simultaneously with the β-phase in a coupled manner, so that PCG established. This growth mode continued for hours, whereby the appearance of the pro-peritectic α-phase changed from patch-like to lamellar and finally to rod-like.

- During growth, the global s/l interface recoiled by approximately 3 mm, which corresponds to a reduction in the interface temperature by approximately 9 K. However, during the 22 h of growth, no steady state was reached. This was evident not only based on the changes in the α-phase growth morphology but also based on the fact that the growth velocity was still lower than the pulling speed.
- After pulling was stopped, the α-phase fibres spread over the s/l interface, at least in some areas. In other areas, the β-phase grew over the α-phase fibres. However, a comprehensive interpretation of these observations is still underway.

Acknowledgements
We thank ESA and QinetiQ Space for aiding in the development of the TA instrument and E-USOC at the Polytechnical University Madrid for their support in performing experiments on the ISS. This research was completed owing to the support provided by ESA for the framework of the project METCOMP and the Austrian Research Promotion Agency (FFG) under grant number 865969.

References

[1] Trivedi R 1995 Theory of layered structure formation in peritectic systems *Metall. Mater. Trans.* **26A** pp 583-90.
[2] Trivedi R 2005 The role of heterogeneous nucleation on microstructure evolution in peritectic systems *Scripta Mater.* **53** pp 47-52.
[3] Boettinger W J 1974 The structure of directionally solidified two-phase Sn-Cd peritectic alloys *Metall.Trans.* **5** pp 2023-2031.
[4] Dobler S, Lo T S , Plapp M, Karma A, Kurz W 2004 Peritectic coupled growth *Acta Mater.* **52**, pp. 2795–2808.
[5] Park J S, Trivedi R 1998 Convection-induced novel oscillating microstructure formation in peritectic systems **187** *J. Cryst. Growth* pp 511-515.
[6] Barrio M, Lopez D O, Tamarit J L, Negrier P, Haget Y 1995 Degree of Miscibility between Non -isomorphous Plastic Phases: Binary System NPG (Neopentyl-glycol)-TRIS [tris(hydroxymethyl)aminomethane] *J. Mater Chem.* **5 (3)** pp 431-439.
[7] Ludwig A, Mogeritsch J P, Grasser M 2009 In situ observation of unsteady peritectic growth modes *Trans. Indian Inst. Met.* **62** pp 433-436.
[8] Mogeritsch J P, Eck S, Grasser M, Ludwig A 2010 In Situ Observation of Solidification in an Organic Peritectic Alloy System *Mater. Sci. Forum* **649** pp 159-164.
[9] Ludwig A, Mogeritsch J P 2011 In situ observation of coupled peritectic growth *John Hunt International Symposium London* pp 233-242.
[10] Ludwig A, Mogeritsch J P, Rettenmayr M 2022 On/off directional solidification of near peritectic TRIS-NPG with a planar but tilted solid/liquid interface under microgravity conditions *Scripta Mater.* **214** p. 114683.
[11] Mogeritsch J P, Sillekens W, Ludwig A 2022 In-situ Observation of Coupled Growth Morphologies in Organic Peritectics under Pure Diffusion Conditions, *TMS annual meeting, Anaheim, California*, ISBN: 978-3-030-92381-5
[12] http://www.sigmaaldrich.com
[13] NPG CAS-No 126-30-7 and TRIS CAS-No 77-86-1.
[14] https://www.qinetiq.com/en/sectors/space
[15] Bottin-Rousseau S, Witusiewicz V T, Hecht U, Fernandez J, Laveron-Simavilla A, Akamatsu S 2022 Coexistence of rod-like and lamellar eutectic growth patterns *Scripta. Mater.* **207** p. 114314

In-situ investigation of the solidification dynamics in an irregular eutectic alloy

S Mohagheghi[1], S Bottin-Rousseau[2] and M Şerefoğlu[3]
[1] Istinye University, Department of Mechanical Engineering, Istanbul, Turkey
[2] Sorbonne Université, CNRS UMR 7588, Institut des NanoSciences de Paris, 4 Place Jussieu, 75252 Paris Cedex 05, France
[3] Marmara University, Department of Metallurgical and Materials Engineering, Maltepe, Istanbul, Turkey

Abstract. We present an in-situ experimental investigation of irregular eutectic growth dynamics observed optically in real time during directional solidification (DS) and rotating directional solidification (RDS) of a model transparent alloy, namely Amino-methyl-propanediol-Succinonitrile (AMPD-SCN). The SCN-rich solid grows with a nonfaceted solid-liquid interface, while AMPD forms thin whisker-like faceted crystals. With RDS, we observed simultaneously four distinct growth dynamics along the solidification front thanks to the rotation induced velocity ramp. At low velocity, AMPD crystals grow individually, with their leading tip ahead of – at a higher temperature than – the SCN-liquid interface which presents a linear undercooling change with V ramp. This regime is defined as a quasi-steady decoupled growth. During RDS, AMPD crystals are found to rotate with the sample in this regime. When they can reach the unsteady coupled-growth regime at higher V, since increasing V favors the formation of AMPD-SCN-liquid trijunctions, they start to branch noncrystallographically. At higher velocity, impurity effects lead to the formation of two-phase fingers. Further increase in the velocity results in the formation of primary SCN dendrites and a fine interdendritic eutectic microstructure. DS experiments confirmed the existence of those regimes at different velocities.

1. Introduction

Irregular eutectics are widely used in industry. In irregular eutectic alloys, at least one of the eutectic solid phases grows in a faceted way, and exhibits large, atomically smooth solid-liquid facets [1]. Many crystals of that kind grow on the shape of thin platelets (e.g., Si crystals [2]) or needles with "blocked" facets, the growth kinetics of which is highly nonlinearly dependent of the undercooling and is complexified by the role of crystal defects acting as elementary step flow sources. This entails nonsteady, irregular growth regimes, with a marked disorder in the solidified microstructures. Despite their practical importance, scientific studies on irregular eutectics remain essentially descriptive, and the simultaneous growth of faceted and nonfaceted crystals in an irregular eutectic alloy has been rarely addressed with the aim of highlighting the solidification processes that would not be specific to the chemistry of the alloy. In a recent paper [3], we showed that a faceted/nonfaceted transparent alloy, *i.e.* Amino-methyl-propanediol/Succinonitrile (AMPD-SCN) at near-eutectic concentration, can be used as an experimental model system for exploring the growth dynamics of irregular-eutectic microstructures in situ. As a main outcome, we identified a microstructural transition between decoupled- and coupled-growth as a function of the solidification velocity. Qualitative features that

[3] Corresponding author

were thus revealed in a model transparent system shed new light on previous observations in metallic irregular-eutectic alloys, such as austenite/graphite eutectic in cast irons [4] and Ni-Ni_3Sn_4 alloy [5].
In this work, we present observations of the growth dynamics in thin AMPD-SCN samples during rotating directional solidification (RDS) experiments. During RDS, a thin sample is slowly rotated at constant angular velocities ω with respect to the axis of the fixed unidirectional temperature gradient G. Previously, the RDS method has been used to vary continuously the orientation of the crystals with respect to G during solidification [6-8]. Here, we took advantage of the imposed velocity ramp for the first time to study the different growth mechanisms observed as a function of growth velocity. As expected, the eutectic spacing is decreasing from center to the sample edge following the increase in velocity [7]. During RDS, the local V value increases from zero at the center of rotation to ωR at the sample edge in the lateral direction, x, where R (≈ 2.5 mm) is the outer radius of a circular sample. This *velocity ramp* along x, enabled us to observe different microstructures and to locate the transitions from one pattern to another in a sample with a given composition and geometry –and impurity content. Our RDS observations confirm the previously established classification into loosely separated categories: low-velocity patterns dominated by decoupled growth; intermediate velocity with transient coupling and noncrystallographic branching; high-velocity, impurity dominated regime with two-phase fingers. In addition, we observed a very high-velocity regime characterized by the formation of single-phase SCN dendrites with interdendritic eutectic. In this paper, we shall compare RDS and ordinary DS patterns, thus evidencing the efficiency of the RDS method at uncovering the complex V dependency of an irregular-eutectic growth dynamics.

2. Experimental procedure

Commercial AMPD and SCN were purified to prepare the eutectic alloy. The eutectic composition and temperature of the AMPD-SCN system are C_E = 97.39 mol% SCN and T_E = 325.7 K [9]. Thin glass-wall samples (with 12 μm nominal thickness) were filled by capillarity with the molten AMPD-SCN alloy under Argon atmosphere and sealed. Rectangular and both circular and rectangular samples were used for DS and RDS experiments, respectively. Samples were solidified with various pulling velocities (0.01 - 4 μm/s) during DS and angular velocities (0.015 and 0.032 °/s) during RDS. In RDS experiments, an average local velocity V_{ave} for each regime is reported since the pulling velocity linearly differs from one location to another. During RDS, solidification occurs in half of the sample while the other half undergoes directional melting – this region will not be shown in this paper, and we only focus on the region where solidification occurs. Additional information about RDS method can be found elsewhere [6, 10]. Similar to the DS experiments, the RDS stage was mounted on the XYZ stage of an optical microscope which is used in transmitted light mode for real-time observation of the solidification front.

3. Results and discussions

Using RDS experiments, we could observe four different types of growth dynamics at the solidification front, as shown Fig. 1. In these examples, the pulling velocity is linearly increasing from left (close to the center of rotation) to the right of the image (sample edge). Close to the center of rotation, where the velocity is low, in Fig. 1 (a), decoupled pattern forms in which AMPD crystals grow at higher temperatures than the SCN front. Coupled patterns, *i.e.* trijunctions where AMPD, SCN, and liquid are at local equilibrium, can be observed as going towards the right side of the image where the imposed velocity increases (Fig. 1 (a)). In addition, going from the left to the right side of this image, the increase in velocity results in a decrease of the AMPD apparent thickness and the distance between AMPD phases, yielding much finer solidification microstructure. The overall solid/liquid interface is highly tilted, compared to RDS experiments performed with regular eutectics [6], due to the fact that decoupled and coupled regimes are growing at different undercooling values (the decoupled regime is at higher temperatures than the coupled regime) which can be measured using Gz.
In an RDS experiment with larger sample, higher velocities are obtained close to the sample edge. The growth front towards the edge of this sample is shown in Fig. 1 (b). Towards the center of this sample,

only limited regions of decoupled and coupled growth dynamics are observed due to the high velocity gradient. At this high V_{ave} regime shown in Fig. 1 (b), two-phase fingers and primary SCN dendrites with interdendritic eutectic were observed. Two-phase fingers in this system are composed of a thin AMPD crystal surrounded by dendritic SCN phase. Two-phase fingers were observed during univariant growth of a non-faceted binary alloy in the presence of a small quantity of a third compound or a foreign impurity [11]. We suspect that the trace of a third compound may remain a few ppm. SCN dendrites with interdendritic eutectic is an expected observation when the system is not at the coupled zone as defined by Kurz and collaborators [12]. Evolution of this structure from two-phase fingers occurs when the velocity increases and, as a result, primary SCN dendrites grow at a higher temperature than the fine interdendritic AMPD – SCN coupled eutectic. From the left to right of the image, although V increases, the temperature of the solidification front with two-phase finger structure does not change much. However, when the transition from the two-phase finger to the primary dendrites with interdendritic eutectic occurs, the tips of the primary dendrites grow at higher temperatures, and the solidification front of the interdendritic eutectic structure is at lower one than the two-phase fingers front temperature. After the transition, the tips of the primary dendrites are growing almost at the same temperature although the velocity increases (Fig. 1 (b)). On the other hand, the overall solidification front temperature of the interdendritic eutectic decreases as V increases, as expected. Primary SCN dendrites shown in the right side of Fig. 1 (b) originate from a single grain as their orientation is identical.

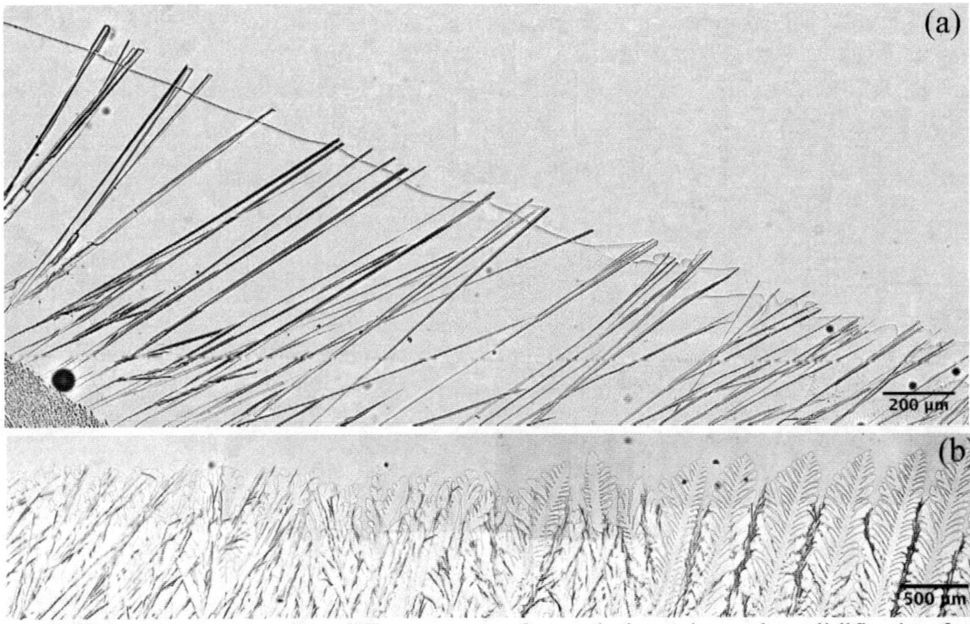

Figure 1. RDS images showing four different types of growth dynamics at the solidification front. As the imposed velocity increases from left to the right side of the images: (a) Transition from decoupled to coupled. $\omega = 0.032°/s$, $G = 5.1 K/mm$, $0 < V_{ave} < 1.3 \mu m/s$. (b) Transition from two-phase fingers to primary SCN dendrites with interdendritic eutectic structure. $\omega = 0.015 °/s$, $1.6 < V_{ave} < 3.9 \mu m/s$.

In the decoupled regime, the faceted phase cannot branch either crystallographically or noncrystallographically. Therefore, further rotation of the sample is accompanied by the rotation of the faceted crystals with the sample. The example shown in Fig. 2 (a) perfectly illustrates the strict rotation of growing faceted crystals for 108° with the sample without branching. It is the extreme case similar to the locked interphases observed in regular eutectics that were studied using RDS stages [8,10]. Absence of both branching and thickening of the crystal indicate that the overall growth direction of the faceted AMPD is in the normal direction of the facet plane at the tip of the crystal in the decoupled regime. When the tip of the crystal shown with the arrow in Fig. 2 (a) reached a higher

V regime (due to growing towards the edge of the sample), coupling occurs. At this moment, the AMPD crystal branches noncrystallographically, as shown by the arrow in Fig. 2 (b). In the coupled regime, the AMPD crystals either continue to rotate with the sample or branch noncrystallographically.

The results of RDS experiments can be confirmed using DS experiments performed at different pulling velocities (Fig. 3). At low velocities, say $V < 0.3$ µm/s, the solidification front comprises a few faceted needle-like AMPD crystals growing mostly ahead of the planar nonfaceted SCN/L interface (Fig. 3 (a)). Therefore, trijunctions, where the growing interfaces of the two solids meet the liquid, are absent, *i.e.* the AMPD-SCN-liquid contact is not at local equilibrium. The tilting angles of AMPD crystals are inherited from the crystallographic orientations in the as-cast solid.

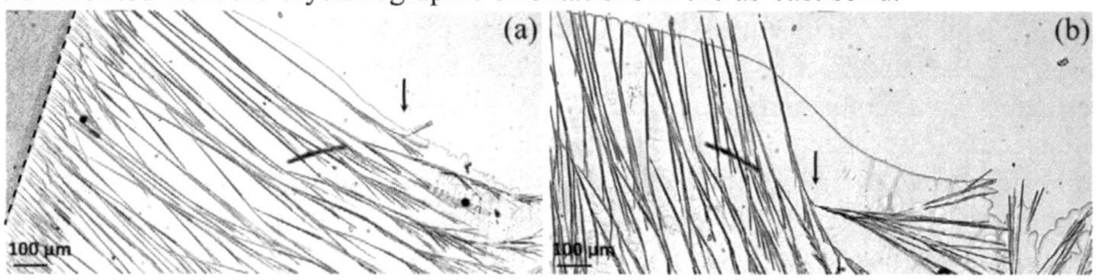

Figure 2. RDS images taken at two different rotation angles, (a) $\Delta\theta = 108°$, (b) $\Delta\theta = 150°$. In this experiment, the initial solid/liquid interface (dashed-line in (a)) was perpendicular to the growth direction at $\theta = 0°$. $\omega = 0.026°$/s, $G = 5.1$ K/mm, $0 < V_{ave} < 0.6$ µm/s. The arrows show the same AMPD crystal.

Figure 3. DS images showing four different types of growth dynamics, namely (a) Decoupled at $V = 0.05$ µm/s, (b) Coupled at $V = 0.6$ µm/s, (c) Two-phase fingers at $V = 1$ µm/s, and (d) Primary dendrite with interdendritic eutectic at $V = 4$ µm/s. For (a) – (c) $G = 7.5$ K/mm, for (d) $G = 8.5$ K/mm.

When the velocity is between $0.3 < V \leq 0.6$ µm/s, unsteady coupled growth is observed (Fig. 3 (b)). In this pattern, many trijunctions form between SCN, AMPD, and liquid phases at the solidification front; however, these trijunctions are frequently pinning and depinning indicating unsteadiness in the growth dynamics. The dynamics was complexified by a phenomenon that is not commonly observed in nonfaceted systems, which was called noncrystallographic branching in reference to spherulitic growth [13]. During noncrystallographic branching, the faceted crystal, in contrast to lamellar and rod splitting in regular eutectics, forms a new crystal with a small misorientation with respect to the parent phase. These frequent branching events lead to a dramatic increase in the number of AMPD crystals and hence reducing the AMPD interphase spacing as shown in Fig. 3 (b).
At high velocity regime ($0.6 < V < 3$ µm/s), two-phase fingers form (Fig. 3 (c)). It is well-established that two-phase fingers form due to impurities in the alloy, as a result of which SCN/L interface is nonplanar and most of the time dendritic. Some of the fingers shown in Fig. 3 (c) are oriented in the

direction of the temperature gradient and some are highly inclined. The inclined blocked AMPD facets that are not exposed to the liquid impose the growth direction, indicating the strong role of the anisotropy of SCN/AMPD interphase boundary. The tip of the interior AMPD crystal can be coupled with the SCN dendrite tip, *i.e.* forming a temporary trijunction. In this case, the blocked facets at the sides of the AMPD crystal are not exposed to the liquid and the two-phase finger can continue to grow for a long time. However, if such a coupled growth configuration is broken and the blocked facets of the AMPD crystal are exposed to liquid, new crystals can be created by noncrystallographic branching from the parent AMPD crystal as shown in Fig. 4. Interestingly, these branching events can occur both at the tip and at locations far away from the tip of the AMPD crystal as shown in Fig. 4 (b) and (d), respectively.

At even higher velocities, primary SCN dendrites with interdendritic eutectic structure are observed (Fig. 3 (d)). Using RDS experiments we already showed that both planar and nonplanar SCN growth microstructures can be observed along the same solidification front. To clarify this point further, we performed another experiment where two major downward velocity jumps were performed. First, the sample was grown at V = 4 μm/s and the resulting microstructure was composed of primary dendrites with interdendritic eutectic. Then, the first downward V jump was employed from 4 μm/s to 1 μm/s. During the transient, the distance between the primary SCN dendrite tips and the eutectic decreased with time. In about 10 minutes, two-phase fingers were established. At the tip of the two-phase fingers, SCN and AMPD grow roughly at the same temperature. After 635 seconds, the second downward V jump was employed from 1 μm/s to 0.1 μm/s. During this second transient, many AMPD crystals were eliminated, SCN/L interface became planar, and the growing tips of the AMPD crystals were ahead of the SCN/L interface. At the end, decoupled microstructure at V = 0.1 μm/s was established after an overnight experiment. Observation of decoupled microstructure where SCN/L is essentially planar after solidifying primary SCN dendrites proves that the transitions between all these four regimes are only due to velocity. It is important to note here that the threshold velocities from the two-phase fingers to the primary dendrites with interdendritic eutectic structure would depend on the impurity level.

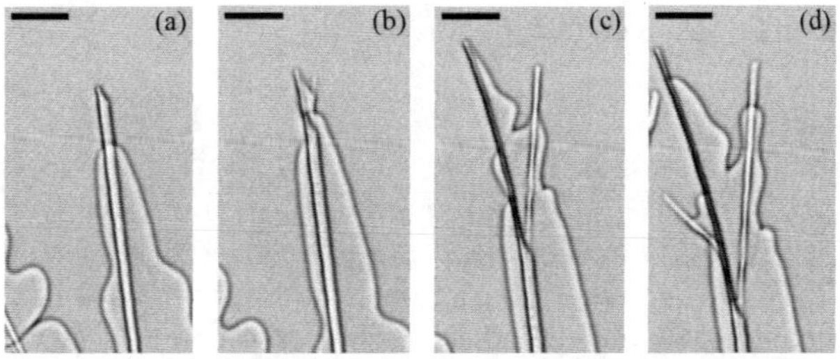

Figure 4. Time sequence showing noncrystallographic branching events of a two-phase finger (a) t = 0, (b) t = 78 s, (c) t = 180 s, (d) t = 234 s. Noncrystallographic branchings occur at the growing tip of AMPD and from the side facet of the same crystal. V = 1 μm/s, G = 7.5 K/mm. The scale bar is 50 μm.

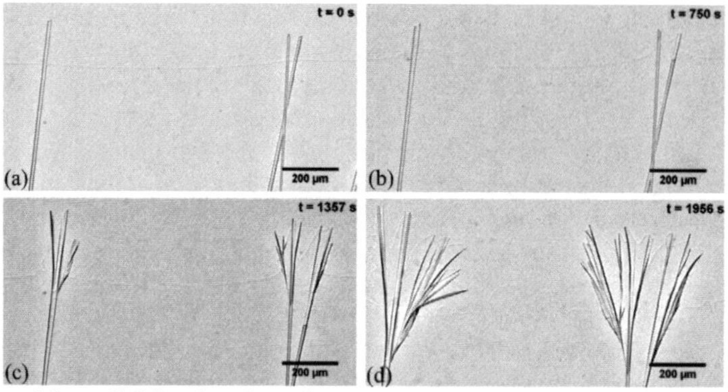

Figure 5. Evolution of the solidification front from decoupled to coupled pattern upon upward V-jump from 0.1 to 0.5 µm/s. G = 7.5 K/mm.

An upward velocity jump experiment was also performed to study the dynamics of the decoupled to coupled transition. After the establishment of the decoupled regime at V = 0.1 µm/s, an upward velocity jump to 0.5 µm/s was applied, and the time is taken as reference, t = 0. As shown in Fig. 5 (a), there are only three crystals in the observation window. As expected from decoupled pattern, the SCN/L interface is planar, and all the crystals grow at higher temperatures than SCN solidification front. After the velocity jump to V = 0.5 µm/s, the SCN close to the AMPD crystals grow faster along the blocked facets of the AMPD crystals which results in a shallow depression at the SCN/L interface between the far away crystal and the two close crystals. Due to this forward jump of SCN, AMPD/SCN/L trijunctions are formed as shown in Fig. 5 (b). At this point, the condition for noncrystallographic branching is satisfied and frequent branching events result in creation of many thin AMPD fibers growing in a coupled manner with SCN (Fig. 5 (c) & (d)). Due to the continuous noncrystallographic branching events, slightly-misoriented branches form over time. As a result of the cumulative effect of these misorientations, new crystals with more than 60° misorientation with respect to the parent phase are observed in Fig. 5 (d).

Summary and Conclusions

We used a model organic transparent alloy system and in-situ observations to investigate the growth dynamics of irregular eutectic alloys. V-ramp imposed by the RDS experiments enabled us to further investigate the various microstructures obtained at different velocities during DS experiments [3] at the same solidification front. Observing all the three previously reported microstructures at the same solidification front confirms that for a given alloy composition, both the microstructure and the growth dynamics, i.e. being coupled vs. decoupled etc., are selected only by the growth velocity. Furthermore, at the very high-V regime, i.e. towards the edge of the sample during RDS experiments, we observed that the system gets out of the coupled-zone region, which results in observation of primary SCN dendrites with eutectic at the interdendritic region. As a result, irregular eutectic growth dynamics can be classified into four main categories as: *(i) Decoupled growth* is mainly observed at low-velocity regime where a steady-state decoupled growth pattern is likely to exist, but its stability is limited by nonlinear interfacial kinetics at the growing tips of the faceted crystal. *(ii) Unsteady coupled growth* is observed at the intermediate velocity regime along with frequent noncrystallographic branching and elimination events. Over long experiments, a balance is reached between the branching and elimination of the AMPD crystals. Hence, the average eutectic spacing is essentially constant. *(iii) Growth of two-phase fingers* is observed at high-velocity regime. Noncrystallographic branching events yield disordered solidification microstructure. *(iv) Growth of primary SCN dendrites with interdendritic eutectic* is observed at even higher velocities. Increase in the velocity drives the system out of the coupled-zone region and favors the formation of primary SCN phase with dendritic interface. The eutectic structure evidently grows at a lower temperature than the tip of the dendritic SCN and in between the dendrites. Unsteady coupled growth with noncrystallographic branching events is observed in the interdendritic eutectic region. The transitions between these growth regimes are reproducible and reversible. Locating a sharp V threshold for these transitions seems to be not feasible. We show the use of RDS stage clearly evidence the V-dependency of the growth dynamics and the selection of the microstructure. By performing RDS experiments, we were able to observe all four patterns in a sample at the same solidification front where the composition and impurity content are constant. Additionally, the RDS results demonstrate the locking dynamics of the AMPD crystals with rotation in the decoupled regime. When reaching the coupled regime, AMPD crystals either continue to rotate or branch noncrystallographically through formation of a triple junction.

The current observations of faceted/nonfaceted eutectic microstructures can assist in explaining some mechanisms of pattern formation in industrial irregular alloy systems. Further theoretical and numerical developments as well as further detailed experimental studies are needed to fully explain the growth dynamics of irregular eutectics.

Acknowledgments

We are deeply grateful to S. Akamatsu for enlightening discussions and to our late colleague G. Faivre for sharing with us his profound thinking at the beginning of this work. This work is funded by TÜBİTAK 2509 (Grant no: 217M089) and Campus France (PHC BOSPHORUS 39706RE).

References

1. Fisher DJ, Kurz W 1980 *Acta Metallurgica* **28** 777-94
2. Hellawell A 1970 *Progress in Materials Science* **15** 3-78
3. Mohagheghi S, Bottin-Rousseau S, Akamatsu S, Şerefoğlu M 2020 *Scripta Materialia* **189** 11-5
4. Lacaze J, Sertucha J 2018 *China Foundry* **15** 457-63
5. Hou N, Belyakov SA, Pay L, Sugiyama A, Yasuda H, Gourlay CM 2018 *Acta Materialia* **149** 119-31
6. Akamatsu S, Bottin-Rousseau S, Şerefoğlu M, Faivre G 2012 *Acta Materialia* **60** 3206-14
7. Mohagheghi S, Şerefoğlu M 2019 *Metallurgical and Materials Transactions A* **50** 5221-33
8. Mohagheghi S, Şerefoğlu M 2018 *Acta Materialia* **151** 432-42
9. Witusiewicz VT, Sturz L, Hecht U, Rex S 2005 *Acta Materialia* **53** 173-83
10. Akamatsu S, Bottin-Rousseau S, Şerefoğlu M, Faivre G 2012 *Acta Materialia* **60** 3199-205
11. Akamatsu S, Faivre G 2000 *Physical Review E* **61** 3757-70
12. Kurz W, Fisher DJ 1979 *International Metals Reviews* **24** 177-204
13. Bisault J, Ryschenkow G, Faivre G 1991 *Journal of Crystal Growth* **110** 889-909

IOP Conf. Series: Materials Science and Engineering 1274 (2023) 012034 doi:10.1088/1757-899X/1274/1/012034

In situ solidification of eutectic Al-33wt%Cu droplets

J Valloton[1,a], N Mahdi[2], L Rabago[1], J Chung[1] and H Henein[1]

[1] Advanced Materials and Processing Laboratory, University of Alberta, Edmonton, Canada
[2] Norcada inc., Edmonton, Canada
[a] Corresponding author: valloton@ualberta.ca

Abstract. Al-33wt%Cu eutectic droplets were rapidly solidified using Impulse Atomization, a drop tube technique. The samples were then processed in situ in an SEM using an HTN-0101 MEMS heating chip manufactured by Norcada. The temperature of the chip, as well as the heating and cooling rates, can be easily set using the chip's control software, allowing for heating and cooling rates as fast as 1000°C/second. The droplets were heated above the eutectic melting temperature. Thanks to the oxide skin, the droplets retained their spherical shape, which allowed in situ solidification experiments. The live feed of the surface of the droplet obtained with the SEM was recorded during the experiments alongside the temperature-time profile. Various cooling rates were imposed and the liquid samples were shown to undercool prior to solidification. The resulting eutectic morphologies and spacings were then analyzed as a function of the cooling rates and undercoolings.

1. Introduction

Solidification of alloys is a complex phenomenon arising in many modern experimental techniques and industrial technologies related to casting and surfaces processing. The variation of different conditions of solidification (such as undercooling or cooling rate) gives a possibility to control the morphology and scale of the microstructure, which substantially influence physical and chemical properties of alloys. Numerous favourable microstructural features can arise from rapid solidification conditions (i.e. solidification occurring at high cooling rates and/or undercoolings). These include microstructure refinement, minimal segregation, solubility extension and the formation of metastable phases [1-6]. Understanding the relationship between solidification parameters and resulting microstructures is therefore critical in order to control the solidification path and resultant mechanical properties. However, relating the observed microstructure with process parameters can be challenging at times due to experimental difficulties. Critical variables such as cooling rate, undercooling and solidification rate, are not always readily measureable. Therefore, to understand the phenomena at play, solidification experiments with controlled parameters are useful. This contribution presents preliminary results of in situ solidification of rapidly solidified Al-33wt%Cu eutectic droplets.

2. Experimental method

Impulse atomization (IA) (a type of drop tube) is a single fluid type of atomization technique (Figure 1) [7]. In this study, 200g of high purity Al (99.99%) and 100g of commercial purity Cu (99.9%) is melted in a graphite crucible using an induction furnace. To ensure full mixing of the alloy, the melt is held for one hour at 750°C (~200°C above the equilibrium eutectic temperature of 548°C) under argon atmosphere. The liquid alloy is then physically pushed through a nozzle plate by the application of a mechanical pressure (impulse) with the help of a plunger. Droplets are created from the destabilisation and subsequent breakup of the liquid ligaments by Rayleigh instabilities. Rapid cooling (and solidification) of the droplets ensues during free fall by heat loss to the surrounding Argon gas. The solidified samples are collected at the bottom of the atomization tower. IA generates a range of droplets sizes per run, giving a range of cooling rates and undercoolings. Thus, the collected droplets are sieved into different size ranges. A detailed description of the process can be found in [7].

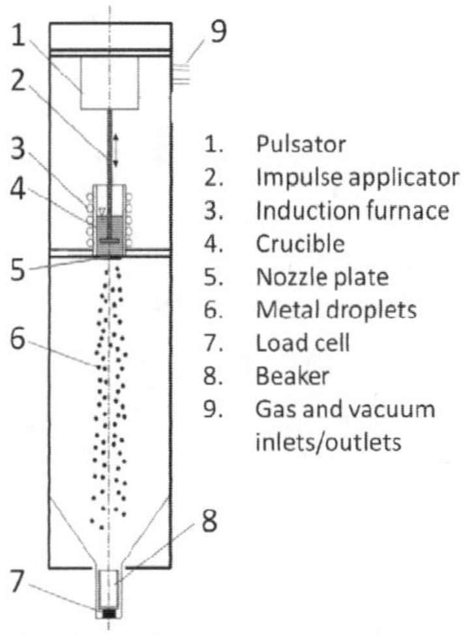

Figure 1: Schematic view of an Impulse Atomization apparatus [1].

One Al-33wt%Cu atomized droplet of diameter less than 90 µm is then mounted on an HTN-0101 MEMS heating chip designed and manufactured by Norcada (Figure 2) [8]. The chip is inserted in a Norcada MEMS in situ heating stage made for ZEISS scanning electron microscopes (NHB-Z-SNL-001). The temperature of the chip and the thermal profiles are controlled via a dedicated software. A constant temperature can be held with a stable and ultra-low drift in the target temperature while heating and cooling rates as rapid as 1000 K/s can be achieved. In this works, the Al-33wt%Cu were heated to 550°C at a rate of 2K/s and subsequently cooled down to 400°C at rates of 5, 10 or 20 K/s. The temperature-time profile is recorded during the in situ experiments alongside the SEM live feed of the surface of the droplet. Changes in microstructure occurring at the surface of the sample can be monitored in situ as a function of time and temperature.

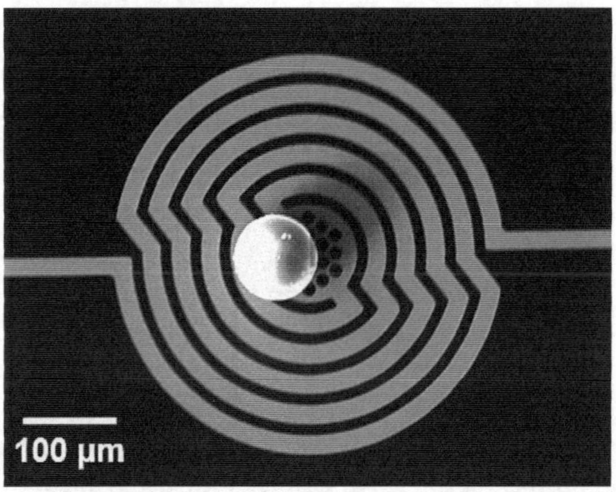

Figure 2: SEM view of Norcada HTN-0101 MEMS heating chip with an Al-33wt%Cu droplet placed at the center for in situ experiments [8].

3. Results and discussion

Figure 3a shows a snapshot of the SEM live feed after heating of the droplet above the eutectic temperature. Thanks to the oxide skin, the liquid does not spill over and the droplet becomes fully spherical. After cooling, the droplet has fully solidified, with visible solidification shrinkage (Figure 3b). The eutectic microstructure at the surface of the sample is clearly apparent, which allows for quantification. The dark grey phase is the α-Al solid solution while the lighter phase is the Al_2Cu intermetallic. Impurities on the surface of the samples are sometimes observed, but they do not seem to act as nucleation points.

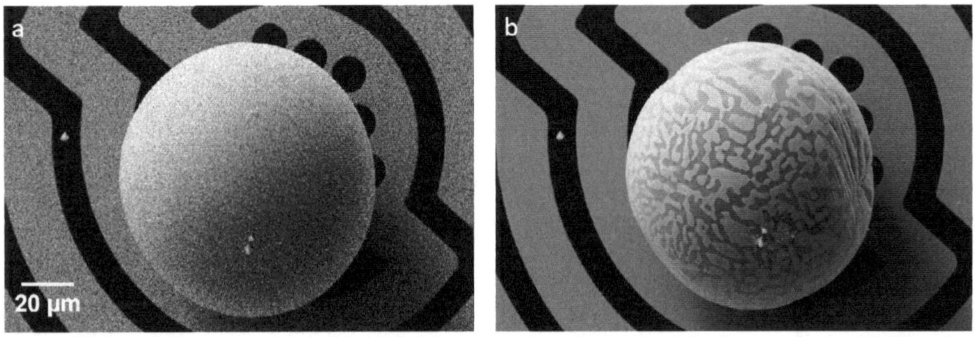

Figure 3: SEM micrographs of an Al-33Cu droplet when fully liquid (a) and after solidification (b).

Figure 4 (left) shows a typical temperature-time profile obtained during an in situ solidification experiment. In this case, heating of the droplet to 550°C was performed at a rate of 2 K/s and cooling to 400°C at a rate of 5 K/s. The temperature acquisition rate of the heating chip is 1 Hz. In all heating cycles, melting of the droplets was observed at 540±2°C. Even though this temperature is slightly lower than the equilibrium eutectic temperature (548°C), it can be used as a reference to see if the samples undercool prior to solidification. On the other hand, the nucleation temperature can only be identified approximately, especially at the higher cooling rates. To estimate this value, the frame from the recorded video feed that contains the first indication of solidification is identified and related to the frames corresponding to the temperature acquisitions directly prior to and following solidification (all the analyzed droplets fully solidified in less than a second). The nucleation temperature is then determined using a linear fit (Figure 4 (right)).

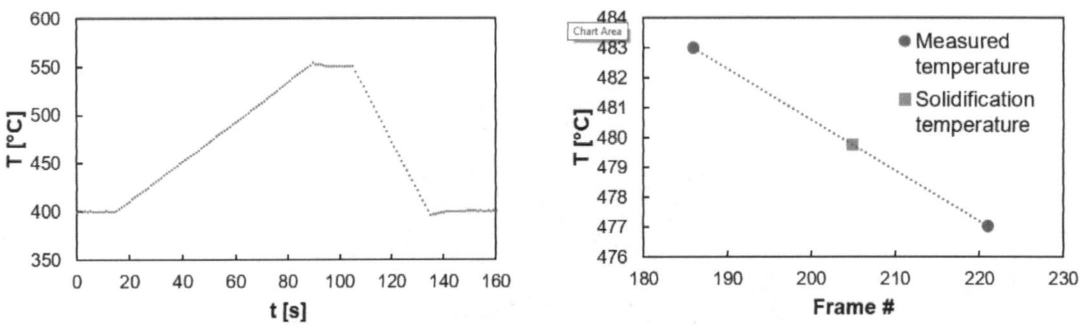

Figure 4: Left: temperature-time profile of a droplet heated at 2 K/s and cooled at 5 K/s. Right: estimation of solidification temperature

Figure 5 shows the surface of Al-33wt%Cu samples that have been solidified in situ with imposed cooling rates of 5, 10 and 20 K/s (a, b and c respectively). Looking at the solidification temperatures, all droplets experienced a certain degree of undercooling prior to nucleation, which increases with the cooling rate (Table 1). However, due to the small size of the droplets, no recalescence event is visible on the temperature-time profiles. Lamellar spacings were measured on the 3 samples shown in Figure 5. To minimize measurement errors due to the near spherical nature of the samples, measurements were only carried out at the center of the samples. As expected, the lamellar spacing decreases as the cooling rate and the undercooling increase. Further experiments will be carried out in order to obtain different undercooling values for a same cooling in order to clarify the role of undercooling on the refinement of the microstructure. Interestingly, the sample with the coarsest eutectic (Figure 5a) does not exhibit a typical lamellar eutectic structure. Some regions see large chunks of Al_2Cu while others are rod-like.

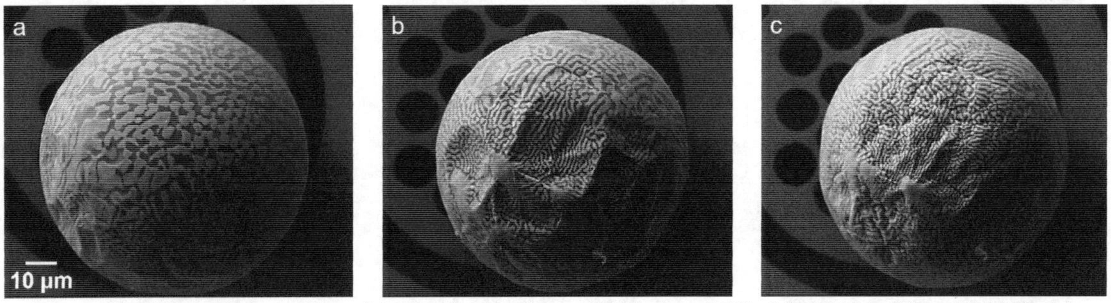

Figure 5: SEM micrographs of Al-33Cu droplets solidified at 5 K/s (a), 10 K/s (b) and 20 K/s (c).

Table 1: Cooling rate, undercooling and lamellar spacing of the samples shown in Figure 5.

Sample	Cooling rate [K/s]	Undercooling [K]	Lamellar spacing [µm]
a	5	68	3.3 ± 0.6
b	10	75	1.5 ± 0.2
c	20	97	1.0 ± 0.2

4. Conclusions

In situ experiments were carried out on Al-33wt%Cu impulse atomized droplets. The samples were processed in a scanning electron microscope using Norcada MEMS heating stage. When the sample is brought over the eutectic temperature, the oxide skin allows the liquid droplet to keep its spherical shape. This enables in situ solidification experiments. First tests show that the droplets undercools prior to solidification. The undercoolings increase as the imposed cooling rates increase. The resulting microstructure is thus refined.

The Norcada stage for MEMS in situ heating and cooling of samples in a scanning electron microscope shows great potential for aging and solidification experiments that will be explored further.

Acknowledgments

The financial support of the Natural Sciences and Engineering Research Council of Canada (NSERC) is gratefully acknowledged.

References

[1] Jones H 1982 *Rapid Solidification of Metals and Alloys* (London: Institute of Metallurgists)
[2] Mehrabian R 1982 *Int. Met. Rev.* **27** 185
[3] Froes FH, Kim YW, Hehmann F 1987 *JOM* **39** 14
[4] Murakami Y 1996 *Mater. Sci. Technol.* **8** 213
[5] Lavernia EJ, Srivatsan TS 2010 *J. Mater. Sci.* **45** 287
[6] DebRoy T, Mukherjee T, Wei HL, Elmer WJ and Milewski JO 2021 *Nat. Rev. Mater.* **6** 48
[7] Henein H 2002 *J. Mater. Sci. Eng. A* **326** 92
[8] Valloton J, Mahdi N, Rabago L, Henein H 2022 Eskin, D. (eds) Light Metals 2022. The Minerals, Metals & Materials Series. Springer, Cham.

Microstructure evolution during directional solidification of off-eutectic Al-Cu-Mg alloys

Dominic Ezemenaka, Samuel Small and Amber Genau[a]

Materials Science and Engineering Department, University of Alabama at Birmingham, Birmingham, AL 35294, USA

Email: [a]genau@uab.edu

Abstract. Building on previous studies of invariant ternary eutectic Al-Al_2Cu-Al_2CuMg, this work aims to deepen our understanding of the microstructural possibilities in the Al-rich region of the Al-Cu-Mg phase diagram by considering the solidification behaviour of several off-eutectic compositions. Four Al-Cu-Mg alloys were directionally solidified at a constant growth rate (1.25 μm/s) and thermal gradient (9-14 K/mm) using a Bridgman-type furnace and analysed using a combination of scanning electron microscopy, energy-dispersive X-ray spectroscopy, X-ray diffraction, and differential scanning calorimetry. Qualitative and quantitative analyses showed a preference for faceted Al_2Cu dendrites surrounded by a ternary eutectic matrix in the two Mg-poor samples, even though one composition was not in the primary Al_2Cu phase region. The two Al-poor samples consisted of large regions of three different intermetallic phases and little eutectic, making these samples very brittle.

1. Introduction

As the microstructural morphologies that evolve during solidification of ternary eutectics can be significantly more complicated than those of binary alloys [1], their study provides valuable opportunities to deepen our understanding of the overarching mechanisms that control microstructure formation in multi-component, multi-phase alloys. Many such alloys are of advanced technological importance for both structural and functional applications [2]. Precise control of the growth of microstructural features of these alloys is often important to achieve the desired mechanical or physical properties. Directional solidification using a Bridgman-type furnace is a valuable tool for interrogating the connections between composition, processing conditions and microstructure formation during freezing, as both velocity of the solidification front and thermal gradient can be independently and precisely controlled.

The growth mechanisms of higher-order eutectic systems have become a subject of interest in recent years, with researchers examining both invariant and univariant coupled growth. Al-Cu-Ag has emerged as a model system of particular interest [3-7], with recent efforts extending to related Al-Cu-Mg alloys [8-10]. Newly published work by the authors showed that in directional solidification of Al-Cu-Mg with nominally the invariant ternary eutectic composition [$L \leftrightarrows \alpha(Al) + \theta(Al_2Cu) + S(Al_2CuMg)$], the system was able to maintain stable, three-phase coupled growth over a range of liquid compositions, solidification velocities and thermal gradients, and to be able to accommodate a range of compositions within the ternary eutectic itself [8]. By adjusting processing conditions, pure ternary eutectic or a bimodal structure consisting of two-phase cells and highly regular ternary eutectic could be produced. These types of hierarchical microstructures show promise for a variety of high-strength, lightweight applications [11].

An early study focusing on directional solidification of monovariant Al-Cu-Mg alloys was carried out by Garmong in 1971 [12] at compositions along the tie line joining the binary Al-Cu and ternary Al-Cu-Mg eutectics, which deployed the constitutional undercooling model to describe the criterion for the breakdown of the planar growth front in the presence of varying Mg content. Additional investigations of the microstructure evolution and solidification paths in Al-Cu-Mg alloys can be found in [13-16]. The goal of the current work is to use similar solidification conditions as [8,9] to compare the microstructure growth behaviour of the eutectic composition alloy with those of off-eutectic alloys having compositions: (a) on the Al-Al_2Cu monovariant line, (b) within Al_2Cu and Al_2CuMg primary crystallization fields, and (c) at the end of Al_2Cu-Al_2CuMg tie line.

2. Experimental Procedure

In this study, four experimental alloys with compositions in the vicinity of the Al-rich ternary eutectic (I_4 of Fig. 1) were prepared from high purity elements: Al (99.99%), Cu (99.9%) and Mg (99.8%). Each 150 g batch of starting material was melted inside a graphite crucible in an electric furnace. Extraction of the melt was done using a suction bulb connected to quartz tubes with 4 mm inner diameter. Phase transition temperatures of the as-cast alloys were measured with a differential scanning calorimeter (TA Instruments Q100 DSC) using a continuous supply of nitrogen purge gas with a flow rate of 50 mL/min. The experimental runs were carried out using sample sizes within the range of 8 - 11 mg in an alumina crucible. The heating and cooling rates were kept constant at 10°C/min.

The samples were directionally solidified using a Bridgman-type furnace at a constant solidification velocity (V) of 1.25 μm/s and thermal gradients (G) of 9.0 K/mm for the first alloy, and 14.4 K/mm for the remaining three alloys. In order to produce these gradients, the temperature of the hot zone within the furnace was set to 942°C and 1200°C, respectively. The reader is directed to ref. [8] for more details about the working principle of this furnace. The processed length of each sample was approximately 75 mm which ensured that the sample had sufficient time to establish a stable microstructure. After processing, the samples were quenched by allowing the crucible to fall into a container of water so that the shape of the solid/liquid interface could be examined.

Metallographic preparation of the samples was carried out by hand using different sizes of SiC grinding papers followed by polishing with diamond pastes of 1.0, 0.5, and 0.1 μm. This was done on the total longitudinal section of each processed sample as well as on transverse sections (parallel to the solidification direction) for different growth heights. The microstructural features of the polished samples were qualitatively examined using both optical microscope (Zeiss Axioplan) and scanning electron microscope (Quanta FEG 650 SEM). Four representative backscatter electron (BSE) images of the transverse sections were taken at each growth height for quantitative microstructure analysis using a modified version of the MATLAB codes described in ref. [18].

The elemental composition of the phases was measured using an energy-dispersive x-ray spectroscopy (EDX) detector attached to the SEM, while phase determination was carried out via x-ray diffraction (XRD) using CoKα radiation. The Bruker Discover D8 diffractometer is equipped with 2D Vantec-500 detector and a 0.8 mm collimator. The matching of the diffraction patterns with the calculated/literature data was carried out using ICDD (International Centre for Diff. Data) database.

3. Results and Discussion

The composition of each of the four alloys as determined by EDX is marked on the partial liquidus surface of the Al-Cu-Mg ternary system in Figure 1b and listed in Table 1. According to the current phase diagram, alloys #1 and #3 are within the primary crystallization fields of $\theta(Al_2Cu)$ and $S(Al_2CuMg)$, respectively. Alloy #2 is a monovariant alloy with composition close to that of the ternary eutectic at I_4. Alloy #4 falls approximately at the end of the tie line joining the ternary eutectic point with transition point II_9 (±0.5 at% compositional difference between Alloy #4 and the point II_9). Transition temperatures for each alloy as determined from DSC are given in Table 1. Three of the alloys show a final solidification temperature of 500.0°C, 3.0°C lower than the accepted ternary eutectic temperature. Although microstructurally, alloy #1 should also end with the ternary eutectic, it shows a higher final solidification temperature of 506.8 °C. Nevertheless, alloys #1 and #2 have similar liquidus temperatures (515.6 °C and 516.5 °C respectively), presumably corresponding to the solidification of primary Al_2Cu dendrites (see micrographs in Figure 2 below). Based on the positions of the compositions on the phase diagram, alloy #2 ought to have the lowest liquidus temperatures and either alloy #3 or #4 the highest, but the opposite is the case. More investigation is needed.

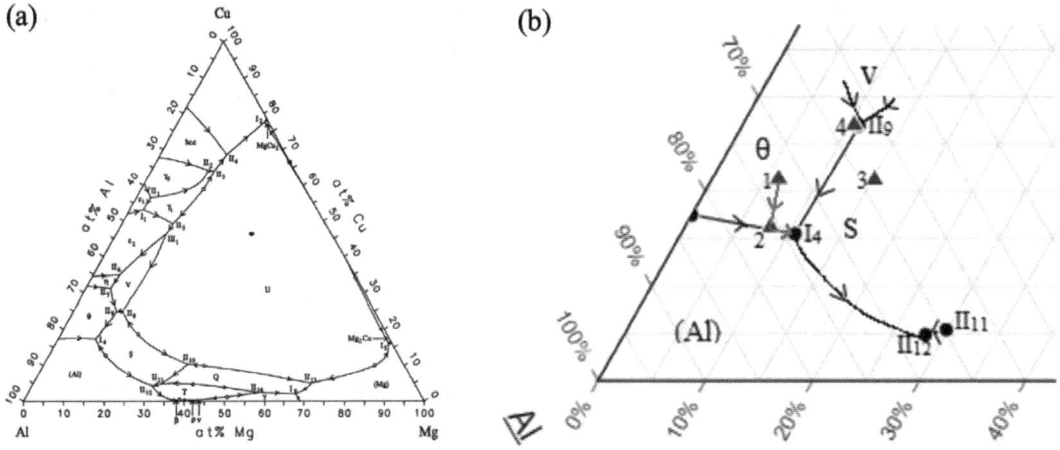

Fig. 1: (a) Calculated liquidus projection of the Al-Cu-Mg system [17], (b) Partial liquidus showing four measured experimental compositions as blue triangles around the ternary eutectic point I_4. Red arrows indicate the presumed solidification path of alloy #1 based on microstructural evidence.

Table 1: Composition of the off-eutectic alloys and their processing conditions

Alloy composition (at. %)				Transition temperature (°C)		Solidification parameters		
#	Al	Cu	Mg	Solidus	Liquidus	G (K/mm)	V (1.25 μm/s)	G/V (K min/mm²)
1	72.5	21.3	6.2	506.8	515.6	9.0	1.25	120
2	75.8	16.2	8.0	500.0	516.5	14.4	1.25	192
3	63.6	21.2	15.2	500.0	510.0	14.4	1.25	192
4	62.6	27.0	10.4	500.0	508.5	14.4	1.25	192

The appearance and evolution of the microstructure in alloys #1 and #2 is similar and shown via transverse cross sections in Figure 2. The solidification distance (growth height) is given in millimeters in the top right of each image. For both alloys, the microstructure is dominated by blocky primary Al_2Cu surrounded by regions of finer eutectic. Near the beginning of solidification, the primary phase consists of relatively large, irregular structures, which cover a large fraction of the surface area. As solidification progresses, the primary dendrite cross sections become smaller and mostly L- or U-shaped, sometimes exhibiting an incomplete square shape, with a clear preferred alignment. Figure 2e shows a lower magnification image of most of the transverse section of alloy #1 just below the quenched interface. At this magnification, the primary Al_2Cu dendrites are all approximately equal in size and are distributed uniformly across the sample. However, the blocky phases can be separated into two distinct regions based on shape and orientation. The upper right third of the sample is covered by L-shaped structures with similar orientation. The remainder of the sample is covered by clusters of U- and rectangular shaped primary phase. It must be assumed that there is a crystallographic relationship between the primary phases in each of these two regions. Small isotropic Al dendrites also begin to appear at this distance.

Closer examination of the eutectic in alloy #1 shows it to be composed of two different morphologies: somewhat coarser regions of intermixed Al and Al_2CuMg and very fine regions of three-phase eutectic. The ternary regions exhibit the strong lamellar tendency between Al and Al_2CuMg observed by [8,19] and others. The three distinct types of microstructure in alloy #1 can be used to determine its apparent solidification pathway. As the sample is cooled, the Al_2Cu primary phase begins to form, changing the melt composition and causing it to move down the slope of the liquidus surface. Because the primary phase is removing Al and Cu atoms from the melt, one would expect that the liquid composition would shift toward the right, causing a binary eutectic of Al_2Cu (θ) and Al_2CuMg (S) to form, but there is no microstructural evidence for this. Instead, solidification seems to follow the path marked by the red arrow in Figure 1 to the Al-θ binary groove, forming irregular regions of semi-coupled Al and Al_2Cu for a short time before the final liquid freezes at the ternary eutectic point I_4.

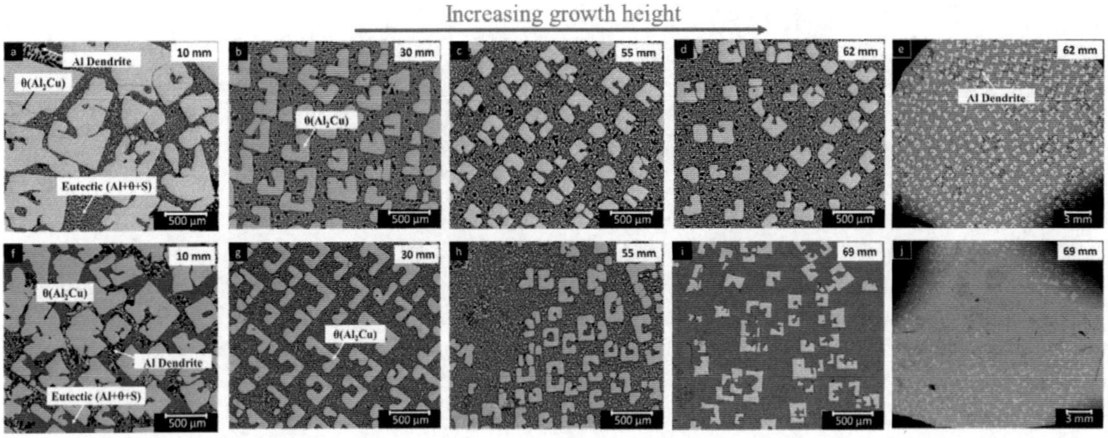

Fig. 2: Microstructure evolution of the alloys as growth height increases. Alloy #1 is (a) to (e). Alloy #2 is (f) to (j), with both quench (i) and (j) showing continued directional growth beyond quenching.

In comparison to alloy #1, the primary phase morphology of alloy #2 displays cross sections that are thinner and somewhat more elongated. Although they are not distributed so uniformly across the sample, they are surrounded by a more uniform three-phase eutectic, with no evidence of two-phase growth. Note that the eutectic in Fig. 2i appears extremely fine because this image was taken from a cross section just past the quenched interface. The phase diagram predicts alloy #2 to fall along the monovariant eutectic groove, so that one would expect no primary phase, but rather two-phase cellular regions surrounded by fine ternary eutectic. Instead, the solidification path appears to go directly from a point in the Al_2Cu primary phase region to the ternary eutectic. The 1971 study by Garmong of alloys along the same univariant groove showed no primary Al_2Cu, but rather primary Al (easily distinguishable by its non-faceted morphology) when velocity and Mg levels were high (that is, compositions close to the ternary eutectic point) [12]. The Garmong experiments used significantly higher velocities and lower thermal gradients in comparison to the current study, so the dendritic phase was formed by destabilization of the solid/liquid interface, a different mechanism than that which formed the primary phase in alloy #1. It may be that very slight changes in composition can shift a sample from just above to just below the univariant eutectic line, from the primary Al_2Cu region to the solid solution Al region. More investigation into this area is needed.

Figure 3 shows longitudinal images of alloy #1 and #2 near the end of controlled solidification as the sample was quenched. In both cases, the Al_2Cu dendrites can be seen growing ahead of the eutectic interface, which is relatively flat. In the manner typical of faceted dendrites, the Al_2Cu grows with a particular crystallographically determined tilt angle, periodically branching at specific angles to maintain approximate alignment with the direction of heat extraction. This behaviour, where θ(Al_2Cu) exhibits faceted behaviour when growing as the primary phase but simultaneous non-faceted behaviour as a couple eutectic phase has also been observed in Al-Cu [19], Al-Cu-Ag [20] and Al-Cu-Ag-Mg [21] alloys. For alloy #1, Al dendrites are also visible. From this viewpoint, it is more apparent that the Al dendrites are growing cooperatively with Al_2Cu dendrites, further supporting the conclusion that this sample follows the solidification path shown in Figure 1. Again, only fine three-phase eutectic is observed surrounding the primary Al_2Cu dendrites in alloy #2.

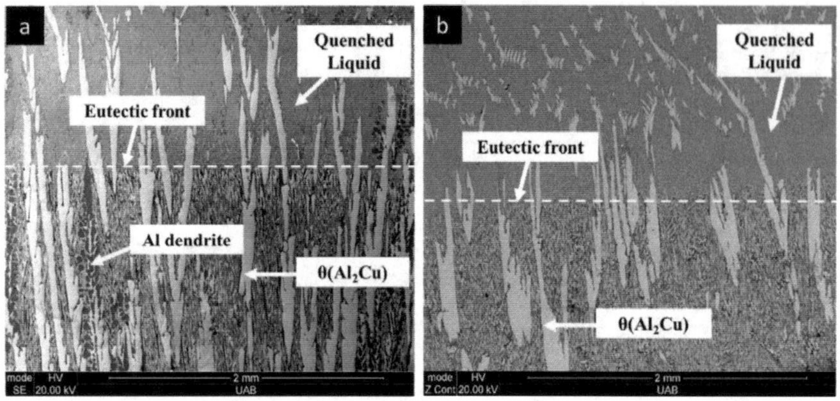

Fig. 3: Longitudinal BSE images showing quenched interfaces of (a) alloy #1, and (b) alloy #2. Solidification direction in the images is upwards.

Quantitative analysis of the primary Al_2Cu in alloy #1 and #2 is shown in Figure 4. In both cases, the amount of primary phase begins very high and drops rapidly within the first 30 mm of solidification. The value then mostly plateaus, although it continues to drop slightly. The decreasing amount of primary phase with increasing solidification length can be attributed to the decreased amount of solute (here, aluminum) in the liquid as the remaining liquid gradually approaches the ternary eutectic composition.

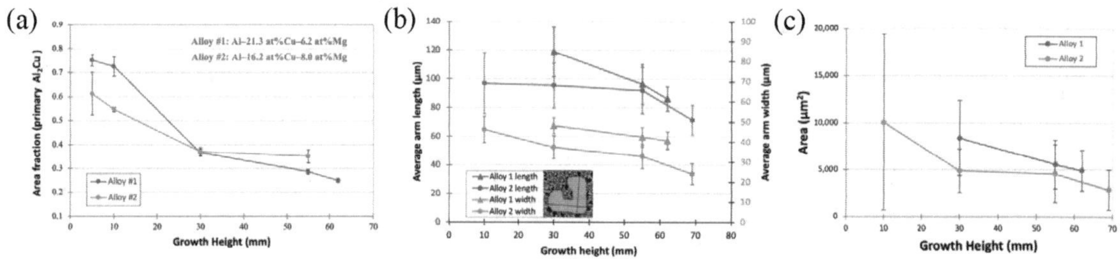

Fig. 4: Variation of amount and dimensions of primary Al_2Cu phase with composition and growth height. Error bars indicate one standard deviation.

As expected, alloy #2 contains more eutectic and less primary phase because the composition is significantly closer to the ternary eutectic point. However, during the period where the microstructure was well established (between 30 and 60 mm), the two alloys contain roughly the same area fraction of bulk Al_2Cu. The reduction in amount of primary phase with growth distance could be accomplished either by a reduction in the number of dendrites or by a reduction in cross-sectional area of existing dendrites. Figure 4b illustrates how the dimensions of the primary Al_2Cu cross sections change with growth length. Both the major axis (length) and thickness (width) of the cross sections decrease as solidification proceeds in both alloys. The dendrites are slightly larger but significantly thicker in alloy #1 compared to alloy #2. Figure 4c shows that the average cross-sectional area of dendrites in both alloys also decrease with growth height, suggesting that the system reduces the amount of primary phase by making the dendrites smaller and/or thinner, rather than termination of dendrites as solidification progresses.

Representative microstructures of alloys #3 and 4 are shown in Fig. 4. These two samples were extremely brittle, making them impossible to remove from the alumina crucibles without multiple fractures. It was therefore not possible to determine the specific growth height of each cross section, although the images are provided in order of increasing growth distance. Both samples are dominated by large, faceted growths of Al_2Cu, which, although more irregular, maintain traces of the characteristic shapes displayed by Al_2Cu in alloys #1 and #2. The Al_2Cu is surrounded by slightly faceted S phase (Al_2CuMg) of a similar scale. Although this microstructure does not meet the strict definition of a divorced eutectic with one massive phase growing in a continuous phase, the two massive phases are clearly growing in a semi-coorperative way, as one would expect from a solidification path following the univariant groove from II_9 to I_4 (see Figure 1). No additional primary S phase could be observed in alloy #4 as one would expect from its position on the phase diagram, although larger regions of Al were present and the V phase ($Al_5Cu_6Mg_2$) was confirmed by XRD to exist in some regions of both samples. The large contiguous intermetallic phase regions account for the brittle behavior of these alloys, which lack the continuous ternary eutectic matrix in which the intermetallic dendrites of #1 and #2 grow.

Fig. 4: Representative microstructures of alloys with increasing growth height. Alloy #3 is (a) to (c). Alloy #4 is (d) to (f). Image (g) is higher magnification of alloy #3 showing a region where the V ($Al_5Cu_6Mg_2$) phase was also observed.

4. Conclusions

In this work, the microstructures produced by four off-eutectic Al-Cu-Mg alloys were investigated. The microstructures of alloys #1 (Al–21.3at%Cu–6.2at%Mg) and #2 (Al–16.2 at%Cu–8.0 at%Mg) are dominated by blocky, primary Al_2Cu dendrites which grow with particular orientations within the finer eutectic matrix. Quantitative analysis shows that area fraction of the primary phase decreases with growth distance by decreasing both the cross-sectional length and width of these features. The microstructures of alloys #3 and #4 are composed primarily of bulky intermetallic phases S(Al_2CuMg), θ(Al_2Cu), and sometimes V ($Al_5Cu_6Mg_2$) with only small amounts of ternary eutectic in between, making the samples extremely brittle. All four samples showed unexpected microstructural features when compared to the predictions of the thermodynamically calculated ternary phase diagram and only one solidification path could be determined. However, the results do show that unpredicted structures which may be of industrial interest (e.g. multimodal Al_2Cu dendrites surrounded by fine, uniform ternary eutectic) can be produced given the right composition and processing conditions. More experimental work, including mechanical testing, on these unique alloys is clearly needed.

Acknowledgements

This work was carried out with funding support from the National Aeronautics and Space Administration (NASA), under Grant No. NNX14AP82G within the Microgravity Research Program and the National Science Foundation (NSF) under Grant No. 1554856. The authors acknowledge Christian Patino for valuable research assistance.

References

[1] Hecht U, Gránásy L, Pusztai T, Böttger B, Apel M, Witusiewicz V, Ratke L, De Wilde J, Froyen L, Camel D, Drevet B, Faivre G, Fries S, Legendre B, Rex S 2004 *Mat Sci Eng R* **46** 1-49

[2] Tiwary CS, Pandey P, Sarkar S, Das R, Samal S, Biswas K, Chattopadhyay K 2022 *Progress in Mat Sci* **123** 100793

[3] Genau A, Ratke L 2012 *Int J Mater Res* **103** 469-475

[4] Hötzer J, Steinmetz P, Dennstedt A, Genau A, Kellner M, Sargin I, Nestler B, 2017 *Acta Mater* **136** 335-346

[5] De Wilde J, Froyen J, Rex S 2004 *Scripta Mater* **51** 533-538

[6] Steinmetz P, Dennstedt A, Şerefoğlu M, Sargin I, Genau A, Hecht U 2018 *Acta Mater* **157** 96-105

[7] Lindemann, GR, Shahani, AJ 2020 *J Alloys and Compounds 827* 154124

[8] Ezemenaka D, Patino C, Genau A 2021 *J Alloys Compd* **883** 160818

[9] Ezemenaka D, Genau A 2022 *J Crystal Growth* **577** 126389

[10] Kaygısız Y, Maraşlı N 2017 *Physics Metals Metallogr* **118** 389-398

[11] Cai Q, Mendis CL, Chang ITH, Fan Z 2021 *Mater Sci Eng A* **800** 140357

[12] Garmong G 1971 *Metall Trans* **2** 2025-30

[13] Chen R, Xu QY, Liu BC 2015 *Acta Metall Sin* **28** 173–181

[14] Vušanović I, Krane MJM 2002 *Int Commun Heat Mass Transf* **29** 1037–1046

[15] Yan EH, Li XZ, Su YQ, Liu DM, Xu DM, Guo JJ, Fu HZ, 2013 *Int J Mater Res* **104** 244–254

[16] Zhao G, Chen J, Ding C, Gu M, Ye Y 2019 *J Mater Eng and Perform* **28** 6980–92

[17] Chen SL, Zuo Y, Liang H, Chang YA 1997 *Metall Mater Trans A* **28** 435

[18] Gauntt E 2018 *Automated image analysis of ternary eutectic microstructures* MSc Dissertation, University of Alabama at Birmingham

[19] Hamar R, Lemaignan C 1981 *J Cryst Growth* **53** 586–90

[20] De Wilde J, Nagels E, Lemoisson F, Froyen L 2005 *Mater Sci Eng A* **413/414** 514-20

[21] Zhou B, Froyen L 2014 *Metallogr Microstruct Anal* **3** 164-73

Pseudo-4D view of the growth and form of crystallographically 'locked' eutectic colonies

P Chao[1], G R Lindemann[1] and A J Shahani[1,2]

1 Dept. Mater. Sci. and Eng., University of Michigan, Ann Arbor, MI 48109, USA
2 Dept. Chem. Eng., University of Michigan, Ann Arbor, MI 48109, USA

E-mail: shahani@umich.edu

Abstract. A number of fundamental questions remain unanswered regarding the interior structure of eutectic cells or colonies growing into a liquid. Here, we observe experimentally the solidification behaviour of an Al-Al_2Cu eutectic as a model system using a novel imaging approach that combines *in situ* X-radiography and *ex situ* X-ray tomography. The former allows us to trace the solid-liquid interfaces while the latter enables us to visualize the solid-solid interfaces. When put together, we can visualize in 4D (*i.e.*, three-dimensional space plus time) the development of a crystallographically 'locked' eutectic microstructure in the presence of an oscillatory eutectic-liquid velocity driven by thermosolutal convection. These details provide direct insight into the mechanism of the lamella-to-rod transition as the eutectic accommodates fluctuations in interfacial composition and growth velocity. We find that this morphological transition reflects a delicate balance between the motion of trijunctions and the stiffness of low-energy, solid-solid interfaces. Our pseudo-4D tomography approach holds broad appeal to the solidification science community, as it can temporally resolve the solidification process on the order of seconds and spatially resolves individual lamellae on the order of micrometres, thereby overcoming the space-time trade-off in conventional *in situ* X-ray tomography.

1. Introduction

In the past decade, synchrotron X-ray imaging has opened new doors to the visualization of solidification processes in metallic systems. Detected phenomena include the nucleation and growth of crystals [1,2,3], dendrite fragmentation [4], dendrite growth and coarsening [5], interface instabilities [6] and flow effects [7]. In the realm of eutectics, investigators have so far focused on the breakdown of a planar front into an array of colonies of a typical length-scale on the order of 10-100 μm.

For example, Wang *et al.* show that this instability is driven by an impurity-induced constitutional undercooling while thermal and convective effects are negligible [8]. Likewise, the colony size is uncorrelated to the distance between dendrites, indicating that a pinning of the eutectic solidification front is not the underlying mechanism. Synchrotron X-radiography demonstrates that the evolution of the colonies is completely unsteady, with eutectic colonies being created and destroyed as solidification proceeds (Figure 1(a)), in agreement with recent phase field simulations [9].

More recently, the same team investigated the evolution of macrofaceted eutectic colonies in a liquid, see Figure 1(b) [10]. That is, the macrofacets extend over distances of many lamellar spacings. They are brought about by the introduction of (at least) one phase in the eutectic that possesses faceted solid-liquid interfaces (in this case, Mg_2Zn_{11}). The eutectic nucleates on this faceted phase, inheriting its crystalline anisotropy (observe in Figure 1(b) that the macrofacets of the eutectic are parallel to the facets of the primary Mg_2Zn_{11} phase, in orange). Two classes of long-wavelength instabilities can be

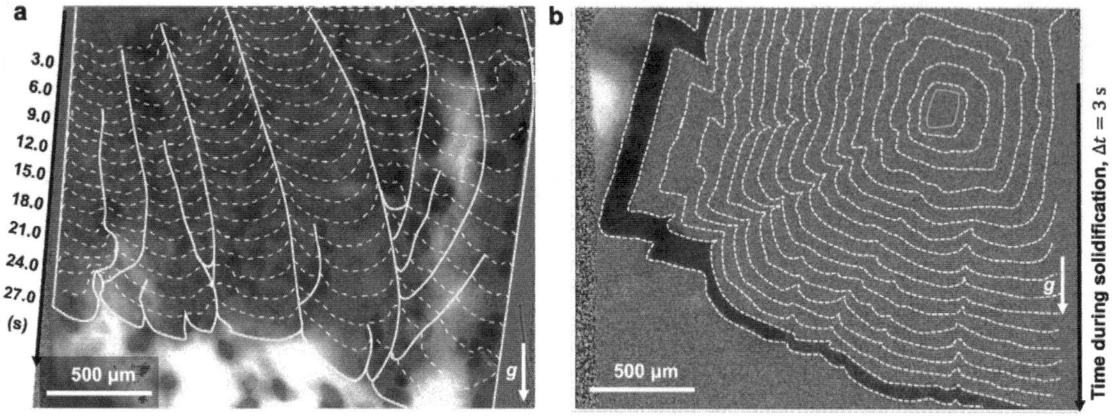

Figure 1: Two types of cellular eutectic microstructures, imaged *via in situ* X-radiography during solidification and shown as spatiotemporal diagrams: (a) one with smoothly curved solid-liquid interfaces (dashed lines), as seen for the Al-Zn eutectic [8], and (b) another with macrofaceted interfaces (dashed), which bound the Al-Zn-Mg_2Zn_{11} eutectic [10]. In (b), the eutectic nucleates and grows heterogeneously from a faceted Mg_2Zn_{11} particle (orange outline), preserving the facet orientations. Direction of gravity indicated by g. Retrieved from Refs. [8,10], respectively.

seen: slower-moving shallow cells (at early times) and faster-moving macrosteps (at later times). Presumably, the latter arises due to an imbalance between the constitutional undercooling and the interfacial stiffness. The macrosteps are unstable, drifting along the eutectic solid-liquid interfaces [10].

Considerably less effort has been devoted to imaging *via* synchrotron-based X-ray tomography the microstructure evolution at *shorter* wavelengths, on the order of the lamellar spacing. Open questions pertain to the scaling relations between interface velocity and lamellar spacing in a cellular eutectic; the mechanisms by which the eutectic microstructure adapts to changes or fluctuations in composition and velocity under non-steady conditions; and the influence of the solid-solid interfacial energy anisotropy on the resulting pattern formation. These questions persist because it is nontrivial to image the solidification dynamics in metals on the scale of 1 μm or less. This is particularly true for regular eutectics, which generally show a finer lamellar spacing compared to irregular eutectics [11].

The challenge in tomographic imaging arises due to the fact that the temporal resolution Δt is limited by the number of projection images N_p per 180° rotation of the sample as well as the exposure time per projection image t_{exp}. That is, $\Delta t \geq t_{exp} \cdot N_p$ in the conventional sense [12]. Reducing either t_{exp} or N_p to keep pace with the dynamics of eutectic solidification leads to a blurring in the spatial domain such that it is difficult to trace the eutectic solid-solid interfaces with reasonable accuracy. Likewise, increasing these two parameters leads to a blurring in the temporal domain such that the eutectic microstructure may have evolved significantly within a span of Δt. This serves to illustrate the fundamental trade-off between spatial and temporal resolutions in *in situ* X-ray tomography.

Here, we circumvent this issue by developing a new method for real-time imaging, pseudo-4D X-ray tomography [13]. We apply this approach to investigate the evolution of microstructure *inside* eutectic colonies, with a focus on the morphological transitions between lamellae and rods.

2. Methods

We summarize the experimental methods given already in Ref. [13]. As a proof-of-concept, we investigate samples of eutectic composition Al-33wt%Cu, cast *via* vacuum arc remelting at the Materials Preparation Centre at Ames National Laboratory (Ames, IA, USA). The sample contains also trace quantities of Si (0.02wt%), as determined *via* inductively coupled plasma mass spectrometry. With a partition coefficient below unity in the eutectic phases Al and Al_2Cu, the impurity Si species will be rejected into the melt, giving rise to a constitutional undercooling and

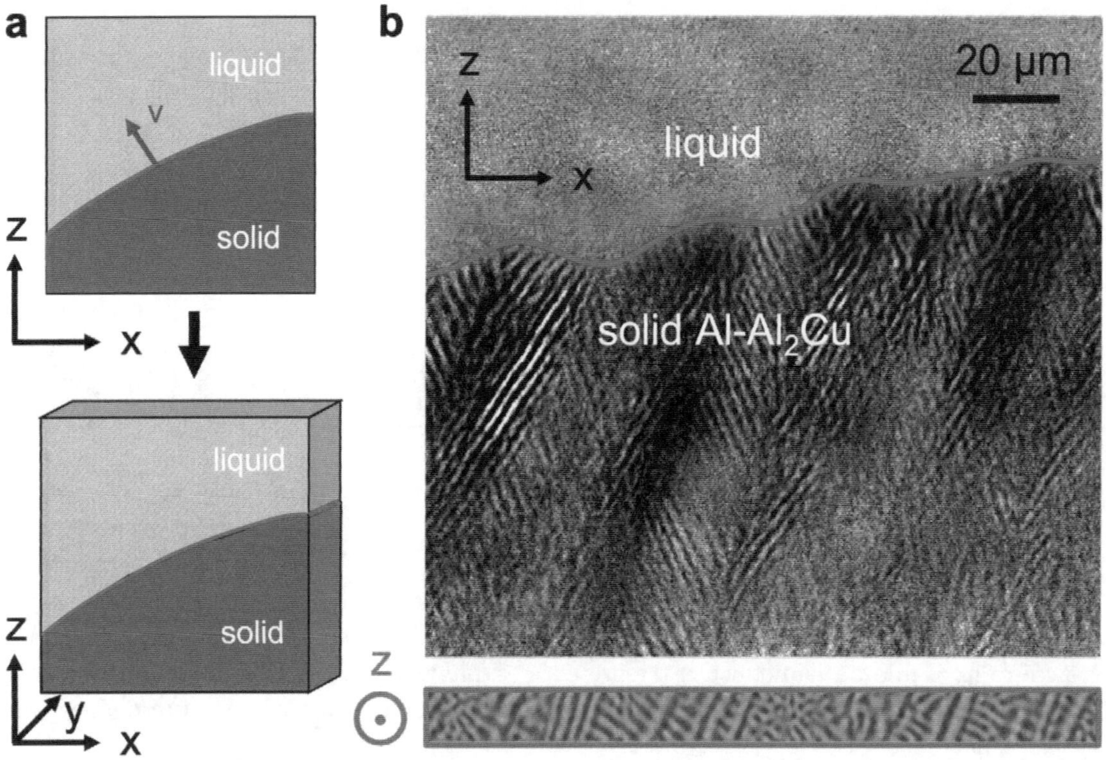

Figure 2: New approach for pseudo-4D X-ray tomography, which combines *in situ* X-radiography (to extract the time-dependent solid-liquid interface position, as in Figure 1) and *ex situ* X-ray tomography of a fully solidified specimen. We can interpolate the interface position (red) on the 3D reconstruction under the assumption that there is no tilt with respect to the incident beam, as shown in (a). With this approach, we can capture in (b) the evolution of the lamellar microstructure within the eutectic colonies. Here the microstructure is shown from two perspectives: digitally enhanced side and bird's eye view.

hence a long wavelength destabilization of the eutectic solid-liquid interfaces above a critical velocity (*vide infra*).

We conducted pseudo-4D X-ray microtomography at Argonne National Laboratory's Advanced Photon Source (Lemont, IL, USA), in two steps, as follows. First, we collected 2D radiographic images on a 30 μm-thick sample *during* the dynamic process, in order to identify the solidification front as time proceeds (with no attention to the eutectic microstructure). This experiment was done under "gradient freeze" by imposing a cooling rate (0.5 K/min) and a constant thermal gradient (~1 K/mm) while the sample stayed in the field-of-view of incident beam. Then, after the sample had fully solidified, we conducted a *post mortem* X-ray tomography scan to identify the eutectic solid-solid interfaces.

On the 3D reconstruction we can interpolate the position of the solid-liquid interface (see Figure 2(a) for a schematic illustration). This can be done by assuming there is no tilt of the solid-liquid interfaces along the direction of the X-ray beam (efforts are underway to overcome this assumption, *e.g.*, by reprojecting the 4D microstructure and comparing it to a 'ground truth' X-radiograph). Secondly, one must assume that the eutectic has not coarsened appreciably between the two scans.

The key advantage of this two-step approach is that we *decouple* the spatial and temporal resolutions in 4D imaging. In other words, *in situ* X-radiography can be conducted at the lowest possible t_{exp} (limited to the frame rate of the camera, here 200 ms) while *post mortem* X-ray tomography can be performed with as high Δt as necessary to ensure high signal-to-noise reconstructions (pixel size of 0.69 μm). Another benefit is that we eliminate the influence of fluid flow caused by rotation of a semisolid sample during *in situ* X-ray microtomography (natural convection is, however, unavoidable).

Further computational details on data segmentation and fusion and given in Ref. [13].

3. Results and discussion

The observation of shallow cells in Figure 2(b) suggests that we are just above the critical threshold for morphological instability due to an impurity-induced constitutional undercooling. Based on the criterion given by Refs. [14,15], we calculate a critical velocity V_c of 1 μm/s for a thermal gradient of 1 K/mm. This value of V_c is on the same order-of-magnitude as our experiment, see also Figure 3(a).

The eutectic Al-Al_2Cu solidifies in a direction antiparallel to gravity g (where $-z \parallel g$ in Figure 2(b)). Since the Si impurity is lighter than the Al-Cu melt, *i.e.*, $\nabla \rho_l \cdot g < 0$ where ρ_l is the density field in the liquid, the liquid tends to be unstable [16,17]. This results in the thermosolutal convection of Si close to the eutectic solid-liquid interface and an unsteady, oscillatory interfacial velocity (Figure 3(a)). The latter is due to the cycles of Si pile-up and convection away from the growth front. While we cannot detect the presence of solutal Si directly from X-radiographs (as in Ref. [7]), the time-dependent interfacial velocity supports the idea of an unsteady impurity concentration.

What is the influence of the velocity transients on microstructure evolution? As a starting point, we calculate the time-dependent lamellar spacing using methods described elsewhere [18]. In response to the oscillations in interfacial velocity V, the lamellar spacing λ fluctuates out-of-phase (Figure 3(b)), as would be predicted by theory on univariant eutectics [19]. We find no substantial lag in λ with respect to V, and thus no hysteresis in the lamellar spacing *vs.* interfacial velocity plane. This would imply that the Al-Al_2Cu eutectic microstructure can readily adapt to velocity transients.

We expect that the evolution in the impurity concentration is coupled also to that of the phase fractions of the eutectic solids at the growth front (Figure 3(c)). Such an effect is also not entirely surprising: As noted by Ref. [20], the impurity species may stimulate a change in the interdiffusivity, solid-liquid interfacial energies, *etc.*, which may increase the undercooling of one eutectic phase. In turn, the interfacial curvature of this phase would need to be reduced, to maintain an isothermal growth front. This leads to a shift in the contact angles at the trijunction. The angles at equilibrium can only be recovered if the trijunction moves laterally along the solid-liquid interface, which would consequently bring about an adjustment in the phase fractions of the eutectic solids, as we see here (Figure 3(c)).

As the areal fraction of Al_2Cu rises, so does the fraction of Al rods at the solid-liquid interface (Figure 3(d)). That is, the microstructure adapts to an evolving composition field by producing (or eliminating) rods. We distinguish between rods and lamellae using a morphological criterion: a two-dimensional connected body is rod-like if it has an eccentricity less than 0.9 and solidity greater than 0.8. Interestingly, the rods are not perfectly circular in cross-section but rather ellipsoidal (eccentricity of ~0.7), suggesting an anisotropy in solid-solid interfacial energies.

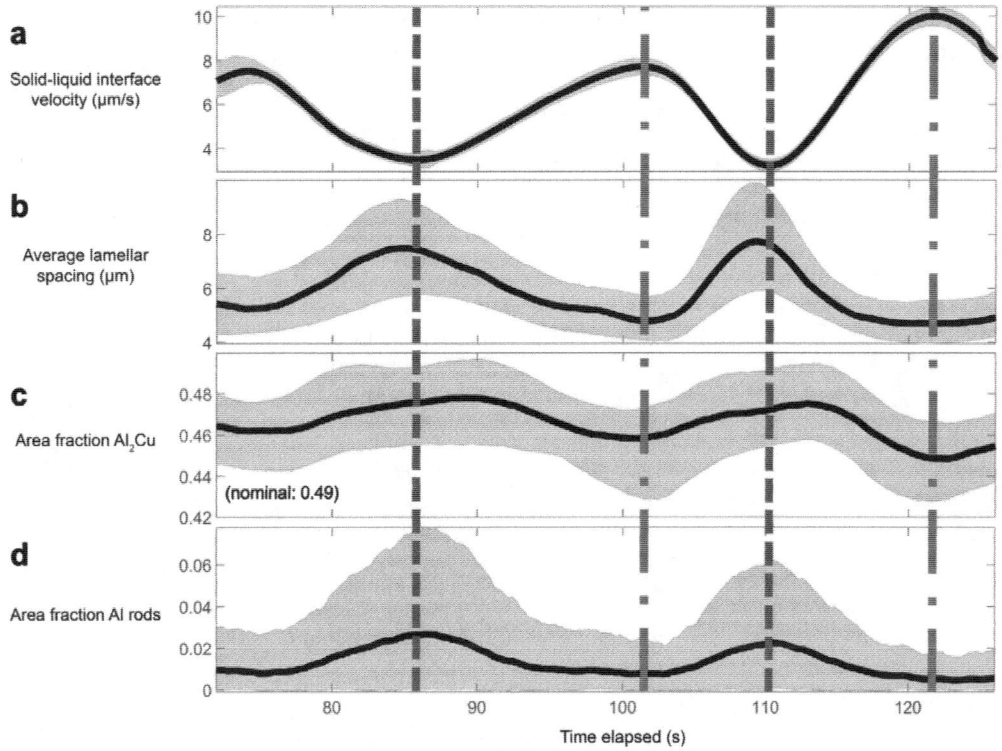

Figure 3: Various time-resolved attributes measured *via* pseudo-4D X-ray tomography: (a) solid-liquid interfacial velocity, (b) lamellar spacing, (c) area fraction of Al_2Cu, and (d) area fraction of Al rods. Data tabulated across a region 1.7 mm along x, 1.4 mm along z and 30 μm along y (*cf.* Figure 2(a)). Grey bands are uncertainty intervals. Extrema in velocity indicated with coloured lines (blue for minima and red for maxima). From Ref. [13].

The pseudo-4D data invites a further inquiry into the mechanism by which Al lamellae transform into Al rods and *vice versa.* In Figure 4(a), we examine the solidification dynamics of one eutectic colony. The lamellar interfaces develop sinusoidal perturbations that amplify in time. Eventually, the one lamella pinches off to yield array of rods, all of which lie within the plane of the lamella. That is, the rods are partially bound by the lamellar interfaces. This process is not exactly a Rayleigh instability since it is brought about by impurity-driven forces acting on the trijunction, as described above.

Our finding also goes against the predictions of Ref. [21]. Hunt argues that the eutectic microstructure must accommodate the solid-liquid interfacial curvature of cells; the implications are that lamellae grow along low energy directions and rods are produced when the local growth direction is furthest from the lamellar plane [21]. That is, rods branch from lamellae. Instead, we see that the lamella and rods are coplanar (Figure 4(a)), which would indicate that the (presumably) low-energy planes do not necessarily perpetuate for all time but may breakdown through an impurity-induced instability.

The morphological transition is not reversible, either. In other words, when the area fraction of Al_2Cu drops, the same rods from above do not recombine to produce the same lamella. Rather, the rods will expand in cross-section (eccentricity $\gg 0.7$) and merge with other rods and/or lamellae. Note the end structure in Figure 4(b) is quite different from the initial structure in Figure 4(a).

4. Conclusions

In this study, we examine univariant solidification of an Al-Al_2Cu eutectic in the presence of trace Si. The influence of the impurity species is twofold: (i) it brings about a constitutional undercooling that destabilizes the eutectic solid-liquid interfaces at long wavelengths ($\gtrsim$10 μm); and (ii) it changes the undercooling of one phase with respect to the other, on local, lamellar scales, which in turn changes the phase fractions and hence the eutectic morphology. These results provide insight into the selection of eutectic patterns that is relevant to a broad class of materials, including eutectic high entropy alloys [22].

Our efforts were made possible by new strides in multimodal, synchrotron-based X-ray imaging. By conducting *in situ* X-radiography and *post mortem* X-ray tomography in series and on the same specimen, we were able to achieve a first-ever, pseudo-4D reconstruction that maximizes spatial and temporal resolutions over that of either technique on its own. This imaging strategy will open the doors to future 4D experiments on solidification, deformation, shock, and related materials processing, that are otherwise out-of-reach for traditional 4D X-ray tomography.

Acknowledgments

We gratefully acknowledge financial support from the National Science Foundation CAREER program under Award No. 1847855. This research used resources of the Advanced Photon Source (APS), a U.S.

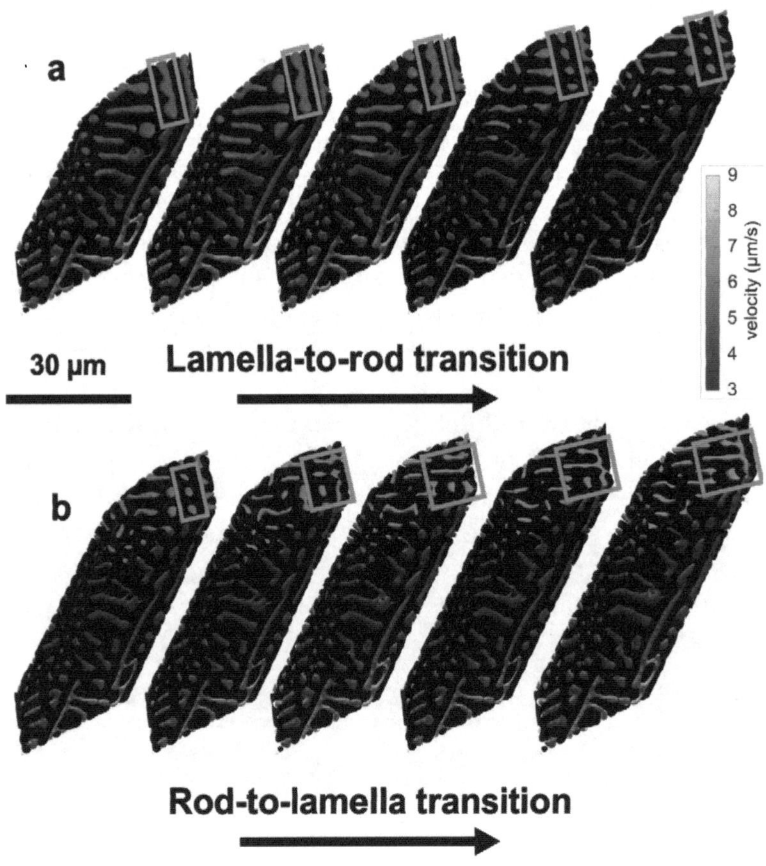

Figure 4: Morphological transitions in eutectic colonies, showing (a) lamella-to-rod and (b) its inverse, uncovered by pseudo-4D X-ray tomography. Al_2Cu is depicted in black, Al is in color (corresponding to the interfacial velocity, see colorbar), and liquid and neighboring colonies are transparent. Arrows indicate the passage of time, with an interval of 1 s between consecutive 3D renderings. In (a), one Al lamella (boxed in green) breaks down to form an array of rods, while in (b) those same rods expand and combine with other rods to form lamellae of a different orientation than in (a). Retrieved from Ref. [13].

Department of Energy (DOE) Office of Science User Facility operated for the DOE Office of Science by Argonne National Laboratory under Contract No. DE-AC02- 06CH11357. We thank P Shevchenko for technical assistance at APS and S Akamatsu and S Bottin-Rousseau for fruitful discussions.

References

[1] Mathiesen R H, Arnberg L, Ramsøskar K, Weitkamp T, Rau C and Snigirev A 2002 *Metall. Mater. Trans.* B **33** 613
[2] Liotti E, Arteta C, Zisserman, A Lui A, Lempitsky V and Grant P S 2018 *Sci. Adv.* **4** eaar4004
[3] Han I, Xiao X, Sun H and Shahani A J 2019 *Acta Cryst.* A **75** 218
[4] Wang F, Eskin D, Mi J, Wang C, Koe B, King A, Reinhard C and Connolley T 2017 *Acta Mater.* **141** 142
[5] Limodin N, Salvo L, Boller E, Suéry M, Felberbaum F, Gaillièguc S and Madi K 2009 *Acta Mater.* **57** 2300
[6] Wang Y, Peng L, Ji Y, Cheng X, Wang C, Wu Y, Fu Y, Chen L Q 2018 *J. Mater. Sci. Tech.* **34** 1142
[7] Reinhart G, Grange D, Abou-Khalil L, Mangelinck-Noël N, Niane N T, Maguin V, Guillemot G, Gandin C A and Nguyen-Thi H 2020 *Acta Mater.* **194** 68
[8] Wang Y, Gao J, Chao P, Muyanja N S, Mathiesen R H and Shahani A J 2021 *Mater. Lett.* **292** 129637
[9] Kellner M, Hötzer J, Schoof E and Nestler B 2020 *Acta Mater.* **182** 267
[10] Wang Y, Gao J, Sun W and Shahani A J 2022 *Scripta Mater.* **206** 114224
[11] Kurz W and Fisher D J 1989 *Fundamentals of Solidification* (Switzerland: Transtech Publications)
[12] Shahani A J, Xiao X, Lauridsen E M and Voorhees P W 2020 *Mater. Res. Lett.* **8** 462
[13] Chao P, Lindemann G and Shahani A J 2022 submitted
[14] Rinaldi M D, Sharp R M and Flemings M C 1972 *Metall. Trans.* **3** 3139
[15] McCartney D G, Hunt J D and Jordan R M 1980 *Metall. Trans.* A **11** 1243
[16] Dantzig J A and Rappaz M 2009 *Solidification* (Switzerland: EPFL Press)
[17] De Wilde J and Froyen L 2005 *Microgravity Sci. Technol.* **16** 40
[18] Aramanda S K, Khanna S, Salapaka S K, Chattopadhyay K and Choudhury A 2020 *Metall. Mater. Trans.* A **51** 6387
[19] De Wilde J, Froyen L, Witusiewicz V and Hecht U 2005 *J. Appl. Phys.* **97** 113515
[20] Lahiri A and Choudhury A 2017 *Acta Mater.* **133** 316
[21] Hunt J D 1966 *J. Inst. Met.* **94** 125
[22] Lu Y *et al.* 2014 *Sci. Rep.* **4** 6200

IOP Conf. Series: Materials Science and Engineering 1274 (2023) 012037 doi:10.1088/1757-899X/1274/1/012037

Formation of locked-lamellar grains in a slightly hypoeutectic Al-Al_2Cu alloy during thin-sample directional solidification

M Medjkoune[1], S Bottin-Rousseau[1], L Carroz[1,2], G Prévot[1], B Croset[1], J S Micha[3] and S Akamatsu[1]

[1]Sorbonne Université, Institut des Nanosciences de Paris, CNRS UMR 7588, 4 place Jussieu, 75005 Paris, France
[2]Currently at ArcelorMittal Industeel, 71201 Le Creusot, France
[3] Univ. Grenoble Alps, CEA-IRIG, 17 avenue des Martyrs, 38054 Grenoble Cedex 9, France

Abstract. We studied the formation and growth of locked-lamellar microstructures in a thin sample of a slightly hypoeutectic Al-Al_2Cu alloy. The coupled-growth dynamics, including early stages and steady-state regimes, was observed optically in real time during directional solidification. The orientation of the α (Al) and θ (Al_2Cu) crystals was measured ex situ in a series of eutectic grains by X-ray Laue microdiffraction. A nucleation event of a θ crystal on a pre-existing α crystal, and the subsequent growth of a eutectic grain with a type-C orientation relationship, that is, with a coincidence of {123}-α and {100}-θ planes, were observed in situ. In type-C eutectic grains, lamellar locking occurred parallel to the low-energy coincidence plane. A regular (*floating*) coupled-growth dynamics was observed in misoriented eutectic grains.

1. Introduction

Composite microstructures in solidified eutectics are a frozen trace of two-phase growth patterns at the moving solid-liquid interface [1]. During directional solidification of a nonfaceted binary eutectic alloy at an imposed velocity V in a fixed temperature gradient G, two solid phases grow simultaneously. In steady-state, they form periodic patterns, the dynamics of which is essentially determined by solute diffusion in the liquid and capillary effects. This is well described by the Jackson-Hunt (JH) theory of regular eutectics [2]. Depending on initial conditions, a coupled-growth pattern smoothly evolves towards a spatially periodic structure, or undergoes symmetry breaking instabilities leading to complex spatio-temporal phenomena [3]. Such a *floating* dynamics is typical of an isotropic system. In practice, a eutectic ingot is often composed of distinct eutectic grains –in a eutectic grain, all the crystals of each phase have the same, fixed orientation with respect to a reference frame. The eutectic-grain dependence of coupled-growth microstructures in eutectic alloys has been extensively documented since the 1970's [4]. The problem has been rigorously formalized much more recently by considering the so-called locking dynamics of tilted-lamellar patterns [5]. The central feature is that the free energy γ of the interphase boundary between the two solids is commonly anisotropic, and depends on the inclination angle ϕ (we consider a 2D geometry) of the interphase boundary. This modifies the theoretical problem via the local equilibrium at the triple junction between the liquid and the two solids (Young-Herring equation) [6]. A marked interfacial-anisotropy effect arises in eutectic grains with an orientation relationship (OR) such that γ presents a deep, possibly singular minimum for an inclination corresponding to that of dense lattice planes in coincidence. Locked lamellae grow (nearly) parallel to such a low-energy plane, independently of control parameters. They can be strongly tilted with respect to the main solidification axis **z**, by an angle ϕ_t that is equal to, or close to that of the coincidence planes. A semi-empirical theory has been proposed, according to which the steady-state tilt angle ϕ_t can be calculated, on a first approach, knowing the function $\gamma(\phi)$ and the orientation of the eutectic grain [5,7]. Recent experimental studies also shed light to the link between coincidence planes and the dynamics of lamellar patterns in anisotropic eutectic grains [8-12]. However, key information on the formation mechanisms of locked-lamellar eutectic grains during early solidification stages in well-characterized alloys, with good knowledge of both the ORs and the coupled-growth dynamics is still lacking.

We present the analysis of a thin hypoeutectic Al-Al_2Cu sample slightly below the eutectic concentration (on the Al side). We combined in situ observations during directional solidification, and ex situ crystallographic analysis. This study brings several pieces of clarification that were lacking so far. First, by monitoring early stages of the solidification process, direct, unequalled evidence was given of the formation of a eutectic grain with a type-C OR (see below), starting from the nucleation of a θ crystal onto a pre-existing α one, to the growth of a locked-lamellar microstructure (α: fcc Al-rich solid solution; θ: tetragonal intermetallic Al_2Cu). Second, the coexistence of several eutectic grains in a thin sample permitted to differentiate crystallographic effects, all other parameters being equal. Third, crystal orientation measurements were made, for the first time in this field, with an X-ray Laue microdiffraction (MD) method at the European synchrotron (ESRF). The MD technique delivers accurate information on the orientation of the crystals on a local scale, and permits a non-invasive mapping of the ORs in the eutectic grains on the scale of the sample, without any degradation of the metallic film.

We recall that a remarkable OR is characterized by a heteroepitaxial interphase boundary, corresponding to a pair of mutually parallel dense lattice planes of the two solids (coincidence planes), and a pair of mutually parallel dense directions (coincidence directions). Eutectic grains with the same coincidence planes, but different coincidence directions, are said to belong to a given family of ORs. In bulk Al-Al_2Cu ingots, previously observed ORs were most commonly of the so-called Alpha $[(001)^{\alpha}//(001)^{\theta}]$ and Beta $[(111)^{\alpha}//(2\text{-}11)^{\theta}]$ OR families [13-15]. The type-C OR family, with $\{123\}^{\alpha}$ and $\{100\}^{\theta}$ coincidence planes, has been recently identified in thin hypoeutectic Al-Al_2Cu alloys [11].

2. Methods

Details on the thin-sample directional solidification method can be found in Ref. [11]. A thin Al-Cu film was prepared by plasma sputtering on a flat single-crystal sapphire plate (also see [16]). Calibrated layers of pure Al and Cu were deposited so that the average (hypoeutectic) concentration C_0 = 16.85 at% of the film (total thickness: 14.3 μm) after melting was slightly below the eutectic point C_E = 17.5 at% (majority phase: α). The metallic film (lateral dimensions $\approx 6\times 50$ mm^2) was designed using a mask during evaporation with a 0.45-mm wide crystal-selector channel and a V-shaped crystal expander. A second sapphire plate was sealed on top of the film. At the beginning of the experiment, the sample was placed in the DS setup, and a rapid, but partial directional melting was performed. A solid seed was left unmolten in the crystal selector channel. After a short holding time, directional solidification was started at V=0.5 $\mu m s^{-1}$ along the axis **z** of the temperature gradient ($G \approx 70$ $K cm^{-1}$) realized between two metallic blocks with thermal regulations. The solidification front was parallel to a planar isotherm close to the eutectic temperature $T_E \approx 548$°C, near the middle of the 1-cm gap separating the two blocks. The structure of the contact surface between the sapphire plate and the metallic film was observed in real-time with a long-distance, reflected-light optics (and a camera). The large-scale microstructure (panorama) was reconstructed ex situ from micrographs recorded at room temperature.

The Laue MD technique was performed at the French CRG-IF BM32 beamline at ESRF [17]. In brief, a small (typically 200×500 nm^2 in size) polychromatic (5–22 keV) beam was sent to the surface of the film (with 40° incident angle). Scattering intensity patterns, so-called Laue patterns, were recorded from illuminated crystals in reflection-mode (deviation angles ≈90 °) on a planar detector. Thanks to the high synchrotron flux and X-ray penetration capability, the X-ray beam could cross twice the upper sapphire plate (previously thinned to about 200 μm by polishing), thus avoiding any mechanical degradation of the metallic film. Regions of interest in sample (ROIs) –typically 100×50 μm^2– were chosen by means of visible light microscopy (Fig. 1a). In a Laue pattern, θ or α diffraction peaks were detected along with those of the sapphire (Fig. 1b). Separate indexation of the phases was unambiguous. A raster scan of each ROI was carried out by displacing the XY stage of the sample holder with 2–3 μm steps –significantly smaller than the lamella width. It typically contained several hundreds of pixels, corresponding to as many Laue patterns. Simultaneously, a chemical map of the actual ROI was obtained with a fluorescence detector selecting a characteristic X-ray emission line of Cu (Fig. 1a). The good quality of the signal is evidenced in Fig. 1c: the scattering θ diffraction peak vanishes when the X-ray beam hits an α crystal, and vice versa. Hence, the optical, chemical and crystallographic maps of the θ/α alternation are in good agreement with each other. After the standard calibration geometry procedure, Laue patterns indexation was made using the LaueTools software [18] with the lattice symmetry and unit-cell parameters as input data (in addition to instrumental ones). While standard EBSD systems automatically deliver crystal-orientation maps, the analysis of the Laue diagrams was made by the operator, pixel per pixel at that time. Consequently, a few Laue patterns per lamella were analysed. The results will be presented in the form of stereographic projections.

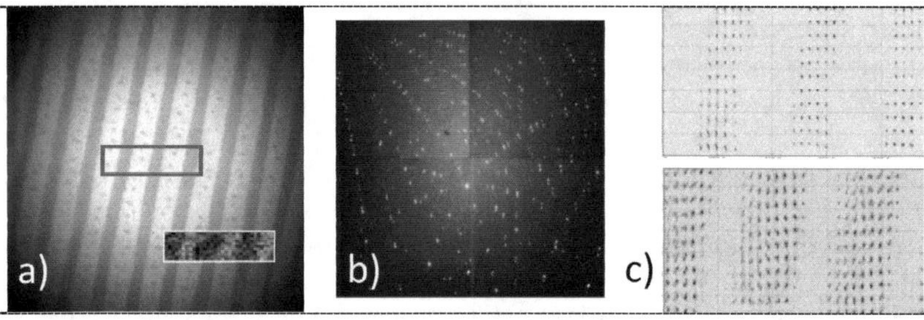

Figure 1. a) Optical view of the lamellar pattern in eutectic grain G6. Horizontal dimension: 200 μm. Lamellae of the θ (α) phase appear dark (bright) –thin θ precipitates are visible in α lamellae. Inset: fluorescence signal over the 62×22 μm^2 ROI (blue frame), with (noisy) contrast between Cu-rich θ and Cu-poor α regions. b) Laue MD diagram in a θ lamella (with diffracted peaks from the sapphire plates and the θ crystal); acquisition time : 3 s. c) Two-dimensional map of the same detector area around a peak of the θ phase (top) and the α phase (bottom).

3. Results

A panorama of the solidification microstructure in the crystal expander of the thin Al-Al_2Cu sample is shown in Fig. 2. Representative details of the microstructures are shown in a series of insets. Most of the eutectic grains (G1 to G6), but not all of them, were oriented, as well as two majority-phase (α) dendrites along the sides of the V-shaped region (see Fig. 2, inset 1). One of the two dendrites, on the left, was an ordinary needle-like dendrite (D), while the other one, on the right, was a seaweed dendrite (SW). Both dendrites originated from two α crystals in the crystal selector channel. Grains G3 and G6 exhibit a locked-lamellar microstructure, G1 and G2 a floating-like one, grain G5 weakly locked (or anisotropic floating) lamellae, and G4 an unsteady, strongly 3D pattern. An in situ snapshot of the coupled-growth pattern (grains G3, G4 and G5) is shown in Fig. 3. The analysis of representative Laue patterns are presented in Fig. 4. The choice was made to display three stereographic projections, within which the $\{123\}^\alpha$ and $\{100\}^\alpha$ poles of eutectic grains (and dendrites) sharing the same α crystal are respectively assembled. The good $(123)^\alpha//(100)^\theta$ coincidence in eutectic grains with a type-C OR was thus highlighted. Let us review the main features. The D dendrite was growing with the $(101)^\alpha$ plane parallel (within 3.4°) to the sample plane (Fig. 4a). The growth direction was about 10° off the $[-101]^\alpha$ direction, and far away from any $<100>^\alpha$ axis. The dendrite was therefore influenced by the edge of the film. The SW dendrite was oriented with a $(111)^\alpha$ plane nearly parallel (within about 11°) to the sample plane (Fig. 4c). Incidentally, in a cubic crystal (here, α), the 4-fold anisotropy of the solid-liquid interface vanishes in a $(111)^\alpha$ plane (3-fold symmetry). In a thin sample, the solidification dynamics is essentially two-dimensional, and needle-like dendritic growth is replaced by a seaweed morphology [19]. Both D and SW dendrites were growing ahead of the eutectic, but were hardly visible in situ because of a very thin, opaque liquid film running between the solid and the sample wall. That film was not uniform in thickness, and liquid pockets surrounded by the crystal contacting the sample wall were visible (Fig. 5). The shape of the dendrites was revealed after encapsulation by the eutectic (Fig. 2).

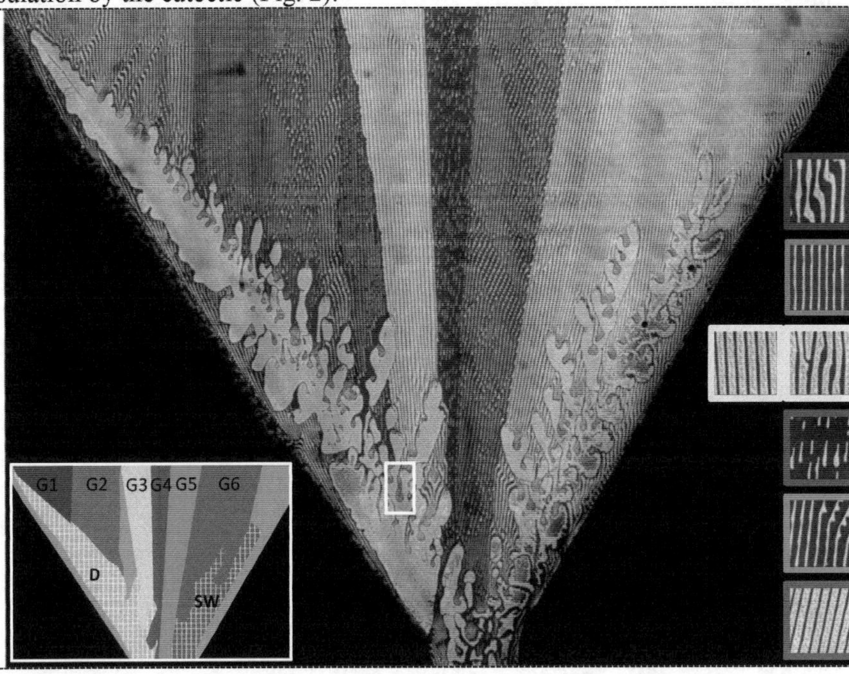

Figure 2. Panorama of the lamellar microstructure in a thin hypoeutectic Al-Al_2Cu sample (horizontal dimension: 5.2 mm). The solidification axis was along the vertical. The α (θ) phase appears bright (dark). The end of the crystal-selector channel can be seen at the bottom of the V-shaped crystal expander. White rectangular frame: region of the θ nucleation event giving rise to grain G3. Inset 1 (on the left): schematic representation of the eutectic-grain structure (G1 to G6 grains were oriented; grey regions were not) and the two α dendrites (D and SW). Insets (on the right): details of the microstructure (from top to bottom: G1 to G6; same color code as in Inset 1).

Figure 3. In situ optical observation of a lamellar pattern. Eutectic grains G3, G4 and G5, from left to right. The apparent variations of the volume phase fraction from one grain to another is related to 3D morphological features of the microstructure in the thickness of the film (see text).

Type-C ORs were found in eutectic grains G3 and G4. Crucially, the nucleation event that gave rise to G3 was captured in real time. In Fig. 5a, one can see the eutectic circumventing primary and secondary branches of the D dendrite. The key feature to be noticed is a shallow pocket of metastable liquid (close to the left hand edge of the image in Fig. 5a) remaining between the dendrite and the sapphire wall, isolated from nearby θ crystals, below the eutectic temperature. A few seconds later (Fig. 5b), a θ crystal nucleated and grew along the triple θ-liquid-sapphire contact line. A fine eutectic microstructure rapidly formed, giving rise to grain G3 during further solidification. In this grain, lamellar locking occurred in the vicinity of the $(123)^{\alpha}//(100)^{\theta}$ coincidence plane, in agreement with Ref. [11]. The coincidence plane was, however, not perpendicular to the sample walls (Fig. 4a). This is thus a typical case of "out-of-plane" locking close to a deep γ minimum: the interphase boundaries are inclined in the thickness of the sample [5]. There is obviously a limit inclination angle for stable out-of-plane locking to occur. This is evidenced in grain G4 (which formed in the crystal selector channel). In that grain, the out-of-plane $(123)^{\alpha}//(100)^{\theta}$ coincidence plane (inclined by an angle of about 35°) was clearly unavailable for steady-state locking. Instead, a complex, unsteady growth, with marked deformations of the interphase boundaries occurred in the thickness of the sample, thus indicating a strong interfacial-anisotropy effect.

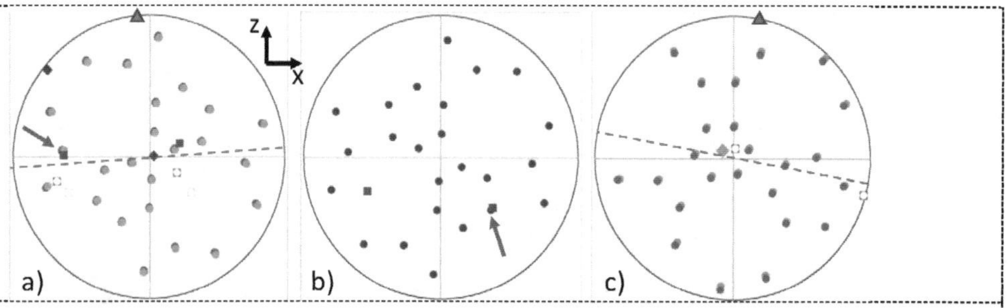

Figure 4. Stereographic projections (central pole: normal to the sample plane) of local crystal orientations obtained by Laue microdiffraction. Disks: $\{123\}^{\alpha}$ poles (same color code as in Fig. 2, inset 1). Filled (empty) squares: $\{100\}^{\theta}$ planes with (without) good coincidence with a $\{123\}^{\alpha}$ plane. a) G1, G2, G3 and dendrite D; b) G4 (grain G5: not shown); c) G6 and dendrite SW. Arrows: coincidence planes (G3 and G5). Triangles and dashed lines: direction of the lamellae and ensemble of possible locking planes (G3 and G6). Diamonds: a) planes $(101)^{\alpha}$ and $(-101)^{\alpha}$ of the D dendrite; c) $(111)^{\alpha}$ plane of the SW dendrite. Directions in coincidence are not shown for clarity.

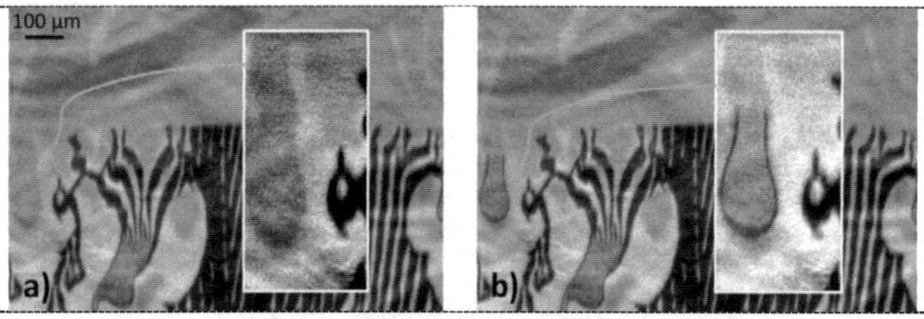

Figure 5. Real-time images (also see Fig. 2). a) On the very left edge of the image a thin liquid pool (with a slightly darker contrast than the surrounding α solid) is enclosed between the sapphire plate and the (hardly visible) D dendrite. b) Nucleation of a θ crystal in the liquid pool leading to the creation of grain G3. Also see the large-magnification insets. Time lapse: 4 s.

No remarkable ORs were found in other (misoriented) grains. They formed from α and θ crystals which did not nucleate one onto the other. The tilted-lamellar pattern in grain G6 is worth considering. The α lamellae originated from the SW dendrite (Fig. 4c). The orientation of the θ crystal was such that the $(100)^{\theta}$ plane was closely perpendicular to the sample walls (to within less than 1°). The lamellae were locked close to that plane. This, again, agrees well with the conclusion in Ref. [11] that, in misoriented grains, the lamellae can lock onto a $(100)^{\theta}$ plane, whatever the orientation of the α crystal relative to θ. In the other misoriented grains, the γ minimum associated to the nearest $(100)^{\theta}$ plane was not deep enough to stabilize an out-of-plane locking. Instead, the lamellae were growing essentially along the main solidification axis (ϕ_t=0), thus falling in a nearly isotropic part of $\gamma(\phi)$. The dynamics, including solitary waves, was typically floating –with a smoothly anisotropic behavior in grain G5.

Let us finally comment on the apparent variation of the volume phase fraction in the solid from one eutectic grain to another. On the basis of previous observations in bulk (near-)eutectic Al-Al_2Cu ingots [12], an approximate 50/50 volume phase fraction was expected. Here, this was essentially the case in grains with a floating dynamics, and interphase boundaries perpendicular to the sample walls. Differences in grains with an out-of-plane locking dynamics is probably due to deformations of the interphase boundaries close to their contact line with the sample walls.

4. Conclusion

Real-time observations during directional solidification of a thin sample of a hypoeutectic Al-Al_2Cu alloy provided a direct confirmation that eutectic grains with type-C ORs form from the nucleation of a θ crystal on a pre-existing α crystal. As in Ref. [11], no ORs belonging to the Alpha and Beta families were found. In the multi-grain structure of the thin sample, it also appeared clearly that steady-state lamellar locking occurred on the $(123)^{\alpha}//(100)^{\theta}$ coincidence plane in type-C grains, and on a $(100)^{\theta}$ plane in misoriented grains, as long as out-of-plane locking was permitted by a strong anisotropy. In other cases, the lamellar growth was either anisotropic, but unsteady, or exhibited a floating dynamics. Similar investigations remain to be made in thin hypereutectic Al-Al_2Cu samples with θ as a majority phase, aiming at establishing whether a (reverse) α-on-θ nucleation scenario can lead to the formation of eutectic grains with Alpha and Beta ORs.

Laue microdiffraction was employed for the first time in this research field. This non-invasive method allowed us to determine the orientation of θ and α crystals on a local scale, and to identify the ORs in distinct eutectic grains. A comparison between a comprehensive Laue microdiffraction analysis and EBSD measurements in a reference sample will be presented in a separate publication. Finally, further experimental information on the anisotropy of the interphase boundaries should be obtained via the so-called rotational directional solidification method [20,21].

5. References

[1] Dantzig J A and Rappaz M 2009 *Solidification* (EPFL Press, Lausanne)
[2] Jackson K A and Hunt J D 1966 *Trans. Metall. Soc. AIME* **236** 1129
[3] Akamatsu S and Plapp M 2016 *Curr. Opin. Solid St M* **20** 46
[4] Hogan L M, Kraft R W and Lemkey F D 1971 *Adv. Mater. Res.* **5** 83
[5] Akamatsu S, Bottin-Rousseau S, Şerefoğlu M and Faivre G 2012 *Acta Mater.* **60** 3199
[6] Hoffmann D W and Cahn J W 1972 *Surf. Sci.* **31** 368
[7] Ghosh S, Karma A, Plapp M, Akamatsu S, Bottin-Rousseau S and Faivre G 2019 *Acta Mater.* **175** 214
[8] Akamatsu S, Bottin-Rousseau S, Şerefoğlu M and Faivre G 2012 *Acta Mater.* **60** 3206
[9] Kokotin V and Hecht U 2014 *Comput. Mater. Sci.* **86** 30
[10] Bottin-Rousseau S, Senninger O, Faivre G and Akamatsu S 2018 *Acta Mater.* **150** 16
[11] Bottin-Rousseau S, Medjkoune M, Senninger O, Carroz L, Soucek R, Hecht U and Akamatsu S 2021 *J. Cryst. Growth* **570** 126203
[12] Hecht U, Eiken J, Akamatsu S and Bottin-Rousseau S 2019 *Acta Mater.* **170** 268
[13] Bonnet R and Durand F 1973, in *Proceedings of the Conference on in Situ Composites* **1** 209
[14] Hecht U, Witusiewicz V and Drevermann A 2012 *IOP Conference Series: Mat Sci Eng* **27** 012029
[15] Wang S J, Liu G, Wang J and Misra A 2018 *Mater. Charact.* **142** 170
[16] Sullivan E J, Tomko J A, Skelton J M, Fitz-Gerald J M, Hopkins P E and Floro J A 2021 *J. Alloy. Compd.* **865** 158800
[17] Ulrich O, Biquard X, Bleuet P, Geaymond O, Gergaud O, Micha J S, Robach O and Rieutord F 2011 *Rev. Sci. Instrum.* **82** 033908
[18] https://gitlab.esrf.fr/micha/lauetools/

[19] Akamatsu S, Faivre G and Ihle T 1995 *Phys. Rev. E* **51** 4751
[20] Akamatsu S, Bottin-Rousseau S, Şerefoğlu M and Faivre G 2012 *Acta Mater.* **60** 3206
[21] Mohagheghi S and Şerefoğlu M 2019 *Metall. Mater. Trans. A* **50A** 5221

Acknowledgments
We thank R. Soucek and L. Becerra for technical help. This work was financially supported by ANPHASES M-ERA.NET project (ANR-14-MERA-004). We also are grateful to CNES for giving us the possibility of a long-term research.

The effect of Sr modification on the eutectic structure of AlSi7Mg alloys at different purity levels

Ozen Gursoy[1] and Giulio Timelli[2]

[1] Corresponding author, Department of Management and Engineering, University of Padova, Vicenza 36100, Italy
[2] Department of Management and Engineering, University of Padova, Vicenza 36100, Italy

Abstract. The effect of different Sr levels on the solidification path and the microstructure of the Al-Si eutectic in AlSi7Mg alloys at 3 different levels of purity (high purity, commercial purity, and secondary alloy) was investigated. Thermal analysis technique was performed to obtain the cooling curves and characteristic temperatures of the eutectic reactions. Microstructural investigations were carried out to study the variation of the eutectic structure due to Sr addition and alloy's purity. The results indicate how the commercially pure alloy shows the highest nucleation and growth temperatures among the unmodified alloys, while the secondary alloy has the lowest ones. Sr addition decreases the characteristic eutectic temperatures. Even though the growth temperature is similar after Sr modification in the commercial and high purity alloys, this last shows a finest eutectic structure. Different levels of Sr are required to fully modify the eutectic of the AlSi7Mg alloy at different purity grade.

Keywords: Modification; high purity alloy; thermal analysis; solidification; eutectic structure

1. Introduction

The presence of impurities in Al alloys occurs as a consequence of nature or additions from the different processes. The role of impurity elements in eutectic solidification and mechanical properties can be crucial and they are difficult and uneconomic to remove [1].

Aluminium phosphide (AlP) particle is known as a pre-existing heterogeneous nucleation site for eutectic Si crystals. In case the P content is lower than 3 ppm in the alloy, eutectic nucleation requires higher undercooling. When the alloy contains P higher than 3 ppm, AlP forms prior to eutectic precipitation and acts as nucleant for eutectic Si [2–4]. Chemical modification of eutectic by Sr addition in the Al-Si alloys is extensively used, and it plays an important role that can improve the mechanical properties by changing the morphology of the eutectic silicon. Strontium addition induces the formation of Al_2Si_2Sr compounds precipitating on AlP; thus, the nucleation frequency is limited during the eutectic reaction in Al-Si alloys and silicon crystals can nucleate independently. A morphology transition also occurs from flake to fibrous [5]. However, the presence of impurity elements at certain levels can change the efficiency of the modification of the eutectic Si.

Iron is one of the most common impurities in aluminium alloys and its detrimental effects on microstructure and mechanical properties become apparent with the increasing Fe content in the alloy.

Vandersluis et al. [6] detected Si flakes in the vicinity of Fe-bearing phases in the Sr-modified Al-6Si alloy. On the contrary, Fracchia et al. [7] found that eutectic particle size decreases when the content of Fe and Mg increases. Although Ca and Bi are well known as eutectic modifiers in Al-Si alloys, Samuel and co-workers [8,9] showed how those elements, cause interaction with Sr and demodification in Sr-modified Al-Si alloys. On the other hand, Huang et al. [10] stated that eutectic silicon shows an incomplete modification in Al-7Si-0.4Mg alloy with respect to Al-7Si alloy. Besides, other impurities, such as Zn, Ni, V, Sn and Pb may also affect the solidification of the eutectic stage [8,11]. Furthermore, the presence of impurity elements may also change the characteristic temperatures of eutectic reactions [12].

Different levels of impurities should require different amounts of Sr modifier mainly due to the variety and density of potent nucleants. In this study, the influence of different levels of Sr on the solidification path and microstructure of hypoeutectic AlSi7Mg alloys at 3 different purity levels was investigated.

2. Experimental Procedure

Three different AlSi7Mg alloys, which were high purity (HP), commercial purity (CP), and secondary alloy (SA), were used as base alloys. The chemical compositions of the alloys were measured by optical emission spectrometry and are shown in Table 1.

Table 1. Chemical composition of the experimental base AlSi7Mg alloys (wt.%).

Alloy	Si	Mg	Fe	Cu	Mn	Zn	Ni	Ti	Pb	Ca	Al
High Purity	7.24	0.282	0.037	0.0421	0.0018	0.0503	0.0026	0.142	-	-	bal.
Commercial Purity	7.29	0.339	0.094	0.0034	0.0020	0.0050	0.0035	0.136	0.0011	0.0005	bal.
Secondary Alloy	7.43	0.329	0.440	0.0528	0.3030	0.0470	0.0112	0.021	0.0077	0.0010	bal.

Each alloy melt was prepared by charging about 1 kg of ingot piece into a SiC crucible using an electrical resistance furnace at 750 ± 5 °C. An AlSr3.5 master alloy was added into the melt to achieve 5 different Sr nominal levels, i.e. 0 (unmodified), 30, 60, 130 and 300 ppm. After each addition, the contact time between the Sr modifier and the melt was about 20 minutes. The melt was then gently skimmed and stirred, and the chemical composition was measured to ensure the proper level of Sr addition. The molten metal was then poured into a BN-coated steel cup (outer diameter 45 mm, height 60 mm, and a uniform wall thickness of 3 mm), preheated at about 550 °C.

Computer-aided two thermocouple thermal analysis technique was used to monitor the solidification path of the experimental alloys and to determine the characteristic eutectic temperatures of the alloys during solidification. After pouring, K-type thermocouples (ø1 mm) covered by stainless steel tube were inserted into the melt. The temperature and time were collected by using a high-speed data acquisition system with a sampling rate of 0.1 s^{-1}, analog-to digital converter accuracy of 0.1 °C and connected to a personal computer. The system was allowed to cool in air to provide a cooling rate of about 0.2 °C/s in the solidification range of the alloy. To determine the characteristic temperatures of eutectic Al-Si reaction, the cooling curves and the corresponding derivative curves (dT/dt) were plotted for each experimental alloy. The eutectic nucleation temperature (T_N), the minimum eutectic temperature (T_{Min}), the eutectic growth temperature (T_G), and the recalescence undercooling of eutectic (ΔT_R) were analysed to estimate the efficiency of eutectic modification.

For the microstructural investigations, samples were drawn close to the tip of the central thermocouple. After conventional mechanical sample preparation, microstructural analyses were performed using an optical microscope (OM).

3. Results and Discussion

3.1 Thermal analysis

The cooling curves of the experimental alloys referring to the Al-Si eutectic reaction are shown in Figure 1 and the characteristic temperatures of the Al-Si eutectic for the experimental alloys are given in Table 2.

In unmodified HP alloy, the eutectic Si nucleated at 574.4 °C showing some degrees of undercooling. According to the predicted phase reactions of Scheil solidification condition, the Al-Si eutectic reaction of HP AlSi7Mg alloy is expected to initiate at 574.6 °C, which is nearly the same as the nucleation temperature from the experimental investigation (Table 2).

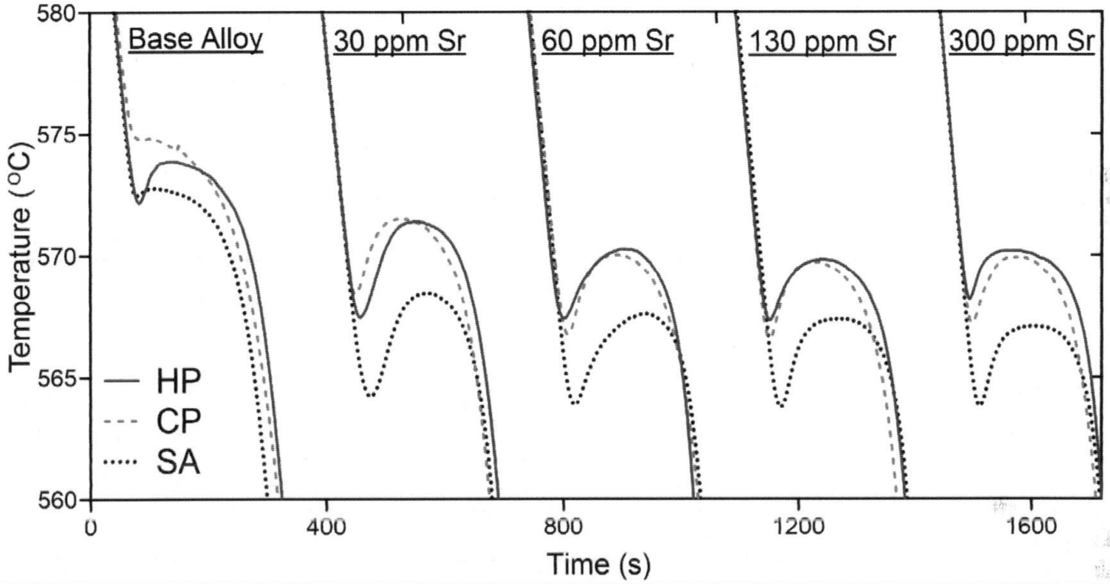

Figure 1. Cooling curves in the region of Al-Si eutectic reaction referred to experimental alloys.

Mondolfo [13] suggested an empirical formula to estimate the eutectic growth temperature, T_G, considering the initial chemical composition (wt.% element) of unmodified commercial Al-Si based alloys:

$$T_G = 577 - (12.5/Si)[4.43(Mg)+1.43(Fe)+1.93(Cu)+1.7(Zn)+3.0(Mn)+4.0(Ni)] \quad (1)$$

Thus, according to eq. (1), the growth temperature of unmodified HP alloy was overestimated (574.6 °C) respect to the experimental one (573.6 °C).

After Sr modification, the eutectic plateau was depressed, and the characteristic eutectic temperatures were decreased in HP alloy. The reduction of T_N and T_G values was observed mainly after 30 ppm Sr, while no significant change in the eutectic temperatures was detected in HP alloy up to 130 ppm Sr additions. A slight increase in the T_G value was detected in the HP alloy modified by 300 ppm Sr. The recalescence undercooling of eutectic Si has been found to have a good correlation in modification level [14]. Since the evaluation of the eutectic modification by thermal analysis is strongly dependent on the impurity concentration, this relashionship cannot be guaranteed for HP Al-Si alloys as observed based on the results of the present study. These results are in agreement with ref. [4].

The nucleation of eutectic Si in unmodified CP alloy was 576.2 °C, which is higher by ~2 °C than that of unmodified HP alloy. According to the predicted phase reactions of Scheil solidification condition, the Al-Si eutectic reaction of CP alloy is expected to initiate at 574.3 °C, which is lower by 1.9 °C than the nucleation temperature from the experimental investigation. The difference in the nucleation temperature from the thermodynamical calculation and the experimental investigation can be evaluated as the indication of the undercooling needed for the nucleation of eutectic Si. The nucleation undercooling of 1.9 °C indicates the existence of potent nucleation sites for the nucleation of eutectic Si at low undercooling. Phosphorus is a common impurity element in Al alloys and as P content is higher than ~ 3 ppm, a pre-eutectic thermodynamically stable AlP phase is favourable to form during the solidification resulting in higher density of heterogeneous nucleants for the precipitation of eutectic Si crystals [2]. It is believed that the unmodified CP AlSi7Mg alloy in the present study contains greater than 3 ppm P content, unlike in the HP alloy. This induces the nucleation of eutectic Si at low undercooling.

The growth temperature of the eutectic Si can be strongly correlated with the modification level and the T_G value of unmodified CP alloy is 574.5 °C. According to eq. (1), the growth temperature of unmodified CP alloy is estimated as 574.1 °C, which is nearly the same as T_G from the experimental investigation. Respect to unmodified HP alloy, a more reliable result in the estimated T_G value by Mondolfo's formulation was detected in the CP alloy. The negligible difference can be attributed to the amount of the impurities in the initial chemical composition other than their calculation in the eq. (1).

Table 2. Characteristic eutectic temperatures of the experimental AlSiMg alloys (°C).

Sr content	High Purity				Commercial Purity				Secondary Alloy			
	T_N	T_{Min}	T_G	ΔT_R	T_N	T_{Min}	T_G	ΔT_R	T_N	T_{Min}	T_G	ΔT_R
Unmodified (Base alloy)	574.4	572.2	573.6	1.4	576.2	574.8	574.5	-0.3	572.6	572.4	572.2	-0.2
30 ppm	569.8	567.5	570.4	2.9	570.9	568.4	571.4	3.0	568.2	564.2	568.4	4.2
60 ppm	570.7	567.4	570.2	2.8	569.0	566.8	570.0	3.2	566.8	563.9	567.4	3.5
130 ppm	570.0	567.3	569.6	2.3	568.7	567.1	569.5	2.4	565.7	563.8	567.3	3.5
300 ppm	569.6	568.2	570.6	2.4	569.1	567.2	569.8	2.6	566.0	563.8	566.8	3.0

Upon increasing the Sr content up to 130 ppm Sr, the T_N and T_G values gradually decreased. The characteristic eutectic temperatures of 300 ppm Sr-containing CP alloy were nearly similar to those of the alloy modified by 130 ppm Sr. It has been studied how Sr neutralizes pre-existing potent nucleation center AlP [15]. Al_2Si_2Sr compounds precipitates on AlP prior to the eutectic reaction in Sr-containing Al-Si hypoeutectic alloys. This results in a drastic reduction of the number of the potent nucleant for eutectic Si. Thus, lower nucleation frequency causes the depression of the eutectic temperatures.

The nucleation of Al-Si reaction in the unmodified secondary alloy was measured at 572.6 °C and the growth temperature of eutectic was at 572.2 °C. According to Scheil calculation and Mondolfo's equation, the initiation of eutectic reaction and the eutectic growth temperature are 572.5 °C and 571.6 °C, respectively. These are nearly the same with T_N and T_G from the experimental investigation of SA. Upon progressive increasing of Sr addition up to 130 ppm, further depression in the nucleation temperature of eutectic Si was observed in SA. However, when the Sr content in the alloy was increased up to 300 ppm, the eutectic reaction initiated at nearly the same temperature as that of SA modified by 130 ppm Sr.

3.2 Microstructure

Figure 2 shows the microstructural evolution of the AlSi7Mg alloys by Sr content. The unmodified eutectic structure of HP alloy exhibits partially refined silicon crystals (Figure 2a). As previously mentioned, nucleation of eutectic Si in unmodified HP alloys required higher undercooling compared to unmodified CP or secondary alloys. When the nucleation rate of eutectic is lower in a higher undercooled melt, the solid-liquid interface is smaller. This causes higher growth rate, consequently refined eutectic structure. The addition of 30 ppm and 60 ppm Sr can partially modify the eutectic structure. While the addition of 130 ppm Sr fully modified the structure (Figure 2h), the microstructure showed over-modification with further Sr addition as shown in Figure 2g.

The eutectic structure of unmodified CP alloy exhibits coarse Si flakes concerning that of unmodified HP alloy (Figure 2b). While adding 30 ppm and 60 ppm Sr induced poor partial modification of eutectic structure, the presence of 130 ppm Sr fully modified the eutectic structure of CP alloy (see Figure 2e). However, the addition of 300 ppm Sr seems unnecessary as the modification ramped down in the microstructure (Figure 2h).

In secondary alloy, a fully modified eutectic structure was provided by the addition of 130 ppm Sr. No alteration was observed with further Sr addition.

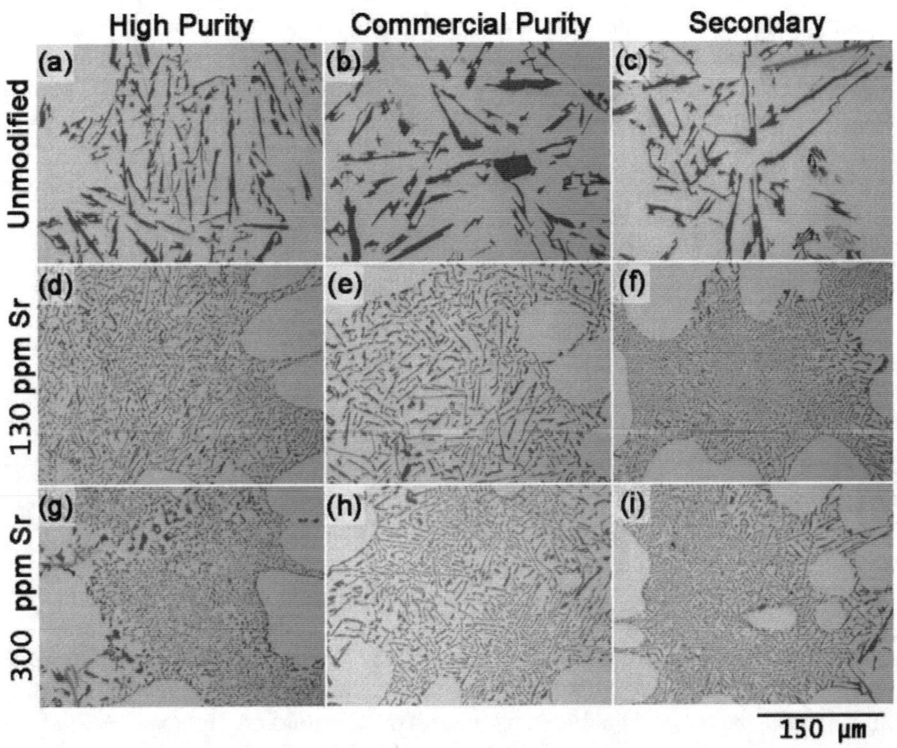

Figure 2. Optical micrographs of the investigated alloys. Images are arranged with increasing Sr content within a column (0 (unmodified), 130, 300 ppm), whereas impurity level increases in a row (High Purity, Commercial Purity, Secondary).

4. Conclusions

In this study, the effect of different levels of Sr contents on the solidification path and microstructure of HP, CP, and secondary AlSi7Mg0.3 hypoeutectic aluminium alloys.

- The growth temperature and recalescence undercooling of eutectic reaction detected by thermal analysis have a strong correlation with actual modification levels in CP alloys.
- Any correlation between characteristic temperatures of eutectic Si and modification level in high purity alloy was not detected.

- Lower Sr content is sufficient for fully eutectic modification when the impurity level is low.

Acknowledgments

The work was developed with the financial support of Fondazione Cassa di Risparmio di Padova e Rovigo (CariPaRo), Padova (2019).

References

[1] Tański T, Labisz K, Krupińska B, Krupiński M, Król M, Maniara R and Borek W 2016 Analysis of crystallization kinetics of cast aluminum–silicon alloy *J. Therm. Anal. Calorim.* **123** 63-74
[2] Ludwig T H, Schaffer P L and Arnberg L 2013 Influence of phosphorus on the nucleation of eutectic silicon in Al-Si alloys *Metall. Mater. Trans.* A **44** 5796-805
[3] Nogita K, McDonald S D, Tsujimoto K, Yasuda K and Dahle A K 2004 Aluminium phosphide as a eutectic grain nucleus in hypoeutectic Al-Si alloys *Microsc* **53** 361-9
[4] Rakhmonov J, Timelli G and Basso G 2018 Interaction of Ca, P trace elements and Sr modification in AlSi5Cu1Mg alloys Journal of *J. Therm. Anal. Calorim.* **133** 123-33
[5] Sigworth G K 2008 The modification of Al-Si casting alloys: important practical and theoretical aspects *Int. J. Met.* **2** 19-40
[6] Vandersluis E, Sediako D, Ravindran C, Elsayed A and Byczynski G 2018 Analysis of eutectic silicon modification during solidification of Al-6Si using in-situ neutron diffraction. *J. Alloys Compd.* **736** 172-80
[7] Fracchia E, Gobber F S and Rosso M 2021 Effect of alloying elements on the Sr modification of Al-Si cast alloys *Metals* **11** 342
[8] Zedan Y, Samuel A M, Doty H W, Songmene V and Samuel F H 2022 Effects of free-cutting elements addition on the microstructure, hardness, and machinability of Al-11% Si–Cu–Mg casting alloys *Int. J. Met.* **5** 1-7
[9] Samuel A M, Ammar H R, Zedan Y, Doty H W and Samuel F H 2022 Effect of Ca–Sr–Mg and Bi–Sr–Mg interactions on the microstructural characterization and tensile properties of B319 alloy *Int. J. Met* **7** 1-20
[10] Huang C, Liu Z and Li J 2022 Influence of alloying element mg on Na and Sr modifying Al-7Si hypoeutectic alloy *Materials* **15** 1537
[11] Ludwig T H, Schaffer P L and Arnberg L 2013 Influence of some trace elements on solidification path and microstructure of Al-Si foundry alloys *Metall. Mater. Trans.* A **44** 3783-96
[12] Gursoy O and Timelli G 2021 Effects of bismuth on the characteristics of eutectic Si particles in Na-modified A356 alloys *Metall. Ital* **1** 73
[13] Mondolfo L F 1976 *Aluminium Alloys: Structure and Properties* (London: Butterworths)
[14] Zamani M and Seifeddine S 2015 Assessment of modification level in EN AC-46000 aluminum casting alloys using thermal analysis and microscopic evaluation *Light metals 2015* 955-60
[15] Li J H, Zarif M Z, Dehm G and Schumacher P 2012 Influence of impurity elements on the nucleation and growth of Si in high purity melt-spun Al–Si-based alloys *Philos. Mag. Lett.* **92** 3789-805

IOP Conf. Series: Materials Science and Engineering 1274 (2023) 012039 doi:10.1088/1757-899X/1274/1/012039

On the role of interfacial segregation on graphite shape in cast irons

J Lacaze

CIRIMAT, Université de Toulouse, CS 44362, 31030 Toulouse, France

Abstract. Understanding spheroidal graphite growth in cast irons is a long standing problem for which several mechanisms have been proposed. During the last decade, observations using transmission electron microscopy added new information about the stacking of graphite layers, leading eventually to propose other mechanisms than previously thought of such as repeated 2D-nucleation of new growth blocks and their lateral extension at the surface of the spheroids. Furthermore, it has been recently realized that elements known to strongly affect graphite shape are also those that are surface active. This is illustrated here in the case of antimony which is known to lead to spheroidal graphite degeneration when added at some tens of ppm. Using atomic resolution microscopy, segregation of antimony could be evidenced at the graphite/matrix interface of both spheroids and protuberances. This strongly suggests that accumulation of antimony repelled magnesium from the graphite surface until growth of spiky protuberances replaced spheroidal growth.

1. Introduction

Graphitic cast irons are casting alloys with a composite microstructure consisting of graphite precipitates in an iron-based matrix. They are near-eutectic Fe-C-Si alloys containing traces of oxygen and sulphur, and also some manganese due to the recycling of steel scrap. Without treatment of the liquid metal, the graphite appears as large plates when primary and as interconnected lamellae that have grown radially from a nucleus of MnS when eutectic (figure 1a). After addition of 300-500 ppm of magnesium (or rare earths that are however less effective), graphite precipitates as individualised spheroids (figure 1b) around nuclei (oxy-sulphides or nitrides) resulting from appropriate inoculation. It is now widely accepted that graphite spheroids nucleate and start growing in the liquid and get encapsulated within an austenite shell at a later stage of solidification. The change in shape from lamellar to spheroidal graphite gives the cast irons mechanical properties that make them suitable for structural and safety applications, which explains why so much work has been done to understand this.

Graphite spheroids can be classified in the family of spherulites which are found in many materials as well as in the plant world. Granasy *et al.* [1] carried out a study of spherulitic growth and the possibility of modelling it by phase field. However, their approach focused on spherulites whose formation is due to the development of a radial and polycrystalline structure, characterised by repetitive branching. This form of growth was described by Ryschenkow and Faivre [2] in the case of selenium, which is tetragonal and has a c-axis of order 3 like graphite. At very low undercooling, selenium forms monocrystalline platelets elongated in the plane perpendicular to the c axis quite similar to plate-like graphite. These platelets become unstable at increasing undercooling and eventually develop as spherulites following a so-called regular-branching mechanism. In agreement with a scheme proposed by Double and Hellawell [3], Faivre [4] has considered the possibility of formation of a graphite spheroid by a similar succession of crystallographic branching of a graphite flake.

The model just mentioned is one of the very many schemes proposed in the literature for the formation of graphite spheroids in cast irons. Rather than attempting to provide a history of these models, we present in section 2 observations by scanning (SEM) and transmission (TEM) electron microscopy that have clarified some of the ambiguities that the only use of optical microscopy has generated. Excluding schemes incompatible with the described observations, a model based on a 2D nucleation and lateral growth mechanism has been elaborated which will be briefly mentioned. This model pre-supposes the role of magnesium which will be discussed in section 3.

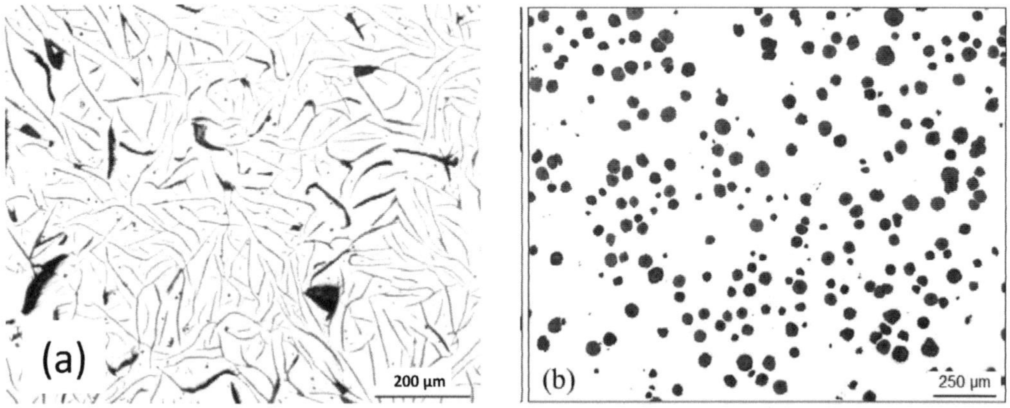

Figure 1. Microstructure of a lamellar graphite (a) and spheroidal graphite (b) iron (courtesy J. Sertucha).

2. The usual structure of graphite spheroids in cast irons

The structure of graphite in cast irons is characterised by the stacking of sheets elongated in the prismatic direction, suggesting that carbon atoms aggregate on these prismatic faces. In the case of flake graphite, these sheets can be several hundred microns long for primary precipitates (figure 2a). Their thickness varies from 50 to 500 nm, and two successive sheets can be rotated around their common c-axis and show low energy configurations for several rotation angles [5]. The observation of this rotation between two successive sheets is a strong indication that a new layer is formed by 2D nucleation on the outer basal surface of the lamella and then grows laterally along the existing surface. This is the mechanism considered by Amini and Abbaschian to model the thickening of primary graphite lamellae [6].

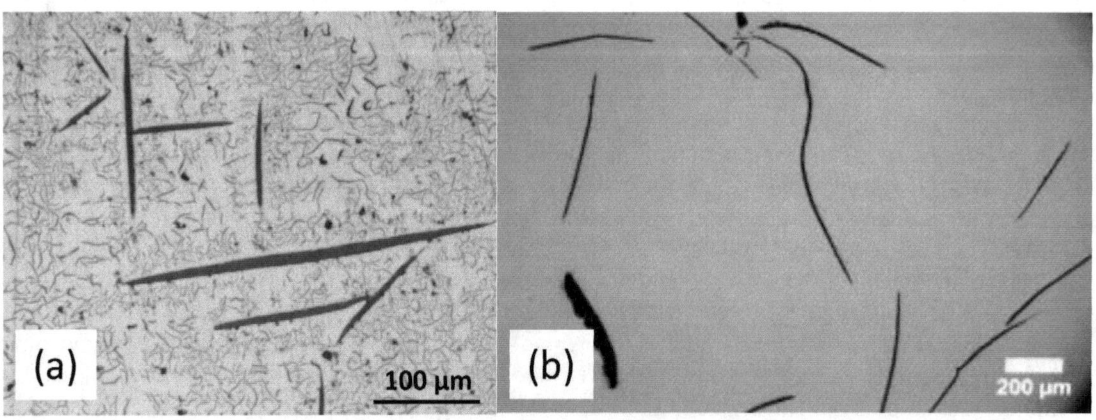

Figure 2. Primary graphite lamellae in a "pure" Fe-C alloy (a) and in a Sb-doped alloy (b). (courtesy of K. Theuwissen)

These possibilities of in-plane rotation should facilitate branching during eutectic growth, but they do not seem to have been reported to influence the shape of the primary precipitates which remain plate-like in "pure" alloys. On the other hand, the addition of certain impurities such as antimony gives a curved appearance to the primary graphite lamellae (see figure 2b) when observed by optical microscopy [7]. TEM observations have shown that such curved lamellae are in fact made up of successive rectilinear blocks that are twisted between each other around an axis parallel to the basal plane. It was then suggested that the antimony accumulates in front of the prismatic planes until it blocks the growth in the initial direction and forces the lamella to take another direction.

Under optical microscopy, and particularly under polarised light, graphite spheroids appear to be formed of adjacent sectors in which the graphite c-axis has a predominantly radial direction (figure 3a). Hillert and Lindholm proposed early on that this arrangement should be associated with growth along screw dislocations [8], a model that has been adopted in many works and is still one of the most frequently presented, together with the slightly modified form of helical growth [3, 9].

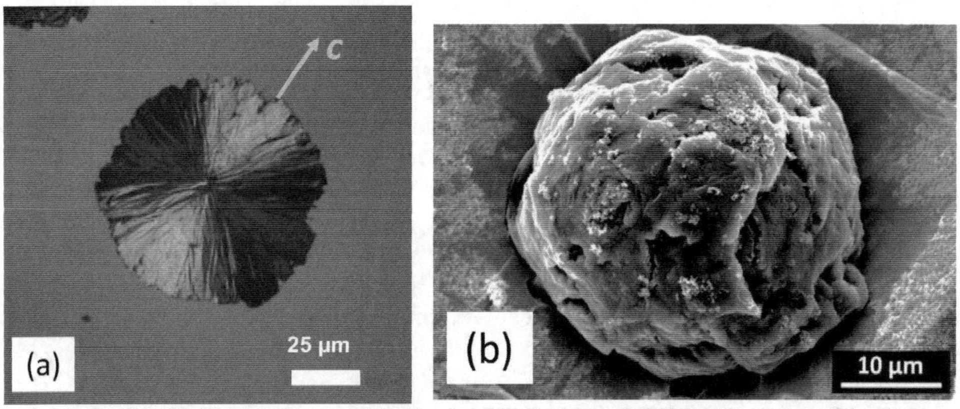

Figure 3. Metallographic section under polarized light (a) and SEM micrograph (b) of a graphite spheroid. The SEM micrograph was taken after deep etching of the matrix. (courtesy of J. Bourdie)

TEM demonstrated that the graphite in spheroids is also made up of blocks elongated in the prismatic direction, but no longer than a couple of microns [10-12]. The very existence of these blocks stacked one on top of the other is contrary to growth around a screw dislocation, unless we assume that the thickness of the blocks corresponds to the height of the growth step. To get to the bottom of this, the relative orientation of successive blocks was followed in TEM along the radius of a spheroid. The blocks showed a slight disorientation around their common c-axis, but without the expected continuity that should appear in the case of screw dislocation [13] or helical growth. In a very detailed work on structural defects in spheroidal graphite, Qing *et al.* [12] came to the same conclusion of no screw dislocation.

Graphite can take an almost infinite number of forms, which Stefanescu has listed [14]. Although some of these forms may be occasionally encountered in cast irons, most of them are not. This is particularly true of dendritic forms emanating from the nucleus of the spheroids and which, through multiple and successive branching, would explain the formation of the spheroids. If such forms existed, they should be easily observed in the 5-7 µm diameter spheroids found in inoculated castings solidified in 2 or 3 seconds, either by quenching small samples or in the products of centrifugal casting. The absence of these structures or any reminiscences excludes this type of growth. The same is true of lamellar branch structures, which have also been invoked as noticed above. It is worth stressing that the radial lines seen in figure 3a are boundaries between sectors or sub-sectors which have been put in relief during polishing [15].

An alternative scheme for spheroidal graphite growth that ensures a compact stacking of the growth blocks while allowing separate growth of the sectors as in exploded graphite (see next section) was recently proposed [16]. It considers repeated epitaxial nucleation of growth blocks at the surface of the sectors while their lateral extension (along this surface, i.e. in the prismatic direction) is limited by adsorption of magnesium atoms. Such a model implies that growth of primary graphite proceeds at some undercooling with respect to the graphite liquidus and it has been found to conveniently explain thermal analysis results [17].

What has not yet been sufficiently studied is the formation of the sectors just after the nucleation of the graphite. Theuwissen *et al.* [7] observed the sectors to appear right in contact with the nucleus in large spheroids corresponding to casting at slow cooling rates. There, the graphite layers are slightly disturbed as also observed by Qing *et al.* [18] for smaller spheroids from samples quenched during solidification. These latter authors associate this disorder with the ability of graphite to bend around the nucleus very much as the circumferential model proposed by Sadocha and Gruzleski [19]. This model is contrary to the existence of sectors, and it has been shown that the observation of enveloping leaves on the outer surface of spheroids as illustrated in figure 3b is a surface artefact and does not imply circumferential growth of graphite spheroids [15] as so often claimed. Finally, a couple of works reported the presence of highly disordered graphite in a 500 nm thick zone around the nucleus; this has been discussed previously [20].

3. Role of magnesium

It was mentioned in the introduction that spheroidisation is achieved by adding magnesium or rare earths. Magnesium is much more effective [21] but rare earths (cerium and/or lanthanum) are often added at the same time to enhance the deoxidising and desulphurising effect. However, it has long been established that 200 ppm of magnesium in solution is required for optimum spheroidisation, with the excess going to capture sulphur and oxygen. In itself, this requirement of free magnesium demonstrates that its role is not simply to lower oxygen and sulphur levels, suggesting that it is adsorbed on the surface of graphite and affects its growth forms. Another observation that makes this interaction evident is the fact that over-treatment with magnesium, in practice above 600-800 ppm of added magnesium to the melt, leads to the degeneration of spheroidal graphite (see reviews by Lux [22] and Lacaze *et al.* [23]). This degeneration manifests itself as graphite that appears "exploded" if primary, and as "spiky" or "mesh" graphite if formed during the eutectic transformation; see figure 4.

Figure 4. a: Exploded graphite spheroid (courtesy of R. Ghergu);
b: overgrowth of spiky graphite onto graphite spheroids [24].

The evidence for this interaction between dissolved magnesium in the liquid alloy and precipitating graphite was defended long ago by Merchant in a comprehensive review of the work on cast irons in the early 1960s [25]. However, it was challenged in the mid-1970s when Auger spectroscopy measurements failed to detect magnesium on graphite spheroids [26]. Advances in microanalysis have revealed the presence of magnesium in some cases but have also confirmed the absence of accumulation of this element at the graphite/matrix interface [27-29]. The lack of detection by Auger mentioned above would therefore be due to the detection limit of this technique (a few %) in the presence of a more diluted spatial distribution of magnesium than imagined.

To further understand the role of magnesium, it appeared interesting to study a spheroidal graphite cast iron doped with antimony. When added at a level of a few tens of ppm to spheroidal graphite cast iron, antimony leads to both a decrease of the diameter of the spheroids and a smoothing of their surface, and then to the development of protuberances. A SEM micrograph of a deep-etched sample showing a spheroid with protuberances is shown in figure 5a. Note the smoothing of the spheroid surface when compared to figure 3b. Further, antimony is known to strongly promote pearlite during the eutectoid transformation and this has been attributed to a Sb-rich layer at the graphite/matrix interface which has been evidenced by Auger analysis [30-32]. This pearlite-promoter effect has been understood to be due to the Sb-rich layer preventing the transfer of carbon to the graphite precipitates, which raises the question of when this layer forms.

As a matter of fact, antimony is known to be surface active in iron melts and, as such, is expected to segregate at the interface between graphite and matrix (either liquid or austenite) as it does at the surface of the melt if there is no more surface active element in the alloy. On the same sample as that illustrated in figure 5a, thin foils were prepared so as to analyse the graphite/matrix interface in an atomic resolution microscope. A 2-4 nm thick Sb-rich layer could be found at the interface of both the spheroids and the protuberances, as seen with the Sb map superimposed to the TEM micrograph in figure 5b. Some lead was also found but no trace of magnesium could be detected. It may be suggested that antimony accumulates progressively at the graphite/matrix interface during early solidification and repels magnesium from the interface, first limiting spheroidal growth and finally stopping it, thus forcing the development of protuberances for graphite to precipitate.

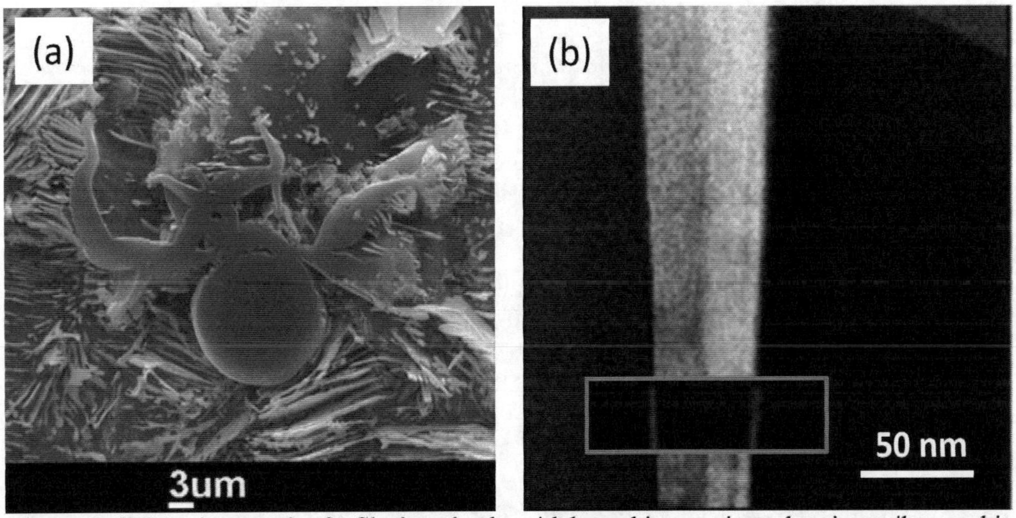

Figure 5. SEM micrograph of a Sb-doped spheroidal graphite cast iron showing spiky graphite growing out from a spheroid (a) and TEM bright field image of a protuberance with superimposed within the red rectangle the map of antimony.

Measurements of the surface tension and contact angle of cast iron drops on graphite have been recently re-analysed [33]. Adding magnesium to a melt decreases the amount of free sulphur and oxygen and thus increases its surface tension, but at the same time strongly decreases the work of adhesion. Turned the way round, melts leading to lamellar graphite are characterized by a high work of adhesion between melt and graphite that relates to the presence of sulphur. It could thus be concluded that the graphite shape is not only determined by the melt surface tension as sometimes stated, but that consideration of the work of adhesion is also needed. Further work is needed to understand how this works: adding Al to an S-bearing alloy increases surface tension while retaining high work of adhesion and the lamellar form, whereas adding antimony to a melt that has been spheroidised decreases surface tension and retains low work of adhesion but leads to a spiky graphite that develops as lamellar graphite.

4. Conclusion

Analysis of the structure of graphite spheroids in cast iron suggested a growth scheme based on a mechanism of epitaxial 2D-nucleation of growth blocks at the surface of the spheroids that then grow parallel to the surface in the prismatic direction. Such a growth is hindered by adsorption of magnesium atoms without any build-up of this element. On the contrary, surface active elements such as sulphur and oxygen in non-spheroidized melts and antimony in magnesium-treated melts do segregate at the graphite/matrix interface, repelling magnesium and then leading to lamellar-type graphite growth.

Acknowledgments
B. Tonn (Uni. Clausthal), A. Undisz and M. Rettenmaier (Uni. Jena) are warmly thanked for the collaborating work illustrated with figure 5.

References

[1] Granasy L, Pusztai T, Tegze G, Warren J A and Douglas J F 2005 *Physical Review E* **72** 011605
[2] Ryschenkow G and Faivre G 1988 *Journal of non-crystalline solids* **87** 221
[3] Double D D and Hellawell A 1974 *Acta Metall.* **22** 481
[4] Faivre G 1997 *Advanced Materials Research* **4-5** 17
[5] Double D D and Hellawell A 1971 *Acta Metall.* 19 1303
[6] Amini S and Abbaschian R 2013 *Carbon* **51** 110
[7] Theuwissen K, Lacaze J and Laffont L 2016 *Carbon* **96** 1120
[8] Hillert M and Lindblom Y 1954 *J. of Iron and Steel Institute* **176** 388
[9] Miao B, Northwood D O, Bian W, Fang K and Fan M H 1994 *J. Materials Science* **29** 255
[10] Mitsche R, Haensel G, Maurer K and Schäffer H 1968 *Fonderie* **270** 367
[11] Monchoux J-P, Verdu C, Tholeet G and Fougères R 2001 *Acta Mater.* **49** 4355
[12] Qing J, Richards V L and Van Aken D C 2017 *Carbon* **116** 456
[13] Theuwissen K, Lafont M-C, Laffont L, Viguier B and Lacaze J 2012 *Transactions of the Indian Institute of Metals* **65 627**
[14] Stefanescu D M, Alonso G, Larranaga P, de la Fuente E and Suarez R 2017 *Acta Mater.* **139** 109
[15] J. Bourdie, J. Lacaze, C. Josse, L. Laffont 2020 *Int. J. Metalcasting* **14** 672
[16] Lacaze J, Bourdie J and Castro Roman M J 2017 *Acta Mater.* **134** 230
[17] Castro-Román M J, Lacaze L, Regordosa A, Sertucha J and del Campo-Castro R 2020 *Metall. Mater. Trans. A* **51** 6373
[18] Qing J, Xu M and Pikhovich V 2020 *Int. J. Metalcasting* **14** 1115
[19] Gruzleski J E 1975 *Carbon* **13** 167
[20] Laffont L, Jday R and Lacaze J 2018 *Metall. Mater. Trans. A* **49** 1287
[21] White C V, Flinn R A and Trojan P K 1983 *AFS Trans.* **91** 549
[22] Lux B 1972 *AFS Cast Metals Research J.* **3** 25
[23] Lacaze J, Connétable D and Castro de Roman M J 2019 *Materialia* **8** 100471
[24] Basutkar P K, Park C S, Miller R E and Loper C R 1973 *AFS Transactions* **81** 180
[25] Merchant H D 1968 Solidification of cast iron – A review of literature *Recent research on cast iron* (Gordon and Breach) pp. 1-100
[26] Johnson W C, Kovacs B V and Clum J A 1974 *Scripta Metall.* **8** 1309
[27] Dierickx P, Verdu C, Reynaud A and Fougères R 1996 *Scripta Mater.* **34** 261
[28] Bourdie J, Bruneseaux F, Parseval P, Gouy S, Laffont L and Lacaze J 2018 *Materials Science Forum* **925** 20
[29] Domengès B, Celis M M, Moisy F and Lacaze J 2021 *Carbon* **172** 532
[30] Kovacs B V 1981 *AFS Transactions* **89** 79
[31] Liu B C, Li T X, Rue Z J, Yang X Y, Huo E Q and Loper C R 1990 *AFS Transactions* **98** 753

[32] Dekker L, Tonn B and Lilienkamp G 2020 *Int. J. Metalcasting* **14** 827
[33] Lacaze J and Dezellus O *Metall. Mater. Trans. B* DOI: 10.1007/s11663-021-02352-x

IOP Conf. Series: Materials Science and Engineering 1274 (2023) 012040 doi:10.1088/1757-899X/1274/1/012040

Preliminary system for data assimilation to infer material parameters from directional solidification experiments: twin experimental study using phase-field method

Y Imai[1], S Sakane[2], M Ohno[3], H Yasuda[4] and T Takaki[2]

[1] Graduate School of Science and Technology, Kyoto Institute of Technology, Matsugasaki, Sakyo-ku, Kyoto 606-8585, Japan.
[2] Faculty of Mechanical Engineering, Kyoto Institute of Technology, Matsugasaki, Sakyo-ku, Kyoto 606-8585, Japan.
[3] Division of Materials Science and Engineering, Faculty of Engineering, Hokkaido University, Kita 13 Nishi 8, Kita-ku, Sapporo, Hokkaido 060-8628, Japan.
[4] Graduate School of Engineering, University of Kyoto, Yoshidahon-machi, Sakyo-ku, Kyoto 606-8501, Japan.

takaki@kit.ac.jp

Abstract. The integration of phase-field (PF) simulations and in situ observations is a promising approach for understanding dendrite growth. In this study, a preliminary data assimilation system is developed to integrate PF simulations and in situ real-time X-ray radiography during the directional solidification of a binary alloy. In this system, only the region around the tip of a primary arm is used for data assimilation. The validity of the developed system is confirmed through twin experiments for columnar dendrite growth with different inclined angles of preferred growth direction.

1. Introduction

A highly accurate prediction of dendrite growth is essential for improving the quality of alloy products [1]. Thus, dendrite growth has been extensively investigated using numerical simulations and experiments.

The phase-field (PF) method is the most accurate model for dendrite growth. The quantitative PF models of dendrite growth [2-4], results of which do not depend on the interface thickness, are generally used for alloy solidification. However, the PF method is computationally expensive because it is a diffuse interface model. To solve this problem, an adaptive mesh refinement method and parallel computing have been applied to PF simulations [5]. Graphics processing units (GPUs) have been actively used in recent years [6], and large-scale computing has been performed using multiple GPUs in parallel [7-10]. As mentioned above, owing to the development of quantitative PF models, computational schemes, and computer hardware, the spatiotemporal domain in which the

PF method can be applied is gradually expanding. Nevertheless, the lack of material properties is a key issue in the PF simulations for dendrite solidification.

As metallic alloys are not transparent and alloy solidification occurs at high temperatures, the direct observation of dendrite growth was impossible. The time-resolved X-ray imaging of dendritic growth was proposed around 2000 [11, 12]. Time-resolved tomography (4D-CT) [13-15] has been developed in recent years, and it has advanced the understanding of dendrite growth. However, 4D-CT observations cannot be applied to fast phenomena owing to their insufficient spatiotemporal resolution, as a 3D image is created by reconstructing hundreds of transmission images obtained during the half rotation of a cylindrical specimen.

Data assimilation [16] has been attracting attention as a promising method for integrating numerical simulations and experiments. Ohno et al. developed a method for estimating interfacial properties by combining molecular dynamics and PF simulations with an ensemble Kalman filter (EnKF) and applied it to pure iron and nickel [17, 18]. Yamanaka et al. employed the local ensemble transform Kalman filter to dendritic solidification to efficiently estimate material parameters [19, 20]. Miyoshi et al. proposed a method of simultaneously inferring multiple grain boundary properties using the EnKF and multi-phase-field method [21]. However, data assimilation using actual experimental data as observation data has not yet been performed. If data assimilation can be performed using in situ observations, it will be possible to simultaneously solve the problems of spatiotemporal resolution in experiments and the material properties required in simulations. However, it is not currently feasible to perform data assimilation using PF simulations for an entire region of in situ observations because of computational costs [8].

The purpose of this study is to construct a preliminary data assimilation system to integrate in situ observations and PF simulations, where observation data is obtained only from the region around the dendrite tip of a primary arm during the directional solidification of a binary alloy to reduce the computational cost. The validity of the system is confirmed through twin experiments and two-dimensional (2D) directional solidification simulations for an Al–Cu alloy. The anisotropy strength of interfacial energy and diffusion coefficient of the liquid are simultaneously estimated.

2. Data assimilation system

In this study, the quantitative PF method for binary alloy solidification [3] was applied to directional solidification, and the EnKF [17] was used for data assimilation.

2.1. Phase-field method

The PF variable, ϕ, is defined as $\phi = 1$ in the solid phase and $\phi = -1$ in the liquid phase, and ϕ varies smoothly from 1 to –1 at the interface. The evolution equation of ϕ is given by

$$t\,[1-(1-k)u']\frac{\partial f}{\partial t} = \nabla\cdot[W^2\nabla f] + \sum_{i=x,y}\left[W\frac{\partial W}{\partial f_i}|\nabla f|^2\right] - f\left(f^2-1\right) - l^*\left(1-f^2\right)^2(u+u') \tag{1}$$

where τ is the relaxation time of ϕ and W is the interface thickness; $\tau = \tau_0\, a_s(\nabla\phi)^2$ and $W = W_0\, a_s(\nabla\phi)$. The anisotropy function is $a_s(\nabla\phi) = 1 - 3\varepsilon_4 + 4\varepsilon_4(\phi_x^4 + \phi_y^4) / |\nabla\phi|^4$, and ε_4 is the anisotropy strength of interfacial energy. ϕ_i denotes the partial derivative of ϕ with respect to i. λ^* is the quantity associated with the thermodynamic driving force. u is dimensionless supersaturation, which is expressed as $u = (c_l - c_l^e)/(c_l^e - c_s^e)$. c_l is the liquid concentration, and c_l^e and c_s^e are the equilibrium concentrations in the liquid and solid, respectively. The temperature change is expressed as $T(y) = T_0 + G(y - Vt)$ by

IOP Conf. Series: Materials Science and Engineering 1274 (2023) 012040 doi:10.1088/1757-899X/1274/1/012040

applying the frozen temperature approximation, where T_0 is the temperature at $y = 0$ and $t = 0$, G is the temperature gradient in the y-direction, and V_p is the pulling velocity. u' in equation (1) is expressed as $u' = (y - Vt)/l_T$ using thermal length l_T. The time evolution equation of u is given by

$$\frac{[1+k-(1-k)\phi)]}{2}\frac{\partial u}{\partial t} = \nabla[D_l q(\phi)\nabla u - j_{AT}] + \frac{1}{2}[1+(1-k)u]\frac{\partial \phi}{\partial t} \quad (2)$$

where $q(\phi) = [kD_s + D_l + (kD_s - D_l)\phi] / (2D_l)$, k is the partition coefficient, and D_l and D_s are the diffusion coefficients of the liquid and solid phases, respectively. j_{AT} is the antitrapping term, which is added to perform quantitative calculations independent of the interface width.

2.2. Data assimilation

A sequential data assimilation system that uses the EnKF was developed in this study. The EnKF can be applied to nonlinear system models such as the PF equation. The probability distribution of state vector $\boldsymbol{x}_t$ containing unknown parameters is represented by a set, $\{\boldsymbol{x}_t^{(i)}\}_{i=1}^N$, of N state vectors. This is referred to as the ensemble approximation, where N is ensemble members. The state vector of i-th ensemble member at time t is expressed by $\boldsymbol{x}_t^{(i)} = \{\boldsymbol{\phi}_t^{(i)}\ \boldsymbol{u}_t^{(i)}\ \varepsilon_{4,t}^{(i)}\ D_{l,t}^{(i)}\}^{\mathrm{T}}$, where $\boldsymbol{\phi}_t^{(i)}$ and $\boldsymbol{u}_t^{(i)}$ are the vectors of ϕ and u at all grid points at time t, respectively. ε_4 and D_l are estimated in the twin experiments and included in the ensemble member as unknown parameters. $\boldsymbol{x}_t^{(i)}$ is updated by repeating the prediction and filtering, and the unknown parameters are expected to gradually approach the true values in this process. The prediction is based on the following system model:

$$\boldsymbol{x}_{t|t-1}^{(i)} = f_t\left(\boldsymbol{x}_{t-1}^{(i)}\right) + \boldsymbol{v}_t^{(i)} \tag{3}$$

where subscript $t|t-1$ in $\boldsymbol{x}_{t|t-1}^{(i)}$ indicates the values at time t obtained using the values at time $t-1$. f_t is a nonlinear map from $\boldsymbol{x}_{t-1}$ to $\boldsymbol{x}_t$, and it corresponds to equations (1) and (2). $\boldsymbol{v}_t^{(i)}$ is system noise, which follows a Gaussian distribution with mean vector $\mathbf{0}$ and variance–covariance matrix $\boldsymbol{Q}_t$. $\boldsymbol{v}_t^{(i)}$ is used to express the uncertainty of numerical simulation. The state vector is updated at every time step in the simulation with time increment Δt on the basis of equation (3). At time $t = n\Delta t_{obs}$, when observed data are available, the state vector is updated by filtering with the following equation:

$$\boldsymbol{x}_{t|t}^{(i)} = \boldsymbol{x}_{t|t-1}^{(i)} + \boldsymbol{K}_t[\boldsymbol{y}_t + \breve{\boldsymbol{w}}_t^{(i)} - \boldsymbol{H}_t\boldsymbol{x}_{t|t-1}^{(i)}] \tag{4}$$

where $\boldsymbol{K}_t$ is the Kalman gain, which determines the degree of correction of the state vector on the basis of the observation data. It is computed as $\boldsymbol{K}_t = \boldsymbol{V}_t\ \boldsymbol{H}_t^{\mathrm{T}}\ (\boldsymbol{H}_t\boldsymbol{V}_t\boldsymbol{H}_t^{\mathrm{T}} + \boldsymbol{R}_t)^{-1}$, where $\boldsymbol{V}_t = (N-1)^{-1}\sum_{j=1}^{N}\breve{\boldsymbol{x}}_t^{(j)}\breve{\boldsymbol{x}}_t^{(j)\mathrm{T}}$ and $\breve{\boldsymbol{x}}_t^{(j)} = \boldsymbol{x}_{t|t-1}^{(j)} - N^{-1}\sum_{j=1}^{N}\boldsymbol{x}_{t|t-1}^{(j)}$. Observation matrix $\boldsymbol{H}_t$ is computed from the observation model, $\boldsymbol{y}_t = \boldsymbol{H}_t\boldsymbol{x}_{t|t-1}^{(i)} + \boldsymbol{w}_t^{(i)}$, and observation noise $\boldsymbol{w}_t^{(i)}$ follows a Gaussian distribution with mean vector $\mathbf{0}$ and variance–covariance matrix $\boldsymbol{R}_t$. In equation (4), $\breve{\boldsymbol{w}}_t^{(i)}$ is expressed as $\breve{\boldsymbol{w}}_t^{(i)} = \boldsymbol{w}_t^{(i)} - N^{-1}\sum_{j=1}^{N}\boldsymbol{w}_t^{(j)}$, and $\boldsymbol{y}_t + \breve{\boldsymbol{w}}_t^{(i)}$ is referred to as the disturbed observation. For simplicity, we consider $\boldsymbol{R}_t = \sigma_t^2\boldsymbol{I}$ with variance σ_t^2, covariance $\mathbf{0}$, and unit matrix $\boldsymbol{I}$.

Table 1. Material properties of Al–3wt%Cu and simulation conditions for obtaining observation data. "*" denotes artificially determined values.

Parameter	Symbol	Value
Initial concentration	C_0	0.013 at.frac.
Melting point of pure Al	T_m	933.25 K
Partition coefficient	k	0.14
Anisotropic strength	ε_4	0.009 *
Gibbs–Thomson coefficient	Γ	0.24×10^{-6} Km
Diffusion coefficient in liquid	D_l	2.0×10^{-9} m²/s *
Diffusion coefficient in solid	D_s	2.0×10^{-13} m²/s *
Liquidus slope	m	-620 K/at.frac
Temperature gradient	G	100 K/mm
Pulling velocity	V_p	100 μm/s

Computational domain size	$L_x \times L_y$	$1000 \times 6000\Delta x^2$
Grid size	Δx	0.4 μm
Interface thickness	W_0	$\Delta x/0.6$
Time increment	Δt	1.6×10^{-5} s

3. Observation data

The 2D directional solidification PF simulations of Al–3wt%Cu were carried out to generate observation data for the twin experiments. The material properties and calculation conditions are listed in table 1. ε_4 and D_l were set to assumed values and these values will be estimated in the EnKF. The simulations were performed for two different preferred growth directions with $\theta = 0°$ and $\theta = 15°$, where θ is the angle between the preferential growth direction and temperature gradient direction (y-axis). As the initial conditions, the entire computational domain was filled with a liquid phase with $u_0 = -0.295$, and a semicircular solid with radius $6\Delta x$ was placed at the center of the bottom of the computational domain. Figures 1(a) and (b) depict the simulation results for $\theta = 0°$ and $\theta = 15°$, respectively. Typical columnar dendrites grew in the temperature gradient direction with secondary arms.

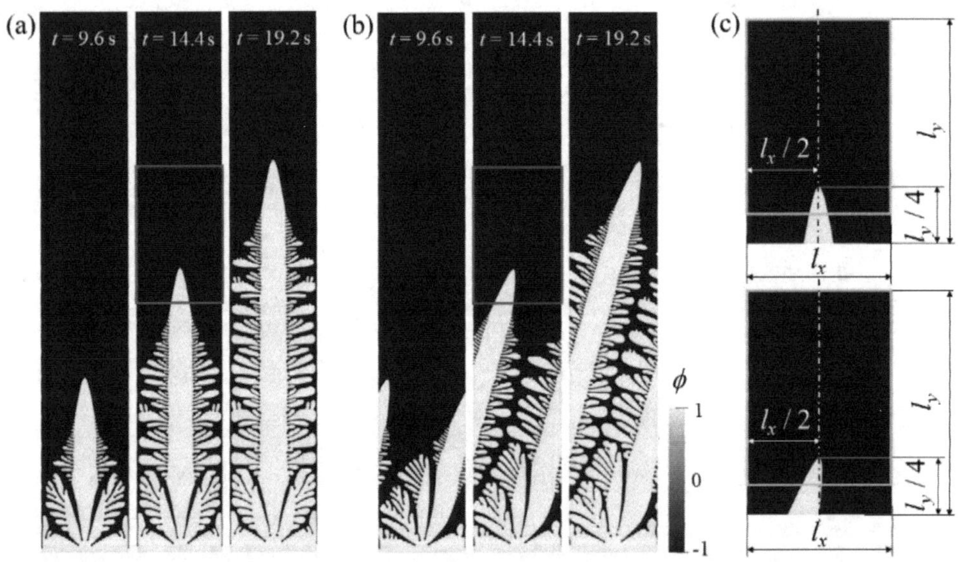

Figure 1. Time evolution of dendrite morphology for (a) $\theta = 0°$ and (b) $\theta = 15°$. (c) Dendrite tip regions enclosed by rectangular red regions in (a) and (b), which correspond to the simulation domain in prediction of data assimilation. The rectangular green regions were used for filtering.

Table 2. Parameters used in the twin experiments.

Parameter	Symbol	Value
Ensemble size	N	100
Variance of observation noise	σ_r^2	1.0
Initial value of ε_4	$\varepsilon_{4,\,init}$	0.005
Initial standard deviation of ε_4	$\sigma_{\varepsilon 4,\,init}$	$0.5 \times \varepsilon_{4,\,init}$
Initial value of D_l	$D_{l,\,init}$	1.0×10^{-9}
Initial standard deviation of D_l	$\sigma_{Dl,\,init}$	$0.25 \times D_{l,\,init}$
Variance of system noise for ϕ	σ_ϕ^2	1.0×10^{-6}
Variance of system noise for u	σ_u^2	1.0×10^{-8}

Variance of noise for ε_4	$\sigma_{\varepsilon 4}^2$	1.0×10^{-11}
Variance of noise for D_l	σ_{Dl}^2	1.0×10^{-23}

4. Twin experiments

ε_4 and D_l were inferred through the twin experiments to confirm the validity of the developed data assimilation system. We used only the region around the dendrite tip of the primary arm to reduce the computational cost of data assimilation. The size of the region was set to $l_x \times l_y = 400 \times 600$ μm^2 ($1000 \times 1500\Delta x^2$) so that the solute diffusion field in front of the dendrite tip did not reach the upper edge of the region. The rectangular red regions were extracted from the results presented in figure 1 so that the dendrite tip was always positioned at ($l_x/2$, $l_y/4$). For $\theta = 15°$, the domain was shifted in the lateral direction such that the tip position was ($l_x/2$, $l_y/4$). We obtained 50 data of microstructure (ϕ profile) during t = 14.4 s–15.2 s as the observation data. Thus, the time increment of the observation data was Δt_{obs} = 0.016 s. As the explicit scheme was used for the time derivatives of equations (1) and (2), the time increment changed depending on D_l. Hence, the maximum value of D_l was set as 4.0×10^{-9} m/s^2, which was twice the true value presented in table 1, and the time increment was set as 0.8×10^{-5} s. The PF simulations during data assimilation were carried out using a moving frame algorithm to maintain the dendrite tip position at ($l_x/2$, $l_y/4$) using the region illustrated in figure 1(c). For $\theta = 15°$, the lateral motion of the region was also introduced to maintain the dendrite tip position at ($l_x/2$, $l_y/4$). The boundary conditions were set as periodic on the left and right sides and zero Neumann on the top and bottom. When the moving frame algorithm is used in a small region, as depicted in figure 1(c), the solidification morphology at the lower end of the region may be destabilized by the zero-Neumann condition. Therefore, data assimilation was performed in the green frame, which did not include the lower edge. Observation vector $\boldsymbol{y}_t$ was set as $\boldsymbol{y}_t = \boldsymbol{\phi}_t^{obs}$. The other conditions for data assimilation are listed in table 2.

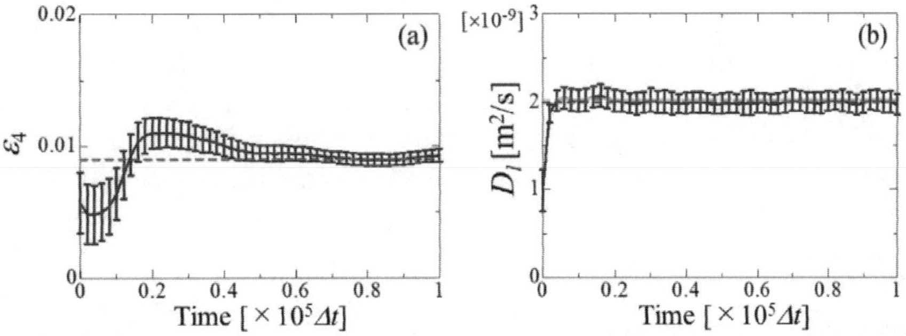

Figure 2. Inference results obtained during data assimilation for (a) anisotropy strength of interfacial energy, ε_4, and (b) diffusion coefficient of liquid, D_l, for the observation data for θ = 0°.

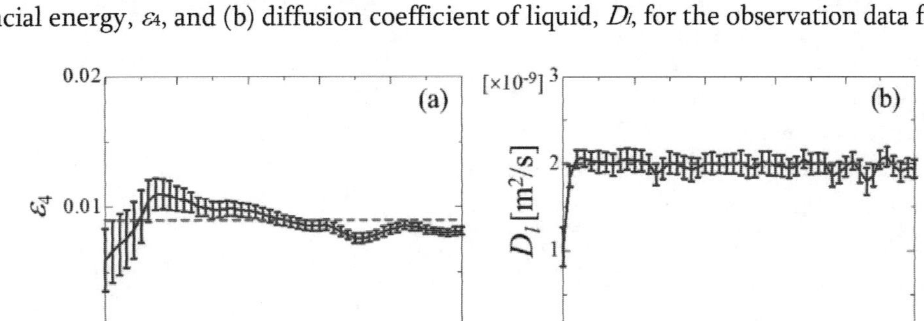

Figure 3. Inference results obtained during data assimilation for (a) anisotropy strength of interfacial energy, ε_4, and (b) diffusion coefficient of liquid, D_l, for the observation data for $\theta = 15°$.

Figure 2 presents the inference results of ε_4 and D_l for $\theta = 0°$. The red dashed line represents the true value, the blue line represents the change in the estimated average value, and the error bars represent the standard deviation of the estimated values. D_l rapidly approached the true value from the initial value, and the approximate true value was estimated in the second filtering. In contrast, the inferred value of ε_4 approached the true value in approximately the 40th filtering. The difference between the convergence behaviors of ε_4 and D_l depended on their contribution to the dendrite tip shape. Although ε_4 and D_l affected the tip morphology of the dendrite, the effect of D_l was stronger than that of ε_4. Nevertheless, we successfully inferred ε_4 and D_l simultaneously using the developed data assimilation system.

Figure 3 presents the inference results of ε_4 and D_l for $\theta = 15°$. The tendency of the convergence behaviors of ε_4 and D_l was similar to that illustrated in figure 2. However, the fluctuations for $\theta = 15°$ were larger than those for $\theta = 0°$. In addition, the convergence of ε_4 to the true value was comparatively slower for $\theta = 15°$. As illustrated in figure 1, when the rectangular red regions were extracted, the data was converted such that the position of the dendrite tip was ($l_x/2$, $l_y/4$). Subsequently, the data was shifted to a unit of the grid size. In addition, the moving frame algorithm in the PF simulations was applied in a unit of the grid size. These migrations in the x-direction caused the error in the inference of ε_4 for $\theta = 15°$. In other words, the effect of the deviation of the interface position between the observation data and simulations was stronger for ε_4 compared to that for D_l. Nevertheless, the estimation accuracy was less than 10%, even for ε_4. Thus, the estimation was highly accurate even when the dendrite was inclined.

5. Conclusions

A preliminary data assimilation system was developed to integrate PF simulations and in situ real-time X-ray radiography during the directional solidification of a binary alloy. Observation data were obtained from the region around the tip of a primary arm to reduce the computational cost. The validity of the developed system was confirmed through twin experiments of the 2D directional solidification of an Al–Cu alloy with different inclined angles of the dendrite. In future work, we will apply the proposed system to the time-resolved X-ray imaging of dendritic growth during directional solidification.

Acknowledgement

This work was supported by KAKENHI, Grant-in-Aid for Scientific Research (A) 20H00217.

References

[1] Dantzig J A and Rappaz M 2009 *Solidification*: EFPL Press)
[2] Karma A and Rappel W J 1998 *Physical Review E - Statistical Physics, Plasmas, Fluids, and Related Interdisciplinary Topics* **57** 4323-49
[3] Ohno M and Matsuura K 2009 *Physical Review E* **79** 031603
[4] Ohno M 2020 *ISIJ International* **60** 2745-54
[5] Sakane S, Takaki T and Aoki T 2022 *Materials Theory* **6** 3
[6] Karma A and Tourret D 2016 *Current Opinion in Solid State and Materials Science* **20** 25-36
[7] Shibuta Y, Ohno M and Takaki T 2015 *JOM* **67** 1793-804
[8] Shibuta Y, Ohno M and Takaki T 2018 *Advanced Theory and Simulations* **1** 1800065

[9] Takaki T, Sakane S, Ohno M, Shibuta Y and Aoki T 2022 *Computational Materials Science* **203** 111143
[10] Takaki T, Sakane S, Ohno M, Shibuta Y and Aoki T 2020 *Computational Materials Science* **171** 109209
[11] Mathiesen R H, Arnberg L, Mo F, Weitkamp T and Snigirev A 1999 *Physical Review Letters* **83** 5062-5
[12] Yasuda H, Ohnaka I, Kawasaki K, Sugiyama A, Ohmichi T, Iwane J and Umetani K 2004 *Journal of Crystal Growth* **262** 645-52
[13] Ludwig O, Dimichiel M, Salvo L, Suéry M and Falus P 2005 *Metallurgical and Materials Transactions A* **36** 1515-23
[14] Puncreobutr C, Phillion A B, Fife J L and Lee P D 2014 *Acta Materialia* **64** 316-25
[15] Yasuda H, Kawarasaki T, Tomiyori Y, Kato Y and Morishita K 2019 Characterization of dendritic growth in Fe-C system using time-resolved X-ray tomography and physics-based filtering. In: *IOP Conference Series: Materials Science and Engineering,*
[16] Houtekamer P L and Zhang F 2016 *Monthly Weather Review* **144** 4489-532
[17] Ohno M, Oka Y, Sakane S, Shibuta Y and Takaki T 2020 *Physical Review E* **101** 052121
[18] Nagatsuma Y, Ohno M, Takaki T and Shibuta Y 2021 *Nanomaterials* **11** 2308
[19] Yamanaka A and Takahashi K 2020 *Materials Today Communications* **25** 101331
[20] Takahashi K and Yamanaka A 2021 *Computational Materials Science* **190** 110296
[21] Miyoshi E, Ohno M, Shibuta Y, Yamanaka A and Takaki T 2021 *Materials & Design* **210** 110089

Simulation of dendritic grain structures with a Cellular Automaton – Parabolic Thick Needle model

Y Wu[1], O Senninger[1], Ch-A Gandin[1]

[1] Mines Paris, PSL University, Centre for material forming (CEMEF), CNRS UMR 7635, Sophia Antipolis, France

Abstract. A multiscale numerical model, coupling the Cellular Automaton (CA) method to the Parabolic Thick Needle (PTN) method, is developed to compute dendritic microstructures. It is applied to directional solidification of bi-crystals under a frozen temperature approximation with fixed temperature gradient and fixed cooling rate. A dendritic branch is modeled as a cylinder headed by a parabolic tip. Its kinetics is computed using the PTN method from the composition field in the liquid in the vicinity of the parabola. This dendritic branch takes part in the definition of a grain envelope by its integration in a CA growth algorithm. The so-called Cellular Automaton – Parabolic Thick Needle (CAPTN) model can simulate the growth of dendritic grains while accounting for non-steady diffusion fields. The PTN method is implemented using Finite Element (FE) method on a heterogeneous mesh. An adaptive FE mesh strategy is developed to optimize computational time. The grain boundary orientation angle of a diverging bi-crystal configuration is investigated at steady state and compared to the Phase-Field (PF) method and the classical CA method. It is shown that contrary to the CA method, the CAPTN model is in good agreement with data deduced from PF simulations, demonstrating the interest of this coupling in the modeling of dendritic growth.

1. Introduction

Microstructures formed during solidification processing have a large influence on the properties of cast metals and alloys. There exist different numerical methods for modeling microstructures of alloys during solidification processes at different scales. The Phase-Field (PF) method allows analyses of dendritic growth at microscopic scale. It can be used in particular to calculate the shape of solid/liquid interface and the evolution of dendrite over time by regarding phase as a continuous field. However, it is very expensive in computation due to the mesh size required at the solid/liquid interface, so its simulation domain is limited to several mm^3 [1]. The Dendritic Needle Network (DNN) method, or the Parabolic Thick Needle (PTN) method, models dendritic growth at a scale between the microscopic scale and the mesoscopic scale. The growing direction of a branch is predefined and fixed, and the dendritic tip is also predefined in parabolic form [2]. To date, it is limited to imposed temperature field.

At a larger scale, the CA method models the grain structure, where only the envelop of dendritic grains is modeled on a cellular grid. An analytical law is used to determine the envelope growth velocity at the scale of automaton cells. This analytical law is however not relevant to describe dendritic kinetics under unsteady state. This CA method can be coupled with the Finite Element (FE) method for thermal evolution, i.e., the Cellular Automaton – Finite Element (CAFE) model [3].

In order to simulate the dendritic microstructure accounting for unsteady diffusion fields with thermal evolution, the CAFE model is coupled with the PTN method. In the PTN method tip radius and velocity are computed from the composition field in the liquid in the vicinity of the tip [2]. The kinetics deduced by the PTN method is relevant for both steady and unsteady state. In this article, the coupling of the CA method to the PTN method is presented in a frozen temperature approximation.

In this article, the Cellular Automaton – Parabolic Thick Needle (CAPTN) model is introduced. This model is applied for investigating the grain boundary orientation angle of a bi-crystal configuration at steady state.

2. Model

2.1. PTN method

In the PTN method, a dendritic tip is represented as a parabola with radius ρ_{tip} truncated by a cylinder of radius r_{cyl} as in Figure 1(a). For computing the kinetics of the tip, the solute diffusion equation is solved in liquid to get the composition field $w^{\text{l}}_{\text{PTN}}$:

$$\frac{\partial w^{\text{l}}_{\text{PTN}}}{\partial t} = D^{\text{l}}\nabla^2 w^{\text{l}}_{\text{PTN}} \tag{1}$$

where D^{l} is the diffusion coefficient in liquid. Dirichlet condition is imposed at the solid/liquid interface. Since the composition field in solid has no influence on the composition in liquid, for the simplicity of implementation, the solid/liquid interface composition $w^{\text{ls}}_{\text{PTN}}$ associated to the local temperature is imposed to all nodes belonging to the parabola.

Tip radius ρ_{tip} and tip velocity v_{tip} are determined using two relations. The first relation is based on the integration of mass conservation along the parabolic solid/liquid interface of the tip included in the integration distance a. This integration is then transferred into

$$\rho_{\text{tip}} v_{\text{tip}}^2 = \frac{2{D^{\text{l}}}^2 \mathcal{F}^2}{d_0} \tag{2}$$

with $d_0 = \frac{\Gamma^{\text{sl}}}{-m(1-k)w^{\text{ls}}}$ the solute capillary length, where Γ^{sl} is the Gibbs-Thomson coefficient, m is the liquidus slope. $\mathcal{F}$ is the Flux Intensity Factor that depends on the composition flux in the vicinity of the tip. Its calculation is converted into two integrals in the integration area, i.e., the yellow part in Figure 1(a): a surface integral in the integration area and a line integral on its arc contour. Both integrals rely on the composition gradient. The formula of Flux Intensity Factor is introduced in [4]. The second relation is the solvability condition

$$\rho_{\text{tip}}^2 v_{\text{tip}} = \frac{D^{\text{l}} d_0}{\sigma} \tag{3}$$

where σ is the marginal stability constant fixed by the strength of crystalline anisotropy [5].

The PTN model is implemented in a finite element library equipped with adaptive heterogeneous mesh. the mesh strategy developed to optimize computational time is illustrated in Figure 1. As the branch has a known trajectory, the idea is to place a box (in blue) with minimum mesh size h_{min} in front of the tip, along its growth direction. This permits to compute the Flux Intensity Factor with a good precision at every time step. The integration area is a circle, and its radius at the time of creation of the box, denoted as $R^{\text{c}}_{\text{integ}}$, is determined by the integration distance a and tip radius at this time $\rho^{\text{c}}_{\text{tip}}$, $R^{\text{c}}_{\text{integ}} = \left(a^2 + 2\rho^{\text{c}}_{\text{tip}} a\right)^{1/2}$. To save computational resources, meshes in solid and liquid have larger sizes. To have good transition between different mesh sizes, two transitional areas TA1 and TA2 are defined based on the distance of nodes to the walls of the blue box. Meshes in these two transitional areas are isotropic with different mesh sizes. If mesh belongs to different areas with different sizes, its size is determined by the smallest one.

The integration area for calculating the Flux Intensity Factor of the tip must always be in the blue box, so this heterogeneous PTN mesh needs to be rebuilt after this dendritic branch grows a certain length. For a single branch, the mesh is necessary to be rebuilt if its integration area is out of the blue box. For simulation with many dendrites, if the accumulated distance passed by the fastest tip at each time step since last remeshing is greater than $2l$, then all these areas will be redefined and the entire PTN mesh is to be rebuilt. Obviously, this criterion is much stricter than the one for single dendrite, and it is used for the simplicity on implementation. If a new dendritic branch appears or the present radius of integration area of a dendritic tip is greater than 0.95 times half of the height of its blue box, then only the blue box and 2 transitional areas of the corresponding tip will be defined and mesh in these areas is to be rebuilt.

(a)

(b)

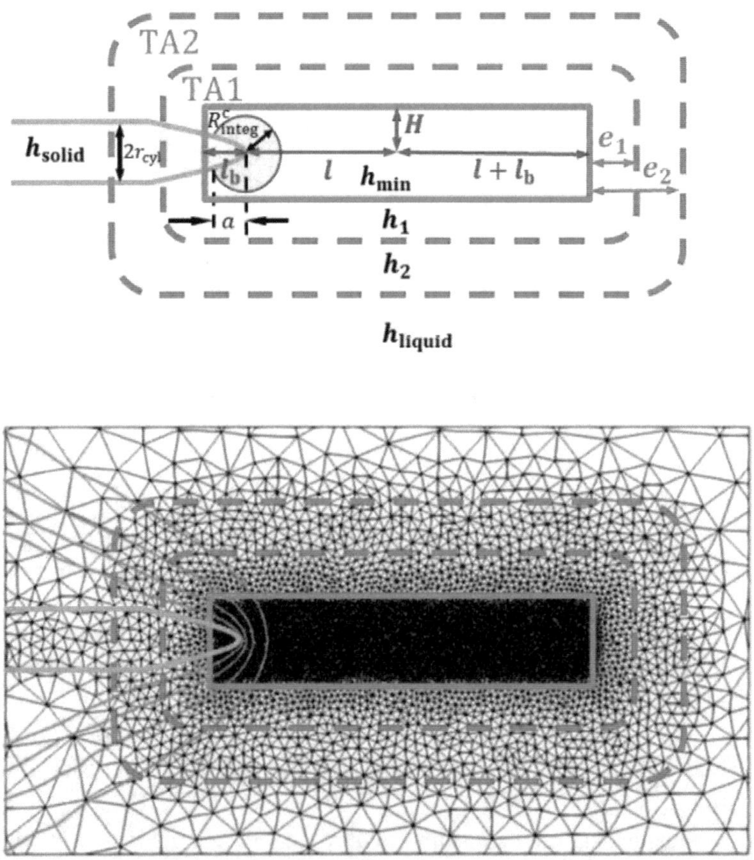

Figure 1. Heterogeneous PTN mesh. (a) Schema of different areas used for adaptive mesh in PTN method. A rectangular area in blue with minimum mesh size h_{min} is defined near the tip, encompassed by two transitional areas TA1 and TA2 with green dashed line boundaries. (b) Mesh corresponding to the schema. Violet lines are the iso-composition lines in the liquid.

Numerical parameters are linked to physical scales of the structure. Consider a single dendritic branch grows in a large isothermal domain. At steady state the composition field in liquid follows Ivantsov law [6]. Its diffusion length $\delta_{d_{Iv}}$ along the axis is defined as twice of the intersection of the tangent line of composition at the tip with the nominal composition. Since the calculation of Flux Intensity Factor relies on composition, the integration distance equals to twice of diffusion length at steady state. The tip radius of the branch at steady state is denoted as $\rho_{tip_{Iv}}$. The truncating cylinder should have no influence on the calculation of kinetics of dendritic branches, so it should not be intersected with the integration area. Consider a dendritic branch growing at steady state, the minimum value of the radius of truncating cylinder is $r_{\text{cyl}}^{\text{min}} = \sqrt{2a\rho_{\text{tip}_{\text{Iv}}}}$. If the radius of truncating cylinder r_{cyl} is greater than $r_{\text{cyl}}^{\text{min}}$, the kinetics of the branch is not influenced by r_{cyl} [2]. Thus, in our application, r_{cyl} is fixed as $r_{\text{cyl}} = 2r_{\text{cyl}}^{\text{min}}$. Verification of our model and choice of parameters are realized by simulation of a single branch of SCN – 1.3 wt% Ace alloy growing along the temperature gradient. The temperature gradient is $G = 0.475\ \text{K/mm}$ and the isotherm velocity is $v_{\text{iso}} = 0.086\ \text{mm/s}$. Theoretical reference is the result of Ivantsov solution and the solvability condition with tip velocity equal to the isotherm velocity. Here $\rho_{\text{tip}_{\text{Iv}}} = 0.0018$ mm and $\delta_{\text{d}_{\text{Iv}}} = 0.0099$ mm. At steady state with parameters in Table 1, the relative error with the theoretical value is about 2% and 12% on tip radius and tip undercooling, respectively.

Table 1. Numerical parameters in our PTN model

Variable	Name	Unit	Value
Minimum PTN mesh size	h_{min}	$\rho_{\text{tip}_{\text{Iv}}}$	1.5
PTN mesh size in TA1	h_1	h_{min}	3
PTN mesh size in TA2	h_2	h_{min}	10
PTN mesh size in solid	h_{solid}	h_{min}	10
PTN mesh size in liquid	h_{liquid}	h_{min}	50
Integration distance	a	$\delta_{\text{d}_{\text{Iv}}}$	2
Distance from the tip to the center of the rectangle with h_{min} at its creation	l	$\delta_{\text{d}_{\text{Iv}}}$	10
Distance from the tip to the back boundary of the rectangle with h_{min} at its creation	l_{b}	a	1.2
Half of the height of the rectangle with h_{min}	H	$R_{\text{integ}}^{\text{c}}$	1.5
Distance from the boundaries of the rectangle with h_{min} to the boundaries of TA1	e_1	l	1
Distance from the boundaries of the rectangle with h_{min} to the boundaries of TA2	e_2	l	1

2.2. Capture in CAPTN model

In the CA method for modeling grain structures, only the envelope of the grain is considered and approximated to growing polygons at cell scale. In the CAPTN model, a parabola is associated to each tip and the PTN method is used as growth law for apices of the envelope. As grains propagate, CA cells will be captured. A CA cell of cell size l_{CA} will be captured if its circumscribed circle whose diameter is $\sqrt{2}l_{\text{CA}}$ is intersected with the envelope, and it can be captured only once.

3. Application

The CAPTN model has been used to investigate the diverging grain boundary orientation between two crystals. Simulations with various cell sizes have been launched for a bi-crystal system under directional solidification with frozen temperature approximation (constant temperature gradient $G = 0.475\ \text{K/mm}$, fixed isotherm velocity $v_{\text{iso}} = 0.086\ \text{mm/s}$) of SCN – 1.3 wt% Ace alloy. The crystal on the left (resp. right) has growing direction of 30° (resp. 10°) with temperature gradient. These simulations correspond to studies already performed with the PF method and the classical CA method [7]. Figure 2 shows the composition field in the PTN method, presented by the wireframe of the mesh, of simulation with $l_{\text{CA}} = 0.1$ mm. Active CA cells (a CA cell is active if at least one of its branch has non-zero velocity) are superposed: black for cells corresponding to the grain with orientation of 30°, and red for cells corresponding to the grain with orientation of 10°. Dendritic microstructures and grain structures of the same simulation are illustrated in Figure 3. The straight light blue dashed lines show the diverging grain boundary at steady state with an angle $\theta_D = 13.9°$ with the temperature gradient.

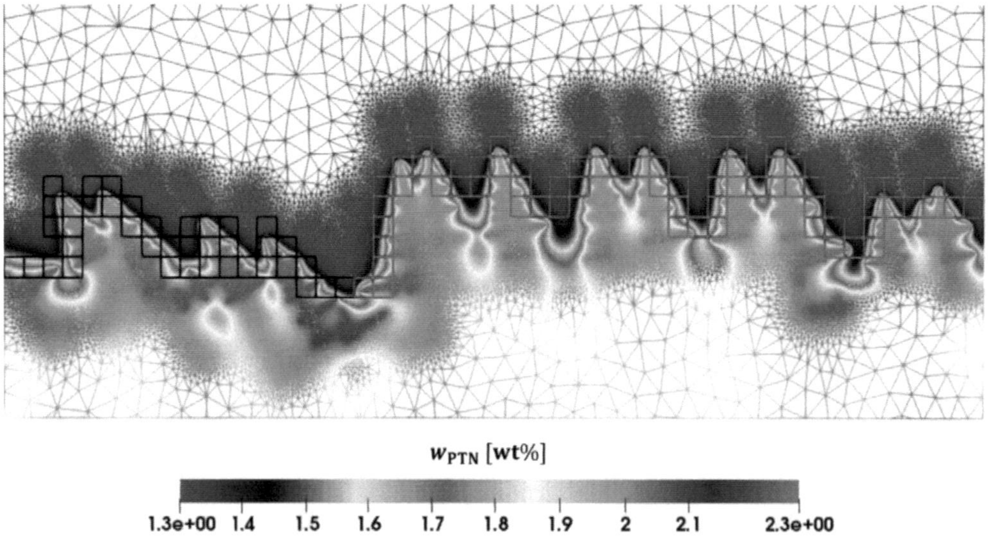

Figure 2. Composition field in the PTN method at the solidification front for the bi-crystal system with $l_{CA} = 0.1$ mm. Active CA cells are superposed. Black CA cells at the left are with orientation of 30°, and red CA cells at the right are with orientation of 10°.

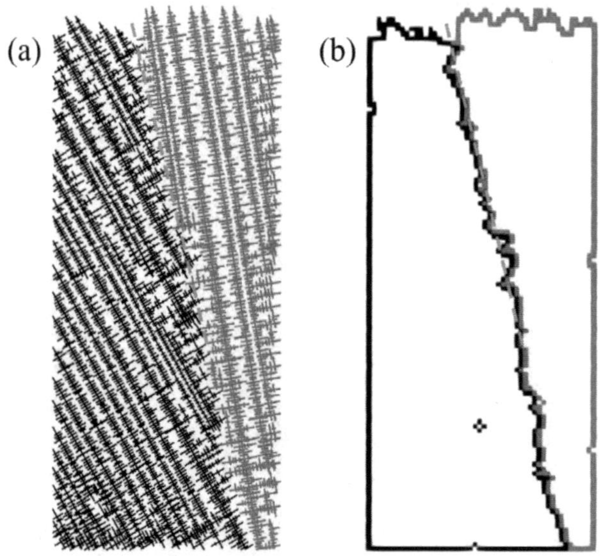

Figure 3. (a) Dendritic microstructure presented by network of needles and (b) grain structure presented by CA cells for the bi-crystal system with $l_{CA} = 0.1$ mm. The light blue dashed lines represent the diverging grain boundary, with an angle of 13.9° with the temperature gradient in vertical direction.

The diverging grain boundary orientation angles from our CAPTN model are presented by green squares in Figure 4. The results from the PF method are presented by horizontal black dashed lines, and the results from the CA method are presented by violet circles. The CA cell size for which CA simulations retrieve the PF angle is indicated by a vertical orange dashed line. In addition, two theoretical criteria, the Geometrical Limit (GL) and the Favorably Orientation Grain (FOG), are shown by the red and blue solid lines, respectively [7]. Compared to the CA model, the CAPTN model produces θ_D in good agreement with the PF model, in a fairly large domain of CA cell size, due to the effectiveness of the PTN method on computing dendritic kinetics at unsteady state.

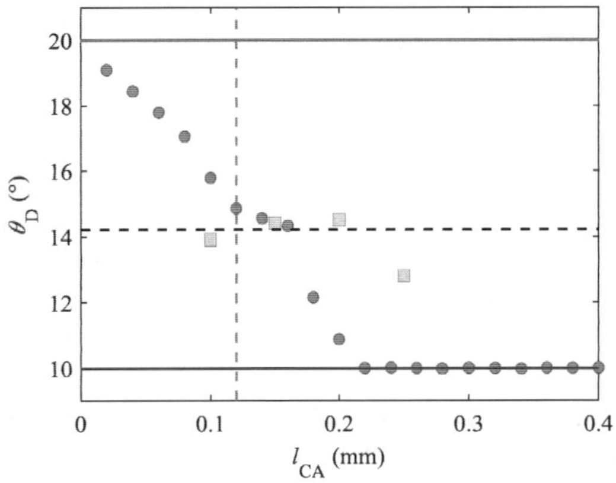

Figure 4. Evolution of diverging grain boundary orientation angle with CA cell size, in CA simulations (violet circles) [7] and CAPTN simulations (green squares).

4. Conclusion

Dendritic grain structures of a bi-crystal system growing under frozen temperature approximation have been investigated by the multiscale CAPTN model implemented in a finite element library. For a certain temperature gradient and isotherm velocity, the diverging grain boundary orientation angles from our simulations are in good agreement with the results of the PF method for a certain range of CA cell sizes and much less dependent on CA cell size than the classical CA method, indicating that our model works well on modeling the dendritic competition at unsteady state. Future investigation should focus on the diverging grain boundary angles of the same bi-crystal system under different temperature gradients and isotherm velocities. Three-dimensional extension is currently under progress. While expressions for the PTN method are modified for a paraboloid of revolution [2], the same numerical framework is being used. This includes the adaptive mesh generation and the finite element methods. Applications to a single dendritic needle show promising results.

5. Acknowledgements

This research work is supported by ESA-MAP program 'CETSOL', contract 14313/01/NL/SH.

Reference

[1] Takaki T, Sakane S, Ohno M, Shibuta Y, Aoki T and Gandin C-A 2018 *Materialia* **1** 104–13
[2] Tourret D and Karma A 2016 *Acta Mater.* **120** 240–54
[3] Carozzani T, Digonnet H and Gandin C-A 2011 *Model. Simul. Mater. Sci. Eng.* **20** 015010
[4] Fleurisson R 2019 *Modélisation multi-échelle parallélisée pour la prédiction de structures de grains dendritiques couplant les éléments finis, un automate cellulaire et un réseau de paraboles* PhD thesis (Paris Sciences et Lettres)
[5] Langer J S 1989 *Science* **243** 1150–6
[6] Ivantsov G P 1947 *Dokl Akad Nauk SSSR* **58** 567–9
[7] Dorari E, Ji K, Guillemot G, Gandin C-A and Karma A 2021 *Acta Mater.* 117395

IOP Conf. Series: Materials Science and Engineering 1274 (2023) 012042 doi:10.1088/1757-899X/1274/1/012042

Time evolution of interface shape distribution of equiaxed dendrite: A phase-field study

Munekazu Ohno[1], Mikihiro Kudo[2], Geunwoo Kim[2], Ryo Yamada[1], Yasushi Shibuta[3] and Tomohiro Takaki[4]

[1] Division of Materials Science and Engineering, Faculty of Engineering, Hokkaido University, Kita 13 Nishi 8, Kita-ku, Sapporo, Hokkaido 060-8628, Japan
[2] Graduate School of Engineering, Hokkaido University, Kita 13 Nishi 8, Kita-ku, Sapporo, Hokkaido 060-8628, Japan
[3] Department of Materials Engineering, The University of Tokyo, 7-3-1 Hongo, Bunkyo-ku, Tokyo 113-8656, Japan
[4] Faculty of Mechanical Engineering, Kyoto Institute of Technology, Matsugasaki, Sakyo-ku, Kyoto 606-8585, Japan

mohno@eng.hokudai.ac.jp

Abstract. An understanding of the morphology of growing dendrites in alloys is needed for an analysis of microsegregation, as well as an estimation of the permeability for macroscopic fluid dynamics. Quantitative phase-field simulations were used to study the growth process of three-dimensional (3D) equiaxed dendrites in an Al-1.0 mass%Cu alloy during continuous cooling. The dendrites were analysed using an interface shape distribution (ISD) map, which provides the probability of the local interface having a morphology with a given curvedness (C) and shape factor (S). Morphological changes in the microstructure can be measured sensitively from the change in the average value of the curvedness $\langle C \rangle$ relative to the solid volume fraction. The ISD map continued to change over time during continuous cooling, implying that it was not time-invariant. Furthermore, when microstructural changes occurred, similarities between the ISD maps were observed, independent of the cooling rates and system sizes.

1. Introduction

Characterisation of the three-dimensional (3D) morphology of growing dendrites in alloys is needed for a detailed analysis of microsegregation [1], as well as an estimation of the permeability in macroscopic fluid dynamics [2, 3]. Time-resolved synchrotron X-ray micro-tomography techniques enable detailed investigation of the time evolution of the 3D morphology of dendrites in alloys [4-7]. These techniques were used to characterise 3D dendrite morphology in terms of the distribution of local solid-liquid interfaces specified by mean and Gaussian curvatures, or equivalently, the curvedness (C) and shape factor (S). The interface shape distribution (ISD) map provides the probability of a given morphology of the local interface in terms of S and C. It is important to investigate and clarify the details of the time evolution of an ISD map under different cooling conditions.

The quantitative phase-field model enables accurate description and prediction of the 3D morphology of dendrites growing under various cooling conditions [8-16]. We previously used a quantitative phase-field simulation to obtain ISD maps that were used to measure the morphological evolution of freely growing equiaxed dendrites during the isothermal solidification of an Al-Cu alloy [17]. The ISD map exhibited time invariance for this solidification process when C was normalised using the average value of C at each time step, $(\langle C \rangle)$. In this study, quantitative phase-field simulations and the resulting ISD maps were used to investigate the time evolution of 3D equiaxed dendrites in an Al-1.0 mass%Cu alloy during continuous cooling at different rates.

2. Computational methods

2.1. Quantitative phase-field simulation

The phase-field model is a diffuse interface model in which the phase field continuously varies inside the interface. The primary advantage of this model is that it avoids the explicit tracking of the moving interface [18]. The quantitative phase-field model was developed based on thin-interface asymptotics, where the diffuse interface model is mapped to the free-boundary problem of interest in the limit of a small, but non-zero interface thickness [19-21]. Therefore, quantitative phase-field models are an effective tool for computational experiments on the formation of microstructures during solidification [9-16]. In this study, we employed a quantitative phase-field model developed previously [22], which recovers the free-boundary problem for solidification in a dilute binary alloy in the thin-interface limit.

We solved the time evolution equations of the phase field and dimensionless supersaturation simultaneously. The details of these equations can be found in previous papers [17, 22]. The time evolution equations were discretised using a second-order finite difference scheme, and time integration was performed using the explicit Euler method. We used these input parameters for an Al-1.0 mass%Cu alloy system: Al melting temperature, $T_m = 933.25$ K; liquidus slope, $m = -620$ K/mole fraction; partition coefficient, $k = 0.14$; molar volume, $V_m = 1.0 \times 10^{-6}$ m^3/mol; liquid diffusivity, $D_L = 3.0 \times 10^{-9}$ m^2/s; solid diffusivity, $D_S = 3.0 \times 10^{-12}$ m^2/s; solid-liquid interfacial energy, $\gamma_0 = 0.15$ J/m^2, and anisotropic strength $\epsilon_1 = 0.08$.

To reduce the computational cost, one-eighth of the dendrite was simulated based on four-fold symmetry. The initial solid seed was placed at the origin of the computational domain occupied by the liquid phase. The mirror (or zero-flux) boundary condition was employed on the *x-y*, *y-z*, and *z-x* planes at $z = 0$ ($z = X/2$), $x = 0$ ($x = X/2$), and $y = 0$ ($y = X/2$), respectively. Here, $X/2$ is the length of the cubic computational domain; hence, X is equivalent to the distance between the equiaxed dendrites during continuous cooling in this investigation. X was set at 128, 256, and 512 μm. The temperature was cooled to the solidus temperature at a constant rate, $\dot{T}$, which was set at 1.0, 5.0, and 10 K/s. All computations were accelerated using a graphics processing unit.

2.2. Calculation of interface shape distribution (ISD) map

The morphology of the local interface is conveniently characterised by C and S, which indicate the degree to which the interface is curved and the type of shape, respectively. These quantities are calculated as [23]:

$$C = \sqrt{\frac{\kappa_1^2 + \kappa_2^2}{2}}$$

and

$$S = \frac{2}{\pi}\tan^{-1}\frac{\kappa_2 + \kappa_1}{\kappa_2 - \kappa_1}$$

where κ_1 and κ_2 are the maximum and minimum curvatures of the local interface, respectively. Interfaces with values of $S = \pm 1$, ± 0.5, and 0 correspond to spherical, cylindrical, and saddle point shapes, respectively. In addition, the sign of S denotes the convexity direction of the interface. Importantly, the morphology of entire dendrites can be determined based on the probabilities of given morphologies of the local interfaces specified by sets of C and S values.

3. Results and discussion

Figure 1 shows the evolution of dendrite structures and the corresponding ISD maps calculated for $X = 512\,\mu m$ and $\dot{T} = 10$ K/s. The colour of the dendrite represents the value of S at the local interface. In the ISD map, C is normalised by $\langle C \rangle$. The dependence of $\langle C \rangle$ on the volume fraction f_v is shown in figure 2, which also shows the minimum and maximum values of $\langle C \rangle$, as indicated by the circles. The dendrites and ISD maps shown in figure 1 correspond to those at these extreme values of $\langle C \rangle$. Although it is not clear in figure 2, $\langle C \rangle$ decreases rapidly at the beginning of solidification, which corresponds to the initial growth of the spherical solid. Subsequently, $\langle C \rangle$ exhibits the minimum where the growth of the primary arms starts, which is immediately followed by the growth of the secondary arms (figure 1(a)). As the arms grow, $\langle C \rangle$ gradually increases, and then begins decreasing after the maximum, where the flattening of the secondary arms starts (figure 1(b)). After decreasing, $\langle C \rangle$ reaches the local minimum because of the onset of growth of the tertiary arms (figure 1(c)), followed by the local maximum due to the flattening of the tertiary arms (figure 1(d)).

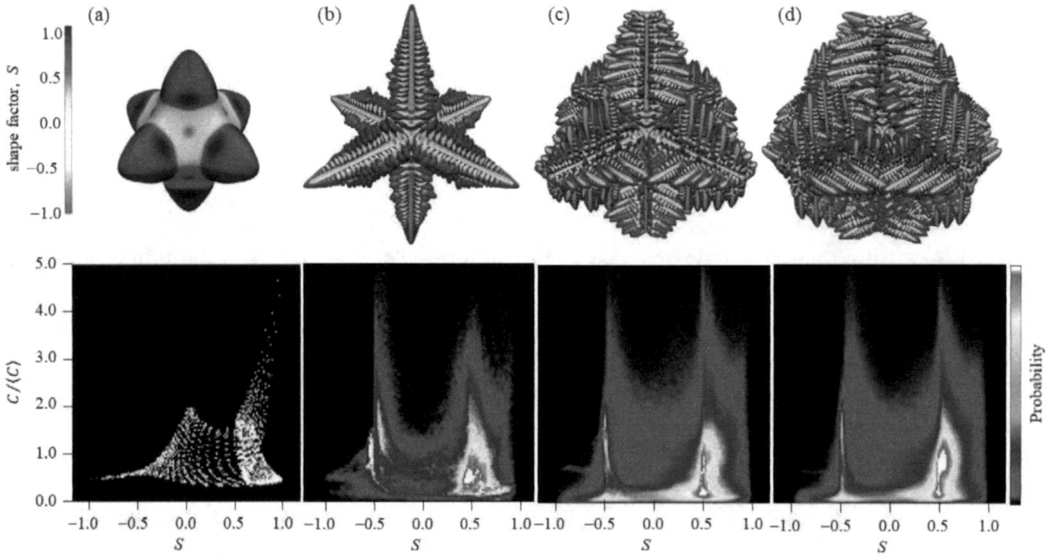

Figure 1. Evolution of dendrite structures characterized by S (upper) and the corresponding ISD maps (lower) calculated for $X = 512\,\mu m$ and $\dot{T} = 10$ K/s. (a) $t = 0.266$ s, (b) $t = 0.633$ s, (c) $t = 0.727$ s, and (d) $t = 0.787$ s.

As shown in figures 1 and 2, $\langle C \rangle$ is sensitive to changes in the morphology of the microstructure. Figure 2 shows the typical behaviour of $\langle C \rangle$. Although an extremum was not clearly observed in some cases, the two maxima and two minima of $\langle C \rangle$ observed in figure 2 generally appear under cooling conditions where the primary and secondary arms develop well. To be more specific, similarities in the behaviour of $\langle C \rangle$ and related microstructural changes were observed for $(X[\mu m], \dot{T}[K/s]) = (256, 5.0)$, $(256, 10)$, $(512, 1.0)$, $(512, 5.0)$, and $(512, 10)$. In the following, our focus is on the results for these conditions, because branching (that is, change in morphology) did not significantly occur under other conditions.

In figure 1, the ISD map represents the morphology of the 3D dendrite in great detail. As mentioned in the introduction, our previous study revealed that the ISD map for free-growing equiaxed dendrites during isothermal solidification becomes time-invariant when C is normalised by $\langle C \rangle$ [17]. However, as shown in figure 1, the ISD map continues to change with time, which means that it does not exhibit time-invariance during continuous cooling. The properties observed in this study are described below.

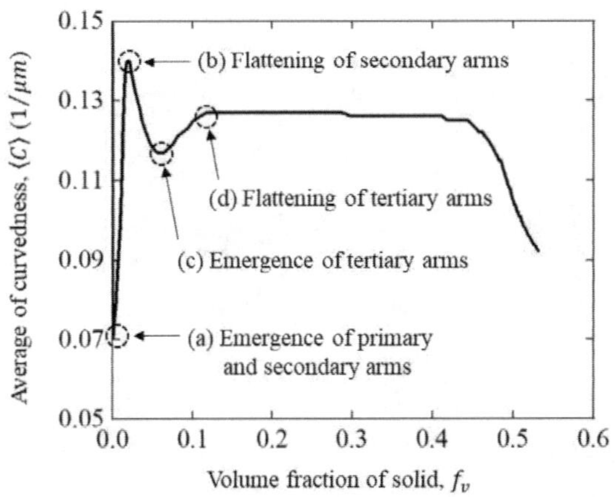

Figure 2. Dependence of average of curvedness on volume fraction of solid calculated for $X = 512\ \mu m$ and $\dot{T} = 10$ K/s.

In the ISD map shown in figure 1(b), a strong probability peak appears at $S \approx 0.5$. This peak is related to high frequency of local interface belonging to edge of the secondary arms (blue colour in the dendrite). Moreover, the peaks at $S \approx -0.4$ and $S \approx -0.5$ are related to the interfaces near the root of the secondary arms and trunk of the primary arms (yellow colour in the dendrite), respectively. These peaks commonly appear in the ISD maps for the other cases when $\langle C \rangle$ reaches its maximum value. Figures 3 and 4 show the dendrite structures and the ISD maps, respectively, at the maximum value of $\langle C \rangle$ where the flattening of the secondary arms begins for different sets of X and $\dot{T}$. Noticeable differences in the dendrite structures appear at this point. For example, the diameters of the primary trunk and the size and number of the secondary arms differ from each other. However, there were similarities in the ISD maps. Specifically, the peaks at $S \approx -0.4$, -0.5, and 0.5 appear in all ISD maps.

The similarities in the ISD maps at the maximum value of $\langle C \rangle$ can be discussed in a more quantitative manner. Figure 5 shows the probability of the interface with respect to S for different values of $C/\langle C \rangle$. At each value of $C/\langle C \rangle$, the probability distributions for different cooling conditions do not match completely, but they are similar in some respects, such as the position of strong peaks and magnitude of probability. Hence, the ISD maps exhibit a unique distribution associated with the flattening of the secondary arms, independent of the size of the microstructure. Similarities in the ISD maps also appear when typical morphological changes occur, as indicated by the circles in figure 2.

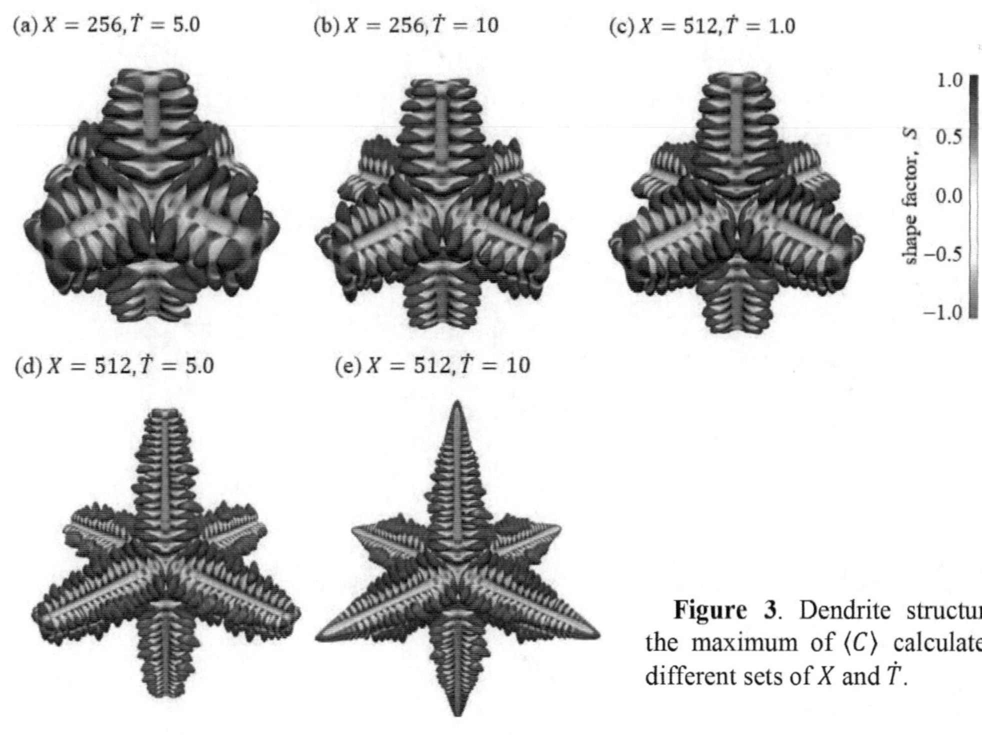

Figure 3. Dendrite structures at the maximum of $\langle C \rangle$ calculated for different sets of X and $\dot{T}$.

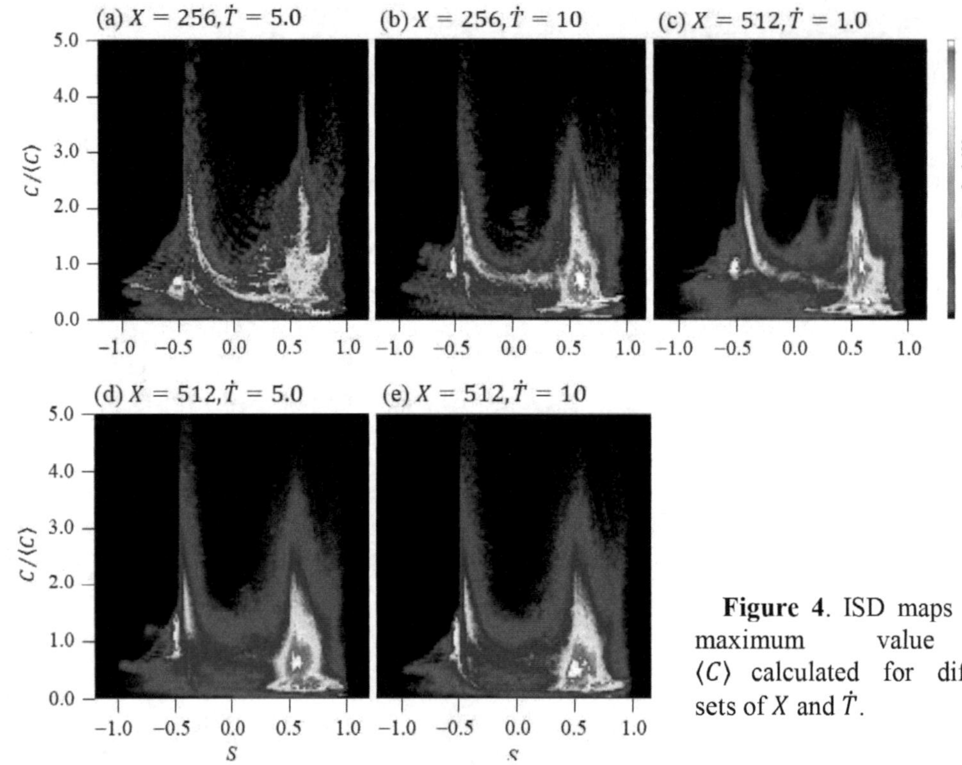

Figure 4. ISD maps at the maximum value of $\langle C\rangle$ calculated for different sets of X and $\dot{T}$.

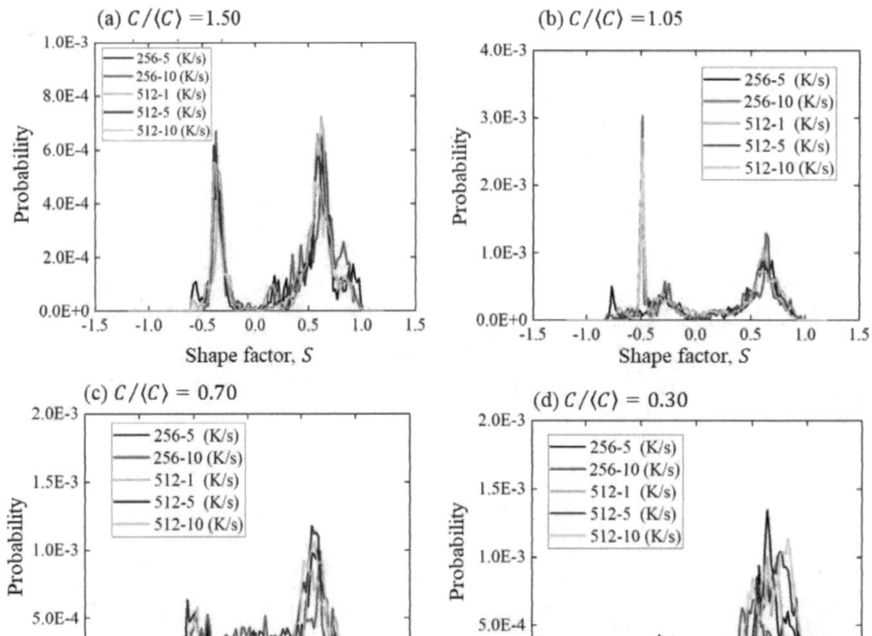

Figure 5. Probability of interface with respect to S for different values of $C/\langle C\rangle$ obtained for different sets of X and $\dot{T}$.

4. Conclusions

Quantitative phase-field simulations and the resultant ISD maps were used to study the time evolution of 3D equiaxed dendrites in an Al-1.0 mass%Cu alloy under different rates of continuous cooling. The value of $\langle C \rangle$ was sensitive to changes in the microstructure, such as the growth of primary and secondary arms, as well as the flattening of the secondary arms. This point should be investigated in more detail in a future work where it is interesting to see the time evolution of specific interface area and compare to its evolution equation developed in Ref. [24]. The ISD map continued to change over time during continuous cooling, in contrast to the previous finding that the ISD map is time-invariant during isothermal solidification. Moreover, when morphological changes occurred, similarities in the ISD maps were observed, independent of the cooling rates and system sizes.

Acknowledgement

This work was partially supported by KAKENHI, Grants-in-Aid for Scientific Research (B) 19H02486 and (A) 20H00217.

References

[1] Ohno M, Yamashita M, Matsuura K 2019 *Int. J. Heat Mass Transf.* **132** 1004
[2] Takaki T, Sakane S, Ohno M, Shibuta Y, Aoki T 2019 *Acta Mater.* **164** 237
[3] Mitsuyama Y, Takaki T, Sakane S, Shibuta Y, Ohno M 2020 *Acta Mater.* **188** 282
[4] Limodin N, Salvo L, Boller E, Suéry M, Felberbaum M, Gailliégue S, Madi K 2009 *Acta Mater.* **57** 2300
[5] Fife J L, Gibbs J W, Gulsoy E B, Park C-L, Thornton K, Voorhees P W 2014 *Acta Mater.* **70** 66
[6] Gibbs J W, Mohan K A, Gulsoy E B, Shahani A J, Xiao X, Bouman C A, De Graef M, Voorhees P W 2015 *Sci. Rep.* **5** 11824
[7] Cai B, Wang J, Kao A, Pericleous K, Phillion A B, Atwood R C, Lee P D 2016 *Acta Mater.* **117** 160
[8] Karma A, Rappel W-J 1998 *Phys. Rev. E* **57** 4323
[9] Takaki T, Sakane S, Ohno M, Shibuta Y, Shimokawabe T, Aoki T 2016 *Acta Mater.* **118** 230
[10] Takaki T, Sakane S, Ohno M, Shibuta Y, Shimokawabe T, Aoki T 2016 *ISIJ Int.* **56** 1427
[11] Clarke A J, Tourret D, Song Y, Imhoff S D, Gibbs P J, Gibbs J W, Fezzaa K, Karma A 2017 *Acta Mater.* **129** 203
[12] Shibuta Y, Ohno M, Takaki T 2018 *Adv. Theory Simul.* **1** 1800065
[13] Takaki T, Sakane S, Ohno M, Shibuta Y, Aoki T, Gandin C-A 2018 *Materialia* **1** 104
[14] Du J, Zhang A, Guo Z, Yang M, Li M, Liu F, Xiong S 2018 *Acta Mater.* **161** 35
[15] Takaki T, Sakane S, Ohno M, Shibuta Y, Aoki T 2022 *Comp. Mater. Sci.* **203** 111143
[16] Ohno M 2020 *ISIJ Int.* **60** 2745
[17] Yamada R, Kudo M, Kim G, Takaki T, Shibuta Y, Ohno M 2022 *Comp. Mater. Sci.* **204** 111173
[18] Takaki T 2014 *ISIJ Int.* **54** 437
[19] Ohno M 2012 *Phys. Rev. E* **86** 051603
[20] Ohno M, Takaki T, Shibuta Y 2016 *Phys. Rev. E* **93** 012802
[21] Ohno M, Takaki T, Shibuta Y 2017 *Phys. Rev. E* **96** 033311
[22] Ohno M, Matsuura K 2009 *Phys. Rev. E* **79** 031603
[23] Koenderink J J, van Doorn A J 1992 *Image Vis. Comput.* **10** 557
[24] Neumann-Heyme H, Eckert K, Beckermann C 2017 *Acta Mater.* **140** 87

Effective evaluation of permeability for interdendritic fluid flow using adaptive mesh refinement: Phase-field lattice Boltzmann study

Y Mitsuyama[1], **S Sakane**[2] **and T Takaki**[2,3]

[1] Graduate School of Science and Technology, Kyoto Institute of Technology, Matsugasaki, Sakyo-ku, Kyoto 606-8585, Japan
[2] Faculty of Mechanical Engineering, Kyoto Institute of Technology, Matsugasaki, Sakyo-ku, Kyoto 606-8585, Japan
[3] Author to whom any correspondence should be addressed.

E-mail: takaki@kit.ac.jp

Abstract. An effective permeability prediction method was developed by applying an adaptive mesh refinement (AMR) scheme to the lattice Boltzmann (LB) simulations to compute interdendritic liquid flow. The developed method can automatically create adaptive meshes for LB simulations based on dendrite morphologies obtained by phase-field (PF) simulations. The AMR scheme drastically reduces the number of grid points and accurately computes the liquid flow around dendrites. We found that the reduction rate of the number of meshes was remarkable when using coarser meshes compared to those used in the PF simulation. This method will enable the systematic evaluation of the permeability for liquid flow in various dendritic structures.

1. Introduction

In macroscopic liquid flow simulations, permeability is an important parameter for predicting casting and macrosegregation [1, 2]. Although the permeability for the flow of interdendritic liquids has been extensively investigated experimentally [3, 4], the accurate experimental measurement of permeability is difficult because dendritic structures change during liquid flow [5]. Thus, a numerical estimation of permeability is indispensable.

Phase-field (PF) simulations can express smooth solid–liquid interfaces in dendritic structures; however, PF simulations of dendrite growth are very expensive. As such, permeability predictions have been limited to a portion of dendrite or single-columnar dendrite structures [6–8]. Recently, we predicted the permeability of columnar structures containing multiple dendrites via large-scale simulations using a graphical processing unit (GPU) supercomputer [9, 10], which clarified that the dimensionless permeability obtained using the specific interface area follows the Kozeny–Carman equation [11] for both normal and parallel flows in columnar dendrites. However, the dendritic structures used in these studies were cellular dendrites with undeveloped secondary arms. Thus, to evaluate the permeability for dendrite structures with developed secondary arms, an efficient computational method for permeability is required.

In this study, we developed an efficient permeability prediction method based on the adaptive mesh refinement (AMR) method [12]. In this method, the dendrite structures are initially computed using the PF method. Subsequently, the adaptive meshes to compute the liquid flow around the dendrites are created automatically for the dendrites computed by the PF method. Finally, the interdendritic liquid flow is computed using the lattice Boltzmann (LB) method with the AMR method, which is hereinafter referred to as the AMR-LB method. The computational accuracy and efficiency of the developed AMR-LB method were confirmed by computing the permeability for simple structures, such as cylinder and sphere, and equiaxed dendrites.

2. Methods

AMR was introduced into a previously constructed permeability prediction method [9, 10]. The two-relaxation-time LB model with the bounce-back method, wherein the solid–liquid interface was treated to be sharp, was employed instead of the previously employed single relaxation-time LB model with dissipative drag forces on the diffuse interface [9, 10]. In the permeability prediction, the dendrite morphology was initially obtained by a PF simulation, during which AMR and multi-GPU parallel computations were applied [12]. Subsequently, for the dendrite morphology computed by the PF simulation, the interdendritic liquid flow was computed by the AMR-LB method. Only the AMR-LB method is explained here.

2.1. *Lattice Boltzmann method*

The LB equation based on the two-relaxation-time model can be expressed as follows [13]:

$$f_i\left(\mathbf{x}+\mathbf{c}_i\Delta t,t+\Delta t\right)=f_i\left(\mathbf{x},t\right)-\frac{f_i^{+}\left(\mathbf{x},t\right)-f_i^{+eq}\left(\mathbf{x},t\right)}{\tau_{LBM}^{+}}-\frac{f_i^{-}\left(\mathbf{x},t\right)-f_i^{-eq}\left(\mathbf{x},t\right)}{\tau_{LBM}^{-}}+F_i\Delta t\,, \quad (1)$$

where f_i ($\mathbf{x}$, t) is the particle velocity distribution function, $\mathbf{x}$ is the position vector, t is the time, $\mathbf{c}_i$ is the discrete particle velocity, f_i^{eq} ($\mathbf{x}$, t) is the local equilibrium distribution function, and τ_{LBM} is the relaxation time. The flow velocity $\boldsymbol{U}$ and density ρ in the liquid phase are obtained by $\rho=\sum_{i=0}^{Q-1}f_i$ and $\rho\boldsymbol{U}=\sum_{i=0}^{Q-1}\mathbf{c}_i f_i$, respectively, where Q represents a number in the discrete direction. In this study, we use the D3Q19 model with Q = 19. The local distribution function is expressed as $f_i^{eq} = w_i\rho[1 + 3(c_i\cdot \boldsymbol{U})/c^2 + 9(c_i\cdot \boldsymbol{U})^2/2c^4 + 3(\boldsymbol{U}\cdot \boldsymbol{U})/2c^2]$. Here, w_i is the weight function and c is the lattice speed, which is equal to $\Delta x/\Delta t$. Furthermore, $f_i^{\pm}(\mathbf{x}, t)$ and $f_i^{\pm eq}(\mathbf{x}, t)$ are defined as $f_i^{\pm} = (f_i(\mathbf{x}, t) \pm f_{\bar{i}}(\mathbf{x}, t))/2$ and $f_i^{\pm eq} = (f_i^{eq}(\mathbf{x}, t) \pm f_{\bar{i}}^{eq}(\mathbf{x}, t))/2$. In these equations, i and $\bar{i}$ are the discrete velocity direction and its reverse direction, respectively. In addition, the kinematic viscosity is expressed as $\nu = (\tau_{LBM}^{+} - 0.5)\Delta x^2/(3\Delta t)$. In this case, $\tau_{LBM}^{-} = (3\tau_{LBM}^{+} + 3)/(4\tau_{LBM}^{+} - 2)$ [14]. The liquid flow is caused by applying a pressure gradient, ∇P, which is expressed by setting $F_i = w_i\rho[3(c_i - \boldsymbol{U})/c^2 + 9(c_i\cdot \boldsymbol{U})c_i/(2c^4)]\cdot\nabla P$ in equation 1, i.e., a second-order accurate discrete external force term. A second-order accurate interpolated bounce-back model was used to apply the no-slip condition at the solid–liquid interface [15].

2.2. *AMR method for LB computation*

For dendrite structures computed by the PF simulation, we created adaptive meshes for use in the LB simulations; these were based on the block-structured AMR method [16]. Initially, the entire computational domain was divided into the coarsest cubic blocks, of which the subdivision level was defined as 0. Subsequently, when the block contained an interface region defined by $\phi_{min} < \phi < \phi_{max}$, where the PF variable ϕ was defined as $\phi = 1$ for the solid phase and $\phi = -1$ for the liquid phase, octree subdivision was performed recursively until the maximum subdivision level n_{max} was reached. Here, the block obtained by dividing the level 0 block by n times was defined as the subdivision level n. All blocks were divided into N^3 grids, and the LB computations were performed on these grids. Setting the minimum grid size at the maximum subdivision level n_{max} to Δx, the length L of one side of the cubic computational domain with size L^3 was expressed as $L = 2^{n_{max}} X_0 N\Delta x$, where X_0 is the number of level 0 blocks on a side with length L. The time increment in a block of level n was given as $\Delta t_n = 2\Delta t_{n+1}$ to maintain the lattice velocity c constant [17]. To keep ν constant, the relationship $\tau^{+}_{LBM,\,n+1} = 2\tau^{+}_{LBM,\,n}$ −

0.5 was used. Furthermore, since the non-equilibrium component of f_i was known to depend on the time increment and the LBM relaxation time as well as on ρ and $\boldsymbol{U}$, upon the movement of fluid particles a lattice with a different resolution, f_i was transformed to maintain continuity of flow [17].

The AMR-LB method developed in this study was coded to allow the parallel computing using multiple GPUs. The following simulations were performed on a GPU supercomputer TSUBAME3.0 (NVIDIA Tesla P100) at Tokyo Institute of Technology.

3. Validation

The validity of the AMR-LB permeability prediction method constructed in this study was confirmed through permeability computations for flow around a cylinder and a sphere [18–20]. As shown in the inset of figure 1, the cylinders were arranged in a square array in the *xy*-plane, and the spheres were arranged in a primitive cubic lattice. The number of grids on a side of cubic computational domain with size L^3 becomes $N_L = L/\Delta x$, when the entire computational domain is uniformly divided by Δx. Here, the computational conditions were set to $N_L = 256$, $n_{max} = 3$, $N = 8$, $\phi_{min} = 0$, $\phi_{max} = 0.4$, and $\nabla P = 1.0\times10^{-6}\rho_L \Delta x/\Delta t^2$ with the initial liquid-phase density ρ_L. The cylinder and sphere were expressed by the PF profile with $\phi(r) = -\tanh\left((r-R)/\sqrt{2}W_0\right)$, where W_0 is the interface thickness set to $\Delta x/W_0 = 1.0$, and r and R are the distance from the center and radius of cylinder and sphere, respectively.

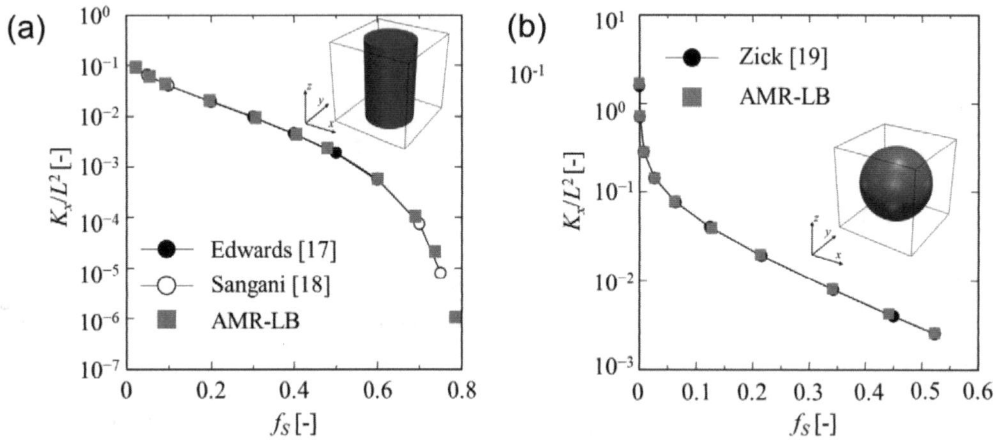

Figure 1. Dimensionless permeabilities computed for periodically arranged (a) cylinders and (b) spheres.

Figure 1 shows the computational results of permeability for (a) cylinders and (b) spheres, wherein the vertical axis represents the dimensionless permeability K_x/L^2 for the flow in the *x*-direction, and the horizontal axis represents the solid fraction f_S. The red plots represent the results obtained using the AMR-LB simulations, while the white and black circles represent the results of previous studies [18–20]. From figure 1, it can be seen that the simulation results obtained using the AMR-LB method are consistent with the results of previous studies over a wide range of solid fractions, which confirmed that the AMR-LB constructed in this study can accurately compute permeability. In our previous work [9], the accuracy of permeability prediction was poor when R became small (i.e., closer to the interface thickness) and large (i.e., touching the boundary) due to the diffuse interface. However, the accuracy of the present AMR-LB appears to be sufficiently high enough regardless of the size of the cylinder and the sphere. This is attributed to incorporation of the bounce-back method with a sharp interface.

Figure 2 shows the accuracy and efficiency of the permeability prediction by the AMR-LB method; the results for the cylinder shown in figure 1(a) are presented, where Δx was varied by changing N_L (i.e., $N_L = 32$, 64, 128, 256, 512, and 1024) under a constant value of L. In addition, the cylinder radius R was varied as follows: $R = 0.15L$ ($f_S = 0.0177$), $0.50L$ ($f_S = 0.197$), and $0.95L$ ($f_S = 0.709$). Figure 2(a) shows the relationship between K_x/L^2 and N_L, wherein the dashed line represents the convergence value. For $R = 0.50L$, the permeability can be calculated with high accuracy over the entire grid size. On the other hand, for $R = 0.15L$, the accuracy declines when N_L is ≤ 128, and, for $R = 0.95L$, the accuracy declines when N_L is ≤ 256. These observations can be attributed to the fact that the cylinder radius for $R = 0.15L$ is smaller relative to the grid size, and there is interference with the domain boundary for $R = 0.95L$. Furthermore, figure 2(b) shows the relationship between the ratio of the number of grids, N_{AMR}/N_L^3, and N_L, where N_{AMR} and N_L^3 are the number of grids used in the AMR-LB simulations and when the entire domain is uniformly divided with Δx, respectively. From figure 2(b), it can be observed that a smaller value of R and a larger value of N_L result in a smaller N_{AMR}/N_L^3 ratio. This indicates that the AMR-LB method is more effective for low solid fraction and high resolution conditions. Figure 2(c) shows the block division of a quarter of the xy cross-section ($z = L/2$) for conditions where $R = 0.95L$ and $N_L = 256$ and 1024. From figure 2(c), we can clearly see that the fine blocks are effectively arranged only around the interface indicated by the thick solid line.

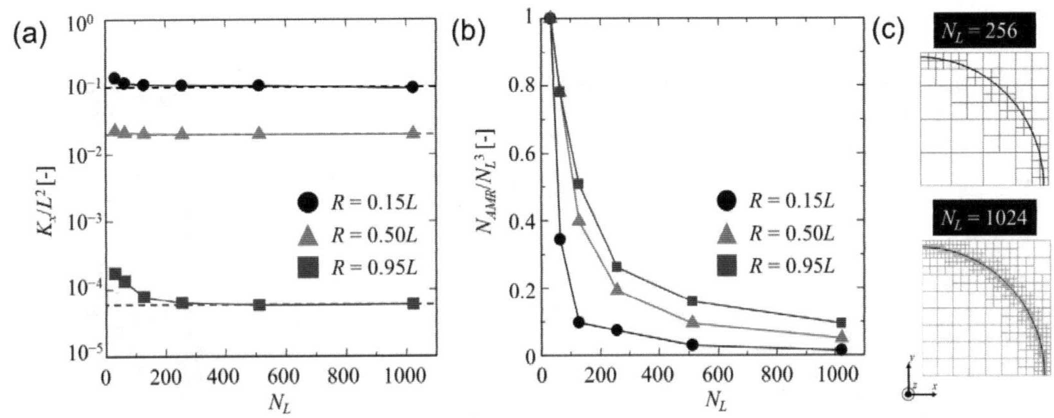

Figure 2. Effect of the minimum grid size Δx on the (a) accuracy and (b) efficiency of the permeability computation using the developed AMR-LB method. (c) Block decomposition on the xy cross-section located at $z = L/2$ for $N_L = 256$ and 1024; the black solid line indicates $\phi = 0$.

4. Permeability computations for equiaxed dendrites

Permeabilities were computed for equiaxed dendrites to evaluate the accuracy and efficiency of the AMR-LB method for dendrite structures. Initially, the growth of a single equiaxed dendrite was simulated for Fe–5.4wt%Si under continuous cooling using the quantitative PF model of a binary alloy [21]. The cubic domain size was set to $L^3 = (2048\Delta x_{PF})^3$ with a minimum grid size of $\Delta x_{PF} = 0.75$ μm. The cooling rate and initial dimensionless undercooling were set to 8.33×10^{-9} K/s and −0.1, respectively. All other conditions were as those described in the literature [12]. Figure 3 shows the computed dendrite morphology at steps 2×10^5, 4×10^5, and 6×10^5, of which solid fractions f_S were $f_S = 0.0010$, 0.0058, and 0.0202, respectively. Subsequently, the liquid flow around the dendrites was computed using the AMR-LB method, where the pressure gradient $\nabla P = \{1.0\times10^{-6}\rho_L\Delta x/\Delta t^2, 0, 0\}^T$ was applied to the domain. For the three dendrite morphologies shown in figure 3, the minimum grid size Δx_{LBM} used in the LB simulations was varied with respect to the value of Δx_{PF} used in the PF simulations, as $\Delta x_{LBM}/\Delta x_{PF} = 1$, 2, and 4. Thus, a total of nine simulations were performed for combinations of three different values of f_S and three different values of $\Delta x_{LBM}/\Delta x_{PF}$. The subdivision level n_{max} was set at 3, 2, and 1 for $\Delta x_{LBM}/\Delta x_{PF} = 1$, 2, and 4, respectively, to fix the maximum block size with $n = 0$. The mesh subdivision conditions were set to $\phi_{min} = -0.8$ and $\phi_{max} = 0.8$ to stably compute the liquid flow for all values of $\Delta x_{LBM}/\Delta x_{PF}$.

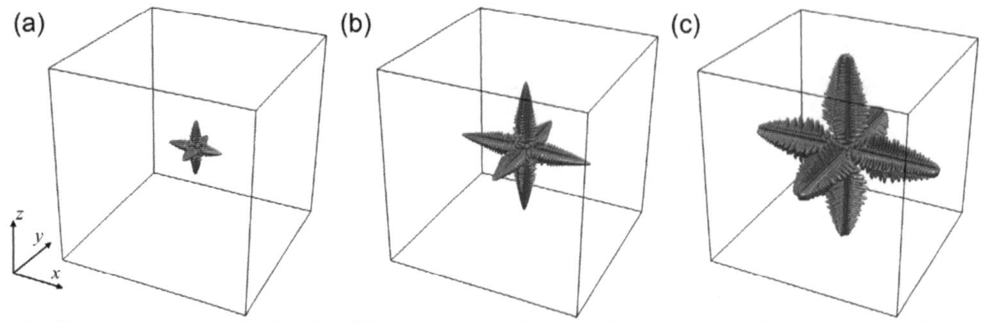

Figure 3. Time increments during the PF growth simulation of an equiaxed dendrite in a cubic domain with a size of $L^3 = (2048\Delta x_{PF})^3$ and a minimum grid size of $\Delta x_{PF} = 0.75$ µm at steps (a) 2×10^5 (f_S = 0.0010), (b) 4×10^5 (f_S = 0.0058), and (c) 6×10^5 (f_S = 0.0202) with a time increment $\Delta t_{PF} = 1.82 \times 10^{-5}$ s.

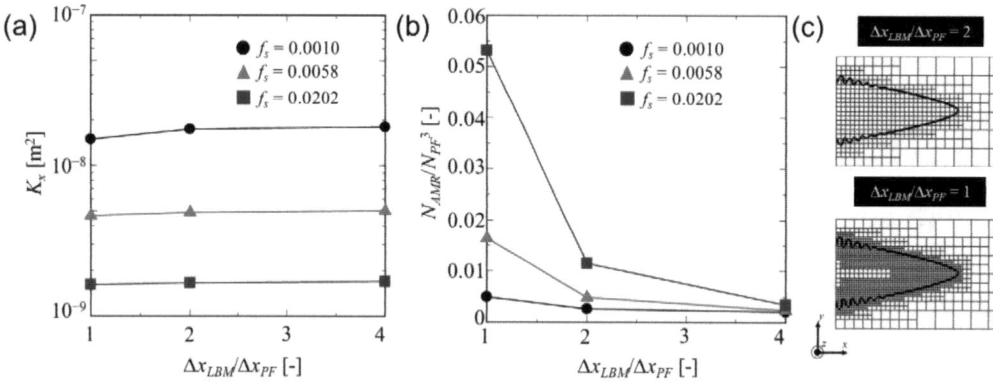

Figure 4. Effect of the solid fraction f_S and the grid size Δx_{LBM} on the permeability predictions by the AMR-LB method. Permeability predictions were performed for the three different equiaxed dendrites shown in figure 3 and three different values of Δx_{LBM}. (a) Permeability K_x for the liquid flow in the x-direction, (b) the number of N_{AMR} grids used in the AMR-LB simulations for $N_{PF}^3 = (L/\Delta x_{PF})^3 = 2048^3$, and (c) block division around the dendrite tip of the xy cross-section ($z = L/2$).

Figure 4(a) shows the relationship between the computed permeability K_x and $\Delta x_{LBM}/\Delta x_{PF}$, while figure 4(b) shows the relationship between the ratio of the number of grids N_{AMR}/N_{PF}^3 and $\Delta x_{LBM}/\Delta x_{PF}$, where N_{AMR} is the number of grids used in the AMR-LB simulations, and $N_{PF}^3 = (L/\Delta x_{PF})^3 = 2048^3$. In addition, figure 4(c) shows block division around the dendrite tip of the xy cross-section ($z = L/2$) in the AMR-LB simulations where $\Delta x_{LBM}/\Delta x_{PF}$ = 1 and 2; the thick black lines represent $\phi = 0$ at the solid–liquid interface.

From figure 4(a), we can see that K_x increases slightly as $\Delta x_{LBM}/\Delta x_{PF}$ increases for all solid fractions. This can be attributed to the reduced resolution, and so fine structures, such as the secondary arm, are not accurately represented. However, the errors were 17% and 21% (f_S = 0.0010), 5.0 and 7.6% (f_S = 0.0058), and 2.6 and 5.2% (f_S = 0.0202) for $\Delta x_{LBM}/\Delta x_{PF}$ = 2 and 4, respectively, corresponding to the results where $\Delta x_{LBM}/\Delta x_{PF}$ = 1; these are sufficiently small on a logarithmic scale. It was therefore confirmed that the permeability prediction for the present dendrite microstructures can be performed with sufficient accuracy using a value of Δx_{LBM} that is four times larger than the value of Δx_{PF} used in the PF simulation.

In figure 4(b), it can be observed that the lower the value of the solid fraction f_S, the smaller the N_{AMR}/N_{PF}^3 value. Even at the condition where $\Delta x_{LBM}/\Delta x_{PF}$ = 1, the number of grids is quite small as N_{AMR}/N_{PF}^3 = 0.49, 1.64, and 5.31%; hence, we can say that efficient permeability predictions were possible. In addition, a remarkable reduction in the grid number was achieved when increasing Δx_{LBM} compared to Δx_{PF}. More specifically, the reduction rates were 47 and 58% (f_S = 0.0010), 70 and 85% (f_S = 0.0058), and 78 and 94% (f_S = 0.0202) for $\Delta x_{LBM}/\Delta x_{PF}$ = 2 and 4, respectively, corresponding to the results where $\Delta x_{LBM}/\Delta x_{PF}$ = 1. Using the previous permeability prediction method [9], the use of uniform grids in $\Delta x_{LBM}/\Delta x_{PF}$ = 1 required approximately 2,720 GB of memory to perform the LB simulations for $N_{PF}^3 = (L/\Delta x_{PF})^3 = 2048^3$ grids. Meanwhile, when f_S = 0.0202, $\Delta x_{LBM}/\Delta x_{PF}$ = 4, and N_{AMR}/N_{PF}^3 = 0.003, and so the required memory was approximately 20 GB, thereby indicating that the simulation can be performed on a single GPU.

5. Conclusions

The adaptive mesh refinement (AMR) scheme was applied to lattice Boltzmann (LB) simulations to develop a new method (i.e., the AMR-LB method) to efficiently compute the permeability for the interdendritic flow using the results of phase-field dendrite growth simulations. This method was found to automatically create adaptive meshes for the AMR-LB simulations and was validated via permeability predictions for periodically arranged cylinders and spheres. Furthermore, through the permeability predictions for equiaxed dendrites, we confirmed that the developed AMR- LB method can efficiently compute the permeability using a small number of meshes. In our future work, we will systematically investigate the permeabilities for various dendritic structures using the AMR-LB method developed in this study.

Acknowledgments

This work was supported by a Grant-in-Aid for JSPS Fellows (Grant No. 21J23371), a KAKENHI Grant-in-Aid for Scientific Research (A) (Grant No. 20H00217), the JFE 21st Century Foundation, and the "Joint Usage/Research Center for Interdisciplinary Large-scale Information Infrastructures" and "High Performance Computing Infrastructure" in Japan (Project ID: jh210010).

References

[1] Beckermann C 2002 *International Materials Reviews* **47** 243-62
[2] Wu M, Ludwig A and Kharicha A 2019 *Metals* **9** 229
[3] Poirier D R 1987 *MTB* **18** 245-55
[4] Duncan A J, Han Q and Viswanathan S 1999 *Metall and Materi Trans B* **30** 745-50
[5] Nielsen Ø and Arnberg L 2000 *Metallurgical and Materials Transactions A* **31** 3149-53
[6] Ludwig A, Kharicha A, Hölzl C, Domitner J, Wu M and Pusztai T 2014 *Engineering Analysis with Boundary Elements* **45** 29-35
[7] Natsume Y, Takahashi D, Kawashima K, Tanigawa E and Ohsasa K 2013 *ISIJ International* **53** 838-47
[8] Böttger B, Haberstroh C and Giesselmann N 2016 *JOM* **68** 27-36
[9] Takaki T, Sakane S, Ohno M, Shibuta Y and Aoki T 2019 *Acta Materialia* **164** 237-49
[10] Mitsuyama Y, Takaki T, Sakane S, Shibuta Y and Ohno M 2020 *Acta Materialia* **188** 282-7
[11] Carman P C 1956 *Flow of gases through porous media* (New York: Academic Press)
[12] Sakane S, Takaki T and Aoki T 2022 *Materials Theory* **6** 3
[13] Sakane S and Takaki T 2021 *Computational Materials Science* **186** 110070
[14] Seta T, Rojas R, Hayashi K and Tomiyama A 2014 *Physical Review E* **89** 023307
[15] Bouzidi M h, Firdaouss M and Lallemand P 2001 *Physics of Fluids* **13** 3452-9
[16] Schive H-Y, ZuHone J A, Goldbaum N J, Turk M J, Gaspari M and Cheng C-Y 2018 *Monthly Notices of the Royal Astronomical Society* **481** 4815-40
[17] Dupuis A and Chopard B 2003 *Physical Review E* **67** 066707
[18] Edwards D A, Shapiro M, Bar- Yoseph P and Shapira M 1990 *Physics of Fluids A* **2** 45-55
[19] Sangani A S and Acrivos A 1982 *International Journal of Multiphase Flow* **8** 193-206
[20] Zick A A and Homsy G M 1982 *Journal of Fluid Mechanics* **115** 13-26
[21] Ohno M and Matsuura K 2009 *Phys. Rev. E* **79** 031603

Data assimilation for dendritic solidification with melt convection: phase-field lattice Boltzmann study

Ayano Yamamura[1], Shinji Sakane[2], Munekazu Ohno[3], Hideyuki Yasuda[4] and Tomohiro Takaki[2]

[1] Graduate School of Science and Technology, Kyoto Institute of Technology, Matsugasaki, Sakyo-ku, Kyoto 606-8585, Japan.
[2] Faculty of Mechanical Engineering, Kyoto Institute of Technology, Matsugasaki, Sakyo-ku, Kyoto 606-8585, Japan.
[3] Division of Materials Science and Engineering, Faculty of Engineering, Hokkaido University, Kita 13 Nishi 8, Kita-ku, Sapporo, Hokkaido 060-8628, Japan.
[4] Department of Materials Science and Engineering, Kyoto University, Yoshidahonmachi, Sakyo-ku, Kyoto 606-8501, Japan.

takaki@kit.ac.jp

Abstract. Time-resolved in-situ X-ray tomography and high-performance phase-field simulations are state-of-the-art approaches to clarifying dendrite solidification. However, major issues persist, such as the insufficiency of spatiotemporal resolution in experiments and lack of material properties in simulations. To overcome these issues, in this study, we developed a data assimilation system using an ensemble Kalman filter based on the phase-field lattice Boltzmann method as a simulation model for the dendrite solidification of binary alloys with liquid flow. The validity of the developed system was confirmed through twin experiments to infer the kinematic viscosity in a two-dimensional dendrite growth problem with forced convection.

1. Introduction

As most microstructures in cast products are formed through dendrite growth, an accurate prediction of dendrite growth during alloy solidification is essential to improve the properties of cast products. As a state-of-the-art technology to investigate dendrite solidification, time-resolved in-situ X-ray tomography (4D-CT) [1-4] and large-scale phase-field (PF) simulations [5-10] have been employed.

Using 4D-CT, the time evolution of solid–liquid interface area and interfacial curvature could be evaluated [3], which has not been achieved by conventional two-dimensional (2D) imaging for thin films [11]. Furthermore, 4D-CT was applied to the growth of equiaxed grains with solid motion [12]. In contrast, the spatiotemporal resolution becomes insufficient for faster phenomena, as in 4D-CT, one three-dimensional (3D) image is created from several hundred transmission images over 180° rotation of a cylindrical specimen [12]. In PF studies, realistic 3D dendrite growth simulations were enabled owing to the development of high-performance computing schemes, such as the adaptive mesh refinement [13], parallel computing, and graphics processing units [14]. Recently, simulations of polycrystalline solidification with liquid flow and solid motion have also become possible [5, 15]. However, the lack of material properties is a major issue in PF simulations [16].

Combining in situ observations and PF simulations is a promising technique to overcome these issues. Data assimilation is a promising method for combining experiments and simulations [17, 18]. Yamanaka et al. [19, 20] applied data assimilation to the 3D dendrite growth problem using the local ensemble Kalman filter. Ohno et al. integrated molecular dynamics and PF simulations using data assimilation to estimate the interfacial properties of pure metals [16, 21]. In this study, we developed a data assimilation system using the phase-field lattice Boltzmann (PF-LB) method [22] for dendrite solidification with liquid flow, which plays a very important role in the solidification microstructure formation of alloys. To confirm the validity of the developed data assimilation system, twin experiments were performed for dendrite growth under the forced convection of a binary alloy.

2. Data assimilation system

In this study, we used the ensemble Kalman filter (EnKF) for data assimilation [16, 17]. In Section 2.1, we describe the overall computational flow of the data assimilation system. The data assimilation system is executed by repeating two processes, that is, a simulation known as *forecasting* and modification of simulation results known as *filtering*, which are explained in Sections 2.2 and 2.3, respectively.

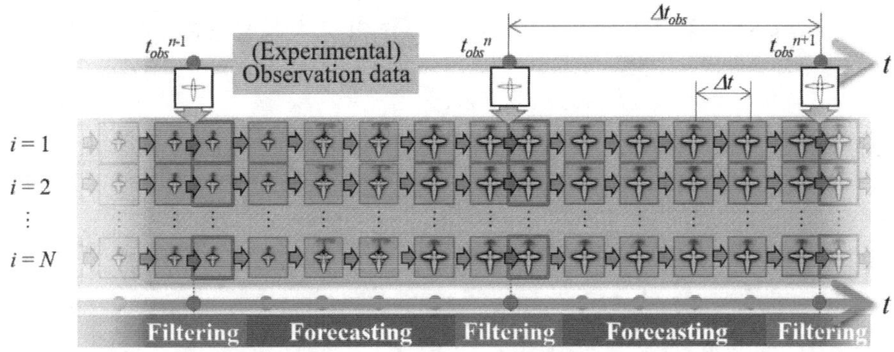

Figure 1. Illustration of data assimilation system developed in this study.

2.1. Computational procedure in the data assimilation system

Figure 1 shows a schematic of the data assimilation system developed in this study. In figure 1, the horizontal axis denotes time, and the experimental results are obtained at the time indicated by the red circle, i.e., t_{obs}^{n-1}, t_{obs}^{n} and t_{obs}^{n+1}. The experimental results are denoted as the observation data $\mathbf{y}_t$. Here, we assume that the observation data are obtained at fixed time intervals Δt_{obs}. The area with the light blue background consists of forecasting (green arrows) and filtering (red arrows), in which N simulations are performed simultaneously to update N state variables $\mathbf{x}_t^{(i)}$ ($i = 1, 2, \ldots, N$), and i specifies an ensemble member. Although the components of $\mathbf{x}_t^{(i)}$ are described later, unknown material properties can be included in $\mathbf{x}_t^{(i)}$. In general, the Δt in the simulation is much smaller than the time increments Δt_{obs} of the observed data, i.e., $\Delta t << \Delta t_{obs}$. $\mathbf{x}_t^{(i)}$ is modified at times corresponding to a multiple of Δt_{obs}, such as t_{obs}^{n-1}, t_{obs}^{n}, and t_{obs}^{n+1}, in figure 1, where the observation data are available, using the EnKF to make the simulated data closer to the observation data. By repeating the two operations, forecasting and filtering, the unknown properties used in the simulations are expected to become closer to the true values.

2.2. Forecasting

In this study, we focused on the isothermal dendrite solidification of a binary alloy with liquid flow. For the simulation, we used the PF-LB model [22], which combines the quantitative phase-field model for a binary alloy [23], LB method with a two-relaxation time model, and interpolation bounce-back method at the solid–liquid interface. In general simulations using the PF-LB model [22], the phase-field equation, advection–diffusion equation of solute, and LB equations are solved numerically up to a given time. This process, referred to as the forecasting in the data assimilation, is represented as

$$\mathbf{x}_{t|t-1}^{(i)} = F_t\left(\mathbf{x}_{t-1}^{(i)}\right) + \mathbf{v}_t^{(i)}, \tag{1}$$

where $F_t(\mathbf{x}_{t-1}^{(i)})$ is the simulation model wherein the PF-LB simulation is performed using $\mathbf{x}_{t-1}^{(i)}$, and $\mathbf{v}_t^{(i)}$ is the system noise. The components of the ensemble member $\mathbf{x}_t^{(i)}$ at time t consist of the PF variable vector $\boldsymbol{\phi}_t^{(i)}$ ($\phi = +1$ in solid and $\phi = -1$ in liquid), the dimensionless supersaturation vector $\boldsymbol{u}_t^{(i)}$, the particle density distribution function vector $\boldsymbol{f}_t^{(i)}$ with nine components in the LB equations, and the unknown property vector $z_t^{(i)}$, as $\mathbf{x}_t^{(i)} = \{\boldsymbol{\phi}_t^{(i)}\ \mathbf{u}_t^{(i)}\ \mathbf{f}_t^{(i)}\ z_t^{(i)}\}^T$. Here, as $\boldsymbol{\phi}_t^{(i)}$, $\mathbf{u}_t^{(i)}$, and $\mathbf{f}_t^{(i)}$ have values on all grid points as components, the number of components of $\mathbf{x}_t^{(i)}$ becomes 11 × (number of grid points) + (number of unknown properties) when the D2Q9 (two-dimensional model with nine directions) model is used in the LB simulation. The subscript $t|t-1$ in $\mathbf{x}_{t|t-1}^{(i)}$ denotes the values at time t computed using the values at time $t-1$. Note that the unknown parameter $z_t^{(i)}$ is not updated during the forecasting but is added to the system noise $\mathbf{v}_t^{(i)}$. The system noise $\mathbf{v}_t^{(i)}$ is represented as $\mathbf{v}_t^{(i)} \sim N(\mathbf{0}, \mathbf{Q}_t)$ with the Gaussian distribution of the mean vector $\mathbf{0}$ and variance–covariance matrix $\mathbf{Q}_t$. The probability distribution of the ensemble $\{\mathbf{x}_{t|\cdot}^{(i)}\}_{i=1}^{N}$ at time t, $p(\mathbf{x}_{t|})$, is approximated by the Gaussian distribution as $p(\mathbf{x}_{t|\cdot}) \doteq \sum_{i=1}^{N} \delta\left(\mathbf{x}_{t|\cdot} - \mathbf{x}_{t|\cdot}^{(i)}\right)/N$. Here, $\doteq$ denotes the ensemble approximation.

2.3. Filtering

At time, t, wherein the observation data $\mathbf{y}_t$ are available, the ensemble member $\mathbf{x}_{t|t}^{(i)}$ is updated using $\mathbf{x}_{t|t-1}^{(i)}$, computed in the forecasting using equation (1), as

$$\mathbf{x}_{t|t}^{(i)} = \mathbf{x}_{t|t-1}^{(i)} + \mathbf{K}_t\left[\mathbf{y}_t + \breve{\mathbf{w}}_t^{(i)} - \mathbf{H}_t \mathbf{x}_{t|t-1}^{(i)}\right], \tag{2}$$

where $\mathbf{H}_t$ is the observation matrix that transforms $\mathbf{x}_{t|t-1}^{(i)}$ into the quantities corresponding to $\mathbf{y}_t$ and is given by

$$\mathbf{H}_t = \begin{Bmatrix} 1 & 0 & \cdots & 0 & 0 & \cdots & 0 \\ 0 & 1 & \cdots & 0 & 0 & \cdots & 0 \\ \vdots & \vdots & \ddots & \vdots & \vdots & \ddots & \vdots \\ 0 & 0 & \cdots & 1 & 0 & \cdots & 0 \end{Bmatrix}. \tag{3}$$

$\mathbf{w}_t^{(i)}$ is the observation noise following a Gaussian distribution with mean vector $\mathbf{0}$ and variance covariance $\mathbf{R}_t$, denoted by $\mathbf{w}_t^{(i)} \sim N(\mathbf{0}, \mathbf{R}_t)$. In this study, $\mathbf{R}_t = \sigma_r^2\mathbf{I}$ is assumed with the identity matrix $\mathbf{I}$ and the variance of the observation noise σ_r^2. In equation (2), $\mathbf{y}_t + \breve{\mathbf{w}}_t^{(i)}$ is the perturbed observation with $\breve{\mathbf{w}}_t^{(i)}$ expressed as $\breve{\mathbf{w}}_t^{(i)} = \mathbf{w}_t^{(i)} - \sum_{j=1}^{N} \mathbf{w}_t^{(j)}/N$, and $\mathbf{K}_t$, the Kalman gain, is expressed as

$$\mathbf{K}_t = \mathbf{V}_t\mathbf{H}_t^T\left[\mathbf{H}_t\mathbf{V}_t\mathbf{H}_t^T + \mathbf{R}_t\right]^{-1}, \tag{4}$$

and the variance-covariance matrix $\mathbf{V}_t$ of the ensemble $\{\mathbf{x}_{t|t-1}^{(i)}\}_{i=1}^{N}$ is given by $\mathbf{V}_t = \sum_{j=1}^{N} \breve{\mathbf{x}}_t^{(j)}\breve{\mathbf{x}}_t^{(j)T}/(N-1)$, where $\breve{\mathbf{x}}_t^{(i)}$ is expressed as $\breve{\mathbf{x}}_t^{(i)} = \mathbf{x}_{t|t-1}^{(i)} - \sum_{j=1}^{N} \mathbf{x}_{t|t-1}^{(j)}/N$.

Figure 2 illustrates the filtering for a single component in the observation vector using the EnKF. The black, red, and yellow circles correspond to the component of $\mathbf{x}_{t|t-1}^{(i)}$, $\mathbf{x}_{t|t}^{(i)}$, and $\mathbf{y}_t$, respectively. The black and red solid lines are the Gaussian distributions, $p(\mathbf{x}_{t|t-1})$ and $p(\mathbf{x}_{t|t})$, respectively. The yellow solid line is the Gaussian distribution of one component of the perturbed observation $\mathbf{y}_t + \breve{\mathbf{w}}_t^{(i)}$, that is, the Gaussian distribution adds a random number to one component of $\mathbf{y}_t$ computed using equation (3).

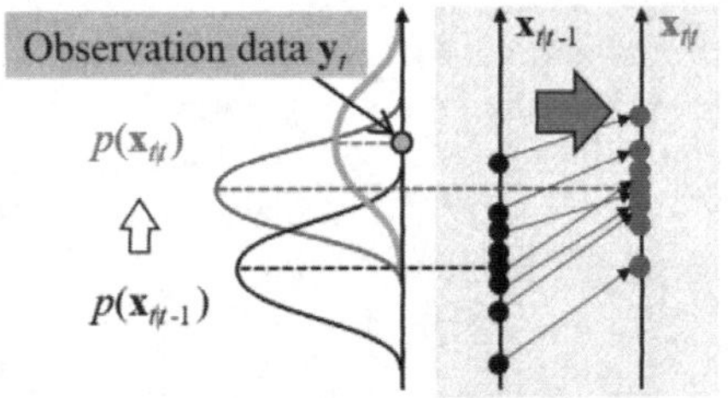

Figure 2. Illustration of filtering using EnKF at time t. The number of black and red circles is N. Here, an example for a single component of the observation data $\mathbf{y}_t$ is shown.

3. Preparation of observation data

To investigate the accuracy of the data assimilation system developed in this study, twin experiments were conducted. In the twin experiment, a simulation result obtained beforehand was used as the observation data instead of the experimental results. In this study, the kinematic viscosity ν was estimated from the twin experiments on the 2D dendrite growth under forced convection. The PF variable vector $\boldsymbol{\phi}_t^{obs}$, obtained from the simulation conducted herein, was set as the observation data, that is, $\mathbf{y}_t = \boldsymbol{\phi}_t^{obs}$. As ν used in the simulation is a known value, we can estimate the accuracy of the data assimilation system.

Figure 3 shows the computational and initial conditions in the simulation used to obtain the observation data. We used Al–3wt%Cu with the following properties [24]: melting point of pure Al T_m = 933.25 K, Gibbs-Thomson coefficient $\Gamma = 0.24 \times 10^{-6}$ K· m, strength of anisotropy $\varepsilon_4 = 0.02$, solute diffusivity in liquid phase $D_l = 3.0 \times 10^{-9}$ m²/s, solute diffusivity in solid phase $D_s = 3.0 \times 10^{-13}$ m²/s, liquidus slope m = -620 K/at. frac., partition coefficient $k = 0.14$, and kinematic viscosity $\nu_L = 5.0 \times 10^{-7}$ m²/s [25]. The grid size and interface thickness were set to Δx = 0.25 μm and W_0 = 0.4167 μm, respectively, in the domain with dimensions of $L_x \times L_y = 512\Delta x \times 512\Delta x$. The initial conditions were as follows: nondimensional supersaturation u_0 = -0.22 (ΔT = 1.88 K), solute concentration C_0 = 0.013 at. frac., and flow velocity $\mathbf{U}_{ini} = \{U_0, 0\}^T$ with U_0 = 50 μm/s. A solid nucleus with a radius of $6\Delta x$ was placed in the center of the domain filled with liquid. Boundary conditions (BC) were set as the inflow condition ($\phi = -1$, $u = u_0$, and $\mathbf{U} = \{U_0, 0\}^T$) on the left side, the outflow condition ($d\phi/dx = du/dx = 0$ and $d\mathbf{U}/dx = \{0, 0\}^T$) on the right side, and the slip condition ($d\phi/dy = du/dy = 0$, $dU_x/dy = 0$, and $U_y = 0$) at the top and bottom sides. The time increment Δt was set to $\Delta t = \Delta x^2/(5D_l) = 4.167 \times 10^{-6}$ s, and simulations were performed for 150000 steps (0.63 s).

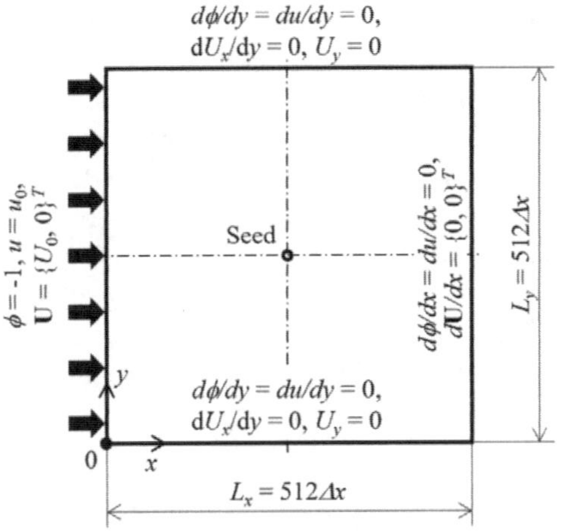

Figure 3. Computational and initial conditions for free growth of a dendrite with forced convection.

Figure 4 shows the simulation results. The initial circular solid grew to a typical cross-shaped dendritic pattern owing to interfacial anisotropy. Then, typical free dendrite growth in forced convection was observed: the upstream arm preferentially grew, the downstream arm delayed in the growth, and the vertical arms slightly tilted toward the upstream side owing to the forced convection [26].

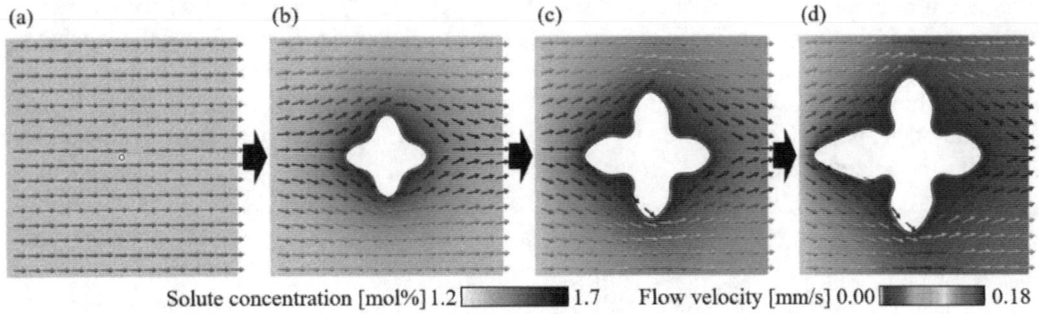

Figure 4. Time evolution of the solid morphology, solute concentration, and liquid flow at t = (a) 0.00 s, (b) 0.21 s, (c) 0.42 s, and (d) 0.63 s. Solid–liquid interface is indicated by the black line.

4. Twin experiments

The PF variable vectors obtained in the simulation shown in figure 4 were used as the observation data $\mathbf{y}_t$ in the twin experiments to infer the kinematic viscosity ν. The following hyperparameters were set in the twin experiments: variance of system noise given to the PF variable ϕ is 4×10^{-8}, variance of system noise given to the supersaturation u is 1×10^{-12}, variance of system noise given to the particle density distribution function f is 1×10^{-20}, variance of the system noise given to the kinematic viscosity ν is 25

$\times 10^{-24}$ (m^2/s)^2 (=($\nu_L \times 10^{-5}$)^2), and variance of the observation noise is 1×10^{-1}. The variance of the initial values of ν was set to 6.25×10^{-14} (m^2/s)^2 (=($0.5\nu_L$)^2). The number of ensembles, N, was set to $N = 128$. To avoid a negative value of ν, ν computed as negative at filtering was set to $\nu = \nu_L \times 10^{-5}$. In the following data assimilation, we investigated the effect of the initial value of ν, the filtering interval, and the value of ν on the accuracy of data assimilation.

Figure 5 shows the inference results in the case of changing the initial value of ν, ν_{ini}, as $\nu_{ini}/\nu_L = 0.2$, 1.0, and 1.8. The black dashed line is ν_{ini}, the red solid line is the true value ν_L, and the black solid line with circles is the time variation of the average of the estimated values. Error bars indicate the standard deviation of the distribution. The number of filters was $n_{filt} = 30$ with $\Delta t_{obs} = 0.021$ s. There is a deviation between ν_{ini} indicated by the dashed line and the average initial values indicated by the black circle. This is because the N is not infinite, and the average value of the ensemble members does not completely match the set value. From figure 5, for the three results with different ν_{ini}, the average of the inferred values gradually approaches the true value ν_L, and the standard deviations become smaller with time. Thus, we confirmed the validity of the developed system. Although the convergence speed was higher for the case in which ν_{ini} was closer to ν_L, the final results were independent of the initial value ν_{ini}.

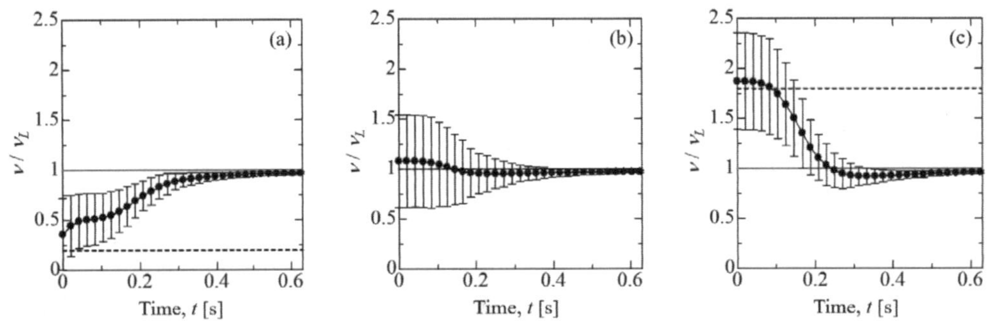

Figure 5. Effect of initial value, ν_{ini}, of kinematic viscosity on convergence behavior: ν_{ini}/ν_L = (a) 0.2, (b) 1.0, and (c) 1.8.

Next, we evaluated the effect of the filtering interval on the inference accuracy by changing $n_{filt} = 15$, 30, and 60 for the inference in figure 5(c). Figure 6 shows the results in the same manner as those in figure 5. From figure 6, it can be observed that the convergence behavior is better for a larger n_{filt}. In contrast, the accuracy is almost independent of n_{filt}; thus, sufficient accurate inference was achieved for all n_{filt}. However, when a much larger n_{filt} is used, the accuracy reduces because for a small Δt_{obs}, the PF variables perturbed in the filtering cannot recover the proper profile in such a small duration. In addition, if the simulation shown in figure 4 could be continued longer with a larger domain, we could obtain sufficient accuracy even using much larger Δt_{obs}.

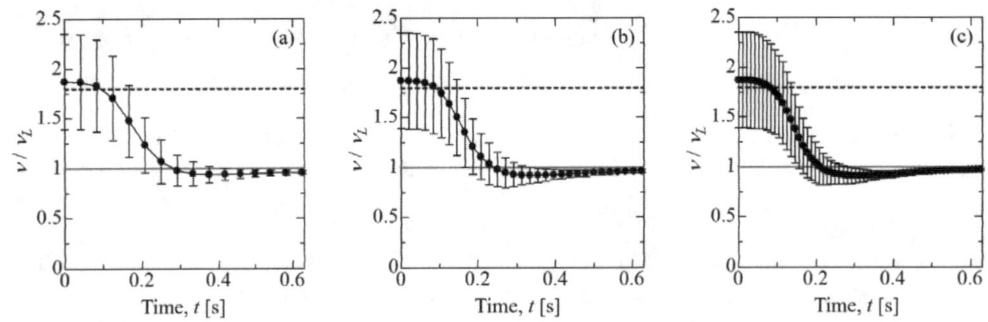

Figure 6. Effect of filtering number, n_{filt}, on convergence behavior: n_{filt} = (a) 15, (b) 30, and (c) 60.

Finally, to confirm the versatility of the data assimilation system developed in this study, twin experiments were conducted by varying the kinematic viscosity as ν_{true}/ν_L = 0.2, 1.0, and 1.8. Figure 7 shows the simulation results of the solid–liquid interface (contour line at $\phi = 0$) at t = 0.63 s for three different ν_{true}. The result with ν_{true}/ν_L = 1.0 is the same as the result shown in figure 4(d). Figure (a) shows the entire domain, and (b) shows an enlarged view around the tips of the upstream arm enclosed by the square in (a). As these are the early stages of dendrite growth, remarkable differences cannot be observed in morphology except for around the tip of the upstream arm. Figure (b) show that the upstream arm grows faster for smaller ν because the liquid flow around the solid–liquid interface is faster for smaller ν. Figure 8 shows the inference results. Three data assimilations started from the same initial value of ν_{ini}/ν_L = 1.0. For all ν_{true}, excellent convergence behaviors were observed in figure 8. The comparison of figures 8(a) and (c) evidences that the convergence behavior is better for a smaller kinematic viscosity (figure 8(a)). This is attributed to the fact that the small viscosity (ratio 0.2) causes a larger morphology change than the large viscosity (ratio 1.8).

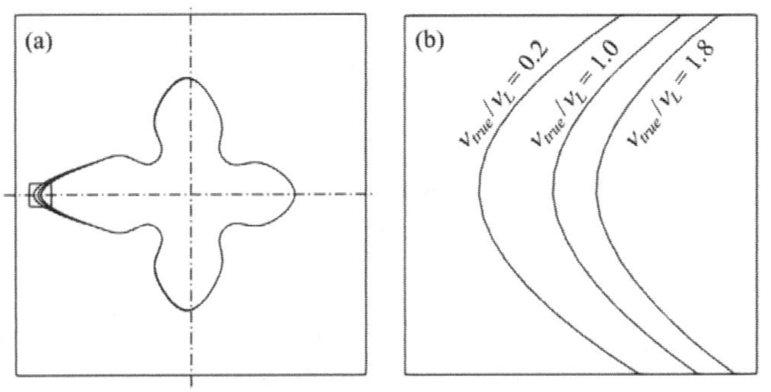

Figure 7. Dendrite morphologies at t = 0.63 s with kinematic viscosities ν_{true}/ν_L = 0.2, 1.0, and 1.8. (a) Entire domain and (b) magnified view of a square domain indicated in (a).

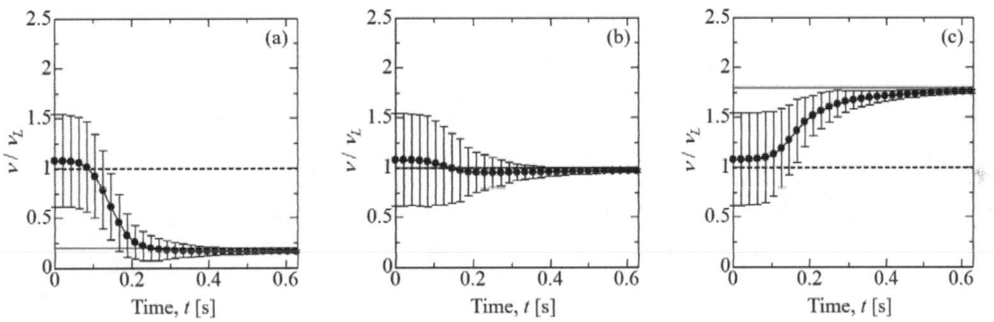

Figure 8. Effect of kinematic viscosity, ν_{true}, on convergence behavior: ν_{true}/ν_L = (a) 0.2, (b) 1.0, and (c) 1.8.

5. Conclusions

In this study, we constructed a data assimilation system using the PF-LB model as a simulation model for the isothermal dendrite solidification of binary alloys with liquid flow. The validity of the system was confirmed by twin experiments to estimate the kinematic viscosity. In the future, we will extend the data assimilation system developed in this study to the problem of motion of multiple equiaxed dendrite grains in 3D and aim to develop a highly accurate solidification microstructure prediction system combining simulation and experiment.

Acknowledgement

This work was supported by a KAKENHI Grant-in-Aid for Scientific Research (A) 20H00217.

References

[1] Cai B, Wang J, Kao A, Pericleous K, Phillion A B, Atwood R C and Lee P D 2016 *Acta Mater.* **117** 160-9
[2] Azeem M A, Lee P D, Phillion A B, Karagadde S, Rockett P, Atwood R C, Courtois L, Rahman K M and Dye D 2017 *Acta Mater.* **128** 241-8
[3] Yasuda H, Kawarasaki T, Tomiyori Y, Kato Y and Morishita K 2019 *IOP Conf. Ser.: Mater. Sci. Eng.* **529**
[4] Nakano K, Narumi T, Morishita K and Yasuda H 2020 *Mater. Trans.* **61** 596-604
[5] Yamanaka N, Sakane S and Takaki T 2021 *Comput. Mater. Sci.* **197** 110658
[6] Takaki T, Sakane S, Ohno M, Shibuta Y and Aoki T 2020 *Comput. Mater. Sci.* **171** 109209
[7] Sakane S, Takaki T, Ohno M, Shibuta Y and Aoki T 2020 *Comput. Mater. Sci.* **178** 109639
[8] Shibuta Y, Ohno M and Takaki T 2018 *Adv. Theory Simul.* **1** 1800065
[9] Shibuta Y, Ohno M and Takaki T 2015 *JOM* **67** 1793-804
[10] Wang N, Korba D, Liu Z, Prabhu R, Priddy M W, Yang S, Chen L and Li L 2021 *Comput. Methods in Appl. Mech. Eng.* **385** 114026
[11] Yasuda H, Morishita K, Nakatsuka N, Nishimura T, Yoshiya M, Sugiyama A, Uesugi K and Takeuchi A 2019 *Nat. Commun.* **10** 3183
[12] Narumi T, Kawarasaki T, Kato Y, Morishita K and Yasuda H 2020 *J. Japan Inst. Light Metals* **70** 339-46
[13] Guo Z and Xiong S M 2015 *Comput. Phys. Commun.* **190** 89-97
[14] Takaki T, Shimokawabe T, Ohno M, Yamanaka A and Aoki T 2013 *J. Cryst. Growth* **382** 21-5
[15] Sakane S, Takaki T, Rojas R, Ohno M, Shibuta Y, Shimokawabe T and Aoki T 2017 *J. Cryst. Growth* **474** 154-9
[16] Ohno M, Oka Y, Sakane S, Shibuta Y and Takaki T 2020 *Phys. Rev. E* **101** 052121
[17] Houtekamer P L and Zhang F 2016 *Monthly Weather Review* **144** 4489-532
[18] Houtekamer P L and Mitchell H L 2005 *Q. J. R. Meteorol. Soc.* **131** 3269-89
[19] Takahashi K and Yamanaka A 2021 *Comput. Mater. Sci.* **190** 110296
[20] Yamanaka A and Takahashi K 2020 *Mater. Today Commun.* **25** 101331
[21] Nagatsuma Y, Ohno M, Takaki T and Shibuta Y 2021 *Nanomaterials* **11** 2308
[22] Sakane S and Takaki T 2021 *Comput. Mater. Sci.* **186** 110070
[23] Ohno M and Matsuura K 2009 *Phys. Rev. E* **79** 031603
[24] Takaki T, Rojas R, Sakane S, Ohno M, Shibuta Y, Shimokawabe T and Aoki T 2017 *J. Cryst. Growth* **474** 146-53
[25] Konstantinova N Y, Popel' P S and Yagodin D A 2009 *High Temp.* **47** 336-41
[26] Tong X, Beckermann C and Karma A 2000 *Phys. Rev. E* **61** R49-R52

IOP Conf. Series: Materials Science and Engineering 1274 (2023) 012045 doi:10.1088/1757-899X/1274/1/012045

2D multi-phase-field lattice Boltzmann simulations of semi-solid shear deformation

N Yamanaka[1], S Sakane[2], and T Takaki[2]

[1] Graduate School of Science and Technology, Kyoto Institute of Technology, Matsugasaki, Sakyo-ku, Kyoto 606-8585, Japan
[2] Faculty of Mechanical Engineering, Kyoto Institute of Technology, Matsugasaki, Sakyo-ku, Kyoto 606-8585, Japan

takaki@kit.ac.jp

Abstract. The deformation of solid–liquid coexistence regions strongly affects the formation of solidification defects in casting applications. Thus, understanding the mechanism of semi-solid deformation is crucial. This study used the multi-phase-field lattice Boltzmann model (MPF-LBM), which can simulate polycrystalline solidification with solid motion and liquid flow, to simulate two-dimensional semi-solid shear deformation with various grain shapes and solid fractions. It was concluded that the MPF-LBM could express the characteristic behaviors of semi-solid deformation, such as the Reynolds' dilatancy and shear band, depending on the grain shape and solid fraction.

1. Introduction

Semi-solid deformation of alloys, which induces severe solidification defects, occurs in die, centrifugal, and continuous casting [1], and its behavior is highly dependent on the solid fraction. When the solid fraction exceeds the dendrite coherency point, a solid network transmitting forces is formed, resulting in rheological deformation behavior similar to that of soil [2]. This causes Reynolds' dilatancy, which leads to band segregation, porosity, and cracks [3-5]. Therefore, the accurate prediction of the semi-solid deformation behavior is crucial to reduce these solidification defects.

Studies on semi-solid deformation have been conducted extensively. Experiments have shown that band segregation occurs through Reynolds' dilatancy, whereby solute-rich liquid flows into spaces due to the interaction between solid grains [6-8]. In-situ observations using a synchrotron radiation facility revealed that grain rearrangement was the dominant deformation mechanism in semi-solid deformation [9-16]; further, by tracking the grain positions during deformation, the local dilatancy behavior was explained based on the arching of crystals [9].

As described above, the experiments provided a considerable understanding of semi-solid deformation. However, numerical simulations are essential to elucidate the involved phenomena in further detail. Su et al. investigated semi-solid deformation with circular grains using the discrete element method and showed that the solid fraction at which dilatancy and cracks occur depends on the deformation rate [17, 18]. In addition, through simulations of semi-solid deformation with dendrite crystals using the discrete element method, Yuan et al. concluded that the dendrite coherency point depends on the solid fraction and morphology of dendrite envelopes [19, 20]. However, these simulations did not include solid–liquid phase transformation in semi-solid deformation, and the solid morphologies were artificial. On the basis of this background, we developed a multi-phase-field lattice Boltzmann model (MPF-LBM) that can simulate polycrystalline solidification by considering solid–liquid phase transformation, liquid flow, solid motion, and solid contact in a previous study [21].

Moreover, we showed that the MPF-LBM has excellent potential as an evaluation model for semi-solid deformation [21, 22].

In this study, simulations of semi-solid shear deformation were performed using the MPF-LBM by changing the grain shape and solid fraction. Through the simulations, we evaluated the versatility of the MPF-LBM as an evaluation method for semi-solid deformation. Note that solid–liquid phase transformation and solute diffusion were not considered in this study in order to confirm the fundamental performance of the MPF-LBM.

2. Numerical model

Solid grains with a constant shape and size were considered to investigate the fundamental performance of the MPF-LBM for semi-solid deformation. Thus, we simply solved the liquid flow and solid motion for multiple grains, expressed by phase-field variables, in two dimensions.

2.1. *Advection of solid grains*

Multiple solid grains are expressed by multiple phase-field variables ϕ_i, which are defined as $\phi_i = +1$ for the i-th grain, and $\phi_i = -1$ in the other grains or liquid, and change smoothly at the interface region. Solid grains were treated as rigid bodies that do not exhibit solid–liquid phase transformation. Thus, only the advection equations were solved for ϕ_i to express the movement of the grains as follows:

$$\frac{\partial \phi_i}{\partial t} + \mathbf{U}_{Si} \nabla \phi_i = 0 , \qquad (1)$$

where $\mathbf{U}_{Si}$ is the solid velocity computed on each grid point.

2.2. *Liquid flow*

The LBM with two relaxation times [23] was used to compute the liquid flow. The lattice Boltzmann equation is expressed as follows:

$$f_k(\mathbf{x}+\mathbf{c}_k\Delta t, t+\Delta t) = f_k(\mathbf{x},t) - \frac{f_k^+(\mathbf{x},t) - f_k^{+eq}(\mathbf{x},t)}{\tau_{LBM}^+} - \frac{f_k^-(\mathbf{x},t) - f_k^{-eq}(\mathbf{x},t)}{\tau_{LBM}^-} , \qquad (2)$$

where $f_k(\mathbf{x},t)$ is the particle distribution function in the k-th discrete direction at position $\mathbf{x}$ and time t; $\mathbf{c}_k$ is the particle velocity in the k-th discrete direction; f_k^{eq} is the equilibrium distribution function; $f_k^{\pm}(\mathbf{x},t)$ and $f_k^{\pm eq}(\mathbf{x},t)$ are expressed by $f_k^{\pm}(\mathbf{x},t) = \left[f_k(\mathbf{x},t) \pm f_{\bar{k}}(\mathbf{x},t)\right]/2$ and $f_k^{\pm eq}(\mathbf{x},t) = \left[f_k^{eq}(\mathbf{x},t) \pm f_{\bar{k}}^{eq}(\mathbf{x},t)\right]/2$, respectively, where $\bar{k}$ indicates the number of directions opposite to the k-th direction; and $\tau_{LBM}^{\pm}$ represents two relaxation times. At the solid–liquid interface, an interpolated bounce-back model considering solid motion [24] was used. Moreover, the fluid reaction force on solids $\mathbf{G}_i$ exerted on the solid by particle reflection at the solid–liquid interface was calculated using the Galilean invariant momentum exchange method [25] as $\mathbf{G}_i = \sum_k \mathbf{c}_k f_k^m - \mathbf{c}_{\bar{k}} f_{\bar{k}}^{m-1}$, where superscripts m and m-1 correspond to the particle distribution functions before and after collision at the solid–liquid interface, respectively. Owing to the introduction of the two-relaxation LBM and bounce-back model, a stable liquid flow simulation with a real kinetic viscosity for a metallic material under the reasonable time increment, difficult in the single-relaxation LBM [21], was achieved [23].

2.3. *Solid velocity and force applied to the solid*

The solid velocity $\mathbf{U}_{Si}$ of the i-th solid grain in equation (1) was computed as $\mathbf{U}_{Si} = \mathbf{U}_{Ti} + \boldsymbol{\omega}_{Ti} \times (\mathbf{x} - \mathbf{X}_i)$, where the translational velocities $\mathbf{U}_{Ti}$ and rotational velocities $\boldsymbol{\omega}_{Ti}$ are updated using the following equations of motion [26]:

$$M_i \frac{d\mathbf{U}_{Ti}}{dt} = \mathbf{F}_i , \quad \mathbf{I}_i \frac{d\boldsymbol{\omega}_i}{dt} = \mathbf{T}_i , \qquad (3)$$

where M_i and $\mathbf{I}_i$ are the mass and moment of inertia, respectively; and $\mathbf{F}_i = -\Sigma[\mathbf{G}_i + \mathbf{F}_{ci}]\Delta V$ and $\mathbf{T}_i = -\Sigma[(\mathbf{x} - \mathbf{X}_i) \times (\mathbf{G}_i + \mathbf{F}_{ci})]\Delta V$ are the force and moment acting on the i-th solid, respectively. Here,

$\mathbf{F}_{ci} = \sum_{j=1}^{n}\left(K_n(\phi_i+1)^2(\phi_j+1)^2\mathbf{n}_{ij}\right)$ with a contact constant K_n and unit interface normal vector $\mathbf{n}_{ij} = \nabla(\phi_i - \phi_j)/|\nabla(\phi_i - \phi_j)|$, is the contact force [22] between the i-th and j-th solids, ΔV is the local volume expressed by Δx^2 in the two-dimensional (2D) problem, and $\mathbf{X}_i$ is the center of gravity of the i-th solid. The contact constant K_n was determined as K_n = 1.5 N/mm^3, which is the maximum value that can be computed stably. Tangential forces were not considered in this study for the sake of simplicity.

3. Computational conditions

Figure 1 illustrates the computational conditions used in this study. Simulations were performed in a square domain with $L \times L = 1.25^2$ mm^2, as shown in figure 1(a), where the black, red, and green lines are solid walls with $\phi_0 = 1$. The black, red, and green lines represent the immobile wall, moving wall, and elongating wall, respectively. The shear force is applied by the upward movement of the red line at a constant velocity V_p = 200 μm/s. The green line was elongated following the upward movement of the red line. We enabled the use of a square computational domain by setting the blue dashed line as the periodic boundary [21]. When visualizing the results, the shaded area at the bottom left of figure 1(a) moved to the top left, as shown in figure 1(b), to easily understand the shear deformation.

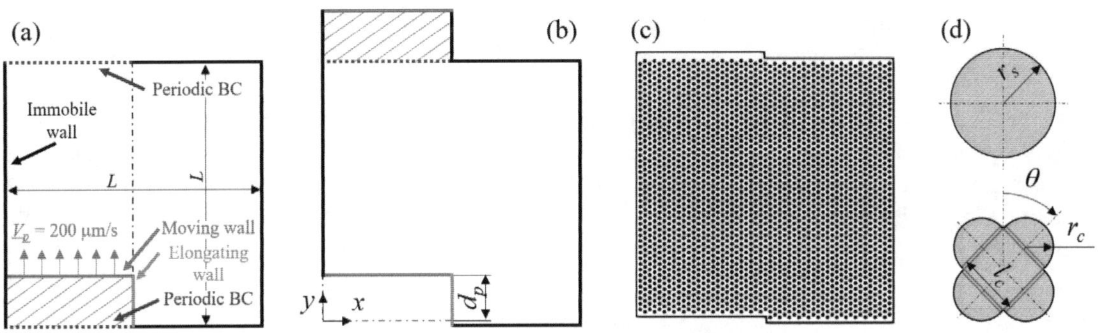

Figure 1. Computational conditions for semi-solid shear deformation: (a) Computational square domain; (b) visualization method of (a); (c) initial grain distribution; and (d) grain shapes. In (a), black, red, and green lines correspond to solid walls expressed by the phase-field variable ϕ_0. To apply the shear force, the moving wall (red line) moves upward with a constant velocity V_p = 200 μm/s. In (b), d_p indicates the displacement of the moving wall compared to the initial state. The initial array of grains was set to an equilateral triangular array, as shown in (c), and the number of grains was fixed to 2,371. In (d), r_s is the radius of a circular grain, where $r_c = 9.5/18r_s$ and $l_c = (\pi r_s^2 - 2\pi r_c^2 + 4r_c^2)/4r_c$ were set such that the cross-shaped grain had the same area as the circular grain. θ indicates the crystal orientation of cross-shaped grains.

The square computational domain was divided into 2,048 × 2,048 grid points (grid size Δx = 0.610 μm). The interface thickness and the time increment were set to W_0 = 2.2 μm and $\Delta t = 3.1 \times 10^{-5}$ s, respectively. Assuming the material to be Al–Cu alloy, the kinematic viscosity and relaxation time were set to $\nu = 5.960 \times 10^{-7}$ m^2/s [27] and $\tau^{+}_{LBM} = 150$, respectively. As shown in figure 1(c), the number of grains was fixed at 2,371, and the distance between the grains was set to 26.7 μm. Two crystal shapes, circular and dendritic cross-shaped, were treated (figure 1(d)), where the solid fraction was changed by changing the grain size. The phase-field profile of the initial grains was given by $\phi_i = -\sin(\eta/W_0)$ [28, 29], where η is the coordinate normal to the interface. The radius of circular grains r_s was set to r_s = 14Δx, 16Δx, and 18Δx (corresponding to 8.51, 9.76, and 10.1 μm) for f_s = 0.40, 0.47, and 0.59, respectively. Thus, the grid size was sufficiently smaller than the grid size.

4. Results

4.1. Circular grains

Figure 2 shows the grain array and pressure distribution with respect to time for solid fraction f_s = (a) 0, (b) 0.40, (c) 0.47, and (d) 0.59 at deformations corresponding to d_p = 30.5, 122, and 366 μm. As shown in figure 2(b), for f_s = 0.40, the liquid pressure in front of the moving wall was high, and this positive pressure area expanded with increasing deformation. This tendency was also observed near the upper-right wall. Meanwhile, a negative liquid pressure was observed in the upper-left region. Such change in pressure distribution as shown in figure 2(b) was similar to that in figure 2(a) where the entire region was only liquid (f_s = 0). This indicates that little contact occurred between the crystals.

Figure 2(c) reveals that for f_s = 0.47, the pressure distribution at d_p = 30.5 μm was similar to that for f_s = 0.40 (figure 2(b)), while the negative pressure regions enclosed by the pink lines were also generated. These negative pressure regions developed vertically at d_p = 122 μm. In such regions, the grains were in contact with each other and arranged in a straight line. This rearrangement causes Reynolds' dilatancy, which has been observed in previous experiments [6-8, 15] and simulations [17]. At d_p = 366 μm, the negative pressure region was not observed. Thus, this solid fraction is considered to approach the threshold for the occurrence of Reynolds' dilatancy.

Furthermore, as shown in figure 2(d) for f_s = 0.59, a shear band with negative pressure was clearly formed at the center because of the Reynolds' dilatancy at d_p = 30.5 μm. This shear band remained even in the late stages of the deformation. In addition, shear bands in the 30° direction—corresponding to the initial grain array—were formed, as indicated by the white arrows. Thus, in the high-solid-fraction condition, the contact between the grains dominates the deformation mechanism.

Figure 2 confirms that the deformation behavior changed depending on the solid fraction, namely from the complete fluid behavior to the grain-contact-dominated behavior with Reynolds' dilatancy with increasing solid fraction. Notably, differences in the deformation behavior due to the solid fraction were confirmed in previous studies [6-8]; thus, it was confirmed that the present method could describe the characteristic behaviors of semi-solid materials.

4.2. Cross-shaped grains

Figure 3 shows the results for cross-shaped grains with f_s = 0.59. The results for f_s = 0.40 and 0.47 are not shown because they were very similar to those for the circular grains shown in figure 2(b) and (c). The initial orientation was set to random. For a small solid fraction, the shape anisotropy of the grains does not affect the deformation behavior. However, the results for f_s = 0.59 shown in figure 3 are different from those for the circular grains shown in figure 2(d). At d_p = 122 μm in figure 3, a negative pressure area is formed at the center, similar to that shown in figure 2(c). However, there was no clear shear band at d_p = 30.5 and 366 μm. Figure 3(b) shows the crystal orientation, where the crystal orientation θ is indicated by the color change every 22.5°, from 0° to 90°. As shown in figure 3(b), the grains were oriented in the y direction near the center of the domain as the deformation progressed, as shown in figure 3(b). Because the grain shape is almost a square, the grains were rearranged such that their flat edges were facing each other. This caused a slip between the flat surfaces, impeding the dilatancy.

As described above, it was shown that the effect of grain shape on deformation behavior is small at low solid fractions, whereas it is significant at high solid fractions. Further, it was confirmed that the present method could express the differences in semi-solid deformation behavior owing to the grain shape.

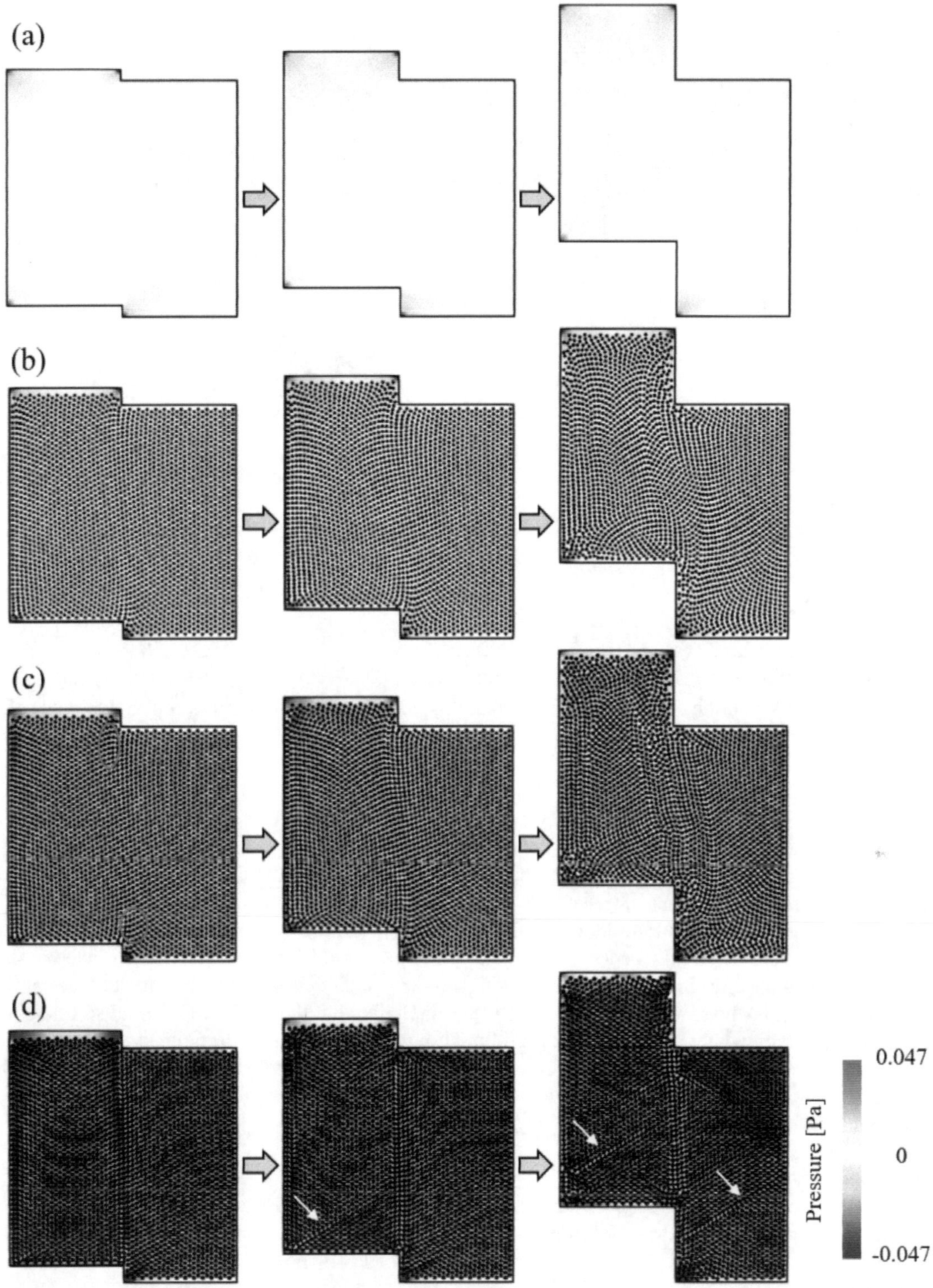

Figure 2. Change in grain array and liquid pressure during shear deformation with circular grains. The solid fraction f_s was changed as f_s = (a) 0, (b) 0.40, (c) 0.47, and (d) 0.59. The deformation proceeded from left to right as d_p = 30.5, 122, and 366 μm. The solid grains with $\phi_i \geq -0.8$ are shown in black. The liquid pressure is a relative value with respect to the initial one.

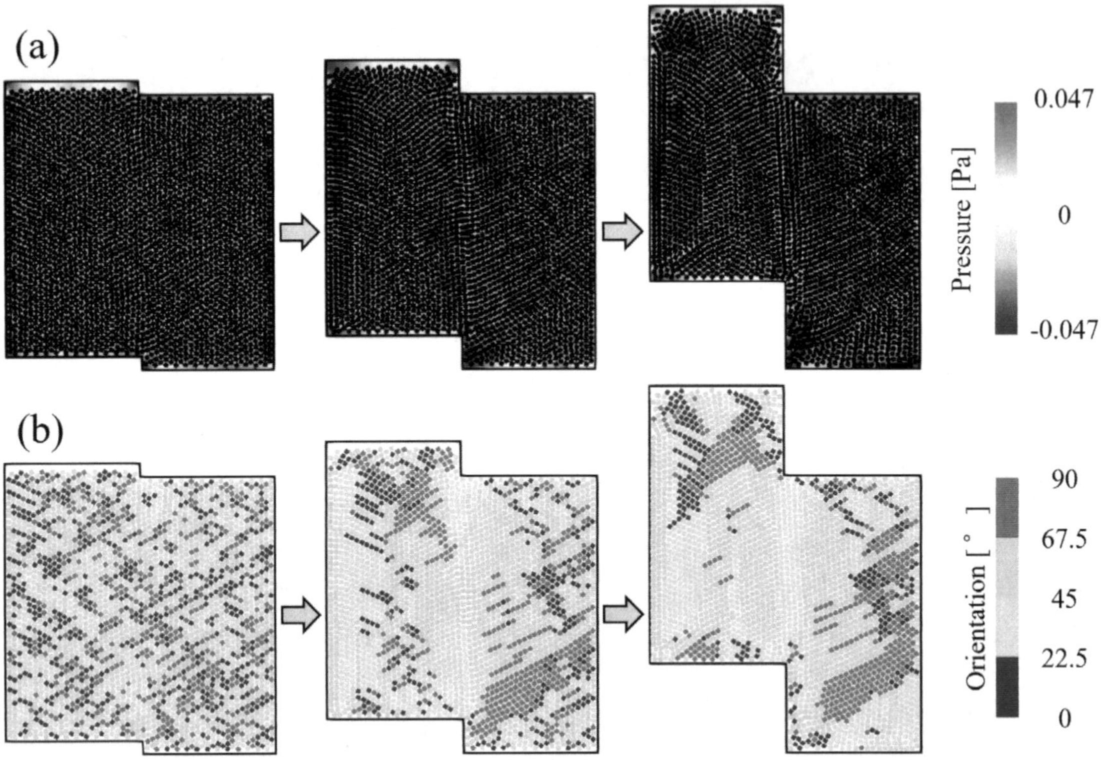

Figure 3. Simulation results for cross-shaped grains with $f_s = 0.59$. (a) Change in grain array and liquid pressure and (b) distribution of crystal orientation at $d_p = 30.5$, 122, and 366 μm (left to right). The solid grains with $\phi_i \geq -0.8$ are shown in black.

5. Conclusions

The MPF-LBM [21], which can simulate polycrystalline solidification with solid motion and liquid flow, was used to simulate 2D semi-solid shear deformations with circular and cross-shaped grains. At low solid fractions, the deformation behavior of the circular grains was almost the same as that of a liquid. However, at high solid fractions, contact between grains occurred, and a shear band due to Reynolds' dilatancy was formed. In addition, the cross-shaped grains showed the same deformation behavior as the circular grains at low solid fractions. Nevertheless, at high solid fractions, the cross-shaped grains were oriented in the y direction near the center of the domain, slip occurred between the flat faces of the grains, and Reynolds' dilatancy did not occur. It was concluded that the MPF-LBM could simulate the semi-solid deformation behavior for various solid fractions and grain shapes. In future work, systematic simulations of real solidification structures will be performed to reveal the semi-solid deformation mechanism. In addition, three-dimensional simulations are expected to reveal the deformation mechanism under practical conditions.

Acknowledgements

This work was supported by a KAKENHI Grant-in-Aid for Scientific Research, (A) 20H00217. This work also used the TSUBAME3.0 computational resources provided by the Tokyo Institute of Technology through the HPCI System Research Project (Project ID: hp210082).

References

[1] Flemings M C 1991 *Metallurgical Transactions B* **22** 269-93
[2] Martin C L, Favier D and Suéry M 1997 *International Journal of Plasticity* **13** 215-35
[3] Kirkwood D H, Suery M, Kapranos P, Atkinson H V and K. P. Young 2010 *Semi-solid Processing of Alloys* vol 124 (Berlin Heidelberg: Springer)
[4] Nafisi S and Ghomashchi R 2019 *Metals* **9**

[5] Karagadde S, Leung C L A and Lee P D 2021 *Materials (Basel)* **14**
[6] Gourlay C M and Dahle A K 2007 *Nature* **445** 70-3
[7] Gourlay C M, Meylan B and Dahle A K 2008 *Acta Metall.* **56** 3403-13
[8] Meylan B, Terzi S, Gourlay C M, Suéry M and Dahle A K 2010 *Scr. Mater.* **63** 1185-8
[9] Gourlay C M, Dahle A K, Nagira T, Nakatsuka N, Nogita K, Uesugi K and Yasuda H 2011 *Acta Metall.* **59** 4933-43
[10] Nagira T, Gourlay C M, Sugiyama A, Uesugi M, Kanzawa Y, Yoshiya M, Uesugi K, Umetani K and Yasuda H 2011 *Scr. Mater.* **64** 1129-32
[11] Gourlay C M, Nagira T, Dahle A K, Nakatsuka N, Uesugi K and Yasuda H 2012 *IOP Conf. Ser. Mater. Sci. Eng.* **27** 012086
[12] Fonseca J, O'Sullivan C, Nagira T, Yasuda H and Gourlay C M 2013 *Acta Metall.* **61** 4169-79
[13] Nagira T, Yokota H, Morita S, Yasuda H, Yoshiya M, Gourlay C M, Sugiyama A, Uesugi K and Umetani K 2013 *ISIJ Int.* **53** 1195-201
[14] Gourlay C M, O'Sullivan C, Fonseca J, Yuan L, Kareh K M, Nagira T and Yasuda H 2014 *JOM* **66** 1415-24
[15] Nagira T, Morita S, Yokota H, Yasuda H, Gourlay C M, Yoshiya M, Sugiyama A, Uesugi K, Takeuchi A and Suzuki Y 2014 *Metall. Mater. Trans.A* **45** 5613-23
[16] Narumi T, Nakata T, Kobayashi R and Yasuda H 2021 *ISIJ Int.* **61** 1567-78
[17] Su T C, O'Sullivan C, Nagira T, Yasuda H and Gourlay C M 2019 *Acta Metall.* **163** 208-25
[18] Su T C, O'Sullivan C, Yasuda H and Gourlay C M 2020 *Acta Materialia* **191** 24-42
[19] Yuan L, O'Sullivan C and Gourlay C M 2012 *IOP Conf. Ser. Mater. Sci. Eng.* **33**
[20] Yuan L, O'Sullivan C and Gourlay C M 2012 *Acta Metall.* **60** 1334-45
[21] Yamanaka N, Sakane S and Takaki T 2021 *Comput. Mater. Sci.* **197** 110658
[22] Yamanaka N, Sakane S and Takaki T 2020 *IOP Conf. Ser. Mater. Sci. Eng.* **861** 012067
[23] Sakane S and Takaki T 2021 *Comput. Mater. Sci.* **186**
[24] Bouzidi M h, Firdaouss M and Lallemand P 2001 *Phys. Fluids* **13** 3452-9
[25] Binghai Wen, Chaoying Zhang, Yusong Tu, Chunlei Wang and Fang H 2014 *J. Comput. Phys.* **266** 161-70
[26] Takaki T, Sato R, Rojas R, Ohno M and Shibuta Y 2018 *Comput. Mater. Sci.* **147** 124-31
[27] Konstantinova N Y, Popel' P S and Yagodin D A 2009 *High Temp.* **47** 336-41
[28] Ohno M, Takaki T and Shibuta Y 2017 *J. Comput. Phys.* **335** 621-36
[29] Ohno M and Matsuura K 2009 *Phys Rev E Stat Nonlin Soft Matter Phys* **79** 031603

A sharp-interface mesoscopic model for dendritic growth

Mitja Jančič[1,2], Miha Založnik[3,4], Gregor Kosec[1]

[1] Institut Jožef Stefan, department of distributed and parallel systems, Jamova cesta 39, 1000 Ljubljana, Slovenia
[2] International postgraduate school Jožef Stefan, Jamova cesta 39, 1000 Ljubljana, Slovenia
[3] Université de Lorraine, CNRS, IJL, F-54000 Nancy, France
[4] Laboratory of Excellence DAMAS, Université de Lorraine, Nancy/Metz, France

E-mail: Gregor.Kosec@ijs.si

Abstract. The grain envelope model (GEM) describes the growth of envelopes of dendritic crystal grains during solidification. Numerically the growing envelopes are usually tracked using an interface capturing method employing a phase field equation on a fixed grid. Such an approach describes the envelope as a diffuse interface, which can lead to numerical artefacts that are possibly detrimental. In this work, we present a sharp-interface formulation of the GEM that eliminates such artefacts and can thus track the envelope with high accuracy. The new formulation uses an adaptive meshless discretization method to solve the diffusion in the liquid around the grains. We use the ability of the meshless method to operate on scattered nodes to accurately describe the interface, i.e., the envelope. The proposed algorithm combines parametric surface reconstruction, meshless discretization of parametric surfaces, global solution construction procedure and partial differential operator approximation using monomials as basis functions. The approach is demonstrated on a two-dimensional *h*-adaptive solution of diffusive growth of dendrites and assessed by comparison to a conventional diffuse-interface fixed-grid GEM.

1. Introduction

Dendritic grains are the most common growth morphology in solidification of metallic alloys. Their growth is to a large extent governed by collective interactions with adjacent grains. The Grain Envelope Model (GEM) [1,2] is a method that can simulate large ensembles of grains and thus fully describe the collective interactions. The GEM is based on the idea that we do not need to represent individual branches of the dendrite but rather represent the dendrite by its envelope, i.e., a smooth surface that connects all growing dendrite tips and has a much simpler shape. The growth of the envelope is obtained from an analytical stagnant film model that relates the growth speed to the solute concentration in the vicinity of the envelope. This formulation allows us to use numerical mesh spacings that are around an order of magnitude coarser than for methods that describe the solid-liquid interface (e.g., phase field).

The accuracy of the GEM simulations [2,3] is principally controlled by the resolution of the sharp variations of solute concentration in the vicinity of the evolving envelopes and by the algorithm used to track the grain envelopes. Presently, diffuse-interface-capturing methods [4] are used for the tracking and fixed meshes are employed for the solution of the transport equations. However, diffuse-interface methods can lead to numerical artifacts that are possibly detrimental in describing highly nonlinear phenomena, such as the formation of new branches of the dendrite envelope or the initial stages of growth after nucleation. The use of sharp-interface tracking combined with an adaptive spatial discretization would be advantageous for two reasons: (1) a sharp-interface description would enable us to eliminate the numerical artifacts related to a diffuse-interface; (2) adaptive methods can automatically refine the spatial discretization where required by the form of the solution, typically in the regions of high gradients surrounding the grains, thus increasing the accuracy of the numerical solution.

Because of their flexibility regarding spatial discretization, meshless methods are particularly suitable for the solution of the moving boundary problems encountered in the GEM. The core idea behind the meshless methods is that they operate on scattered nodes [5], in contrast to traditional mesh-based methods that require internodal connectivity. Generally speaking, node placement is simpler than mesh generation, however, not trivial [6]. Dedicated node positioning algorithms for meshless discretization support variable nodal density [7] that can also be applied to parametric surfaces, e.g., a dendrite envelope. In the context of surface reconstruction from a cloud of points, important advances have been just recently made with a two dimensional algorithm using splines [8] and radial basis functions based three-dimensional algorithms [9,10].

In this paper we develop a sharp-interface formulation of the GEM that uses an adaptive meshless discretization method to solve the diffusion in the liquid around the grains. We use the ability of the meshless method to operate on scattered nodes to accurately describe the interface, i.e., the envelope. We demonstrate the ability of the new approach on a two-dimensional h-adaptive solution of diffusive growth of dendrites and compare it to a conventional diffuse-interface fixed-grid GEM simulation.

2. Mesoscopic Grain Envelope Model for equiaxed isothermal solidification

The Mesoscopic Grain Envelope Model (GEM) [1,2] represents a dendritic grain by its envelope. The growth speed of the envelope, v_n, is obtained from the speed of the tips of the dendrite branches (which are not represented in the GEM), v, by the relation $v_n = v\cos\theta$, where θ is the angle between the outward normal to the envelope, $\mathbf{n}$, and the branch growth direction. In the GEM we assume that the dendrite branches grow in predefined growth directions and θ is the smallest of the angles between $\mathbf{n}$ and the individual directions. In the present study the branches are given $\langle 10 \rangle$ directions, i.e., four possible directions. Note that all equations are written in dimensionless quantities, where the characteristic scales for normalization are v_{iv} for velocity, D / v_{iv} for length and D / v_{iv}^2 for time[1]. The tip speed is calculated from a stagnant-film formulation of the 2D *Ivantsov*

[1] $v_{iv} = 4\sigma^* D Pe_{iv}^2 / d_0$ is the steady-state velocity of a free (Ivantsov) tip growing into an infinite liquid at supersaturation Ω_0, σ^* is the tip selection parameter, D is the diffusion coefficient, d_0 is the capillary

solution that relates the *Péclet number* of the tip, Pe, to u_δ, the normalized concentration difference between the liquid at the tip and that at a finite distance δ from the tip [11] as

$$u_\delta = \sqrt{\pi Pe}\exp(Pe)\left[\operatorname{erfc}\left(\sqrt{Pe}\right) - \operatorname{erfc}\left(\sqrt{Pe(1+\delta Pe)}\right)\right] \tag{1}$$

and from a tip selection criterion that reads

$$v = (Pe / Pe_{iv})^2. \tag{2}$$

Equations (1) and (2) are used to solve for v at any given point on the envelope. The concentration u_δ is obtained from the concentration field in the liquid around the envelope, which is resolved numerically from the diffusion equation

$$\frac{\partial u}{\partial t} = \nabla^2 u, \tag{3}$$

subject to boundary conditions $u = 0$ on the envelope and $\mathbf{n} \cdot \nabla u = 0$ on the outer boundary, where $\mathbf{n}$ is the normal vector to the boundary. The initial condition is $u = \Omega_0$, i.e., the liquid is homogeneous and has a dimensionless supersaturation of Ω_0. In this paper we used a stagnant film of $\delta = 1.0$.

3. Diffuse-interface GEM on a fixed mesh

The established formulation of the GEM [1,2] uses a phase-field-like interface capturing method [4] to track the grain envelope on a fixed mesh. An indicator function, α, is used to distinguish between the interior of the grain ($\alpha > 0.95$) and the liquid ($\alpha < 0.95$). The interface is diffuse since the transition of α between 1 (grain) and 0 (liquid) is smooth with a width w_α. The evolution of α is given by

$$\frac{\partial \alpha}{\partial t} + v_n \mathbf{n} \cdot \nabla \alpha = -b\left[\nabla^2 \alpha - \frac{\alpha(1-\alpha)(1-2\alpha)}{w_\alpha^2} - |\nabla \alpha| \nabla \cdot \left(\frac{\nabla \alpha}{|\nabla \alpha|}\right)\right], \tag{4}$$

where v_n is the envelope growth speed. The parameters w_α and b, as well as the grid size and timestep are selected following the guidelines of Souhar et al. [2]. The advantage of this method is that we can avoid explicit tracking of the envelope and instead simply solve the PDE (4) on a fixed mesh. The drawback is that its accuracy is limited when it needs to describe radii of curvature smaller than around 5 times the mesh spacing [2,4].

4. Sharp-interface GEM solved with an adaptive meshless method

In the scope of meshless methods, a linear differential operator L (for example, ∇^2 from diffusion equation (3)) is approximated from a set of nearby nodes, also referred to as *support nodes* or *stencil*. In this paper, the closest n nodes are chosen as stencil of a node $\mathbf{x}_c$. The approximation of operator L in node $\boldsymbol{x}_c$ is then sought using an ansatz

$$(\mathrm{L}u)(\mathbf{x}_c) \approx \sum_{i=1}^{n} w_i u_i(\mathbf{x}_i) \tag{5}$$

---

length, and Pe_{iv} is the Péclet number of the free tip, given by the solution of

$\Omega_0 = \sqrt{\pi Pe}\exp(Pe_{iv})\operatorname{erfc}\left(\sqrt{Pe_{iv}}\right)$.

for support nodes $\mathbf{x}_i$, support weights w_i, an arbitrary function u and nodal values u_i. To determine the weights w_i, the equality of approximation (5) is enforced for a given set of m basis functions $\{\phi_j\}_{j=1}^{m}$. This allows us to write a linear system

$$\begin{bmatrix} \phi_1(\mathbf{x}_1) & \cdots & \phi_1(\mathbf{x}_n) \\ \vdots & \ddots & \vdots \\ \phi_m(\mathbf{x}_1) & \cdots & \phi_m(\mathbf{x}_n) \end{bmatrix} \begin{bmatrix} w_1 \\ \vdots \\ w_n \end{bmatrix} = \begin{bmatrix} (\mathrm{L}\phi_1)(\mathbf{x}_c) \\ \vdots \\ (\mathrm{L}\phi_m)(\mathbf{x}_c) \end{bmatrix}, \tag{6}$$

where only the weights w_i are unknown. In this paper monomials with order up to including 2 are used. The support size has been set to $n = 12$, resulting in an overdetermined linear system (6), which is solved using the Householder QR decomposition.

The approximation (5) is general in the sense that it can also be used to interpolate a global solution. In this paper, partition-of-unity (PU) based interpolation has been employed to obtain a global solution approximation [12,13], which is essential for interpolation of concentration or concentration difference (1) at positions that do not correspond to discrete nodal positions.

The stand-alone C++ implementation of the meshless methods, including PU approximators and node positioning algorithms, is available as part of *the Medusa library* [14].

4.1. *Algorithm for the sharp-interface GEM*

The sharp-interface GEM simulation procedure consists of multiple steps shown in figure *1*.

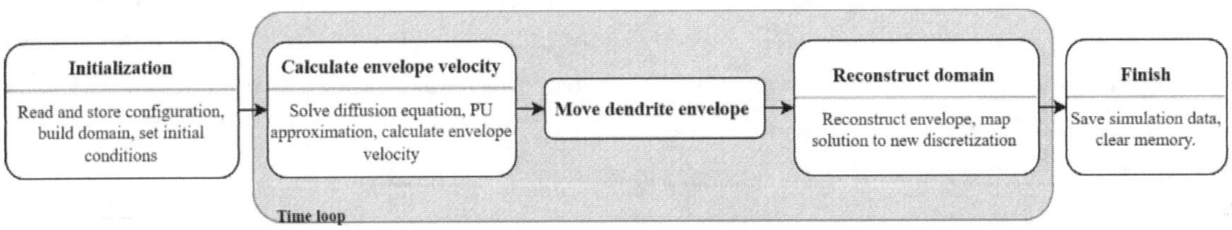

Figure 1: Simulation workflow scheme for the meshless sharp-interface GEM.

The initial liquid domain is a difference between the square domain with dimensionless side length $a_m = 20$ and the initial grain envelope represented as a circle with dimensionless radius $r_d = 0.22$. This is the smallest radius of curvature that can be accurately represented with the diffuse-interface method using a grid spacing of 0.05, which is used for comparison in Section 5. The domain is discretized with a linear nodal density distribution between the grain envelope and the outer domain boundary, ranging from $h_d = 0.05$ on the envelope to $h_m = 3h_d$ on the outer boundary. The initial concentration in the liquid is homogeneous at $u = \Omega_0 = 0.18$.

The first step within the explicit time-marching (blue box in figure *1*) is solving the diffusion equation (3). Afterwards, the envelope velocity is computed for each node on the envelope. In addition to the model parameters, the GEM also requires an accurate concentration field value $u_\delta(\mathbf{x}_i^\delta)$, where $\mathbf{x}_i^\delta = \mathbf{x}_i + \delta\mathbf{n}_i$ for dendrite boundary node $\mathbf{x}_i$, normal $\mathbf{n}_i$ and stagnant-film thickness δ. PU approximation is employed to interpolate a global concentration field $u(\mathbf{x})$ and thus enables us to obtain $u_\delta(\mathbf{x}_i^\delta)$ anywhere in the domain. As a result, the GEM model returns the nodal velocity v_{tip} for the dendrite boundary node $\mathbf{x}_i$ and the boundary node is moved to a new location

$$\mathbf{x}_i^{'} = \mathbf{x}_i + \cos\theta_i v_{tip} dt \mathbf{n}_i \tag{7}$$

where θ_i is the minimal angle between the normal $\mathbf{n}_i$ and 4 unit directional vectors, i.e., $\mathbf{e}_x$, $-\mathbf{e}_x$, $\mathbf{e}_y$ and $-\mathbf{e}_y$.

However, moving the envelope boundary nodes in the domain space introduces several problems. Firstly, the internodal distance on the dendritic envelope is generally no longer locally constant and approximately h, and secondly, some of the nodes that were originally in the interior of the domain, can fall outside of the computational domain, i.e., in the interior of the dendritic envelope. For the sake of simplicity, we omit these problems by choosing to build a completely new domain discretization with the new dendritic envelope, whose shape is now exactly prescribed with the new boundary nodes from equation (7). To accurately build the domain discretization and assure minimal distortion to the dendritic envelope, a surface reconstruction algorithm is required. In this paper we reconstruct the two-dimensional envelope shape using splines, as in [8]. After the discretization is completed, the concentration values in the new nodal positions are obtained by employing the PU interpolation constructed over the old domain discretization.

The entire process, i.e., the blue rectangle on figure 1, is then repeated, until the total simulation time $t_{tot} = 12$ is reached.

5. Results

The initial domain has been filled with 21945 nodes and computational time of approximately 35 hours on a computer with AMD EPYC 7702 64-Core processor with 512GB of DDR4 memory for simulation time $t_{tot} = 12$ with timestep $dt = 10^{-4}$. The code[2] was compiled using g++ (GCC) 9.3.0 for Linux with -03 -std=c++17 -fopenmp -DNDEBUG flags.

A timelapse of the growing grain is shown in figure 2.

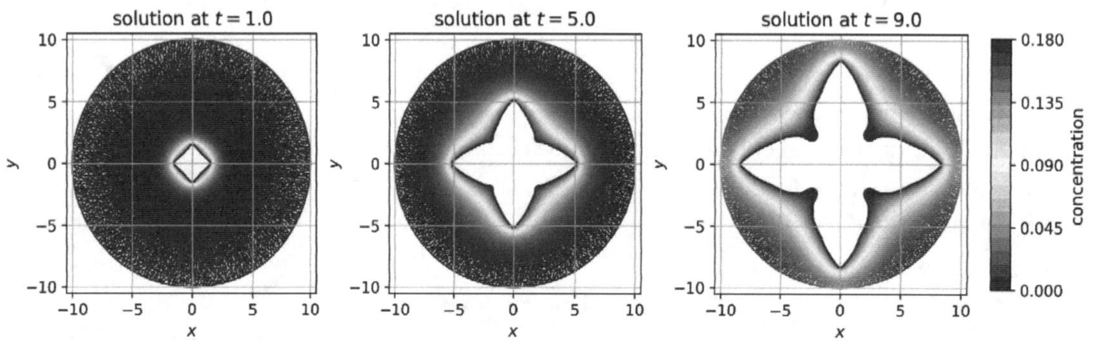

Figure 2: Dendrite growth timelapse.

As seen in figure *2*, we are able to obtain a reasonable solution using the sharp-interface meshless method. To assess its quality, we compare it to the diffuse-interface fixed-mesh solution obtained

[2] Source code available at: https://gitlab.com/e62Lab/public/2022_cp_icasp_dendrite_growth under tag *v1.0.*

with identical physical and model parameters and a mesh spacing of 0.05. The primary tip speed as a function of the primary tip position is compared in figure *3* (left) and the grain envelope shape at $t = 6$ in figure *3* (right).

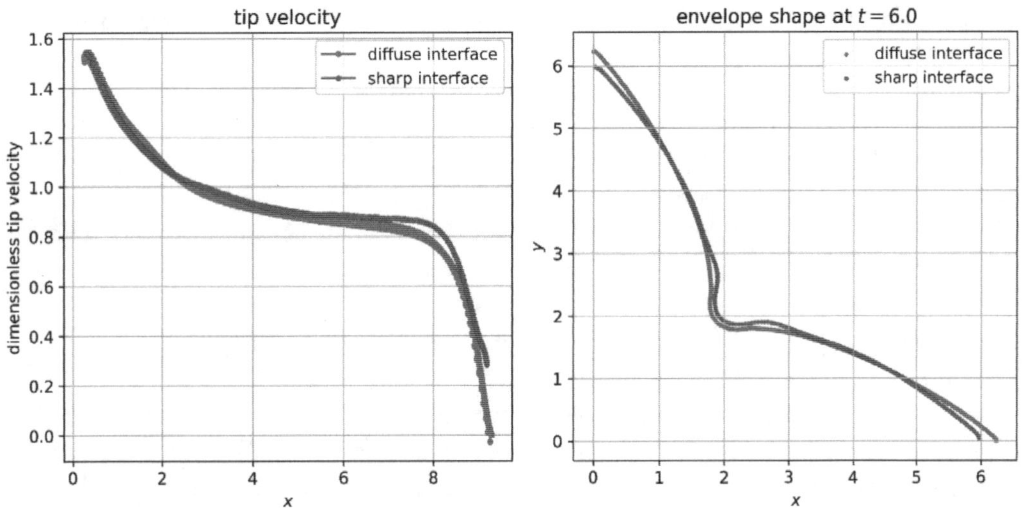

Figure 3: Dimensionless tip velocity vs. tip position (left) and dendrite envelope shape at selected simulation time (right).

The evolution of the tip velocity shows the characteristic initial transient form a high initial velocity towards a velocity of 1. In the present case the tip velocity does not reach a steady state since the diffusion field interacts with the outer domain boundary before steady state is reached. The final transient consists of a slowdown of the tip that finally stops at a distance δ from the boundary, i.e., $x_{tip} \approx 9$.

6. Conclusions

We have shown that the *h*-adaptive meshless method can be successfully employed to simulate the growth of a dendrite envelope during solidification, using a sharp-interface representation of the envelope. The meshless numerical solution has been compared to the conventional diffuse-interface fixed-grid GEM solution in terms of maximum tip velocity for all the simulation steps and in terms of tip shape at a selected time step. We have shown that both solutions obtained by different numerical methods are comparable. Future work could include extension to the more realistic three-dimensional dendrites and full exploitation of parallelism to reduce the computational times of computationally more expensive meshless methods.

Acknowledgments. The authors would like to acknowledge the financial support of the Slovenian Research Agency (ARRS) research core funding No. P2-0095 and the World Federation of Scientists. A part of the required high performance computing resources was provided by the EXPLOR center hosted by the University of Lorraine.

References

[1] Steinbach I, Diepers H-J and Beckermann C 2005 Transient growth and interaction of equiaxed dendrites *J. Cryst. Growth* **275** 624–38

[2] Souhar Y, De Felice V F, Beckermann C, Combeau H and Založnik M 2016 Three-dimensional mesoscopic modeling of equiaxed dendritic solidification of a binary alloy *Comput. Mater. Sci.* **112** 304–17

[3] Tourret D, Sturz L, Viardin A and Založnik M 2020 Comparing mesoscopic models for dendritic growth *IOP Conference Series: Materials Science and Engineering* vol 861 (IOP Publishing) p 012002

[4] Sun Y and Beckermann C 2007 Sharp interface tracking using the phase-field equation *J. Comput. Phys.* **220** 626–53

[5] Wang H and Qin Q-H 2019 Overview of meshless methods *Methods of Fundamental Solutions in Solid Mechanics* (Elsevier) pp 3–51

[6] Shankar V, Wright G B, Kirby R M and Fogelson A L 2015 A Radial Basis Function (RBF)-Finite Difference (FD) Method for Diffusion and Reaction–Diffusion Equations on Surfaces *J. Sci. Comput.* **63** 745–68

[7] Slak J and Kosec G 2019 On Generation of Node Distributions for Meshless PDE Discretizations *SIAM J. Sci. Comput.* **41** A3202–29

[8] Jancic M, Cvrtila V and Kosec G 2021 Discretized Boundary Surface Reconstruction *2021 44th International Convention on Information, Communication and Electronic Technology (MIPRO)* 2021 44th International Convention on Information, Communication and Electronic Technology (MIPRO) (Opatija, Croatia: IEEE) pp 278–83

[9] Drake K P, Fuselier E J and Wright G B 2022 Implicit Surface Reconstruction with a Curl-free Radial Basis Function Partition of Unity Method *ArXiv210105940 Cs Math*

[10] Liu S, Liu T, Hu L, Shang Y and Liu X 2021 Variational progressive-iterative approximation for RBF-based surface reconstruction *Vis. Comput.* **37** 2485–97

[11] Cantor B and Vogel A 1977 Dendritic solidification and fluid flow *J. Cryst. Growth* **41** 109–23

[12] Slak J 2021 Partition-of-Unity Based Error Indicator for Local Collocation Meshless Methods *2021 44th International Convention on Information, Communication and Electronic Technology (MIPRO)* 2021 44th International Convention on Information, Communication and Electronic Technology (MIPRO) (Opatija, Croatia: IEEE) pp 254–8

[13] De Marchi S, Martínez A and Perracchione E 2019 Fast and stable rational RBF-based partition of unity interpolation *J. Comput. Appl. Math.* **349** 331–43

[14] Slak J and Kosec G 2021 Medusa: A C++ Library for Solving PDEs Using Strong Form Mesh-free Methods *ACM Trans. Math. Softw.* **47** 1–25

IOP Conf. Series: Materials Science and Engineering 1274 (2023) 012047 doi:10.1088/1757-899X/1274/1/012047

Phase-field lattice Boltzmann model of growth and buoyancy driven motion of dendritic particles: the effect of Coriolis force

L Rátkai[1], L Gránásy[1,2] and T Pusztai[1]

[1] Wigner Research Centre for Physics, P.O. Box 49, H-1525, Budapest, Hungary
[2] BCAST, Brunel University, Uxbridge, Middlesex UB8 3PH, United Kingdom

Abstract. A phase-field lattice Boltzmann (PFLB) model we developed recently (Rátkai L, Pusztai T and Gránásy L 2019 *npj Comput. Mater.* **5** 113) was adapted to investigate the effect of Coriolis force on the solidification morphology forming in a centrifuge. The moving particles are handled using an overlapping multi-grid scheme, assigning a separate moving grid to each individual particle, with attached local fields. Relying on this approach, solidification problems, in which solute diffusion, melt flow, solid motion, the effect of gravity and collision of the particles occur simultaneously can be modelled. Using relevant properties of an Al-Ti alloy, the interaction of columnar and equiaxed solidification was studied. It was found that the Coriolis force influences the path of the falling dendritic particles, which in turn governs the position, at which the transition between columnar and equiaxed morphologies takes place.

1. Introduction

Buoyancy driven melt flow and the motion of growing crystals have an essential role in determining the cast microstructure [1]. The local temperature gradient and the velocity of the solidification front are known to govern a morphological transition between the elongated "columnar" crystals growing inwards from the surface of the mould and the equiaxed structure dominated by nucleation: columnar to equiaxed transition (CET). This morphological transition influences the properties and failure characteristics of the cast components. A detailed understanding of such processes are of practical importance. *In situ* synchrotron studies indicate that buoyancy forces may influence the microstructure of cast metals considerably [2]: buoyancy driven jets, "plumes", drag dendrite fragments of the columnar crystals into the bulk melt, which growing further and heavier fall on top of the columnar layer, leading to CET [3]. Modelling of these phenomena requires the coupling of fluid flow, gravity and particle growth and motion. This task has been addressed in detail only recently. Earlier studies of solidification in flow combine either the lattice Boltzmann method of fluid flow or the Navier-Stokes equations with the phase-field/level set/cellular automaton models of solidification [4-10]. The majority of these studies refer to the growth of static particles in fluid flow. Motion of the growing solid particles was addressed by recent studies by Do-Quang and Amberg [5], Steinbach [6], Takaki et al. [7], and Qi et al. [8]. In the latter works, the phase-field approach is coupled either to the Navier-Stokes equations or to lattice Boltzmann hydrodynamics. Do-Quang and Amberg proposed a semi-sharp interface phase-field model combined with the distributed Lagrangian method that realizes rigid body motion of the solid particle [5]. The phase-field approach put forward by Steinbach and co-workers was used to describe the motion of a single dendritic particle growing in shear flow in two and three dimensions [6]. Takaki et al. addressed a descending single dendritic particle and the motion of several dendritic particles in shear flow [7]. In the multi-particle case, "collision-coalescence" was assumed; i.e., that the solid particles in contact with each other stick together forming a single solid body. In a recent work, Qi et al. [8] addressed the buoyancy induced motion of several dendrites, performing the simulations at realistic viscosity contrast ($\mu_S/\mu_L = 10^6$, where subscripts S and L refer to the solid and the liquid, respectively), however, retaining exclusively hydrodynamic interaction between the dendritic particles, i.e., the model cannot handle colliding particles. In a recent paper, we published a PFLB method that relies on an overlapping grid approach, when handling the solidifying particles [9]. This enables us to treat several dendritic particles (typically up to 100 on a GPU card) that tumble over each other when colliding, a technique complementary to other methods [8]. A similar particle-particle interaction was adopted by Yamanaka et al. in large scale simulations for the sedimentation of equiaxed dendrites [10]. Recently, some interest was shown regarding the effect of another external force, the Coriolis force, expected to act in the case of centrifugal experiments of solidification [11] as illustrated in figure 1. In the present work, we employ the approach described in [9] to model the effect of Coriolis force acting on dendritic particles in a centrifuge, moving in the melt due to buoyancy forces.

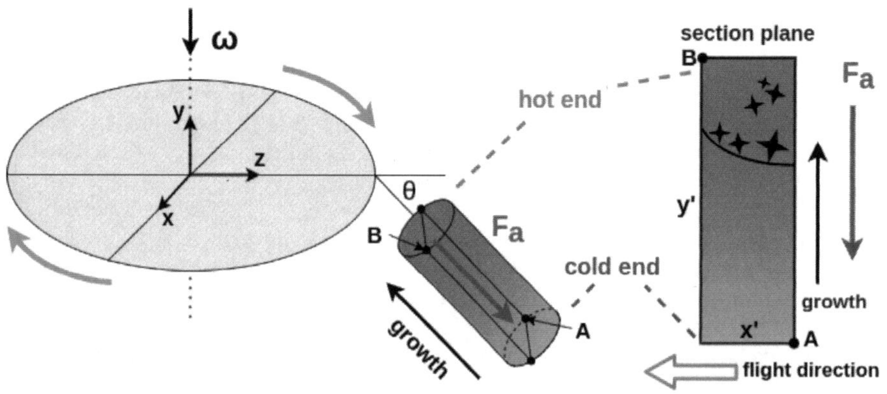

Figure 1. Schematic view of solidification in a container hanging on a centrifuge. Notations: **ω** – angular velocity vector; $\mathbf{F}_a$ – sum of gravitational and centrifugal forces; θ – angle between the axis of the container and the axis of rotation. The 2D simulations are to be performed within a rectangular domain of the plane defined by the axis of the container and the tangential velocity, whose *y'* edges are parallel with $\mathbf{F}_a$, whereas the *x'* edges are parallel with the tangential velocity, while a pair of opposite corners are denoted by A and B. (Reproduced after a drawing by U. Hecht.)

2. Model description

Simulation of CET incorporating mobile particles requires a model that can handle all relevant processes and phenomena simultaneously: solidification, diffusion, melt flow, convection, particle movement and particle collision. Coupling of models that address one or more of these phenomena may be challenging due to numerical problems arising from different length- and time scales. Furthermore, when particles move relative to the grid, distortion of their shape due to numerical issues associated with the advective terms may occur. We address multi-particle interaction using an overlapping grids scheme, where to each individual particle a local grid is assigned that moves with the velocity of the particle.

2.1. Overlapping grids scheme

In the overlapping grids scheme there is a global grid and several local grids (see figure 2). The global grid is a fixed, uniformly spaced rectangular grid that defines the simulation box. The particle distribution functions required by the lattice Boltzmann method and the *global* concentration field are defined on the global grid. The local grids are also rectangular and their linear size grows as the size of the particle defined on it requires. The fields defined on the local grids are the phase field corresponding to the given particle ($\phi = 0$: liquid, $\phi = 1$: solid) and the *local* concentration field. The local grids overlap with and may even extend beyond the global grid, and may also overlap with each other (see the red dashed rectangles in figure 2). The boundaries of the local grids are not physical boundaries of the domain, the boundary conditions applied to the local fields are based on the values interpolated from the global grid.

For the αth local grid, in addition to the fields mentioned above, the following values are calculated and stored in each time step:

- M_α : total mass of the solid
- r_α^l : position of the centre of mass in local grid coordinates
- r_α^g : position of centre of mass in global grid coordinates
- θ_α : rotation angle
- I_α : moment of inertia
- u_α : velocity
- ω_α : angular velocity

The total mass, local center of mass and the moment of inertia can be evaluated from the phase field distribution. First the forces acting on the solid on the local grid are computed, which are used in the equation of motion for the center of mass. The relevant forces are: (i) the fluid force that acts at the boundary of a particle, (ii) particle-particle interaction force that acts at the contact point between neighboring particles, and (iii) the external forces, such as gravity or the Coriolis-force. The fluid force acting on the particle is evaluated after projecting the phase-field values on the local grids to an auxiliary field ϕ^g defined on the global grid using bilinear interpolation. The velocity of solid points is computed on the global grid.

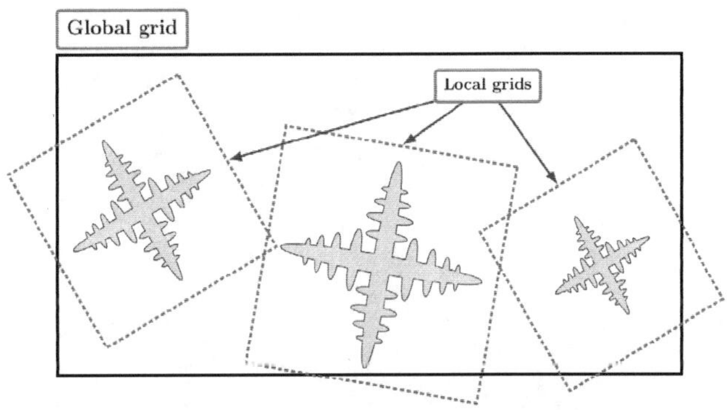

Figure 2. Global and local grids. The global grid (black rectangle) defines the simulation domain. The local grids (red dashed rectangles) may have different sizes as they are resized since the corresponding particles grow. Note that the local grids may extend beyond the border of the simulation box and they may overlap other local grids.

In the hydrodynamic description, sharp moving boundaries defined by $\phi^g = 0.5$ were used. Having all forces and the motion of particles calculated, the phase-field equations are solved on the local grid, whereas the concentration equations are solved on both the global and local grids. The two solutions (global or local) for the concentration equation are considered in updating the values for each point. The effect of fluid flow on the concentration field is described by an advection term in the equation of motion (EOM) for concentration. An advantage of this approach is that the local grid moves with the particle, so no advection terms act inside the particle. In turn, numerical error emerges due to interpolating between the local and global grids, which error is, however, expected to be small as the respective fields are fairly smooth.

2.2. Fluid flow in the lattice Boltzmann method

In simulating the fluid flow, the lattice Boltzmann method [12] was used with the multiple-relaxation-time collision operator [13]. In our model the particles are considered as sharp, arbitrarily shaped, moving walls during the streaming step of the lattice Boltzmann method. To incorporate these moving walls during the fluid calculation, we chose the approach by Mei et al. [14]. The fluid force exerted on the solid particles has been determined using the Galilean invariant momentum exchange method by Wen et al. [15]. For each α particle we calculate a total hydrodynamic force F_α^h and torque T_α^h.

2.3. Particle collisions in the Diffuse Interface Field Approach

In the case of large particle density, collisions do occur. Mobile dendrites may collide with other dendrites (columnar or equiaxed), and the walls of the simulation box (container). In the case of spherical particles, contact events can be determined from the position of the particles. The short range repulsion is often modelled by hard-sphere or soft-sphere potentials. Prediction of contact events between particles of complex shape such as the dendrites is, however, non-trivial and it can be very time consuming. Following [9], we adapted the Diffuse Interface Field Approach by Wang et al. [16], in which a short-range repulsive force emerges from overlapping diffuse fields (in our case, phase fields). It has the advantage that it takes into account the arbitrary shapes and sizes of individual particles automatically, i.e., without explicitly tracking the particle surfaces. The method produces the total short-range interaction force F_α^{sr} and torque T_α^{sr} acting on the center of mass of the αth particle.

2.4. Particle motion

The total force and torque acting on the αth particle include contributions from fluid flow, the short-range particle interaction and the external force: $F_\alpha = F_\alpha^h + F_\alpha^{sr} + F_\alpha^{ext}$, and $T_\alpha = T_\alpha^h + T_\alpha^{sr}$. The Coriolis force acting on a particle is computed as $F_\alpha^{Cori} = -2M_\alpha(\boldsymbol{\omega} \times \mathbf{v}_\alpha)$, where $\boldsymbol{\omega}$ is the angular velocity vector of the centrifuge, whereas $\mathbf{v}_\alpha$ is the velocity vector of the αth particle. The simulations with F_α^{Cori} were performed in the plane defined by $\mathbf{F}_a$ (the sum of gravitational and centrifugal forces) and the tangential velocity of the container, $\mathbf{v}_t$ (see x'– y' plane in figure 1).

2.5. Phase-Field Method

Binary solidification is described by the phase-field model of Kim and co-workers [17]. The respective EOMs were solved on the respective local (equiaxed dendrites) and global (columnar dendrites) grids. For the αth local grid the EOM for the phase-field reads as

$$\frac{\partial \phi_\alpha}{\partial t} = M_\phi\big[\nabla \cdot \epsilon_\phi^2 \nabla \phi_\alpha + p'(\phi_\alpha)\big[f_L(c_{L,\alpha}) - f_s(c_{S,\alpha})\big] \\ -(c_{L,\alpha} - c_{S,\alpha}) f'_L(c_{L,\alpha}) - wg'(\phi)\big]. \tag{1}$$

Here f' stands for the derivative of f with respect to its argument, M_ϕ denotes the phase-field mobility, and ϵ_ϕ^2 is the coefficient of the square-gradient term for the phase field in the free energy functional. $g(\phi_\alpha) = \phi_\alpha^2(1-\phi_\alpha)^2$ and $p(\phi_\alpha) = \phi_\alpha^3(10 - 15\phi_\alpha + 6\phi_\alpha^2)$ are the double well and interpolation functions, f_S and f_L are the free energy densities of the solid and liquid phases, w is the free energy scale, whereas $c_{S,\alpha}$ and $c_{L,\alpha}$ are the compositions of the solid and liquid phases.
The EOM of the concentration field on the local grid α is

$$\frac{\partial c_\alpha}{\partial t} = \nabla\big[D(\phi_\alpha)\nabla c_{L,\alpha}\big] + v_\alpha[1 - p(\phi_\alpha)]\nabla c_\alpha, \tag{2}$$

where D is a phase field dependent diffusion coefficient. The last term on the RHS is the advection current that takes into account the fluid flow. The local flow velocity v_α is determined from the global fluid velocity v by subtracting the velocity of the given volume element of the moving local grid.
The EOM for the global concentration c^g has a similar form

$$\frac{\partial c^g}{\partial t} = \nabla\big[D(\phi^g)\nabla c_L^g\big] + v[1 - p(\phi^g)]\nabla c^g \tag{3}$$

As the EOM for concentration is solved on the global and on all the local grids, several solutions exist for the same volume elements (with the same global coordinates) where the grids overlap. Interpolations of the respective solutions are used to obtain the solution (for details see [9]).

2.6. Computations

The EOMs for the phase and concentration fields were solved numerically using the finite difference method combined with forward Euler time stepping, while prescribing no-flux boundary condition at the borders of the simulation box (global grid). In LB hydrodynamics, no-slip boundary condition was used. The computations were performed on high end GPU cards. The materials properties were taken from [9], including a viscosity of 3 orders of magnitude smaller than the physically realistic one, making the numerical solution tractable. In the computations a temperature gradient of $dT/dy' = 4000$ K/m and a cooling rate of 0.6 K/s was assumed. The time and spatial steps were $\Delta t = 10$ μs and $\Delta x = 0.5$ μm.

2.7. Validation

Motion of the inert particles was validated in a number of test cases [9] such as flow over a cylinder asymmetrically placed into a channel [18], a descending circular particle [15], two sedimenting particles showing drafting, kissing and tumbling motions [19]. Our simulations were in a good agreement with these benchmarks [9]. We also studied the sedimentation of individual and interacting dendritic particles [9]: the predictions are in a reasonable agreement with recent work [7, 20].

3. Results and discussion

Herein, we investigate, how the Coriolis force influences of the sedimentation of equiaxed dendrites on top of columnar dendritic structures. Three cases will be investigated: (i) single descending dendrite, (ii) point source of dendritic particles, and (iii) randomly nucleating equiaxed dendrites.

3.1. Single descending equiaxed dendrite

We compared the behaviour of a single equiaxed dendrite descending in the presence of gravitational force (dimensionless angular velocity: $\omega = 0$ rad/s) with that of a similar arrangement, however, with Coriolis force acting on the particle ($\omega = -2.5$ rad/s). 2D simulations were performed in a coordinate system (x', y') as defined in figure 1. Note that the Coriolis force acts in the $x' - y'$ plane. While in the first case (only gravity) the growing particle descends approximately vertically, the path of the particle bends to the right in the presence of the Coriolis force (a typical ratio of the Coriolis force to $\mathbf{F_a}$ is ~ 0.8).

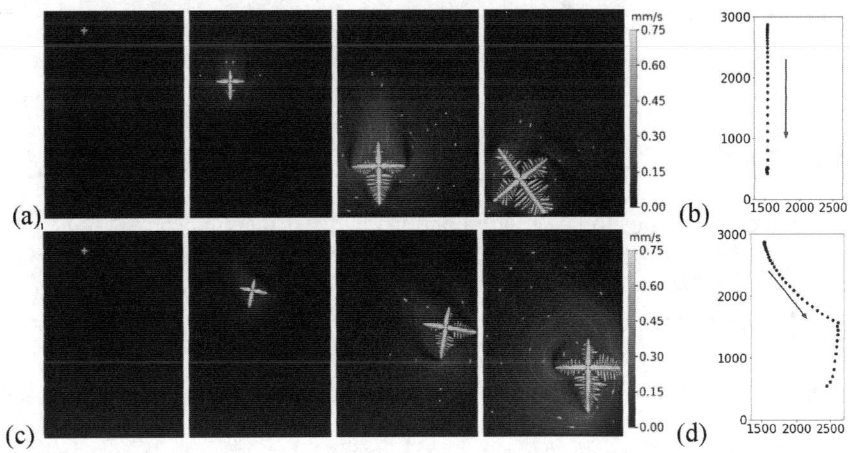

Figure 3. Dendritic particle descending in the melt under gravity: (a), (b) without ($\omega = 0$ rad/s) and (c), (d) with Coriolis force ($\omega = -2.5$ rad/s). (a), (c): Velocity maps are shown (the solid domains are coloured white, whereas colouring in the melt shows the magnitude of the local velocity). The simulations were performed on a 3000 × 3000 (1.5 mm × 1.5 mm) uniform rectangular grid (only a fraction is shown). Time elapses from left to right: Snapshots taken at 1.2×10^5, 3.6×10^5, 5.6×10^5

and 7.0 $\times 10^5$ time steps are shown. (b), (d): The respective particle trajectories (the frame shows a 1500 $\times$ 3000 section of the simulation box). Note the bending of the particle trajectory in panel (d) due to the Coriolis force. The particle bounces back after colliding the wall on the right.

3.2. Effect of Coriolis force on CET in case of point source of nucleation

Next, we explore how the Coriolis force influences the path of equiaxed particles periodically nucleating at a point source above the columnar front. The point source of nucleation was realized by releasing periodically supercritical Al-Ti clusters at a point located well above the columnar front. As expected on the basis of the previous simulation, it was found that in the presence of Coriolis force, the otherwise vertical trajectory of the growing particles bends to the right, though their trajectory is certainly influenced by impingement on the previously nucleated particles (see figure 4).

(a)

(b)

Figure 4. The effect of Coriolis force on sedimenting dendrites that nucleate periodically in a point source in the presence of columnar dendrites growing from the bottom wall of the simulation box. (a) Only $\mathbf{F}_a$ (i.e., $\omega = 0$ rad/s), (b) $\mathbf{F}_a$ and Coriolis force ($\omega = -1.5$ rad/s): note the accumulation of the equiaxed dendritic particles on the right. Velocity maps (1st and 3rd line) and concentration maps (2nd and 4th line) are shown. The simulations were performed on a 6000 $\times$ 4000 grid. Time elapses from left to right: Snapshots taken at 7.2 $\times 10^5$, 1.2 $\times 10^6$ and 1.6 $\times 10^6$ time steps are shown.

3.3. Effect of Coriolis force on the columnar to equiaxed transition with randomly placed nuclei

To investigate how the Coriolis force present in the centrifugal arrangement shown in figure 1 influences the morphological transition between the columnar and equiaxed domains, we performed simulation for $\omega = 0$ rad/s and in dimensionless units. Columnar dendrites of the same orientation were initiated to grow from the bottom surface of the simulation box, whereas a spatially random nucleation process was emulated by placing supercritical Al clusters to the upper part of the simulation domain. As expected on the basis of the previous simulations, it was found that in the presence of the Coriolis force, the otherwise vertical trajectory of the growing particles bends to the right while descending, leading to an accumulation of the equiaxed particles at the right wall of the container (see figure 5).

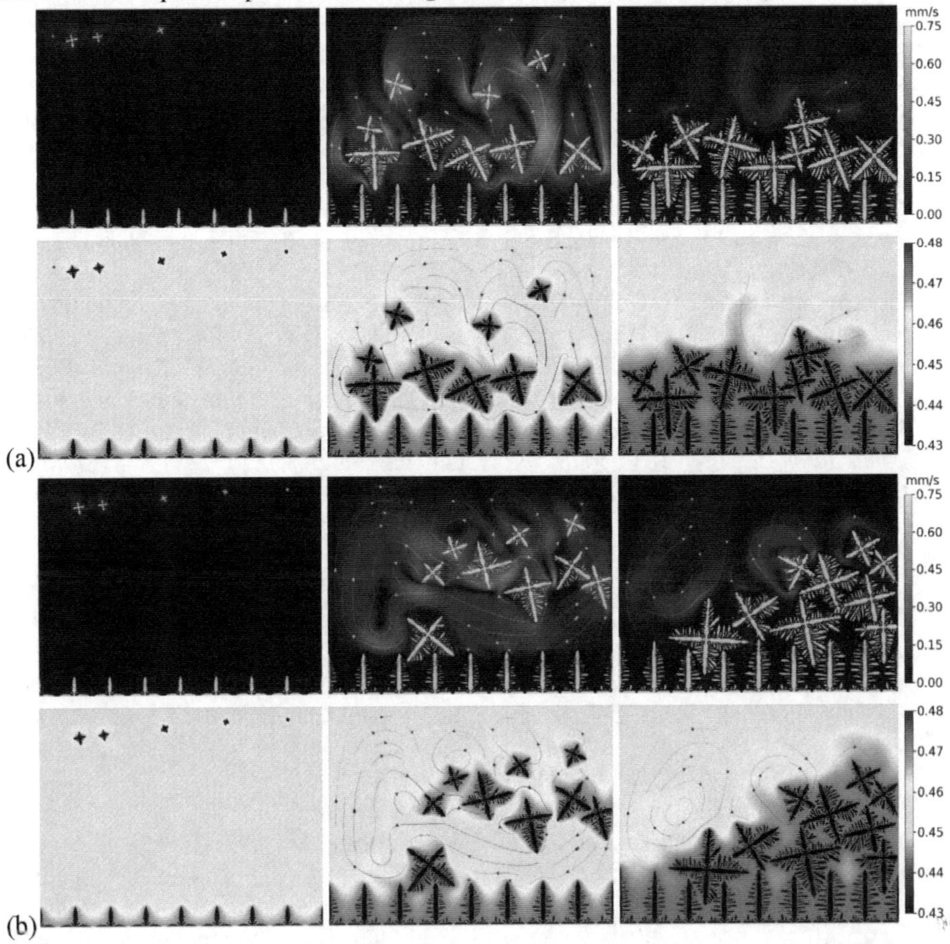

Figure 5. The effect of Coriolis force on columnar to equiaxed transition during the solidification of an Al-Ti alloy under buoyancy forces. (a) Only $\mathbf{F}_a$ (i.e., $\omega = 0$ rad/s), (b) $\mathbf{F}_a$ and Coriolis force ($\omega = -3$ rad/s) note the accumulation of the equiaxed dendritic particles on the right. Velocity (1st and 3rd line) and concentration maps (2nd and 4th line) are shown. The simulations were performed on 4000×3000 uniform rectangular grids. Time elapses from left to right: Snapshots taken at 3×10^5, 6×10^5 and 8×10^5 time steps are shown.

4. Summary

A PFLB model, we developed for simulating the columnar to equiaxed transition (CET), was adapted to solidification in a centrifuge, in which Coriolis force acts on the sedimenting equiaxed dendritic particles. We have demonstrated that this effect may substantially modify the solidification microstructure via moving the equiaxed particles in the direction of the force, leading to the accumulation of equiaxed grains on that side of the container, a phenomenon that influences the location of CET. This result seems to be in accord with preliminary experimental results obtained in the framework of the ESA MAP project "GRADECET" [21]. A detailed study of this phenomenon is underway.

Acknowledgments

This work was supported by the National Agency for Research, Development and Innovation, Hungary (NKFIH) under Contract No. KKP-126749 and by the ESA MAP/PECS project "GRADECET"

(ESTEC Contract No. 40000110759/11/NL/KML). The research infrastructure was provided by the Hungarian Academy of Sciences.

References

[1] Dantzig J A and Rappaz M 2009 *Solidification*, (EPFL Press, Lausanne).

[2] Mathiesen R and Arnberg L 2005 *Mater. Sci. Eng. A* **413-414** 283–7; Mathiesen R and Arnberg L 2005 *Acta Mater.* **53** 947–56; Ruvalcaba D, Mathiesen R, Eskin D, Arnberg L and Katgerman L 2007 *Acta Mater.* **55** 4287–92; Bogno A, Nguyen-Thi H, Reinhart G, Billia B and Baruchel J 2013 *Acta Mater.* **61** 1303–15

[3] Hellawell A, Sarazin J R and Steube R S 1993 *Philos. Trans.* **345** 507–44; Chen C 1995 J. Fluid Mech. **293** 81–98

[4] See for example: Jeong J-H, Goldenfeld N and Dantzig J A 2001 *Phys. Rev. E* **64** 041602; Nourgaliev R, Dinh T, Theofanous T and Joseph D 2003 *Int. J. Multiphase Flow* **29** 117–69; Conti M 2004 *Phys. Rev. E* **69** 022601; Tan L and Zabaras N 2007 *J. Comput. Phys.* **221** 9–40; Selzer M, Jainta M and Nestler B 2009 *Physica Status Solidi (b)* **246** 1197–205; Sun D, Zhu M, Pan S and Raabe D 2009 *Acta Mater.* **57** 1755–67; Yuan L and Lee P D 2010 *Model. Simul. Mater. Sci. Eng.* **18** 055008; Yin H, Felicelli S and Wang L 2011 *Acta Mater.* **59** 3124–36; Zhang S-J, Wang Y-M, Wang Q, Li C-Y and Li R 2021 *Acta Phys. Sin.* **70** 238101; Sakane S and Takaki T 2021 *Comput. Mater. Sci.* ***186*** 110070

[5] Do-Quang M and Amberg G 2008 *J. Comput. Phys.* **227** 1772–89

[6] Medvedev D, Varnik F and Steinbach I 2013 *Procedia Comput. Sci.* **18** 2512–20; Steinbach I 2013 *JOM* **65** 1096–102

[7] Takaki T, Rojas R, Ohno M, Shimokawabe T and Aoki T 2015 *IOP Conf. Ser.: Mater. Sci. Eng.* **84** 012066; Takaki T, Sato R, Rojas R, Ohno M and Shibuta Y 2018 *Comput. Mater. Sci.* **147** 124–31; Sakane S, Takaki T, Ohno M, Shibuta Y and Aoki T 2020 *Comput. Mater. Sci.* **178** 109639

[8] Qi X B, Chen Y, Kang X H, Li D Z and Gong T Z 2017 *Sci. Reports* **7** 45770

[9] Rátkai L, Pusztai T and Gránásy L 2019 *npj Comput. Mater.* **5** 113

[10] Yamanaka N, Sakane S and Takaki T 2021 *Comput. Mater. Sci.* **197** 110658

[11] Hecht U, Huang C, Zollinger J, Droz D, Zaloznik M, Cisternas M, Viardin A, McFadden S, Gránásy L, Lapin J, Leriche N and Kargl F 2018 *Proc. 7th Int. Conf on Solidification and Gravity (Miskolc)*, ed A Roósz, Z Veres, M Svéda and G Karacs (Mikolc: Hungarian Academy of Sciences/University of Miskolc) pp 27–36.

[12] Iglberger K, Thürey N and Rüde U 2008 *Comput. Math. Appl.* **55** 1461–8

[13] Lallemand P and Luo L 2000 *Phys. Rev. E* **61** 6546–62

[14] Mei R, Luo L-S, Shyy W. 1999 *J. of Comput. Phys.* **155** 307–30

[15] Wen B, Zhang C, Tu Y, Wang C and Fang H 2014 *J. Comput. Phys.* **266** 161–70

[16] Wang Y U 2007 *Acta Mater.* **55** 3835–44

[17] Kim S G 2007 Acta Mater. **55** 4391–9

[18] Schäfer M, Turek S, Durst F, Krause E and Rannacher R 1996 *Benchmark Computations of Laminar Flow Around a Cylinder*, (Leipzig/Wiesbaden: Vieweg+Teubner Verlag) pp 547–66

[19] Glowinski R, Pan T W, Hesla T I, Joseph D D and Périaux J 2001 *J. Comput. Phys.* **169** 363–426; Wang L, Guo Z L and Mi J C 2014 *Comput. Fluids* **96**, 20–34

[20] Luo S, Wang P, Wang W and Zhu M 2020 *Metall. Mater. Trans. B* **51** 2268–84; Meng S, Zhang A, Guo Z. and Wang Q 2020 *Comput. Mater. Sci.* **184** 109784

[21] Hecht U 2018 personal communication

The macrosegregation behaviour of Fe-C-Cr-Mo type steel in laboratory scale model ingot.

Y Sumi[1a], S Narita[1] and M Yamashita[1]
[1] Daido steel co., ltd., Aichi, Japan

E-mail: [a]y-sumi@ac.daido.co.jp

Abstract. In ingot casting, it is important to predict and control segregation and casting defects that occur during the solidification process, because they have a significant effect on the quality of the steel ingot. In some special steels, in typical Fe-C-Cr-Mo type high carbon tool steels, macro-segregation tends to be more pronounced due to their wide solid-liquid coexistence temperature range and molten steel properties such as relatively higher liquid viscosity than basic carbon steels. In order to investigate segregation and defect formation phenomena during solidification, many experiments have been conducted using specially designed laboratory scale model ingots that can induce partial difference of solidification to cause macrosegregation phenomena. However, these kind of experimental approaches for special steels are less conducted and influence of alloy types on the macrosegregation behaviour is still not well understood. In this study, a model ingot which have a steel chill ring in the middle region of a mullite mold was prepared, then Fe-0.45C(mass%) carbon steel, Fe-1C-1.8Cr(mass%) bearing steel and Fe-1C-8Cr-2Mo(mass%) high carbon tool steel were cast into 30kg ingots and the macrosegregation behaviour was investigated.

1. Introduction

Ingot casting process is conventional manufacturing method to make large scale steel product. Generally ingot casting is selected for large diameter products, or for special steels with high alloy content those are difficult to manufacture by continuous casting. It is important to predict and control segregation and casting defects that occur during the solidification process, because they have a significant effect on the quality of the steel ingot. Therefore there has been long research works done on macrosegregation phenomena and theoretically well understood [1]. Generally, such macrosegregation and flaw formation are highly affected by the dimension of the ingot case and design of feeder head. Therefore, there has been several attempts to predict the defect formation from the solidification profile in the large scale ingot [2, 3]. These are very useful in practical way, however it needs huge efforts to handle large scale samples, it is beneficial to use small scale mock ingot for more basic research purpose. It is also known that the macrosegregation formation is affected by alloying element and molten metal properties [4]. In some special steels, for instance Fe-C-Cr-Mo type high carbon tool steels, macrosegregation tends to be more pronounced due to their wide solid-liquid coexistence temperature range and molten steel properties such as relatively higher liquid viscosity than basic carbon steels. It is commonly known that the macrosegregation is highly related to suction of liquid phase, therefore there are long historical approaches to simulate suction phenomena using experimental model [5, 6]. Recently, Sato et al. have developed a laboratory-scale small ingot (Sato-mold) [7]. The original Sato-mold experiment using Al-Cu alloy showed the liquid suction through the mushy zone in the bridged region of primary solidified shell near the chill plates and formation of macrosegregation below the bridged region. Natsume performed numerical simulation on the Sato-mold experiments and confirmed that the relationship of bridging and suction liquid phase on macrosegregation [8]. Thus the Sato-mold is good simple experimental method to simulate macrosegregation phenomena. However, these kind of experimental approaches are based on Al based alloys or low alloy steels, studies for special steels with high alloy content are less conducted and influence of alloy composition on the macrosegregation is yet not known enough. In this study, Sato-mold is modified and adapted for experiment on special steels, then various steels with different alloy composition are cast into modified Sato-mold and the effect of alloy types on macrosegregation is investigated.

2. Experimental procedure

Table 1 shows the chemical composition of steels used for experiment. About 150kg of the material was melt in a vacuum induction melting furnace (VIM), then cast by bottom pouring into 4 ingot molds. Figure 1 shows schematic drawing of the mold, which have a steel chill ring in the middle region of mullite rings stacked vertically. The total height of stacked mold is 340 mm and inner diameter is 100 mm. The weight of each ingot is approximately 30 kg, including about 6 kg of feeder head on the top. The liquidus, solidus for each alloy was calculated by commercial phase diagram calculation software Thermo-Calc (TM), and the target cast temperature was 200 degree plus the liquidus temperature for each alloy. After the casting, the vertical slice at the center plane of ingot was prepared, and macrostructures and microstructures were observed by an optical microscope. To investigate segregation, about 5 g of samples were cut from several positions along the vertical center line of the ingot, then chemical composition for each sample were measured by ICP. The distribution of alloying elements in the microstructure was observed by EPMA.

Table 1. Chemical composition of steels used for experiment and calculated liquidus and solidus temperature for each steel.

	Nominal alloy composition	Liquidus [°C]	Solidus [°C]	Casting super heat [°C]
Steel A	Fe-0.45 mass %C	1490	1415	200
Steel B	Fe-1C-1.8 mass %Cr	1445	1310	200
Steel C	Fe-1C-8Cr-1.8 mass %Mo	1418	1220	200

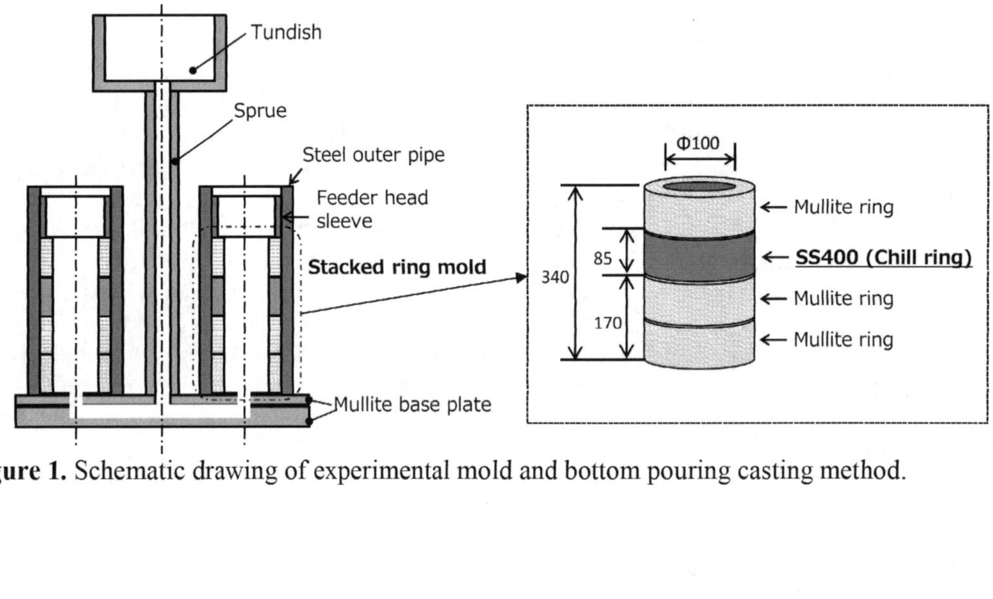

Figure 1. Schematic drawing of experimental mold and bottom pouring casting method.

3. Results and discussions

Figure 2 shows the macrostructures of vertical samples. In every sample, shrinkage cavities were generated in the lower part below the chill ring. In the steel A, smooth spindle-shaped cavity was formed and its top end starts from the center of the chill ring. In the steel B, fine cavities were complexly aggregated in the lower part of the cavity, and upper part had smooth inner surface. Sample of steel C had more complexed cavity, which was composed of relatively simple shape of upper part and complexed shape of lower part. The position of cavity in the steel C was the lowest of three samples, which starts from the lower end of the chill ring. Figure 3 shows the distribution of alloy element along the vertical center line of the ingot In every steel sample, enriched carbon concentration was observed slightly below the center of the chill ring where near the top of the shrinkage cavity. On the contrast, the carbon concentration in the upper part of the chill ring was lower than these in other position. In the bottom side region, carbon distribution in steel A showed different behaviour than other two steels. Steel A showed carbon enrichment near the bottom of the cavity, whereas steel B, C showed lean carbon concentration at the bottom region below the cavity. These results suggest that chill ring induces partial increase of cooling rate at the chill and prevent liquid feeding into the lower part of ingot, results in the cavity formation during solidification. Cavity formation behaviour is highly affected by the alloy element. In this report, steel C was selected for further investigation.

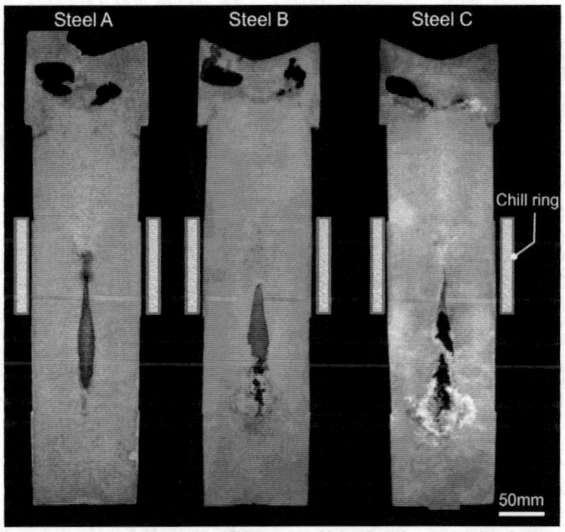

Figure 2. Macrostructures of ingot slice in the vertical center plane.

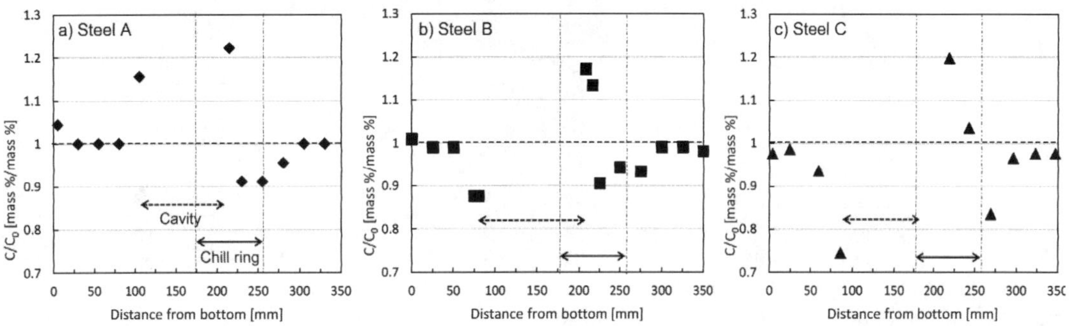

Figure 3. Carbon concentration along the vertical center line of ingot. C/C_0 denotes concentration ratio of carbon, whereas C: Concentration at each sampling position, C_0: Concentration in the ladle.

Figure 4 shows microstructures of each position in vertical slice of steel C ingot. After pouring, solidification of molten steel is considered to have progressed from the inner surface of the mold inward the center. Dendritic growth with side branches was observed at most of the observation positions except the lower center in the chill ring. The dendrite arm spacing near the chill ring were finer than those in other region, which indicates that cooling rate in the chill ring is large due to the larger heat conductivity of steel chill ring than that of mullite mold. In the lower center area in the chill ring, where a large concentration of carbon was observed, eutectic carbides were significantly increased between the very fine dendrite structures. In the bottom center area below the large center cavity, fine shrinkage cavities were dispersed in inter-dendrite region. Figure 5 shows the area fraction of eutectic carbides at the different distance from bottom of the ingot. At 190 mm from the bottom and 50 mm from the surface, which is just above the center cavity, significant increase of eutectic carbide was observed. From microscopic carbide morphology, these carbides were considered to be formed in inter-dendritic region by microsegregation of solute element during the dendritic growth. To confirm that, phase diagram calculation was conducted. Table 2 shows the chemical composition for each observation position. At position a where just above the center cavity, solute elements were concentrated. On the contrast, at the lower end of center cavity, solute concentration was leaner than

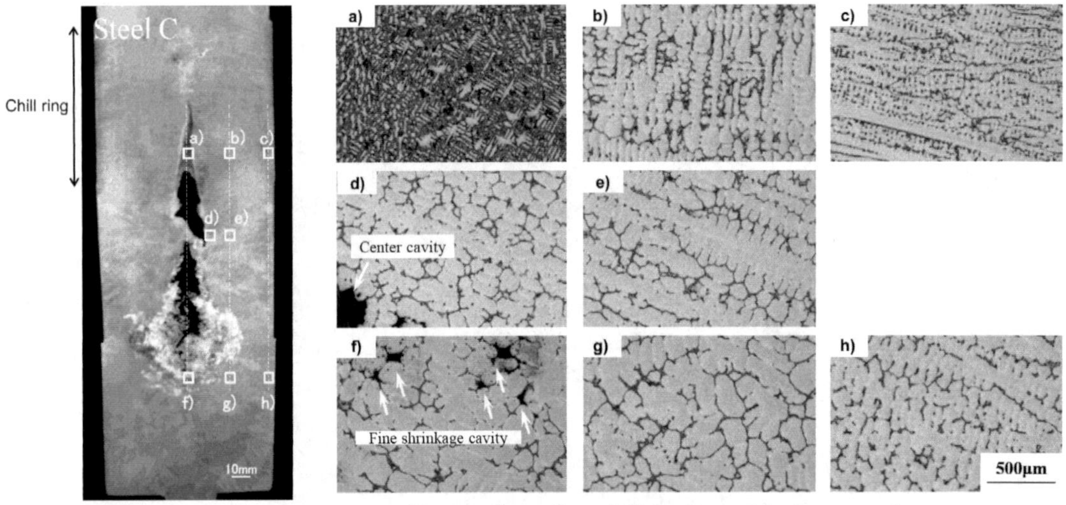

Figure 4. Microstructures of ingot slice of steel C, in the vertical center plane.

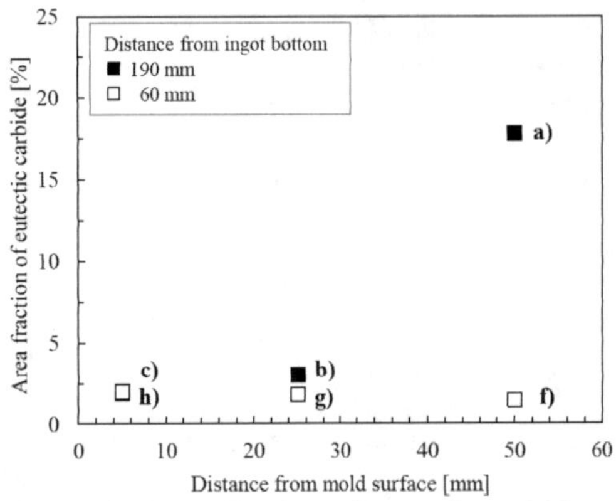

Figure 5. Area fraction of eutectic carbides of steel C. The alphabets (a ~ f) are correspond to the each position in the figure 4.

Table 2. Chemical composition for each observation position (mass %)

Position	Fe	C	Cr	Mo
a	Bal.	1.19	8.70	2.21
b	Bal.	0.99	8.14	1.85
c	Bal.	1.00	8.14	1.86
d	Bal.	1.00	8.19	1.89
e	Bal.	0.99	7.97	1.74
f	Bal.	0.74	7.02	1.31

the nominal composition. Other positions seems to be nearly homogeneous. Figure 6 shows equilibrium phase diagram calculation and Scheil-Gulliver module calculation with each composition in table 2. For position b to e, representative calculation with nominal composition is shown. For nominal composition and position f, eutectic carbides were not predicted to be formed in equilibrium solidification, whereas Scheil-Gulliver calculation showed that microsegregation induces eutectic carbide formation. In position a, where the apparent increase of fraction and morphology change of carbides were observed, carbides were considered to be formed by eutectic reaction. The measured area fraction of carbides were larger than calculated by Sheil-Gulliver model. This might be because microscopic observation was more magnified into final solidification area than sample for chemical analysis. This implies that liquid phase with high solute concentration was entrapped in this region, consequently high amount of eutectic carbide were crystalized in the final solidification. These results suggests that chill ring induces the inhomogeneous solidification in the bottom part of the ingot, consequently generates pressure gradient, thereafter suction of concentrated liquid phase was occurred.

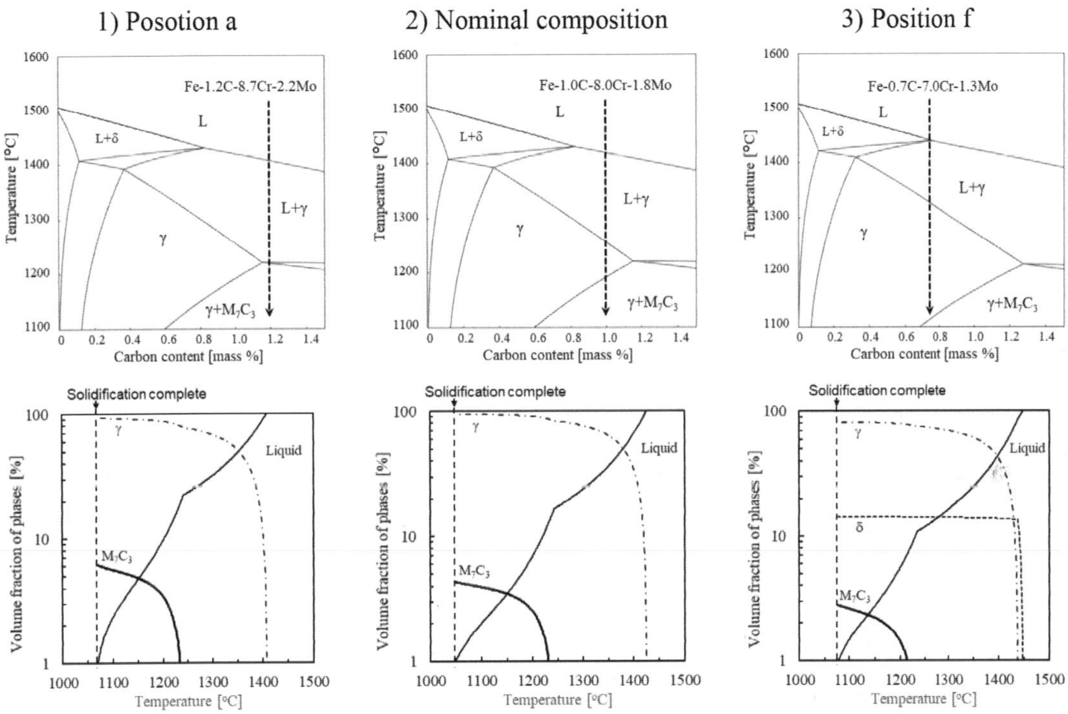

Figure 6. Equilibrium phase diagrams (upper) and Sheil-Gulliver calculation (lower) for position a 1), nominal composition (represents position b ~ e) 2), position f 3)

4. Conclusions

The effect of alloy type on the formation of macrosegregation was investigated by casting special steels into experimental ingot mold which have chill ring in the middle of mullite mold. Shrinkage cavities were formed in the lower region below the chill. In the Fe-0.45C(mass%) carbon steel, the cavity was simple and smooth spindle shape, whereas Fe-1C-1.8Cr(mass%) bearing steel and Fe-1C-8Cr-2Mo(mass%) high carbon tool steel created more complexed shape cavities. In every steel sample, enriched carbon concentration was observed slightly below the center of the chill ring where near the top of the shrinkage cavity. This macrosegregation is considered to be created by the suction of concentrated liquid phase from the upper region of chill, by the negative pressure occurred during solidification in the bottom part of the ingot.

References

[1] M C Flemings, 2000 *ISIJ international* 40 No.9 833
[2] Tashiro K, Watanabe S, Kitagawa I and Tamura I 1981 *Tetsu to Hagane* 67 No.1 104
[3] Yamada H, Sakurai T, Takenouchi T and Iwanami Y 1987 *Tetsu to Hagane* 73 No.14 1706
[4] Ishida H, Sakamoto K, Nohara M, Oki H and Ohsasa K 2009 *Tetsu to Hagane* 795 No.1 26
[5] Flemings M C, Nereo G E 1968 *Trans. TMS* 242 41
[6] Suzuki K, Miyamoto T, 1973 *Tetsu to Hagane* 59 No.12 1549
[7] Sato F, Esaka H and Shinozuka K 2016 *Tetsu to Hagane* 99 No.2 101
[8] Natsume S, 2017 *Tetsu to Hagane* 103 No.12 738

Characterisation of mesosegregations in large steel ingots

L Gutman[1,2], J R Kennedy[2], F Roch[1], A Marceaux dit Clément[1], L Salsi[3], J Cauzid[3], B Rouat[2], H Combeau[2], M Založnik[2] and J Zollinger[2]

[1] Framatome DTI, F-92400 Courbevoie
[2] Université de Lorraine, CNRS, IJL, F-54000 Nancy
[3] Université de Lorraine, CNRS, GeoRessources, F-54000 Nancy

Abstract. The chemical composition mapping of low-alloyed steel ingots used for the nuclear industry is crucial in the manufacturing of forgings and their final quality mastering. Mechanical properties of forged and hot-rolled steels may be affected by chemically segregated bands. These bands arise from segregations that appear at the scale of a few grains in the as-cast structure: the so-called mesosegregations. While segregation at the scale of dendrite arms (microsegregation) and the scale of the casting (macrosegregation) is well understood and can be readily characterized, only little is known about the formation of mesosegregation. The first step towards understanding the cause behind mesosegregation formation can be brought through comprehensive chemical characterisation at the scale of several grains (mesoscopic scale), which requires using different characterisation techniques compared to micro- or macrosegregation characterisation. We developed a sampling and characterisation methodology that allows segregations to be mapped at the mesoscopic scale using micro X-ray fluorescence (µXRF). Characterisation technique, sampling methodology, and sample size must be adapted to consider the different solidification structures; both at smaller (dendrite arms, grains) and larger (macrostructure) scales. Segregations were characterised on a 113 x 98 mm² steel plate extracted from a low-alloyed steel large ingot.

1. Introduction

Segregation of alloying elements occurs as steel solidifies. In steel, the common alloying elements (C, Mn, Mo for instance) are more soluble in the liquid than in the solid [1]. As steel solidifies, alloying elements are rejected into the liquid phase. This solute enrichment of the liquid phase leads to chemical heterogeneity in the solid-state. Such heterogeneity is called segregation and exists at different scales. Segregation is a departure from the nominal chemical composition of the steel, local enrichments cause positive segregations whereas local depletions cause negative segregations. *Microsegregation* is segregation observed at the dendrite scale (from µm to a few hundreds of µm), due to solute rejection during the growth of dendrites. It can be influenced by solute diffusion difference between the liquid and solid-state [2]. Microsegregation creates depleted solid and enriched liquid that move at the ingot scale (m) through convection, movement of grains, solidification shrinkage and mushy zone deformation forming *macrosegregation* [3]. While heat treatment can reduce microsegregation through diffusion, macrosegregation cannot be reduced at the industrial scale since diffusion is too slow of a process. Thus macrosegregation cannot be reduced after solidification, it may remain in the semi-products and even in the forged part thus affecting final mechanical properties [4].

Segregations that exist in the ingot influence solid-state transformations, which cause heterogeneities in microstructures and final mechanical properties. Forging steps change the size, shape, and distribution of segregations but do not decrease their intensity or number [5]. The change in shape and size of segregations causes a banding phenomenon in forged and hot-rolled steels: bands of different microstructures and compositions are observed [6]. Banding in steel leads to anisotropic mechanical properties. This is especially true for impact toughness properties which can be sensitive to orientation and depth of specimen sampling [7, 8]. Banding is caused by segregations larger than the dendrite scale (microsegregations), but smaller than channels (macrosegregation) [9]. Such intermediate scale segregations, probably at the grain-scale, are called *mesosegregations*. To understand and influence banding phenomena, understanding the formation of mesosegregations is key.

A comprehensive characterisation of mesosegregations is the first step towards understanding their formation. Different characterisation techniques exist to track chemical heterogeneities at different scales. Investigation of microsegregation is commonly carried out with an Electron Probe Micro Analyser (EPMA) which enables the quantification of the chemical composition with an interaction volume of roughly 1 μm^3. In large steel ingots, macrosegregation can be tracked with Optical Emission Spectrometry (OES) or infrared combustion analysis for carbon and sulphur investigation. Macrosegregation can also be detected by X-Ray Fluorescence (XRF) spectroscopy. Manganese (Mn) and Molybdenum (Mo) macrosegregation patterns can easily be distinguished, with a 5 mm analysis step and minimal sample preparation [10]. However, these techniques cannot be adapted to intermediate-scale segregation in the case of large ingots due to their lack of spatial resolution. Large ingots used in the nuclear industry have long solidification times and so present large structures: investigated samples present secondary arm spacing of 800 µm. Hence to investigate the occurrence of mesosegregations in such ingots, chemical composition should be mapped at the centimetric grain scale. At this scale, EPMA analysis is impractical: too detailed and too time consuming. In contrast, the OES and combustion analysis techniques are too coarse to be implemented at a centimetric scale.

In this paper we developed a new sampling and characterisation methodology, suitable for the chemical analysis of mesosegregations in steel ingots. This new method is based on micro-X-Ray Fluorescence (µXRF), which has begun to be investigated as a tool to analyse centreline segregation during continuous casting [11, 12]. First, the effects of sampling on the observed structures and recorded data were analysed in order to define sampling parameters applicable to a centimetric scale mapping. Then, calibration was conducted by comparison with EPMA to convert X-ray counts to quantitative concentration. Finally, a centimetric scale quantitative map was produced combining µXRF and EPMA calibration.

2. Material and methods

2.1. Samples

The low-alloy steel 16MND5 (0.16 %wt. C, 1.45 %wt. Mn, 0.725 %wt. Ni and 0.5 %wt. Mo) was selected for analysis. Samples were taken from the equiaxed region of an industrially produced 116 ton directionally solidified ingot. Samples were cut using wire electron discharge machining.

2.2. Micro-X-Ray Fluorescence (µXRF)

Micro-X-Ray Fluorescence (µXRF) analyses were performed with a Bruker M4 Tornado device equipped with a Rh X-ray source (30 W power). Two EDS (Energy Dispersive Spectroscopy) detectors are located on either side of the primary X-ray source. Both the source and the detectors are fixed in the vacuum chamber (20 mbar vacuum) whereas the samples are placed on a motorized stage. The spot size focused on the sample is 20 µm.

Low-alloyed steel samples of different sizes were used to investigate the effects of experimental conditions on the results. For µXRF measurements, a metallographic surface finish is not required, samples were ground flat with Si-C paper down to P800 (21.8 µm granulometry).

Concerning the primary X-ray source conditions, acceleration voltage, current, and the use of filters on the beam enable improvement of the signal and minimisation of artefacts. An indicator for artefact minimisation is the dead time, as dead time approaches 20 %, artefacts are minimised. Dead time can be managed by balancing current, acceleration voltage, and filters. Acceleration voltage was set to 50 keV for all analyses. Filters used on the X-ray beam are high-pass filters, they enable suppression of a part of the signal to avoid saturation and artefacts. However, strong filters can suppress peaks that correspond to elements of interest, for instance, the Manganese signal is affected by such strong filters. The use of filters can be optimised to approach 20 % of dead time without decreasing the current. This is beneficial as a low current leads to a weak outcoming signal.

2.3. Electron Probe Micro Analyzer (EPMA)
The Electron Probe Micro Analyser (EPMA) device used is a JEOL JXA-8530F equipped with a FEG source, an EDS and 5 WDS (Energy and Wavelength Dispersive Spectroscopy) detectors. The electron beam operates at 20 kV and 100 nA for the analysis of Mn and Mo on LIF and PET crystals respectively, calibrated with standards Pel6080-Mn and Pel6080-Mo. Samples analysed with EPMA require a mirror-like surface and were polished with a 1 µm diamond slurry.

2.4. Cumulative solid fraction curves
As solidification occurs the solute concentration in liquid increases (the partition coefficients of the studied elements are smaller than one). One way to compare the different characterisation techniques is to associate a solid fraction with data points and to plot the evolution of concentration (for EPMA analysis) or peak intensity (for µXRF analysis) as a function of the increasing solid fraction. The first solid to form is the most depleted and the last solid to form is the richest in solute. Intermediate data points are then ranked. Different ranking techniques exist and are used for microsegregation analysis, several of them are compared by Ganesan [13]. In the present study, the data points are arranged monotonically into ascending order according to a single element (Flemings-Grungor method).

3. Experimental methodology development

3.1. Sampling effect
The X-ray beam of the µXRF device used in this study has a 20 µm width with a fixed focal point. With µXRF the sampling is not homogenous in both directions. The surface is scanned continuously in the *x*-direction, a single spectrum is obtained by collecting X-rays over a given time, as the stage translates at a constant speed this corresponds to a specific length, called the analysis step. In the *y*-direction, the scanning is discrete since the lines are spaced by a distance equal to the analysis step in order to maintain square pixels. A certain area between the lines, corresponding to the difference between the analysis step and the beam width (20 µm) is not analysed, as shown in the schematic in figure 1. The only way to map the whole surface is to use an analysis step equal to the beam width, 20 µm, in which case, there is no space between the analysed lines.

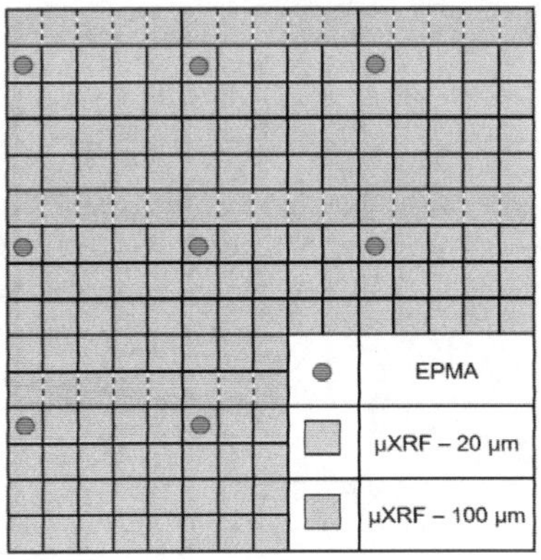

Figure 1. Schematic of the surface scanning during µXRF analysis with different analysis steps. When the analysis step is 100 µm, a band of 80 µm is not analysed between each analysed line of 20 µm width. The EPMA grid discussed in paragraph 3.2 is also represented.

To investigate the effect of the analysis step and time on segregation mapping, the same sample was mapped in similar experimental conditions but with different analysis steps sizes. The tests were performed on a 4.4 x 5.6 mm² zone. The stage speed was kept constant for each test, the primary X-ray source was unfiltered with a 50 keV acceleration voltage and 200 µA current, only the analysis step and the associated analysis time changed. Five different configurations were tested (notation: analysis step_analysis time): 20µm_20ms, 50µm_50ms, 100µm_100ms, 150µm_150ms and 200µm_200ms.

Three representative maps are presented in figure 2. The visual aspects of the maps are affected by the sampling. With a 20 µm step, the surface is completely mapped, the edges of the structures are sharp, small microstructure details are detected. As the acquisition step increases, the structures seem to spread out and their shape can be distorted. With acquisition steps bigger than 100 µm the shape of the structure is no longer representative, as such, acquisition steps smaller or equal to 100 µm should be favoured.

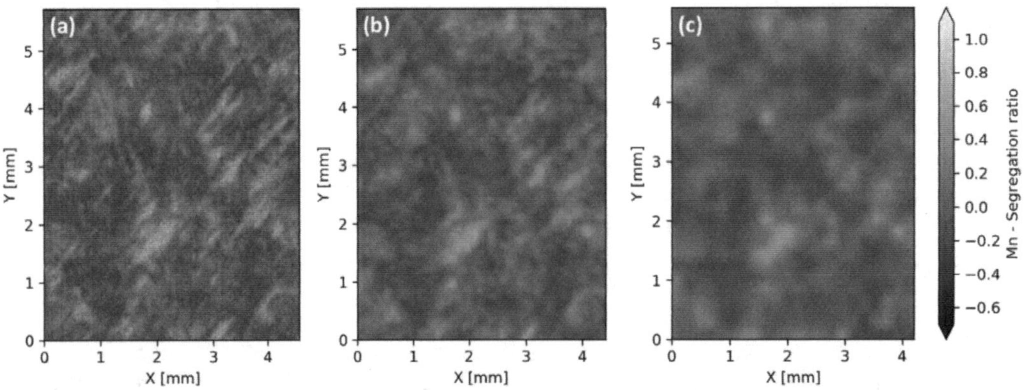

Figure 2. Maps of Mn segregation ratio of a sample 4.4 x 5.6 mm² analysed with µXRF (X-ray beam: unfiltered, 50 keV, 200 µA) with different analysis step and time ((a) 20µm_20ms, (b) 100µm_100ms (c) 200µm_200ms).

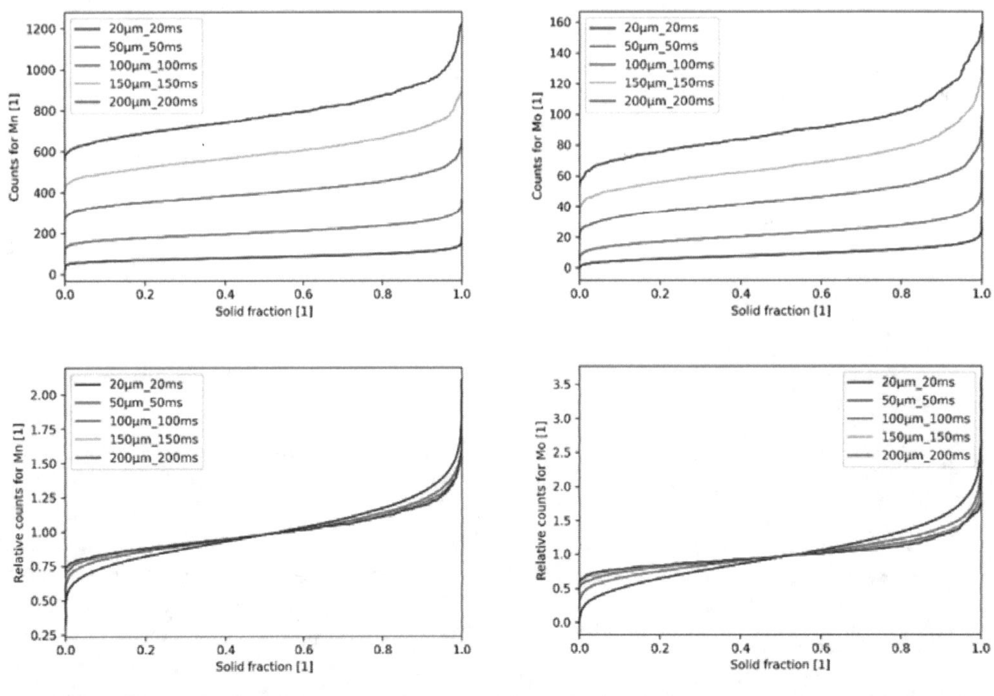

Figure 3. Classic (top) and relative (bottom) cumulative solid fraction curves (peak intensity as a function of the solid fraction) for µXRF data collection on a sample 4.4 x 5.6 mm², in the same experimental conditions (unfiltered, 50 keV, 200 µA) but with different analysis step and time (20µm_20ms, 50µm_50ms, 100µm_100ms, 150µm_150ms and 200µm_200ms). On the left is Mn peak intensity, on the right is Mo peak intensity.

The cumulative solid fraction curves are presented for Mn and Mo in figure 3. The collected peak intensity increases with the analysis step for both elements. It is linked to an increase in acquisition time: the longer the detector collects the fluorescent X-rays, the greater the peak intensity. The intensity of the peaks corresponds to the acquisition time, but the curve shape is linked to the acquisition step. To isolate the influence of the acquisition step relative cumulative solid fraction curves are used. With a step of 20 μm, the whole surface is analysed whereas in the other cases it is not. The curves associated with analysis steps of 100 μm, 150 μm, and 200 μm are very similar. Increasing the analysis step does not seem to decrease the quality of the information needed for solid fraction curves. The benefit of small analysis steps such as 100 μm is questionable: it is time-consuming, and a 200 μm step can provide the same information. Using a small step can however be justified if mapping is of interest rather than cumulative segregation curves.

3.2. Quantification

μXRF analysis can provide chemical composition mapping at a centimetric scale. However, the initial map is expressed in counts, which correspond to the intensity of the peak associated with the element, and not its concentration itself. EPMA analysis, on the other hand, can provide quantified local concentrations, but cannot be used at a large scale. By combining these techniques, the μXRF centimetric scale mapping can be quantified.

To quantify the μXRF analysis, a 400-point grid of EPMA spot analyses were conducted on 3 different samples. A step size of 100 μm was used, hence the surface analysed was 1.9 x 1.9 mm². The same samples were then analysed with μXRF using the same experimental conditions as the centimetric mapping to be quantified. EPMA and μXRF data points were ranked in ascending order, then assuming that the greatest peak intensity corresponds to the greatest concentration and vice versa, EPMA data was plotted against μXRF data and a linear regression was performed on the curve. The regression equation can be applied to convert the peak intensities to concentrations.

3.3. Quantitative centimetric scale mapping

Based on previous observations, experimental conditions were defined to perform centimetric scale mapping. The analysed surface was 113 x 98 mm², the acquisition steps were 100 μm in both directions (resulting in 1 109 511 analysis points) and the analysis time was 45 ms per spectra. The X-ray beam operated at 50 keV and 200 μA, and a 100 μm thick Al filter was used to reduce dead time to 21 %. Once the data was collected, the quantification method was applied and quantitative centimetric scale maps obtained. The quantified map for Mn is presented in figure 4.

Mn is rejected into the liquid phase during solidification. Dendrite trunks and secondary arms can be distinguished on the quantified map. At the centimetric scale, the structures do not appear distorted confirming that our developed analysis protocol is suitable. The microsegregation is visible on the centimetric scale map. At a larger scale, two round spots of a few millimetres can be observed near the centre and at the bottom left corner of the map on both maps, which may be channel segregates. Also, the interdendritic spaces on the top right side of the map seem richer in Mn than in the centre of the map.

4. Conclusion

The understanding of mesosegregation formation requires chemical analysis at a centimetric scale. The objective of this work was the development of a new sampling and characterisation methodology that relies on μXRF and allows centimetric scale mapping. First, the effect of sampling on the observed structures was investigated to select suitable experimental conditions for centimetric scale mapping. As the surface scanning with μXRF techniques is anisotropic, a small acquisition step is employed to avoid shape distortion of the structures (it should be smaller or equal to 100 μm). The chemical composition of a 113 x 98 mm² sample was mapped with a 100 μm analysis step with the μXRF technique and quantified by calibration with EPMA. The chemical composition mapping reveals microsegregations and chemical heterogeneities at a larger scale. A geostatistical approach should be used on centimetric scale mappings to investigate mesosegregations patterns.

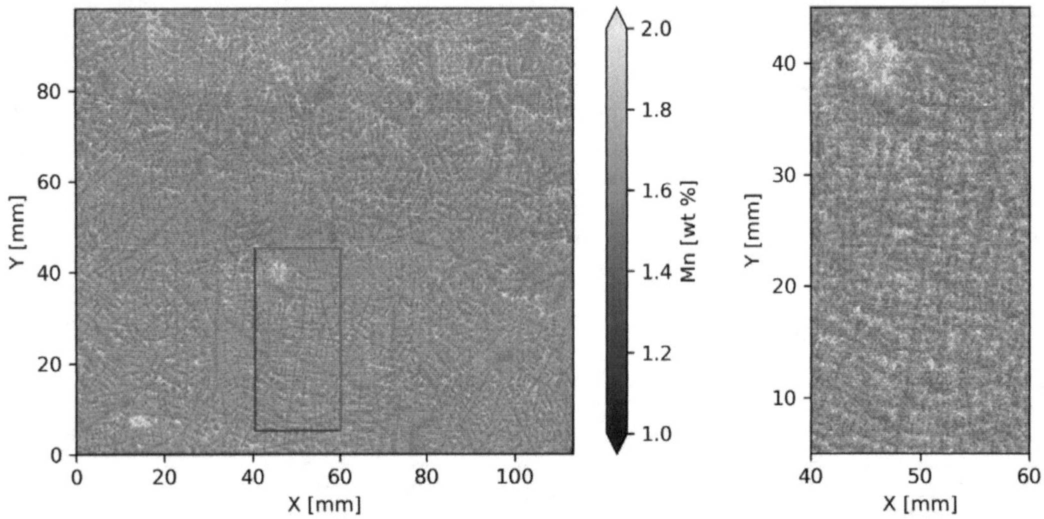

Figure 4. Map of Mn realised on a 113 x 98 mm² sample analysed with µXRF (X-ray beam: Al-100 µm, 50 keV, 200 µA), with a 100 µm analysis step and 45 ms of acquisition time per spectra. An enlarged view of the area delimited in red is shown on the right.

Acknowledgments
This research work is supported by ANRT France under a CIFRE Ph.D. fellowship (Grant Number 2020/1234) and through the Industrial chair *Solidification* at Institut Jean Lamour, supported by ArcelorMittal, ArcelorMittal Industeel, CEA, EDF and Framatome.

References

[1] Lesoult G 1986 *Techniques de l'ingénieur. Matériaux métalliques.* **M58** 1
[2] Dantzig J A and Rappaz M 2009 *Solidification.* (Lausanne: EPFL Press) p 394
[3] Pickering E J 2013 *ISIJ international.* **53** 935–49
[4] Ludwig A, Wu M and Kharicha A 2015 *Metall. Mater. Trans. A* **46** 4854–67
[5] Thornthon P A and Colangelo V J 1976 *Metall. Mater. Trans. B.* **7** 425–33
[6] Pickering E J and Bhadeshia H K D H 2014 *J. Press. Vessel Technol. Trans. ASME.* **136**
[7] Salimi A, Zadeh H M, Toroghinejad M R, Asefi D and Ansaripour A 1996 *Materiali in tehnologije.* **47** 385
[8] Hong S, Song, J, Kim MC, Choi KJ and Lee BS 2015 *Met. Mater. Int.* **22** 196–203
[9] Yamashita T, Torizuka S and Nagai K 2003 *ISIJ international.* **43** 1833–41
[10] Pickering E J and Holland M 2014 *Ironmaking & Steelmaking.* **41** 493–9
[11] Sengupta J, Leung J and Noorafkan A 2017 *AISTech 2017 Proceedings.* (Warrendale: AIST) p 1925
[12] Sengupta J and Noorafkan A 2018 *AISTech 2018 Proceedings.* (Warrendale: AIST) p 2637
[13] Ganesan M, Dye D and Lee P D 2005 *Metall. Mater. Trans. A.* **36** 2191–204

Macrosegregation maps in industrial continuous casting low alloyed steel ingots

M Gennesson[1,3]**, S Evrard**[1] **and Y Amhoud**[2]

[1] Ascometal CREAS, Hagondange, France
[2] Master of advanced metallurgy, Nancy, France
[3] corresponding author: marvin.gennesson@ascometal.com

Abstract. Macrosegregation is a critical defect in industrial steel ingots in both conventional ingot castings and continuous castings. Improvements in terms of macrosegregation require to be able to precisely quantify the intensity and the repartition of macrosegregation along the cast product. However, traditional macrosegregation measurements are a tedious and time consuming process. In this study, three different product lengths of the same caster strand were taken during the same heat. Each sample was cut along the longitudinal central plane of the cast product. One additional transverse section was also cut for one of the sample. Drilling with constant spacing was performed on 50% of the area of longitudinal sections and 25% of the transverse section. Carbon and sulfur contents were then measured on each drilled location. Plotting methodologies to produce chemical maps are carefully described regarding the different parameters such as color map choices and normalization methods. Data for all the characterized sections is compared with each other and the differences are commented with the help of the process parameters.

1. Introduction

Modern steel making always requires cleaner and more homogeneous products in terms of chemical segregations. Chemical segregations at the product scale cannot be decreased once the product is solid and can induce anisotropy and heterogeneities in mechanical properties after rolling. This kind of segregations, i.e. macrosegregations, arises because of the relative motion between the moving solid grains and the thermosolutally convected liquid phase. [1]

Macrosegregation characterization is a key step to be able to assess the effect of any process change. Large distances at stake (at the cm scale) in those phenomenon prevent from using well known techniques for microsegregation characterization such as electron microprobe analysis (EMPA) or electron back-scattered diffraction within a scanning electron microscope (SEM/EDS).

Most of the literature regarding steelmaking have been focused on 1D linear profiles (with emphasis on product central axis) for both conventional ingot casting, e.g. [2-5] and continuous cast products, e.g. [6-8]. This kind of assessment is very popular because, it gives a way to easily create numerical benchmark cases for validation, e.g. [9]. Nevertheless, such data is limited as it seems dangerous to extrapolate the solute repartition within the whole product.

To give a broader picture of the repartition of macrosegregation, few studies have extended the 1D characterization method to 2D. Old studies usually give point values [10] and/or interpolated iso-concentration lines, e.g. [11]. Modern plotting methods allow to color code each point. Element concentration can then directly be read on a color scale. Such 2D maps were only produced so far in the literature for conventional ingot castings but not for continuously cast products.

As said before color map plotting is easily done but each study has its own set of given parameter. Point locations as well as plotting parameters are rarely discussed and few details can be found in the corresponding studies. Maps can also be provided without [12-14] or with interpolation between points [15]. Furthermore, values can also be given raw as weight percent of element [13, 16] or in relative variation with respect to the first composition of the liquid e.g. [9]. In this last option, one can directly see if the composition is below (negative macrosegregation) or above (positive macrosegregation) the initial composition of the liquid. This initial liquid composition usually refers to the composition which has been measured in the ladle during the steelmaking process [9, 16]. Recently, Falkus et al. [7] have discussed the best reference composition to choose for relative variation calculation. They proposed to directly calculate it on the existing measuring points. They do not take directly the measured average composition but correct it with the mass of metal that each point accounts for with symmetry relationships. This seems particularly important for ingot casting where the teeming powder may be a carbon source but also in continuous casting where the composition of the ladle may change with time.

Regarding chemical concentration measurements, most of the previously quoted studies used drilling (or sometimes milling, punching…) at chosen locations on cut sections of the steel product. Carbon content materials was then measured thanks to spectroscopy for each drilling. Spectroscopy is usually combustion infrared carbon-sulfur analyser (LECO) but other techniques such as optical emission spectroscopy (OES) (e.g. [5, 8, 13, 17]) or even titration [13] can also be found. Such techniques are tedious and require a lot manual intervention which generally prevent from characterizing large number of points. Nevertheless, some studies [13, 16] clearly demonstrate that extensive data (above 1000 points) can be acquired through those kind of protocols. It has to be noted that Pickering et al. [18] presented an automated technique based on the use of X-ray fluorescence (XRF) combined with a robotized arm. This technique gives a very high number of characterized points (up to 71000 points) but the entry cost of the needed robotized arm may be too high for existing industrial actors.

In the present study, a 2D characterization method will be given with emphasis on the details about the plotting methodology. The produced 2D macrosegregation data for continuous casting products will also be shortly presented.

2. Materials and methods

2.1. Steel grade and process

A low alloyed steel was cast by continuous casting. One single ladle was used to realize this heat. The measured composition in the ladle during casting is given below in Table 1. The continuous caster machine is made of several vertical strands with each strand being of square section (bloom). An electromagnetic stirring device is positioned at the end of the CC mold. Its intensity is set constant during the whole heat.

Table 1. Composition of steel grade (in weight percent).

C	Cr	Mn	Mo	S	Fe
0.27	1.0	1.3	0.09	0.08	balance

Three samples were taken from one of the internal strands corresponding to the start, middle and end of casting. Their respective positions with respect to the last cast bloom (16th) are: 1/16, 7/16 and 12/16.

2.2. Characterization method

The following steps were applied:

- One longitudinal slice containing the product central axis was cut for each sample. For the mid casting sample, one additional slice was taken transversally.

- Dry drilling was performed on each slice on the side containing the product central axis. The drilling positions were regularly spaced on the characterized surfaces. The drill bit was 0.7 cm in diameter. Additionally, the mid-casting sample was also drilled on the other side but only on half the length of the product. The transverse slice was only drilled on 25% of the surface supposing that square symmetry is respected. On the central axis of each longitudinal slice, the distance between the points was divided by 2. It was also the case on the transverse section axis. A total of roughly 700 points (central axis and body) was analysed during the whole study.
- Drilled material (roughly 1 g per grid point) was packed in labelled plastic bags. Half of the content of each bag was then analysed (roughly 0.5 g taken after shaking plastic bags to ensure random sampling) with a calibrated LECO CS 744 analyser to obtain carbon/sulfur contents. Measurements were repeated only for points where the composition was very different from other nearest neighbour points.

The dimension of the grid and the diameter of the drill bit were chosen for a good compromise between analysis time and the desired level of details in the final map. As an order of magnitude, the typical grain size in the equiaxed area is way below the drilled hole diameter. Each drilled material is thus believed to be representative of the chemical composition at this scale.

2.3. Map plotting

From the pure plotting perspective, Python was used to take advantage of its advanced library. It allows to easily plot 3D data (2D positions plus chemical concentrations) as a 2D plot with a color scale. The chosen color scale was Spectral_r discretized to 15 colors and numerically centered on 0. Those parameters were selected thanks to a parametric study.

Linear interpolation was carried in between measured points. The interpolation grid was also a square grid. In between two nearest drilled holes, there are roughly 20 interpolated points. There was no extrapolation outside of data points. In other words, concentration values outside the convex Hull of drilled holes were set as null.

Following literature guidelines, it was chosen to plot the relative difference between the measured composition and the initial composition of the liquid. Similarly to [7], it was decided to calculate an average value directly based on the measured data points. It is corrected with the weight of all the points that should have the same composition in the final product if the symmetry relationships are correct.

The volume of metal, V, that should be at the same composition within one bloom starting from a generic drilled location (x, y) on a longitudinal section can be written as:

$$V(x, y) = (2x + \Delta x) * \Delta x^2 \quad (1)$$

with Δx, being the interspacing between two drilled holes.

It follows that on the axis of the cast product the volume associated with each point would only be the cubic volume Δx^3. Anywhere else, it would turn into the volume of a torus of square section with side Δx. The average composition within the metal, C_0 was then calculated as a weighted average:

$$C_0 = \frac{\sum_{(x,y)} C(x,y) * (2x+\Delta x) * \Delta x^2}{\sum_{(x,y)} (2x+\Delta x) * \Delta x^2} = \frac{\sum_{(x,y)} C(x,y) * (2x+\Delta x)}{\sum_{(x,y)} (2x+\Delta x)} \quad (2)$$

with $C(x, y)$, being the measured concentration on the drilled material.

Equation (2) is only valid when applied to longitudinal sections. On transverse sections, the product is of infinite length in the cast direction. As only 25% of the section was studied and supposing square symmetry, each point should have a weight of 1 except at the center of the section and on the axis of the transverse section (which must not be confused with the cast product axis) where it should be 0.25 and 0.5 respectively.

The macrosegregation intensity, I, at a location (x, y) can then be defined as:

$$I(x,y) = \frac{C(x,y)-C_0}{C_0} \tag{3}$$

A short parametric study showed that best readability of macrosegregation maps was attained when the color map was only limited to macrosegregation intensities in between -0.15 and +0.15. Everything outside this interval would then be colored with the bottom/top color of the color scale.

3. Results and discussion

3.1. Calculated average composition

Normalized calculated average compositions following the description of 2.3 are given in Table 2. All values are normalized with respect to the ladle composition measured during the casting.

Table 2. Normalized calculated average C_0 values

Casting time	Start	Middle	Middle	Middle	End
Cutting plane	Longitudinal	Longitudinal	Longitudinal off-center	Transverse	Longitudinal
C_0	1.000	0.992	0.982	0.992	1.02

It is straightforward to see that average carbon value is varying along with casting time. One explanation could be that casting parameters such as superheat and casting speed can be subject to small variations within the casting process. Such differences could then affect the depth of the liquid pool. It could even modify the secondary arm spacing of the columnar dendritic network and the liquid circulation within the mushy zone, [19, 20]. In another direction, it could also be the sign that the liquid pool composition is changing during casting. Possible closing of liquid pool known as mini-ingot formation could happen. This phenomenon is already well known to affect the central segregation. [21, 22]

Another result from Table 2 is that even at approximately the same casting time (middle in that case), different average values can be found. The off-center plane gives a slightly lower value than the axial longitudinal plan and the transverse section. It shows that small variations can exist at a lower distance scale (and so casting time scale).

In the following results, a common value – equal to the mean of the raw value C_0 line in Table 2 – will be taken as C_0 value to be able to compare the plotted macrosegregation maps. This fact should be kept in mind when discussing each map individually.

3.2. Macrosegregation characterization on different planes at middle casting time

As said earlier, the middle casting time sample can be used to assess if the symmetry hypothesis are satisfactory. Maps for carbon macrosegregation intensity, I_C, for all the different sections of that sample are given in Figure 1.

Actual drilled hole centers are represented as over-imposed red dots. Thanks to the calculation of relative difference between C_0 and the measured composition at drilled location coupled to the use of a zero centered color map, it is worth noting that blue colored areas correspond to negative macrosegregation whereas red areas are devoted to positive macrosegregation.

Starting from the external side, a negatively segregated band parallel (to the product side) is followed by a positive band and again a negative band just before the product center. Those bands are not perfect and several variations in terms of macrosegregation intensity exist within them. It is of interest noting that the columnar-to-equiaxed transition typically happens shortly before reaching mid-radius in this type of product. Finally, it can be seen that in this very specific case, there is no real trend on the product axis. Macrosegregation intensity takes both negative and positive values.

By comparing the first longitudinal map (Figure 1(a)) to the off-center longitudinal map (Figure 1(b)), it can be seen that there is a good correspondence between the pattern shapes of macrosegregation intensity. Nevertheless, there is a difference in composition particularly for the positively segregated band. It is less intense on the off-center section.

IOP Conf. Series: Materials Science and Engineering 1274 (2023) 012050 doi:10.1088/1757-899X/1274/1/012050

The two longitudinal slices can be positioned relatively to the transverse section. The longitudinal slice cut one of the horizontal axis of the transverse slice whereas the longitudinal off-center slice is one of the two first drilled lines adjacent to one of the axis of the transverse slice.

On the transverse slice, it can be seen that macrosegregation symmetry is satisfactory. The small observed variations were expected regarding the intrinsic complex and dynamic nature of continuous casting.

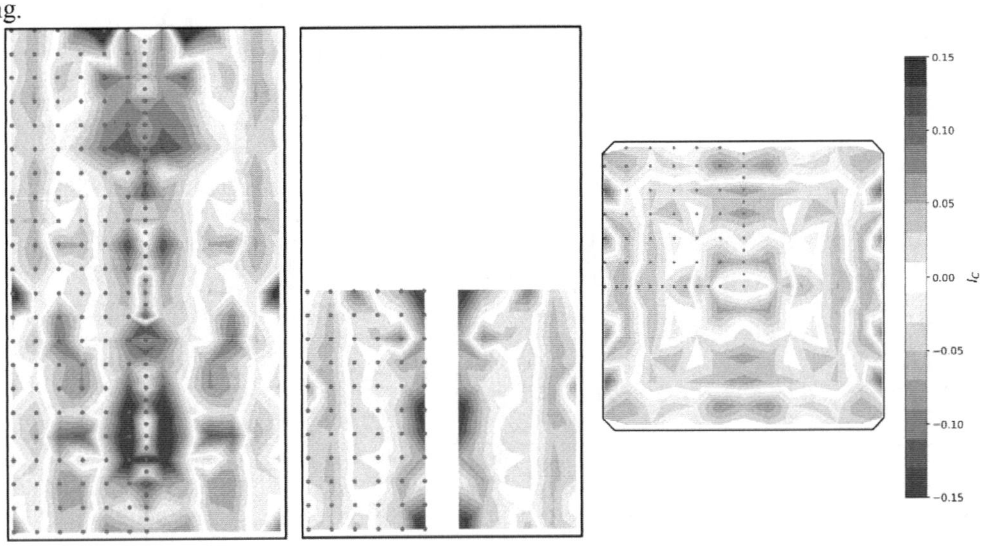

(a) (b) (c)

Figure 1. Carbon macrosegregation intensity maps for sample at mid casting time: longitudinal (a), longitudinal off-center (b) and transverse (c) planes

3.3. Macrosegregation maps at different casting times

Maps at different casting times are provided in Figure 2. It can be observed that the sampling time is of importance when measuring macrosegregation. The severity of segregation in the positive band, in the axis and in between them is changing between the three maps. Extreme points (either of negative or positive macrosegregation) are also occurring on the three samples. It can happen on the side of the product or closer from the product axis.

As mentioned earlier, the variation of conditions during casting can explain the different segregation maps observed within the presented samples. During casting, the metal temperature is decreasing in the ladle. In the meantime, the casting speed is also slightly modified. The effect of those two parameters is already well known in the literature. The smaller overheat and casting speed are, the less intense axial segregation should be, [19-22]. It is worth noting that casting speed gradually increases from Figure 2(a) to Figure 2(c). This change could then explain why the segregation is more intense on the axis of the end casting time sample. This last assertion should be verified with the help of the available numerical models.

4. Conclusion

Macrosegregation map plotting methodology was carefully described. In future work, the same methodology could be applied on more complex shapes such as conventional ingots. This methodology would also be very useful when the composition of the liquid is different from the initial liquid composition (for example, when there are several ladles or when there is carbon enrichment due to the teeming flux).

Different maps were obtained on different planes at the same casting time. It proves that segregation repartitions are respecting the different expected symmetry relationships within the product both longitudinally and transversally.

Other maps at different casting times confirms that there seems to be a close link between casting conditions and observed segregations. Numerical simulations with different conditions (overheat and casting speed) could be used to verify this link.

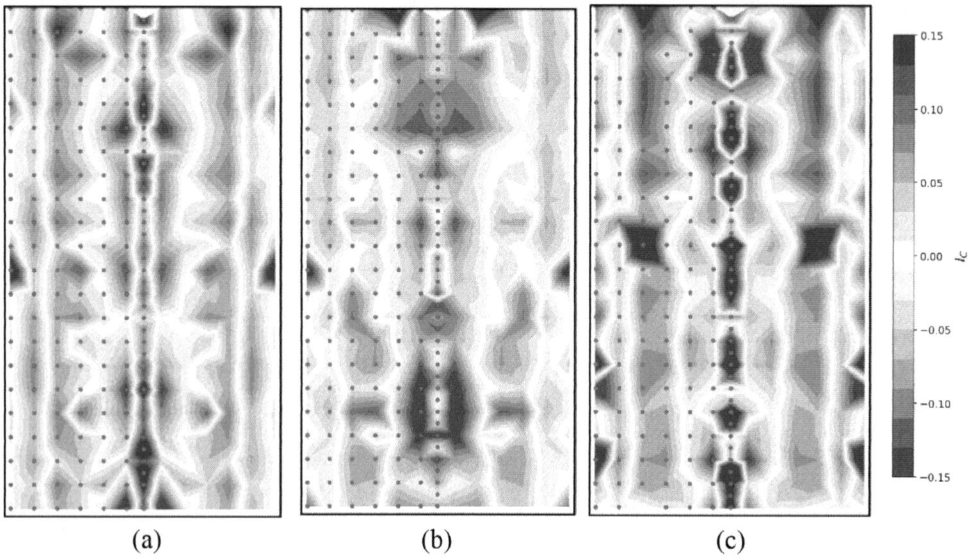

Figure 2. Carbon macrosegregation intensity maps for different casting times: start (a), middle (b) and end (c)

5. References

[1] Rappaz M and Dantzig J 2009 *Solidification* (Lausanne: EPFL Press)
[2] Comon J, Delorme J and Bastien P 1973 *Rev. Metall.* **70** 4 251-8
[3] Kajikawa K, Suzuki S, Takahashi F, Yamamoto S, Suzuki T, Ueda S, Shibata T and Yoshida H 2012 *1st Int. Conf. on Ingot Casting Rolling and Forging (Aachen)*
[4] Chen Z, Senk D and Firsbach F 2018 *Metall. Mater. Trans.* B **49** 5 2514-32
[5] Zhang C, Jahazi M and Tremblay R 2020 *Int. J. Adv. Manuf. Tech.* **107** 1 167-75
[6] Guo D, Hou Z, Cao J, Guo Z, Chang Y and Wen G 2020 *J. Iron Steel Res. Int.* **27** 10 1163-69
[7] Falkus J, Mickowska-Piszczek K, Krajewski P and Ropka T 2020 *Metals* **10** 4 499
[8] Pikkarainen T, Vuorenmaa V, Rentola I, Leinonen M and Porter D 2016 *4th Int. Conf. on Advances in Solidification Processes (Windsor)* 117 (Bristol: IOP Publishing) 1 12064
[9] Kumar A, Založnik M and Combeau H 2010 *Int. J. Adv. Eng. Sci. Appl. Math.* **2** 4 140-8
[10] Olsson A, West R and Fredriksson H 1986 *Scand. J. Metall.* **15** 2 104-12
[11] Maidorn C and Blind D 1985 *Nucl. Eng. Des.* **84** 2 285-96
[12] Tanzer R, Schützenhöfer W, Reiter G, Fauland H-P, Könözsy L, Ishmurzin A, Wu M and Ludwig A 2009 *Metall. Mater. Trans.* B 40 3 305-11
[13] Machovcak P, Opler A, Tkadleckova M, Michalek K, Gryc K, Krutis V and Kovac M 2012 *1st Int. Conf. on Ingot Casting Rolling and Forging (Aachen)*
[14] Loucif A, Shahriari D, Zhang C, Jahazi M, Lapierre-Boire L-P and Tremblay R 2017 *Mater Sci Forum* **879** 1176-81
[15] Wang T, Wang E, Delannoy Y, Fautrelle Y and Budenkova O 2021 *Metals* **11** 2 262
[16] Duan Z, Tu W, Shen B, Shen H and Liu B 2016 *Metall. Mater. Trans.* A **47** 7 3597-606
[17] Chen Z, Schweinichen PV, Senk D and Lob A 2012 *1st Int. Conf. on Ingot Casting Rolling and Forging (Aachen)*
[18] Pickering EJ, Chesman C, Al-Bermani S, Holland M, Davies P and Talamantes-Silva J 2015 *Metall. Mater. Trans.* B **46** 4 1860-74
[19] Gao X, Yang S and Li J 2016 *Mat. Design* **110** 284-95
[20] Zou L, Zhang J, Liu Q, Zeng F, Chen J and Guan M 2019 *Metals* **9** 12 1312
[21] Saleil J and Le Coze J 2018 *Mat. Tech.* **106** 505

[22] Moore JJ 1980 *Iron Steel.* **7** 10 8-16

Simulation of macrosegregation in continuous casting of steel based on large-eddy simulation turbulence model and meshless solution procedure

K Mramor[1], R Vertnik[1,2] and B Šarler[1,3]

[1] Department of Fluid Dynamics and Thermodynamics, Faculty of Mechanical Engineering, University of Ljubljana, Aškerčeva 6, Ljubljana, SI-1000, Slovenia
[2] Štore Steel d.o.o., Železarska cesta 3, SI-3220 Štore, Slovenia
[3] Laboratory for Simulation of Materials and Processes, Institute of Metals and Technology, Lepi pot 11, SI-1000 Ljubljana, Slovenia

Abstract. The formation of macrosegregation pattern in steel is explored for the continuous casting process. The related solidification problem is described by incompressible turbulent fluid flow, governed by mass, momentum, energy, and species conservation equations. The solid-liquid system is designed in a mixture continuum model formulation, and the mushy region is approximated with the Darcy model. The lever rule describes the formation of microsegregation of carbon. Turbulence is modelled with a large-eddy simulation approach implemented with the Smagorinsky model and van Driest correction. The synthetic isotropic fluctuations of the inlet velocity are generated by the von Karman Pao spectrum. The solution of partial differential equations is implemented locally by using collocation with radial basis functions for spatial discretization and the explicit Euler method for time discretization. The two-dimensional large eddy simulation model solutions are compared to the previously calculated examples obtained with the low-Re $k-\varepsilon$ turbulence model.

1. Introduction

One of the most frequently used process in steel production is the continuous casting (CC) process. The molten steel is poured into the mould with relatively high velocities resulting in a chaotic and random fluid motion below the submerged entry nozzle (SEN). A consequent turbulent motion can significantly affect the final macrosegregation and thus influence the product's quality. It is therefore essential to accurately model the turbulent flow.

Turbulence models are, according to the extent of modelling, divided into three groups: the direct numerical simulation (DNS) where the whole energy cascade is calculated, the large eddy simulation (LES) models where the larger eddies are directly calculated and the smaller ones are modelled, and the Reynolds averaged Navier Stokes (RANS) equations where the whole energy cascade is modelled. This article focuses on the LES models as they provide a reasonably good transient solution.

The continuous casting process, including the simulation of macrosegregation [1–3], has been previously modelled with various numerical models, ranging from the classical finite volume method (FVM) [4–8] to more advanced meshless methods [9,10]. In this article, the LES turbulence model and macrosegregation of carbon is for the first time solved by the local radial basis function collocation method (LRBFCM) [11].

The paper compares the new results with the reference results computed with Low-Re $k-\varepsilon$ turbulence model [10].

2. Numerical model

The physical model represents a solidification of steel in a simplified 2D CC system, posed in the Cartesian coordinate system. This concept demonstration model, in which the densities of the liquid and solid steel are constant and equal $\rho = \rho_S = \rho_L$, is formulated within the mixture continuum framework [12]. The corresponding mass conservation equation is defined as

$$\nabla \cdot \boldsymbol{u} = 0; \qquad \boldsymbol{u} = f_S \boldsymbol{u}_S + f_L \boldsymbol{u}_L, \tag{1}$$

where mixture velocity $\boldsymbol{u}$ is a linear combination of phase velocities $\boldsymbol{u}_\ell$ and volume fractions f_ℓ with ℓ standing for $\ell = \mathrm{L}$ liquid phase and $\ell = \mathrm{S}$ solid phase. The macroscopic momentum conservation equation is given as

$$\frac{\partial(\rho \boldsymbol{u})}{\partial t} + \nabla \cdot (\rho \boldsymbol{u}\boldsymbol{u}) = -\nabla p + \nabla \cdot [(\mu_L + \mu_t)(\nabla \boldsymbol{u} + (\nabla \boldsymbol{u})^T)] - \boldsymbol{F}_D + \boldsymbol{F}_B, \tag{2}$$

where t is time, p is pressure, μ_L is dynamic viscosity, μ_t is turbulent viscosity, $\boldsymbol{F}_D = \mu_L[K_0(1-f_L)^2 / f_L^3](\boldsymbol{u} - \boldsymbol{u}_S)$ is Darcy term that describes flow through porous media with permeability K_0 and $\boldsymbol{F}_B = \rho \boldsymbol{g}[\beta_T(T - T_{ref}) + \beta_C(C_L - C_{ref})]$ is Boussinesq approximation used to describe the buoyancy of molten steel where $\boldsymbol{g}$ is gravitational acceleration, β_T is thermal expansion coefficient, T is temperature, T_{ref} is reference temperature, β_C is solute expansion coefficient, C is the concentration of alloying element, and C_{ref} is the nominal concentration of the alloying element. The macroscopic equation describing energy is written as

$$\frac{\partial(\rho h)}{\partial t} + \nabla \cdot (\rho \boldsymbol{u} h) = \nabla \cdot (\lambda \nabla T) + \nabla \cdot [\rho f_S (h_L - h_S)(\boldsymbol{u} - \boldsymbol{u}_S)] + \nabla \cdot (f_L \frac{\mu_t}{\mathrm{Pr}_t} \nabla h_L), \tag{3}$$

where the mass-specific mixture enthalpy h is a linear combination of the mass-specific phase enthalpies h_ℓ and volume fractions f_ℓ, λ is thermal conductivity, and Pr_t is turbulent Prandtl number. The temperature-enthalpy relation is precisely the same as in [13]. The macroscopic equation for species conservation is described as

$$\frac{\partial(\rho C)}{\partial t} + \nabla \cdot (\rho C \boldsymbol{u}) = \nabla \cdot (D \nabla C) + \nabla \cdot [\rho (C_L - C)(\boldsymbol{u} - \boldsymbol{u}_S)] + \nabla \cdot (f_L \frac{\mu_t}{\sigma_C} \nabla C_L), \tag{4}$$

where the mixture concentration of alloying element C is a linear combination of phase concentrations C_ℓ and volume fractions f_ℓ, D is diffusivity, and σ_C is closure coefficient set to 1.0.

The liquid fraction, liquidus temperature, and concentration of solute in the liquid phase are coupled through the lever rule microsegregation model

$$f_L = 1 - \frac{1}{1-k_p} \frac{T - T_L}{T - T_m}; \qquad C_L = \frac{C}{1 + (1 - f_L)(k_p - 1)}; \qquad T_L = T_m + (T_e - T_m)\frac{C}{C_e} \tag{5}$$

with partition coefficient $k_p = C_S / C_L$, melting T_m and eutectic T_e temperatures, and eutectic concentration C_e, determined from the iron-carbon phase diagram.

The turbulent viscosity is modelled by the Smagorinsky model [14] and is presented as a sub-grid scale viscosity

$$\mu_t = \rho L_S^2 (2 S_{ij} S_{ij})^{1/2}; \qquad L_S = \min(\kappa_{vK} d,\ f_\mu C_{SL} \Delta); \qquad S_{ij} = 1/2(\partial u_j / \partial x_i + \partial u_i / \partial x_j) \tag{6}$$

with von Karman constant $\kappa_{vK} = 0.4$, distance to the nearest wall d, van Driest damping function $f_\mu = 1 - \exp(y^+ / A^+)$ [15,16], Smagorinsky-Lilly constant $C_{SL} = 0.168$, and filter width $\Delta = V^{1/2}$. y^+ is non-dimensional wall distance and $A^+ = 26$.

3. Boundary conditions

The initial and boundary conditions determine the solution to the problem. Four different boundary types are set in the present case: inlet, free surface, moving wall, and outlet. The Dirichlet boundary conditions are prescribed for temperature $T = T_{\text{cast}}$, velocity $\boldsymbol{u}(x, y) = (0 \text{ m/s}, u_{\text{cast}}\, a_{\text{b}}/d_2)$, and concentration $C = C_{\text{ref}}$ at the inlet. The casting velocity is scaled with a factor associated with SEN a_{b} and slab width d_2 to maintain a constant flux. The boundary condition for the velocity is further adjusted by generating the isotropic fluctuations. A detailed description of the procedure is given in [17,18]. Pressure at the inlet is calculated from Neumann boundary condition $\partial p / \partial \mathbf{n} = 0\,\text{Pa/m}$. The boundary conditions at the free surface are defined as

$$\partial u_x / \partial \boldsymbol{n} = 0\ \text{s}^{-1}; \quad u_y = 0\ \text{s}^{-1}; \quad \partial T / \partial \boldsymbol{n} = 0\ \text{K/m}; \quad \partial C / \partial \boldsymbol{n} = 0\ \text{m}^{-1}; \quad \partial p / \partial \boldsymbol{n} = 0\ \text{Pa/m}. \quad (7)$$

The boundary conditions on the outlet are

$$\partial \boldsymbol{u} / \partial \boldsymbol{n} = 0\ \text{s}^{-1}; \quad \partial T / \partial \boldsymbol{n} = 0\ \text{K/m}; \quad \partial C / \partial \boldsymbol{n} = 0\ \text{m}^{-1}; \quad p = 0\ \text{Pa}. \quad (8)$$

The boundary conditions on the moving walls are

$$u_x = 0\ \text{m/s}; \quad u_y = u_{\text{cast}}; \quad q_{\text{cz}} = h_{\text{cz}}(T - T_{\text{cw}}); \quad \partial C / \partial \boldsymbol{n} = 0\ \text{m}^{-1}; \quad \partial p / \partial \boldsymbol{n} = 0\ \text{Pa/m}. \quad (9)$$

The heat transfer coefficient h_{cz} is set to 2000 W/(m^2 K) at the mould and to 800 W/(m^2 K) at the secondary cooling system and is used to calculate the heat flux q_{cz}. The temperature of the cooling water T_{cw} is set to 293 K.

4. Numerical method and solution procedure

4.1. Local radial basis function collocation method

The meshless method applied in this paper uses radial basis functions on a local sub-domain and apply collocation to calculate the approximation function and explicit Euler method for time advancement. The multiquadric radial basis functions with shape parameter set to 32 and augmented by a constant are applied in five-nodded subdomains. The pressure and velocity are coupled with the fractional step method. The numerical method is the same as described in [5]. The volume associated with the LES filter width is due to the two-dimensionality of the problem in a meshless context calculated as $V = d_{x/2} d_{y/2}$, where $d_{x/2} = \max(|x_i - x_j|)/2$ and $d_{y/2} = \max(|y_i - y_j|)/2$ are the maximum half distances between nodes in subdomain in each coordinate direction of the Cartesian coordinate system.

4.2. Solution procedure

The first step in the solution procedure is the calculation of wall distances and minimum and maximum wavenumbers, which serve as parameters for the generation of synthetic turbulence at the inlet. The time loop begins by generating the isotropic turbulence at the inlet. The solution procedure for generating synthetic isotropic turbulence at the inlet reproduces the procedure suggested by [18]. In the next step, the momentum equation is solved without the pressure gradient, and intermediate velocity is obtained. We then solve the pressure Poisson's equation to obtain pressure and correct the velocity. The solution of energy and species conservation equations follows by using the temperature-enthalpy relation and lever rule microsegregation model. Finally, the turbulent viscosity is calculated. The variables are updated, and if the solution is converged within a timestep, we move to the next time step, using the final values of the previous timestep as the initial values of the next timestep.

5. Numerical example

5.1. Geometry

The numerical example considered in the present article is a simplified illustration of a CC process. It is represented in simplified 2D geometry and consist of a vertical longitudinal slab that is 1.5 m heigh. The thickness of the computational domain is set to 0.18 m and accounts for the whole width of the slab. The SEN with a diameter of 0.035 m, is set at the top of the slab in such a way that the nozzle entry coincides with the free surface of the slab. The height of the mould is 0.8 m.

IOP Conf. Series: Materials Science and Engineering 1274 (2023) 012051 doi:10.1088/1757-899X/1274/1/012051

5.2. Material Properties

The simplified material properties ($\rho = 7200\,\text{kg/m}^3$, $\lambda = 30$ W/(m K), $c_p = 700$ J/(kg K), $\mu = 0.006$ kg/(m s), $\beta_T = 10^{-4}\,\text{K}^{-1}$, $\beta_C = 4.0 \cdot 10^{-3}$, $K_0 = 1.6 \cdot 10^{-8}$, $D_S = 1.6 \cdot 10^{-4}$ m²/s, $D_L = 10^{-8}$ m²/s, $T_{\text{cast}} = 1793$ K, $u_{\text{cast}} = 1.5\,\text{m/min}$, $T_S = 1680$ K, $T_L = 1760$ K) [19] are used in the present concept demonstration model for carbon steel with carbon content $C_{\text{ref}} = 0.51$ wt%.

5.3. Numerical implementation

The LRBFCM solution of the described physical model has been coded in modern Fortran and compiled with Intel(R) Visual Fortran Compiler version 19.1.1. for applications running on 64-bit Windows. The details of the numerical implementation can be found in [19].

6. Results

The velocity fields are compared for two different turbulence models. The effect of the turbulence model on velocity is shown in figures 1 and 2 where the time-averaged velocity magnitude field of the LES model is compared to the steady-state solution of the RANS type Low-Re $k-\varepsilon$ turbulence model. The time average was calculated for a period of 4000 s starting at t = 1000 s. The time-averaged LES results are being reasonably (but not entirely) symmetric (see figure 4). The RANS solution results in higher velocities in the centre of the slab and lower velocities at the mushy zone. The higher LES velocities in the mushy zone are due to the primary jet oscillations, as shown in figure 3 at t = 3500 s. The shorter average inlet jet is also a consequence of these fluctuations.

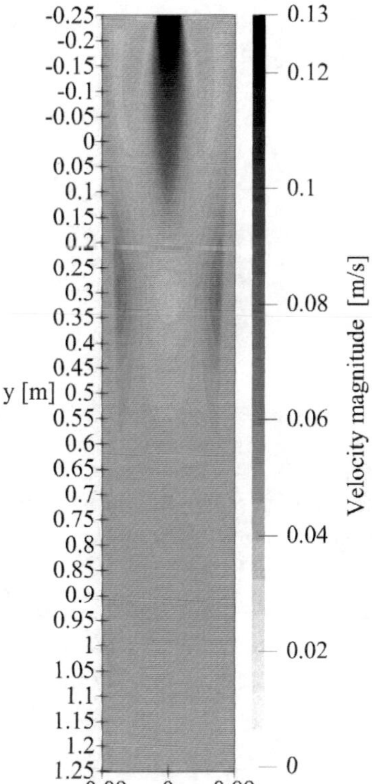

Figure 1. The steady-state velocity map of Low-Re $k-\varepsilon$ model and legend for figure 3.

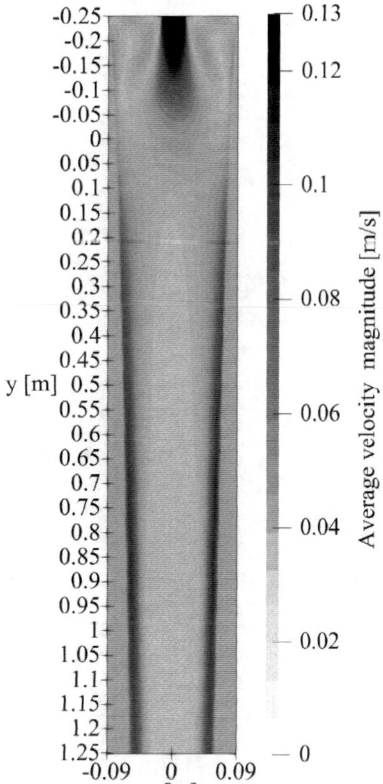

Figure 2. The time-averaged velocity map of LES model.

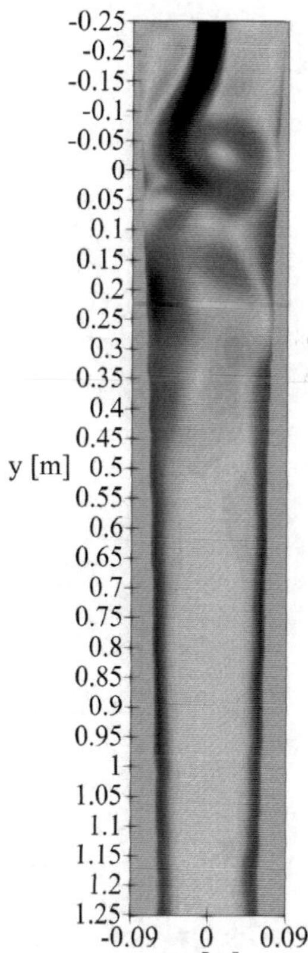

Figure 3. The velocity map for LES model at t = 3500 s.

Next, the cross-section at $y=1.25\,\text{m}$ is examined in figure 4. As seen in figures 2 and 3, the velocity magnitude modelled with LES results in higher values outside the solidified shell and lower values in the centre. Emerging from the SEN, the primary jet is whipping around the vertical direction with an approximate time of oscillations of 200 s.

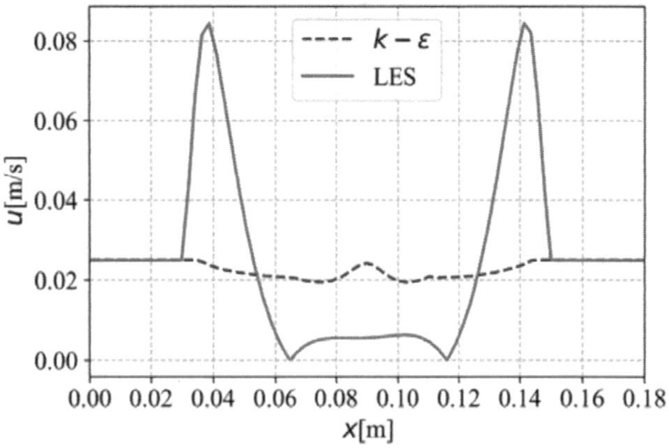

Figure 4. Comparison of velocity magnitude over the cross-section of the slab at height 1.25 m for Low-Re $k-\varepsilon$ and LES models.

The effect of the turbulence model on macrosegregation is shown in figure 5 where time-averaged concentration cross-section at $y=1.25\,\text{m}$ of LES model is compared to the steady-state solution of the RANS type Low-Re $k-\varepsilon$ turbulence model. The concentration in the centre of the slab is slightly higher than the nominal concentration C_0 and approximately the same for both turbulence models. The concentration in the solidified shell is lower. However, the depletion is more significant in the case of Low-Re $k-\varepsilon$ model. Immediately after the mushy region, the solute concentration is enriched. The increase is higher with the LES model. The behaviour of the depleted/enriched regions of the concentration corresponds to the velocity responses.

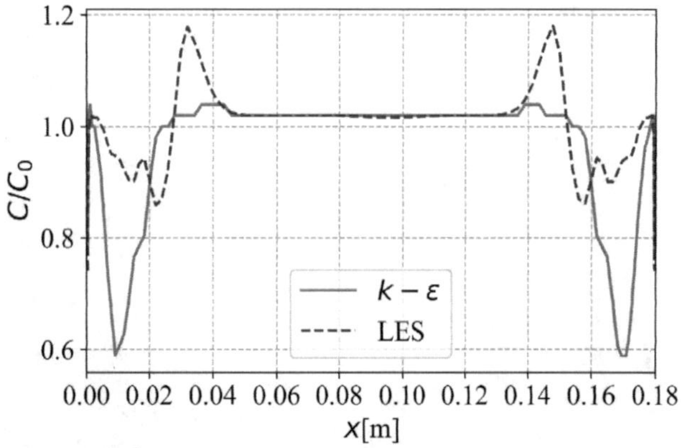

Figure 5. Comparison of concentration cross-sections for Low-Re $k-\varepsilon$ and LES models.

The transient nature of the LES model requires a higher number of nodes in a calculational domain and, consequently, longer computational times. The Low-Re $k-\varepsilon$ model requires 4 h CPU time to reach the steady-state and 137 h to calculate 5000 s, and the LES example requires 507 h CPU time (for t = 5000 s) on 2 Intel Xeon E2-2650 processors with 16 cores running at 2.00 GHz. The cases presented above were calculated on a minimal node arrangement containing 37696 nodes.

7. Conclusions

The present study evaluates the effect of the two turbulence models on macrosegregation. The results, calculated with the meshless LRBFCM method, show that the application of the LES turbulence model can be alternatively used to RANS models for predicting macrosegregation in the CC of steel. The most significant advantage of the LES model is its ability to predict the transient nature of the problem under consideration correctly. It also provides a more detailed description of the recirculation zones and, consequently the species concentration in a fluid and mushy region. The conceptually represented LES model is currently in extension to three dimensions for assessing the realistic macrosegregation pattern of a billet caster in Štore-Steel company.

Acknowledgments
This work was funded by the Slovenian Research Agency (ARRS) and Štore-Steel company in the framework of applied research project L2-9246 Multiscale and Multiphysics Numerical Modelling for Competitive Continuous Casting, and program P2-0162 Multiphase Systems.

References
[1] Vakhrushev A, Wu M, Ludwig A, Tang Y, Hackl G and Nitzl G 2012 Modeling of the flow-solidification interaction in thin slab casting *IOP Conf. Ser. Mater. Sci. Eng.* **33**
[2] Wu M, Ludwig A and Kharicha A 2017 A four phase model for the macrosegregation and shrinkage cavity during solidification of steel ingot *Appl. Math. Model.* **41** 102–20
[3] Wu M, Ludwig A and Kharicha A 2019 Volume-averaged modeling of multiphase flow phenomena during alloy solidification *Metals (Basel).* **9**
[4] Aboutalebi M R, Hasan M and Guthrie R I L 1995 Coupled turbulent flow, heat, and solute transport in continuous casting processes *Metall. Mater. Trans. B* **26** 731–44
[5] Sivaramakrishnan S, Thomas B G and Vanka S P 2000 Large Eddy Simulation of Turbulent Flow in Continuous Casting of Steel *Mater. Process. Comput. Age III* 189–98
[6] Chaudhary R, Ji C, Thomas B G and Vanka S P 2011 Transient turbulent flow in a liquid-metal model of continuous casting, including comparison of six different methods *Metall. Mater. Trans. B Process Metall. Mater. Process. Sci.* **42** 987–1007
[7] Qian S, Hu X, Cao Y, Kang X and Li D 2015 Hot top design and its influence on feeder channel segregates in 100-ton steel ingots *Mater. Des.* **87** 205–14
[8] Vakhrushev A, Liu Z, Wu M, Kharicha A, Ludwig A, Nitzl G, Tang Y and Hackl G 2018 Numerical modelling of the MHD flow in continuous casting mold by two CFD platforms ANSYS Fluent and OpenFOAM *ICS 2018 - 7th Int. Congr. Sci. Technol. Steelmak. Chall. Ind. 4.0*
[9] Vertnik R, Mramor K and Šarler B 2019 Solution of three-dimensional temperature and turbulent velocity field in continuously cast steel billets with electromagnetic stirring by a meshless method *Eng. Anal. Bound. Elem.* **104** 347–63
[10] Mramor K, Vertnik R and Šarler B 2015 A meshless model of electromagnetic braking for the continuous casting of steel *Mater. Tehnol.* **49**
[11] Šarler B and Vertnik R 2006 Meshfree explicit local radial basis function collocation method for diffusion problems *Comput. Math. with Appl.* **51** 1269–82
[12] Bennon W D and Incropera F P 1987 A continuum model for momentum, heat and species transport in binary solid-liquid phase change systems-I. Model formulation *Int. J. Heat Mass Transf.* **30** 2161–70
[13] Vertnik R and Šarler B 2014 Solution of a continuous casting of steel benchmark test by a meshless method *Eng. Anal. Bound. Elem.* **45** 45–61
[14] Smagorinsky J 1963 General circulation experiments with the primitive equations *Mon. Weather Rev.* **91** 99–164
[15] Driest E R van 1956 On Turbulent Flow Near a Wall *J. Aeronaut. Sci.* **23** 1007–11
[16] Pakzad A 2017 Damping functions correct over-dissipation of the Smagorinsky Model *Math. Methods Appl. Sci.* **40** 5933–45
[17] Davidson L and Billson M 2006 Hybrid LES-RANS using synthesized turbulent fluctuations for forcing in the interface region *Int. J. Heat Fluid Flow* **27** 1028–42
[18] Saad T, Cline D, Stoll R and Sutherland J C 2017 Scalable tools for generating synthetic isotropic turbulence with arbitrary spectra *AIAA J.* **55** 327–31
[19] Vertnik R, Mramor K and Šarler B 2019 Solution of three-dimensional temperature and turbulent velocity field in continuously cast steel billets with electromagnetic stirring by a meshless

method *Eng. Anal. Bound. Elem.* **104** 347–63

IOP Conf. Series: Materials Science and Engineering 1274 (2023) 012052 doi:10.1088/1757-899X/1274/1/012052

Computational fluid dynamics calculations of binary representation of Ni-base superalloy solidification in frustum geometries

J D T Allen[1], N Green[2] and N Warnken[1,3]

[1]University of Birmingham, School of Metallurgy and Materials, Edgbaston, Birmingham, England, B15 2TT
[2]High Temperature Research Centre, Airfield Drive, Coventry CV7 9BF

Abstract. CFD calculations were carried out on a number of frustum geometries from the experimental literature on freckling of superalloys. The effects of two imposed thermal fields were investigated, flat and inclined. Experiments document that when thermal fields are inclined the cold side is more prone to formation of freckles; our results conclude that rising liquid plumes in advance of the solidification front, which can correlate to freckles, are more abundant and intense on the cold side. Also, modelled plume activity is greater in inward sloping frustum geometries as opposed to outward sloping ones, which also agrees with established literature. Complex flow patterns were observed in inclined geometries and there is a clear interplay and competition between both the thermal and solutal aspects of flow, with flow being significantly changed, sometimes reversed in direction in some capacity. Our results also offer insight into the incubation length for plume and resultant freckle formation after the transition between frustums, with some critical distance of advance of the solidification front necessary to initiate and establish other plumes that feed into the main one circulating liquid flow which is locally intense enough to potentially create a freckle.

1. Introduction

The phenomenon of the freckle has been known about in the process of directional solidification for approximately 50 years and is significant to many areas of materials science including austenitic stainless steels, superalloys and in geothermal settings [1, 2, 3, 4, 5]. This paper is an investigation of the phenomenon in nickel-based superalloys. A freckle is a chain of small eqiuiaxed grains of random orientation on the surface of a casting [2, 6, 7]. In the scientific literature freckling is widely understood to be caused by the process of thermosolutal convection [7, 8, 9] which is caused when, during thermally constrained directional solidification, heavy elements (such as Re and W in superalloys) segregate to the dendrites, resulting in a layer of liquid of reduced density in the mushy zone [7, 8, 9]. In the mushy zone there are frictional and buoyancy forces active and should the buoyancy forces overcome the frictional forces plumes of liquid can emerge from the mushy zone [7]. These plumes can cause remelting of the secondary dendrite arms forming stable channels of intense and persistent flow allowing for the initiation of freckles during solidification [10, 11]. Freckles are highly significant in turbine applications. For nickel-based superalloys a single freckle can significantly reduce the mechanical properties, resulting in parts which are unfit for service as they cannot be post-hoc removed via mechanical or thermal treatments [1, 2, 12, 13, 14].

[3] Corresponding Author: N.Warnken@bham.ac.uk

There have been many factors identified to affect the possibility and severity of freckling, the first is alloy composition, adding heavy elements that are known to partition to the dendrites, for example Re and W, would worsen freckling, whereas adding heavy elements that are known to segregate to the interdendritic regions such as Ta, is known to reduce freckling [1, 7, 15]. In addition to this, the use of carbon to from carbides close to the liquidus also suppresses freckling [7, 8]. The second is thermal conditions, for example Schadt et. al. [16] found in general for an array of geometries for CMSX-4 that if the product of the two variables thermal gradient, G, and withdrawal velocity, V, was less than 0.14 K/s then freckles would likely occur. On the other hand, the work of Ma et. al. [17] analysed a single cylindrical geometry and concluded that there is a region of freckling at intermediate values of G and V with no freckling outside of that. The third is the thickness of the sample; it is known from the literature that a thicker sample has a longer mushy zone which increases the likelihood of freckling [7, 8, 12]. Finally, and most important to this paper, the geometric features; in many castings freckles are associated with a rapid increase or decrease in the thickness of a sample, or a surface that inclines inwards or outwards, the latter chiefly being the effect investigated in this paper [1, 12, 18, 19].

In the research of Ma et. al. [1, 12] several experiments and thermal field simulations have been carried out on geometries which have sloping surfaces either inwards or outwards [1, 12]. It was determined experimentally that inward sloping faces are more prone to formation of freckles than outward sloping faces [1, 12]. Ma et. al. made the determination that this was due to the reservoir theory [1, 12, 19]. In this reservoir theory a sufficient mushy zone reservoir of enriched liquid below a plume is required to feed it. The size of the reservoir is enhanced when the surface slopes inwards [1, 12]. In Ma's experiments, freckles were found in 2 scenarios, in the outward sloping cases some plumes were found at the transition between 2 frustums, and in the inward sloping cases more prominent plumes were found leading up to the transition points after an incubation height [1, 12]. The reason for the short plumes after the transition to a smaller diameter has been reasoned to be due to the step effect. Immediately after the transition, there is an existing reservoir (the mushy zone) and a newly obtained surface, allowing for a short freckle [1, 12, 20]. The incubation height is caused by the fact that after a cross-section transition to a larger diameter it requires a certain solidification distance to establish said reservoir again [1, 12, 18, 19]. In most castings radiation effects dictate that the thermal field is not perfectly horizontal. In the literature it is documented that freckles are more likely to appear on the cold side of the geometry, *i.e.* the most advanced portion of the advancing solidification front [20].

In this study CFD computer simulations are conducted on inward sloping and outward sloping geometry frustrums using both non-inclined and inclined thermal fields for the purpose of seeing where plumes local to the solidification front occur and whether they correspond to freckles in the experimental results [1, 12]. Both the composition field and flow field are analysed to this end.

2. Simulation

The model used is implemented in this study using the routine created by Crane and Coleman named solidificationFoam, that is implemented within the CFD package OpenFOAM. (https://github.com/OpenFOAM/OpenFOAM-Solidification). The model uses the pressure corrector algorithm SIMPLE, the details of which are explained elsewhere [24, 25]. The model uses 2 iteration loops, one for the temperature and composition and the other for flow velocity and pressure. One was set to be converged when the difference in solid fraction is less than $1 \cdot 10^{-4}$ and the other one was set to be converged when the difference in pressure is less than $1 \cdot 10^{-8}$ Pa. Non-slip velocity boundary conditions were used in all calculations.

Two geometries have been created based on those which feature in the research of Ma et.al. [12] which shall be called geometries G1 and G2 respectively, see Figure 1. In order to accurately assess the flow patterns at and near the surfaces of said geometries, the meshes were constructed as follows: within 4 mm of the surfaces, the maximum cell element volume is set to $6.4 \cdot 10^{-11} m^3$ ($0.4^3 mm^3$), the rationale behind this was it was necessary to have a cell size with lengths comparable to the primary dendrite arm spacing associate with thermally constrained directional solidification of nickel base superalloys. The meshes have larger elements in the core of the geometries to reduce the computational expense.

IOP Conf. Series: Materials Science and Engineering 1274 (2023) 012052 doi:10.1088/1757-899X/1274/1/012052

Two isotherm conditions were applied, horizontal and inclined at 15 degrees to the horizontal. Values for G and V were selected based on the literature, wherein previous research using these geometries selected a fixed V of 1 mm/min but various G values [1, 12]. These G values vary between 1000 K/m and 2900 K/m, the range of reported conditions having been determined by simulations or thermocouple measurements of the researchers [12]. In the end, the intermediate G value 1900 K/m was selected. G and V were then used to calculate the temperature on the surface of the metal as function of time and z-coordinate using the equations $min\{z_{dist} = z - (t \cdot V), 0\}$ and $T = T_{furn} + (G \cdot z_{dist})$ where t is the time, z is the z-coordinate of a point in a vertical frame of reference and T_{furn} is the furnace temperature 1773 K. This was then used a boundary condition for transient thermal fields inside the metal, as obtained from heat transport through conduction and convection.

The primary dendrite arm spacing was determined using the following equation $\lambda_1 = 2.83 \cdot (k \cdot \Gamma \cdot D_{liq} \cdot \Delta T)\ 2.83 \cdot G^{-0.5} \cdot V^{-0.25}$ where ΔT is the difference between T_{liq} and T_{sol}, k is the partition coefficient, Γ the Gibbs Thomson coefficient and D_{liq} the solutal diffusion coefficient in the liquid. This equation was created by Hunt by considering the geometry of cell/dendrite tip as a hemispherical cap and detailed in [21, 22]. Simulations were done using a pseudobinary representation of the commercial single crystal superalloys CMSX-4.

The second generation single crystal superalloy CMSX-4 is a widely used commercial single crystal superalloy, consisting of nine mayor elements. This alloy is well known as a freckle prone alloy and quite widely studied [1,16-20]. CFD simulations with this level of compositional complexity still provide a significant computational challenge. For this reason, a pseudo binary representation is commonly used in the superalloy solidification community, Ren et.al. [23] recently published work on CMSX-4 using this approach. A binary alloy consisting of nickel plus a fictitious second element is created, represented by a liquidus slope and partition coefficient – see Table 1 for data used in the present study, giving a useful representation on properties especially segregation behaviour, solidification range, effect of solute on liquidus temperature.

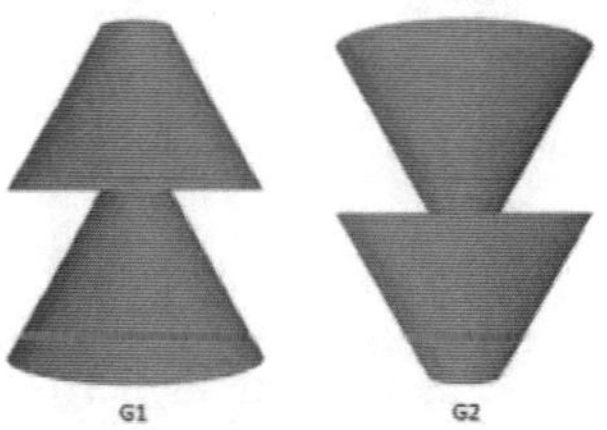

Figure 1: The two geometries used in the simulations, each individual frustum has a diameter of 30mm at its thickest point, a height of 15mm and a sloping angle of 30 degrees.

Table 1. Relevant quantities used in the simulations taken from [1]

Quantity	Value	Units	Quantity	Value	Units
Latent Heat of Fusion	240000	Jkg^{-1}	Dynamic Viscosity (μ)	$4.93*10^{-3}$	$kgm^{-1}s^{-1}$
Reference Density(ρ_{ref})	7646	kgm^{-3}	Liquidus Temperature Nominal (T_{liq})	1660.973	K
Nominal Composition of Solute(C_0)	38.2	wt%	Solidus Temperature Nominal (T_{sol})	1631.523	K
Partition Coefficient(k)	0.6883	-	Thermal Expansion Coefficient (β_T)	$1.16*10^{-4}$	K^{-1}
Solvent Melting Temperature(T_s)	1726	K	Solutal Expansion Coefficient (β_C)	$2.51*10^{-1}$	$wt\%^{-1}$
Heat Capacity(C_p)	575	$kJkg^{-1}$	Solidus Slope (m_S)	-2.4732	$Kwt\%^{-1}$
Solid Thermal Conductivity (K_{ts})	24.6	WmK^{-1}	Liquidus Slope (m_L)	-1.7023	$Kwt\%^{-1}$
Liquid Thermal Conductivity (K_{tl})	33.5	WmK^{-1}	Gibbs Thompson Coefficient (Γ)	$2.84*10^{-7}$	Km
Reference Temperature(T_{ref})	1773	K	Diffusion Coefficient in Liquid (D_{liq})	$3.60*10^{-9}$	m^2s^{-1}
Eutectic Temperature (T_E)	1593	K			

3. Results and Discussion

Figure 2 documents the composition profile at or very close to the end of solidification. Segregation arises as a consequence of, and denotes, flow of partitioned solute enriched liquid plumes. Our results seem to align with the results of the previous literature, in that G1 with contracting frustums has a lot more plume activity than G2 with expanding ones [1, 12]. In all cases the composition of the alloying element increases relative to the nominal composition at and near the very top of the geometry. This is simply the result of the phenomenon of macrosegregation. With regard to the conditions with thermal fields inclined by 15 degrees, stronger plume activity on the cold side of the specimen accords with literature [20]. A series of plumes proceed upwards along the inclined front to feed a singular vertical plume at the point furthest away from the heating element/cooler face in the case of G1, this also aligns with literature data [1, 12]. In the case of G2 less intense plume activity is detected, although notably on the cold side a prominent plume appears just above the transition, this appears consistent with the experimental results where a freckle is often found just above the transition due to the step effect [1, 12].

Flow activity during solidification is summarised in Figure 3. Three snapshots at 3 different points during solidification for each calculation were taken when the solidification was half way up each frustum and at the union. The first important observation, that applies mostly to the results with inclined thermal fields, is that the intensity of flow is reduced as solidification proceeds, apparently due to reducing free liquid volume ahead of the solidification front. In cases without inclined isotherms this is less relevant, as flow is dominated by solutal effects.

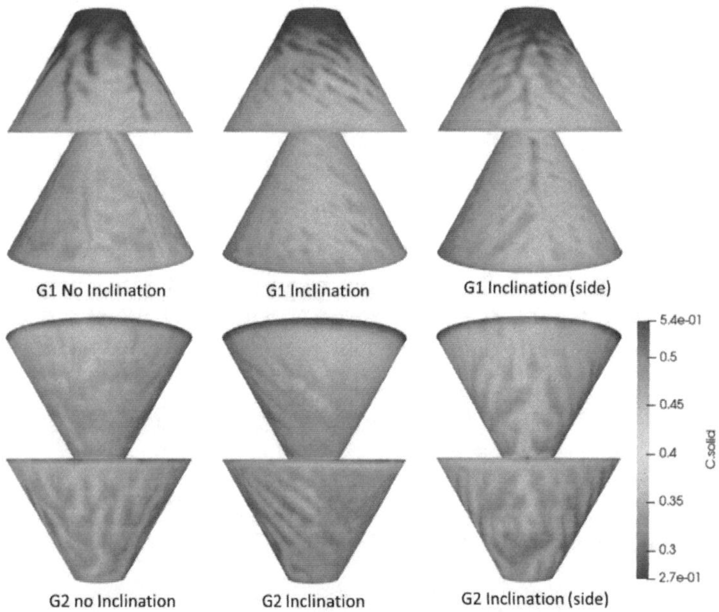

Figure 2. Solid composition profiles in the two geometries for both inclined and non-inclined temperature profiles. In inclined profiles the cold side is always to the left. In the side view the cold side faces the camera.

The second important observation arising from these results is that overall, there is more flow in the far field from the solidification front for cases where the thermal field is inclined. This is not only due to convection effects caused due to the thermal field, liquid on the hot side rising and that on the cold side falling but also as a consequence of more complex interactions arising from the differential distribution of the rising low density liquid ahead of the solidification front. In the lower frustrum of G1 geometry with the inclined front the flow is reversed based on what would be expected with thermal flow alone, indicating conditions where solutal flow is intense and likely to be favourable for freckle formation. However this does not occur in the G2 geometry in the bottom frustum due to the dissipative action of the expanding section leading to much weaker plume flow that could not overwhelm the thermal flow in this way. In the top frustum, due to the fact that the thermal flow is much weaker, we have a reversal effect. At the union also, due to the large plume there is also a reversal effect arising from momentum, but there is also a pocket of flow in the centre that follows the thermal direction, indicating that the solutal flow is not fully yet dominant. In the cases with no inclination we typically see that there is very limited thermal flow (which would be expected to be down at surfaces up in the centre) allowing for a series of weaker plumes at the surface to proceed which have no factor driving them to merge like inclined cases.

A further key observation arising from this study is that the appearance of plumes does not necessarily correspond to the appearance of freckles, for example we see some plume activity in G2 inclined but no freckles experimentally (except for on the cold side just above transition) [12]. In the experimental literature freckles only occur in G1 after an incubation height of *5mm* from the transition [12]. In the example of G1 inclination (where the inclined thermal field serves as a first degree approximation of the experimental curved profile), there is no delay, but there is an incubation period before the main plume is being fed by a lot of others, of about the same length as observed experimentally, this could explain why there is an incubation zone. This observation does counteract the aforementioned reservoir theory, however it is clear that more research needs to be done in this regard.

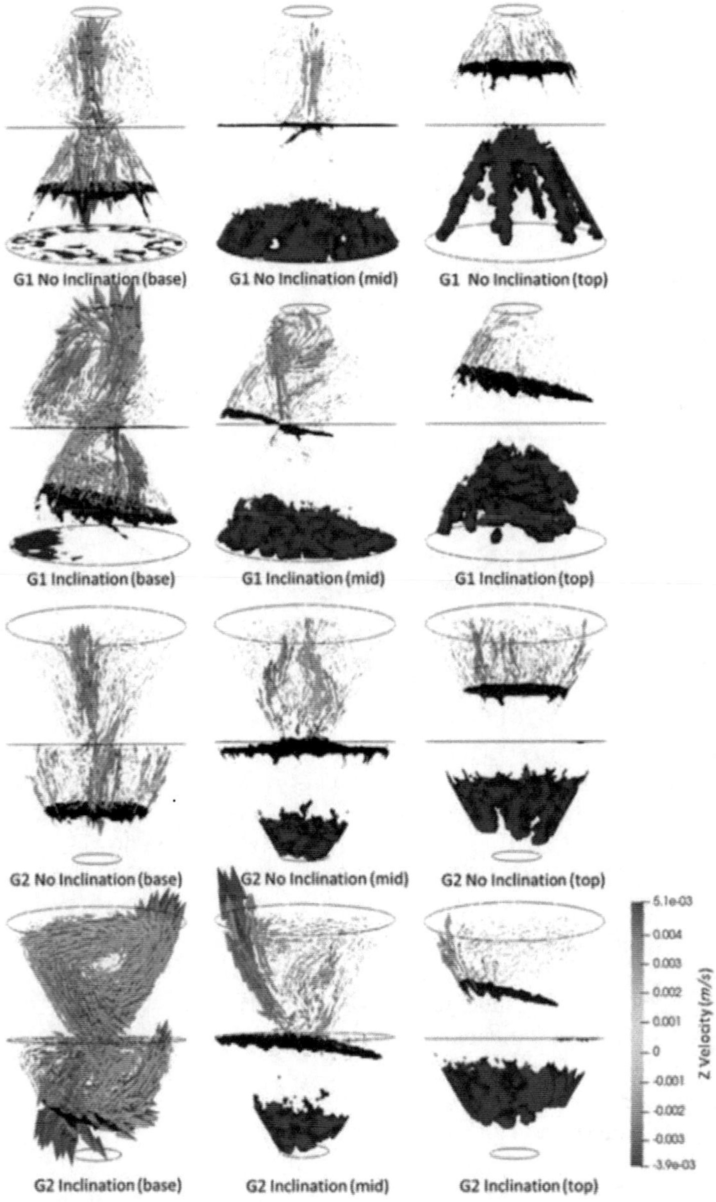

Figure 3. Velocity profiles in the two geometries as described in figure 1 for both inclined and non-inclined temperature profiles. In inclined profiles the cold side is always to the left. The green contour corresponds to full solidification and the black contour corresponds to 1% solidification. The coloring of the arrows is determined by the z-component of the flow vector the redder the more upwards the flow and the blue the more downwards.

4. Conclusions

- CFD computer simulations were designed to assess the effects of geometry and inclination of temperature profile on plume activity. Results showed that outward sloping geometries are less prone to plume occurrence than inward sloping geometries.
- In the case of an inclined temperature field the model indicated that the cold side is more susceptible to plume occurrence than the hot side.In inward sloping geometries the model predicted a series of plumes would combine into a single intense channel at the uppermost point of the advancing grain envelope.
- Computed flow patterns match the occurrence of freckles in the experimental data to which comparison was made.
- Complex flow patterns arising from coupled thermal and a solutal effects were observed in geometries with inclined profiles, with the solutal flow strongly impacting the overall flow in high plume situations, notably the inward sloping geometry.
- The appearance of plumes does not completely correspond to the occurrence of freckles, due to presence of an incubation height.

5. References

[1] Hong J, Ma D, Wang J, Wang F, Dong A, Sun B and Bührig-Polaczek A 2015 *J. Alloys Compd.* **648** 1076–82.
[2] Giamei A F and Kear B 1970 *Met. Trans.* **1** 2185–92.
[3] Fowler A C 1985 *IMA J. Appl. Math.* **35** 159–74.
[4] McKenzie D 1984 *J. Petrol.* **25** 713–65.
[5] Jacobs J A 1953 *Nature* **172** 297–98.
[6] Ren N, Panwisawas C, Li J, Xia M, Dong H and Li J 2021 *Acta Mater.* **215** 117043.
[7] Reed R C 2006 *The Superalloys*, (Cambridge University Press-Cambridge).
[8] Tin S and Pollock T 2003 *Metall. and Mater. Trans. A* **34** 1953–67.
[9] Tin S, Pollock T and Murphy W *Metall. and Mater. Trans. A* 2001 **32** 1743–53.
[10] Yuan Y and Lee P D 2012 *Acta Mater.* **60** 4917–26.
[11] Kao A, Shevchenko N, Alexandrakis M, Krastins I, Eckert S and Pericleous K 2019 *Philos. Trans. Royal Soc. A* **377** 20180206.
[12] Ma D and B¨uhrig-Polaczek A 2014 *Metall. and Mater. Trans. A* **45** 1435–44
[13] Yu K O, Domingue J A, Maurer G E and Flanders H D 1986 *JOM* **38** 46–50.
[14] Amouyal Y and Seidman D 2011 *Acta Mater.* **59** 6729–42.
[15] Amouyal Y and Seidman D 2011 *Acta Mater.* **59** 3321–33.
[16] Schadt R, Wagner I, Preuhs J and Sahm P 2000 *Superalloys 2000: Proc. of the 9th Int. Symp. on Superalloys* T M Pollock (Minerals, Metals and Materials Society- Pennsylvania) 211– 8
[17] Ma D, Zhou B and Bührig-Polaczek A 2011 *Adv. Mat. Res.* **278** 428–33.
[18] Ma D X, Mathes M, Zhou B and Bührig-Polaczek A 2011 *Adv. Mat. Res.* **278** 114–9.
[19] Ma D, Ziehm J, Wang W and Bührig-Polaczek A 2012 *Solidification and Casting of Metals: Proc. of an Int. Conf. on Solidification.* (Rolduc Abbey-Aachen) vol 27 012034.
[20] Ma D, Wu Q and Bührig-Polaczek A 2012 *Metall. and Mater. Trans. B* **43** 344–53.
[21] J. Hunt 1979 *Solidification and Casting of Metals: Proc. of an Int. Conf. on Solidification* (Metals Society-London) 3–9.
[22] Ghosh S, Ma L, Ofori-Opoku N and Guyer J E 2017 *Model. Simul. Mat. Sci. Eng.* **25** 065002.
[23] Ren N, Li J, Panwisawas C, Xia M, Dong H and Li J 2021 *Acta Mater.* **206** 116620
[24] Caretto L S, Gosman A D, Patankar S V and Spalding D B 1973 H Cabannes and R Temam (Springer-Berlin) *Proc. of the 3rd Int. Conf. on Numerical Methods in Fluid Mechanics* vol 2 60–8.
[25] Patankar S V and Spalding D B 1973 *14th Symp. (Int.) on Combustion.* (Pennsylvania State University-Pennsylvania) vol 14 605–14.

IOP Conf. Series: Materials Science and Engineering 1274 (2023) 012053 doi:10.1088/1757-899X/1274/1/012053

Observation of grain motion during semisolid deformation by using 4D-CT and 3DXRD

T Narumi, K Ohta, M Ohta, T Numata, K Asahi, R Katsube, H Yasuda

Department of Materials Science and Engineering, Kyoto University, Yoshida-honmachi, Sakyo, Kyoto, 606-8501 Japan

E-mail: narumi.taka.6n@kyoto-u.ac.jp

Abstract. A technique combining time-resolved computed tomography (4D-CT) and three-dimensional X-ray diffraction (3DXRD) was employed to investigate semisolid deformation in an Al-10mass%Cu alloy with the equiaxed grain structure. The diffraction patterns and reconstructed 3D images were used to quantitatively analyse the translational and rotational motions of solid grains in the semisolid alloy. Finally, the motion of solid grains and the expansion of porosity during the compression tests are discussed on the basis of in situ observations.

1. Introduction

Semisolid deformation often occurs during casting processes of metallic alloys and contributes to casting defects such as macrosegregation, porosity, and hot tears. To improve the quality of casting materials and enhance the efficiency of casting processes, it is essential to understand the mechanism of semisolid deformation in metallic alloys. The semisolid deformation is typically induced by solidification shrinkage and external forces such as gravitational and electromagnetic forces. The appearance of a segregation band during semisolid processing and centrifugal casting is one example of macrosegregation induced by the semisolid deformation [1–3]. Rheological behaviours of semisolid alloys (e.g., apparent volume expansion, also called Reynolds dilatancy) can trigger macrosegregation and the formation of porosities [1,2]. From a grain-scale perspective, Reynolds dilatancy is induced by the rearrangement of solid grains, which changes the local configuration of solid grains through translational and rotational motions. To understand defect formation related to the semisolid deformation, the motion of solid grains during the semisolid deformation process must be elucidated.

Recently, time-resolved computed tomography (4D-CT) using synchrotron radiation X-rays has been employed to observe microstructure evolution in bulk metallic-alloy specimens [4–10]. The 4D-CT technique has also been used to observe the translational motion of solid grains and changes in local liquid fractions of semisolid Al-Cu alloys during the tensile tests [11,12]. Compression test of semisolid Al-Cu alloys was also observed in-situ using the 4D-CT technique, which revealed the local configuration of globular solid grains formed by compression [13]. In addition, the void generation in a semisolid Al-Cu alloy and the plastic deformation of dendritic solid grains were evaluated [14]. Furthermore, transgranular liquation cracking during indentation of a semisolid Al-Cu alloy was detected using the 4D-CT technique, which is expected to lead to structural refinement [15]. These studies provide insights regarding the translational motion of solid grains during the semisolid deformation; however, analysis of the rotational motion is challenging because it is difficult to distinguish the rotational motion and the plastic deformation from the tomographic images. The rotational motion can be quantified by analysing changes in the crystallographic orientation of each solid grain during the semisolid deformation process.

Three-dimensional X-ray diffraction microscopy (3DXRD) is an X-ray imaging technique that can be used for 3D crystallographic orientation mapping in polycrystalline specimens [16–21]. 3DXRD measurements are performed using the CT setup; thus, 3DXRD measurements are often combined with CT measurements. Our group's previous studies revealed the translational and rotational motions of dendritic solid grains during the compression of a semisolid Al-10mass%Cu alloy using a combination technique of 4D-CT and 3DXRD [22,23]. These results also showed that the non-uniform deformation induced by the rearrangement of solid grains was initiated before the compression reached a strain of –0.04. Therefore, it is of interest to understand how solid grains move during the initial stage of the compression test. In this work, the combination technique of 4D-CT and 3DXRD was used to analyse the rearrangement of globular solid grains in a semisolid Al-10mass%Cu alloy during the compression test. We discuss the expansion of porosity and the gap between solid grains until the compression reaches a strain of –0.04.

2. Experimental method

CT and 3DXRD measurements were performed at the imaging beamline BL20XU of the SPring-8 synchrotron radiation facility in Hyogo, Japan, using a monochromatised X-ray (37.7 keV). Figure 1 shows a schematic illustration of the setup. A beam monitor (ORCA-Flash; format: 1024 × 1024 pixels; pixel size: 6.5 μm × 6.5 μm; Hamamatsu Photonics) was used to observe the projection images (transmission images), and a panel-type X-ray detector (format: 992 × 682 pixels; pixel size: 100 μm × 100 μm; Hamamatsu Photonics) was placed along the X-ray beam. The distance between the sample and the panel-type detector was about 0.2 m, and the distance between the sample and the beam monitor was about 0.5 m. For CT measurements, the X-ray beam (width = 5 mm, height = 3 mm) was applied to a specimen rotated at 4 s/360°-rotation. The projection images were obtained at a frame rate of 200 fps. Figure 1(b) shows a projection image of an Al–10mass%Cu alloy. The 3D images were reconstructed using a convolution back-projection method involving 400 projected images over a 180°-rotation. The voxel size of the reconstructed images was 6.5 μm × 6.5 μm × 6.5 μm. 3DXRD measurements followed the CT measurements. The height of the X-ray beam was adjusted to 100 μm (referred to as narrow X-ray). The narrow X-ray beam was applied to the specimen rotated at 10 s/360°-rotation, and the diffraction spots were detected by the panel-type X-ray detector. The diffraction measurements were repeated from the top to the bottom of the specimen for a total of 20 layers. Completing a sequence of CT and 3DXRD measurements took approximately 300 s. In the present setup, a monochromatised X-ray with an energy of 37.7 keV allows measurement of the 111 and 200 diffraction spots of α-Al over a 360°-rotation, which is required to analyse the crystallographic orientations of solid grains. Figure 1(c) shows the diffraction images of an Al–10mass%Cu alloy, where the diffraction images were integrated over a 360°-rotation. It should be noted that the X-ray beam size used for 3DXRD measurements was larger than that used in previous 3DXRD studies [18,19,21], which allows scanning of a larger area in a single diffraction measurement. Namely, 3DXRD measurement was completed in fewer scans, resulting in shorter time measurement. Details of 3D image processing and 3DXRD analysis have been given in our group's previous studies [22,23].

An Al-10mass%Cu alloy with the equiaxed grain structure was used as a test specimen, which had a rod-like shape (diameter = 2 mm, length = 5 mm) with a reduced section in the middle (diameter = 1.5 mm, length = 2 mm). The specimen was inserted into a sintered Al_2O_3 tube (inner diameter = 2.0 mm, outer diameter = 3.0 mm) and heated until the eutectic structures of α-Al and θ-Al_2Cu melted; melting was detected according to the disappearance of the θ-Al_2Cu diffraction spots. The temperature of the semisolid specimen was maintained for 15 min under vacuum (<1 Pa) to create isothermal conditions for the semisolid specimen before the compression test. The semisolid specimen was then compressed by placing an Al_2O_3 rod on the top of the specimen and pushing down with a strain rate of approximately 10^{-4} s^{-1} (calculated from the reduced section length and the moving velocity of the

Al_2O_3 rod). The nominal strain (ε) of each compression was calculated according to the deformation length and the reduced section length. These compression tests were conducted intermittently three times.

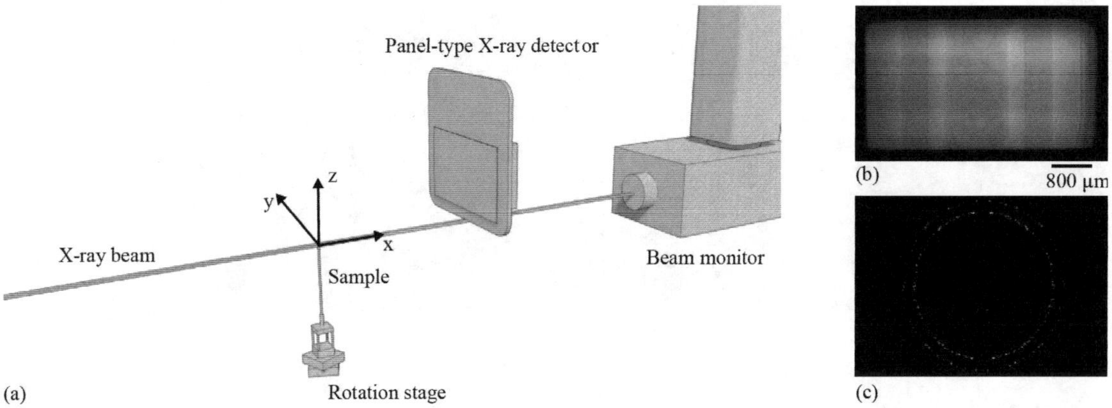

Figure 1. (a) Schematic illustration of the experimental setup of 4D-CT and 3DXRD; (b) A projection image and (c) the diffraction spots from an Al–10mass%Cu alloy.

3. Results and discussion

Figure 2 presents the 3D reconstructed images of solid grains in the semisolid specimen. Discrete solid grains are represented by greyscale variations. There were 212 solid grains in the observation field, and the average sphere-equivalent diameter was 220 μm. The solid fraction before compression was 0.79. Solid grains were uniformly distributed throughout the structure and separated by the liquid phase. The compression tests were performed in triplicate, where the final strain reached –0.04.

Figure 2. 3D reconstructed images of Al-10mass%Cu alloys at strains of (a) ε = 0 and (b) ε = –0.04.

Figure 3 shows the *y-z* slices and *x-y* slices of the reconstructed images of the semisolid specimen, where the dark grey and light grey regions correspond to the solid and the liquid phases, respectively, and the internal black regions indicate porosity. In compression at a strain of –0.01, the gap between grains P and Q expanded, as indicated by red arrows. As the compression increased to a strain of –0.04, the gap continued to expand. The sizes of the solid grains did not change during the compression tests; however, the melt flowing into the expanded gap region reduced the local solid fraction. Therefore, this observation demonstrated that the repulsive force was generated by the interaction between solid grains and consequently non-uniform deformation was initiated at the grain-scale during the initial stage of semisolid deformation, when the strain reached –0.01. As the compression test proceeded, the porosity also expanded, as indicated by yellow arrows. In the present study, the temperature of the specimen was maintained during the compression test, and as a result, the solid fraction remained constant. Thus, the expansion of the porosity occurred when the local dilation reduced the pressure inside the semisolid specimen and the melt was drawn into the dilated regions, such as the gap between grains P and Q.

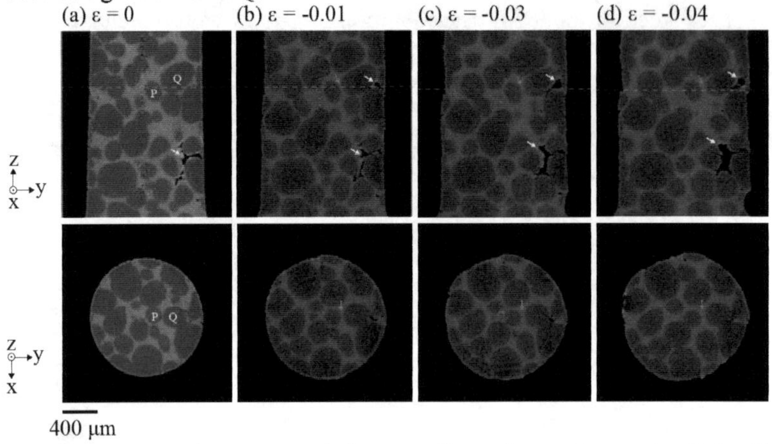

Figure 3. The *y-z* slices (upper images) and the *x-y* slices (lower images) of the 3D reconstructed images of the Al-10mass%Cu alloy at strains of (a) ε = 0, (b) ε = –0.01, (c) ε = –0.03, and (c) ε = –0.04.

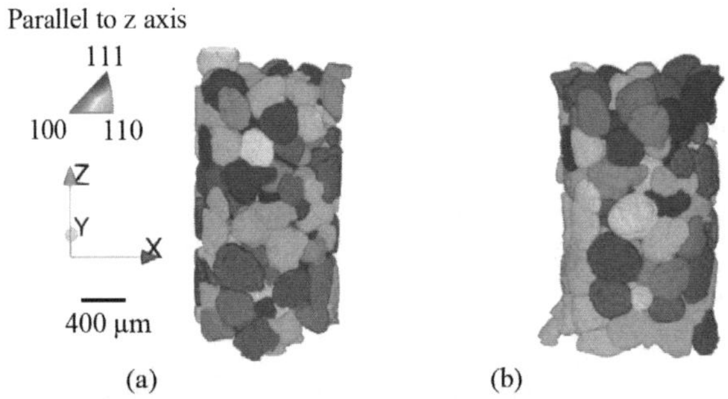

Figure 4. 3D grain configurations in terms of crystallographic orientation at strains of (a) ε = 0 and (b) ε = –0.04.

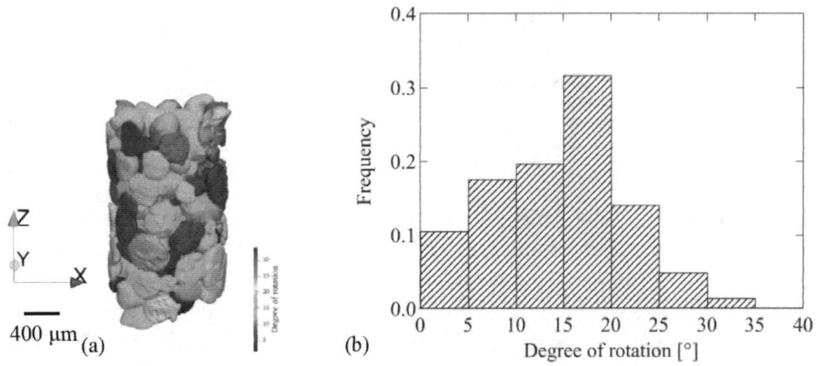

Figure 5. (a) 3D grain configurations in terms of the degree of rotation from a strain of ε = 0 to ε = –0.04 and (b) histogram of the degree of rotation.

The grain configuration obtained by CT was combined with the normal vectors obtained from 3DXRD measurements performed after each compression test. Figure 4 shows the 3D grain configuration, where the colour variations indicate crystallographic orientations. The number of the solid grains in figure 4 was smaller than figure 2, because the crystallographic orientations of several solid grains were not detected by 3DXRD measurement. Each solid grain in figure 4 adopted a unique crystallographic orientation. The solid grains were randomly orientated from a crystallographic perspective and separated by the liquid phase. These results show that the relationship $2\sigma_{SL} < \sigma_{GB}$ (σ_{SL}: solid/liquid interfacial energy; σ_{GB}: grain boundary energy) produced the liquid phase at the grain boundary region, and consequently, the solid grains remained discrete even at a solid fraction of ~0.8. Thus, the liquid phase was generated at the grain boundary regions, and consequently, solid grains were separated by the liquid phase.

The rotational motion of solid grains was assessed according to changes in the crystallographic orientation during compression from a strain of 0 to –0.04. Figures 5(a) and 5(b) respectively show the 3D grain configuration where the colour variations indicate the degree of rotation, and histogram of the degree of rotation. The globular shape of solid grains reduced their resistance to motion relative to the dendritic shape, and consequently, the degree of rotation was as high as 30° even at a strain of –0.04, which was larger than that observed in previous studies [22,23].

Figures 6(a) and 6(b) respectively show the *y-z* slices of the reconstructed images of the semisolid specimen, and the translational motion vectors (arrows) and the rotational axis directions (cylinders) of grains P and Q. The translational motion vectors were obtained by tracking the centre of mass of the solid grains from the tomographic images, and the rotational axis directions were obtained by using changes in the crystallographic orientations of the solid grains. Details of the analysis have been given in the previous studies [22,23]. According to the translational motion vectors, grains P and Q moved apart each other. In addition, their rotational axis directions were nearly parallel. The previous studies revealed that the resistance to motion of solid grains was reduced near the region where the solid grains tended to possess parallel rotational axis directions, and consequently, the gap between the solid grains expanded owing to the repulsive force generated by the interaction between solid grains [22,23]. Thus, the rotational analysis by 3DXRD has shown that the rotational axis directions can influence the interaction between solid grains. To elucidate the non-uniform deformation which causes the gap expansion between solid grains, it is necessary to measure the motion of large number of solid grains during the semisolid deformation and to statistically analyse the interaction between solid grains.

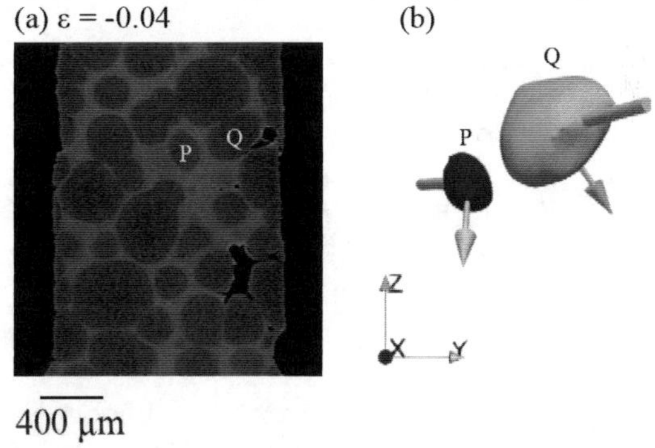

Figure 6. (a) *y-z* slices of the 3D reconstructed images of the Al-10mass%Cu alloy at a strain of ε = –0.04; (b) the translational motion vectors (arrows) and the rotational axis directions (cylinders) of grains P and Q.

4. Conclusion

4D-CT and 3DXRD were combined to observe the motion of solid grains in a bulk Al-10mass%Cu alloy specimen at the solid fraction of 0.79 during the compression test. The semisolid specimen contained 212 globular grains, where the average sphere-equivalent diameter was 220 μm. The translational and rotational motions of the globular solid grains were quantitatively analysed. According to the 3D grain configurations determined by CT measurements after each compression test, the gaps between a few solid grains expanded following compression at a strain as low as –0.01. These gaps continued to expand as the compression proceeded. Simultaneously, the porosity expanded around the expanded gap region. These observations confirmed that the rearrangement of solid grains expanded the gaps and contributed to the expansion of the porosity because the reduced pressure promoted the melt flow into the gaps. The crystallographic orientation of each solid grain measured by 3DXRD enabled rotational analysis. The globular shape of solid grains reduced the resistance of their motion, and consequently, the degree of rotation was as high as 30° when the compression reached a strain of –0.04. The rotation locally induced by the interaction between the solid grains was closely linked to the translational motion. Besides, the rotational axis directions can influence the interaction between solid grains. Therefore, we conclude that it is useful to consider the rotational motion of solid grains in addition to the translational motion to construct a physical model.

Acknowledgements

A portion of this study is supported by a Grant-in-Aid for Scientific Research (S) (No. 17H06155) and a Grant-in-Aid for Early-Career Scientists (No. 20K15071) from the Ministry of Education, Culture, Sports, Science and Technology (MEXT), Japan, and a research grant from Japan Aluminum Association. The synchrotron radiation experiments were performed as general projects at a beamline of the BL20XU at SPring-8 (JASRI) 2019B1478, 2020A0737, 2020A1558. We also thank Suzanne Adam, PhD, from Edanz (https://jp.edanz.com/ac) for editing a draft of this manuscript.

References

[1] Gourlay C M and Dahle A K 2007 *Nature* **445** 70–3
[2] Gourlay C M, Meylan B and Dahle A K 2008 *Acta Mater.* **56** 3403–13
[3] Kang I and Ohnaka I 1997 *J. Japan Foundry Eng. Soc.* **69** 119–26
[4] Ludwig O, Dimichiel M, Salvo L, Suéry M and Falus P 2005 *Metall. Mater. Trans. A* **36** 1515–23
[5] Aagesen L K K, Fife J L L, Lauridsen E M M and Voorhees P W W 2011 *Scr. Mater.* **64** 394–7
[6] Gibbs J W, Mohan K A, Gulsoy E B, Shahani A J, Xiao X, Bouman C A, De Graef M and Voorhees P W 2015 *Sci. Rep.* **5** 11824
[7] Puncreobutr C, Phillion A B, Fife J L and Lee P D 2014 *Acta Mater.* **64** 316–25
[8] Yasuda H, Kawarasaki T, Tomiyori Y, Kato Y and Morishita K 2019 *IOP Conf. Ser. Mater. Sci. Eng.* **529** 012023
[9] Nakano K, Narumi T, Morishita K and Yasuda H 2020 *Mater. Trans.* **61** 596–604
[10] Narumi T, Kawarasaki T, Kato Y, Morishita K and Yasuda H 2020 *J. Japan Inst. Light Met.* **70** 339–46
[11] Terzi S, Salvo L, Suéry M, Limodin N, Adrien J, Maire E, Pannier Y, Bornert M, Bernard D, Felberbaum M, Rappaz M and Boller E 2009 *Scr. Mater.* **61** 449–52
[12] Sistaninia M, Terzi S, Phillion A B, Drezet J M and Rappaz M 2013 *Acta Mater.* **61** 3831–41
[13] Kareh K M, Lee P D, Atwood R C, Connolley T and Gourlay C M 2014 *Nat. Commun.* **5** 4464
[14] Cai B, Karagadde S, Yuan L, Marrow T J, Connolley T and Lee P D 2014 *Acta Mater.* **76** 371–80
[15] Karagadde S, Lee P D, Cai B, Fife J L, Azeem M A, Kareh K M, Puncreobutr C, Tsivoulas D, Connolley T and Atwood R C 2015 *Nat. Commun.* **6** 8300
[16] Fu X, Poulsen H F, Schmidt S, Nielsen S F, Lauridsen E M and Jensen D J 2003 *Scr. Mater.* **49** 1093–6
[17] Lauridsen E M, Schmidt S, Nielsen S F, Margulies L, Poulsen H F and Jensen D J 2006 *Scr. Mater.* **55** 51–6
[18] Offerman S E, van Dijk N H, Sietsma J, Lauridsen E M, Margulies L, Grigull S, Poulsen H F and van der Zwaag S 2006 *Nucl. Instruments Methods Phys. Res. Sect. B Beam Interact. with Mater. Atoms* **246** 194–200
[19] Jensen D J, Lauridsen E M, Margulies L, Poulsen H F, Schmidt S, Sørensen H O and Vaughan G B M 2006 *Mater. Today* **9** 18–25
[20] Poulsen H F 2012 *J. Appl. Crystallogr.* **45** 1084–97
[21] Toda H, Ohkawa Y, Kamiko T, Naganuma T, Uesugi K, Takeuchi A, Suzuki Y and Kobayashi M 2013 *Acta Mater.* **61** 5535–48
[22] Narumi T, Nakata T and Yasuda H 2020 *IOP Conf. Ser. Mater. Sci. Eng.* **861** 012065
[23] Narumi T, Nakata T, Kobayashi R and Yasuda H 2021 *ISIJ Int.* **61** 1567–78

Investigation of Al-20Sn-10Cu alloy directional solidification by laboratory X-radiography

G Reinhart[1], F Ngomesse[1], F. Bertelli[2], P Benigni[1], A Campos[3], and H Nguyen-Thi[1]

[1] Aix-Marseille Univ, Université de Toulon, CNRS, IM2NP UMR 7334, 13397 Marseille, France
[2] Sea Science Department, Federal University of São Paulo – UNIFESP, 11030-400 Santos, SP, Brazil
[3] Aix Marseille Univ, CNRS, Centrale Marseille, FSCM (FR1739), CP2M, 13397 Marseille, France

Abstract. Al-based alloys with a soft phase such as Sn are extensively used for bearing components due to their self-lubricating properties. Al-Sn alloys lack the ability to support heavy loads so the alloying with Cu as a third element provides solution strengthening of the aluminium matrix. A crucial issue in the manufacturing of Al-Sn-Cu alloys is the miscibility gap in the phase diagram of the system. Liquid immiscibility is responsible for severe segregation during the solidification process, due to the large density difference between the Al-rich and Sn-rich liquids, which limits their utilization in industry. It is therefore both scientifically and technically important to accurately understand their solidification path. In the present work, the solidification of a ternary Al-20wt.%Sn-10wt.%Cu alloys was investigated in-situ by using X-radiography. Directional solidification experiments were performed on sheet-like samples in the laboratory device SFINX (Solidification Furnace with IN-situ X-radiography), which consists of a Bridgman-type gradient furnace and an X-radiography system. The solidification sequence was determined based on the observation of the recorded images, enlightening the successive steps of the solidification path. These observations were compared to predictions obtained from thermodynamic calculations. Complementary post-mortem microscopic analyses showed that the dendrite primary trunk and secondary arms developed along <110> crystallographic axes instead of the usually expected <100>.

1. Introduction

The mechanical properties of a bearing material are a balance between a high strength for load bearing and resistance to fatigue, and a low friction coefficient for tribological applications [1]. Aluminium-tin based alloys are suitable for such applications because the presence of Sn as a soft additive provides good anti-frictional and self-lubricating properties. In addition, the alloying with a third element such as Cu can provide solid solution strengthening of the Aluminium matrix [2]. Besides the chemical composition, the microstructure forming during the solidification phase is an important feature, especially relevant for intimate surface contact where the use of lubricants is not possible [3]. A key issue in the processing of Al-Sn-Cu alloys is the liquid miscibility gap in the system, which causes the parent liquid to decompose into two distinct immiscible liquid phases during cooling. This phase separation occurring in immiscible alloys leads to severe segregation at the microstructure level, due to the large density difference between the liquid phases that causes droplets to sink or float [4,5]. The liquid droplets can also grow and coarsen by solute diffusion or be transported by convective flows in the melt which can increase even more the detrimental impact of segregation. Research under microgravity conditions has also demonstrated that significant droplet migration is induced by thermocapillary (Marangoni) motion in the presence of a temperature gradient [6,7].

It is now well established that in-situ and time-resolved X-radiography is a method of choice for unveiling the dynamical evolution of dendritic microstructures and grain structure formation during the solidification of metal alloys [8]. However, studies of immiscible alloy solidification using this powerful observation technique are hitherto very scarce. First investigations have been carried out using synchrotron radiation sources. For example, Schaffer *et al.* investigated spinodal decomposition in hyper-monotectic Al-Bi and Al-Bi-Zn alloy [9] and highlighted the complex interactions between droplets. Lu *et al.* also studied liquid phase separation and the impact of bubbles on the segregation in Al-Bi alloys [10]. Recently, new opportunities have arisen with the improvement of compact micro-focus sources and X-ray sensitive detectors that enable *in-situ* and time-resolved radiography to be used in laboratory devices, with a sufficient spatial and time resolutions to distinguish the microstructure features [11], enabling different orientations of the sample with respect to gravity [12] and compatible with microgravity platforms [13,14]. Such a laboratory device was successfully used by Xavier *et al.* to analyse the morphology of the solidification interface at different cooling rates for an Al-Bi-Cu monotectic alloy [15].

The present work reports on the solidification of an Al-20wt.%Sn-10wt.%Cu sample observed *in situ* using the SFINX (Solidification Furnace with IN-situ X-radiography) laboratory device. The solidification sequence will be described based on the observation of the recorded images. These observations will be compared to predictions obtained from thermodynamic calculations. Complementary post-mortem microscopic analyses will be used to determine the growth orientation of the dendrite primary trunk and secondary arms.

2. Experimental details

The SFINX facility available at IM2NP laboratory was used and was described in detail in [16]. It has been developed within the framework of the ESA MAP named XRMON (In-situ X-ray monitoring of advanced metallurgical processes under microgravity and terrestrial conditions), devoted to the application of X-radiography during microgravity experiments [17]. The core of the facility is a solidification furnace consisting of two heaters separated by a gap that applies a longitudinal temperature gradient G_{app}. The gradient furnace enables directional solidification with cooling rates R within the range of 0.01–1.5 K/s and applied temperature gradient within the range of 5–15 K/mm.

The X-radiography system figures a micro-focus X-ray source with a molybdenum target (3 μm focal spot) providing enough photon flux and has two peaks of energy 17.4 keV and 19.6 keV ensuring a good image contrast to study Al–Cu based alloys. The camera system consists of a scintillator plate and a digital camera with a CCD-sensor. Due to the beam divergence, a geometric magnification of the object is observed at the detector, corresponding to the ratio of the source-detector by source-sample distances. Here, a magnification of approximately 5 was used for a Field-of-View (FoV) of about 5 × 5 mm^2 leading to an effective pixel size of approximately 4 μm. The acquisition rate was 2 frames per second. The studied alloy was Al-20wt.%Sn-10wt.%Cu. The three elements have significantly different absorption contrast so that the liquid and solid phases that are expected to form during the solidification sequence will be clearly distinguishable. Sheet-like rectangular sample of 5 mm × 50 mm in area and about 250 μm in thickness were used. The sample was placed in the middle of stainless-steel spacers sandwiched between two glassy carbon sheets sewn together with a silica thread. The crucible was inserted inside the furnace and meets both sides to achieve the expected thermal profile. The furnace

was set in vertical position, with the sample length parallel to the gravity vector **g**. This configuration is chosen to highlight gravity-related phenomena such as buoyancy.
The X-radiography system recorded a stack of raw images during the sample solidification. The grey level variations in the recorded images are related to the attenuation coefficient of each different part of the sample. Image legibility is enhanced by applying a "flat-field" correction, which consists in dividing each frame by a reference image recorded just before solidification. This procedure reduces the noise and removes defects related to the detector and crucible [18]. After applying the image processing, the bright regions in the processed radiograph are solid Al-grains whereas the dark regions are the Cu-rich and/or Sn-rich liquid regions.

3. Results and discussion

In the present study, we will consider an experiment carried out with a cooling rate, $R = 0.45$ K/s, and an applied temperature gradient G_{app} = 5.5 K/mm. These parameters were chosen to obtain a microstructure with features (dendrite arm spacing, interdendritic liquid fraction) that are distinguishable with respect to the spatial resolution.

3.1. Theoretical solidification sequence

The solidification paths of hypo-eutectic binary Al-Cu and Al-Sn alloys are rather standard with the formation of aluminium as primary phase and then of eutectic from the residual liquid. However, the comprehension of the solidification path in the ternary Al-Cu-Sn system is more complex, in part due to the presence of a liquid miscibility gap. In a first approach, the solidification path of the Al-20wt.%Sn-10wt.%Cu alloy has been determined by the use of the Thermo-Calc software [19] and the COST 507 database [20]. Figure 1 depicts the successive steps during cooling (from the right side to the left side) in the form of the calculated phase fraction as a function of temperature.

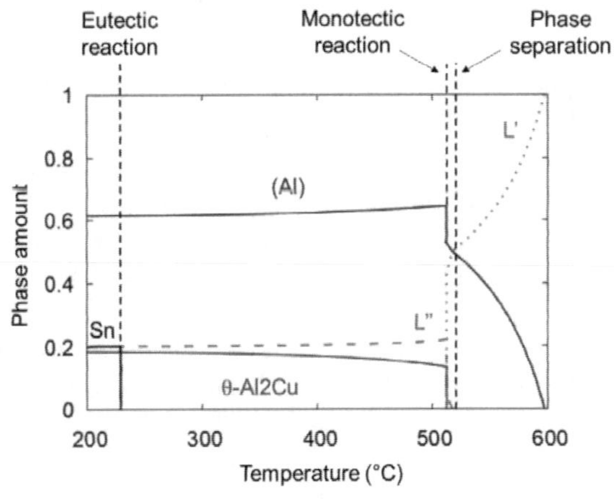

Figure 1. Thermodynamic calculation of the phase evolution during cooling in Al-20wt.%Sn-10wt.%Cu.

The predicted solidification sequence is in agreement with previous studies found in the literature [4,21] and takes place as follows. It commences with the solidification of the α-Al phase and the gradual increase of the liquid phase L′ composition in Sn and Cu due to solute rejection. When the liquid phase L' reaches the boundary of the miscibility gap, phase separation occurs with the formation of Sn-rich liquid L″. At lower temperature, the remaining liquid L′ is transformed during a monotectic reaction that induces the concomitant formation of a large quantity of α-Al and θ-Al_2Cu solid phases. Finally, the remaining liquid L″ is consumed through a eutectic reaction that forms the solid Sn phase.

3.2. In-situ observation of the solidification sequence

X-radiography has been used to visualise in-situ the solidification sequence of an Al-20wt.%Sn-10wt.%Cu sample. The sample was firstly directionally melted from the top and towards the bottom part of the sample. A smooth and planar solid-liquid interface was obtained after a stabilization time of approximately 30 minutes. Then, solidification was triggered by cooling down the two heaters. A selection of radiographs recorded during the solidification is shown in Figure 2. At first, a columnar dendritic microstructure developed (Figure 2a), which corresponds to the α-Al phase. These dendrites appear in light grey while the surrounding liquid appears darker due to the rejection of Cu and Sn solutes that have higher X-ray absorption coefficients. It is already possible to notice that secondary dendrite arms do not develop perpendicular to the primary trunks as usually expected in aluminum alloys, but with a lower angle.

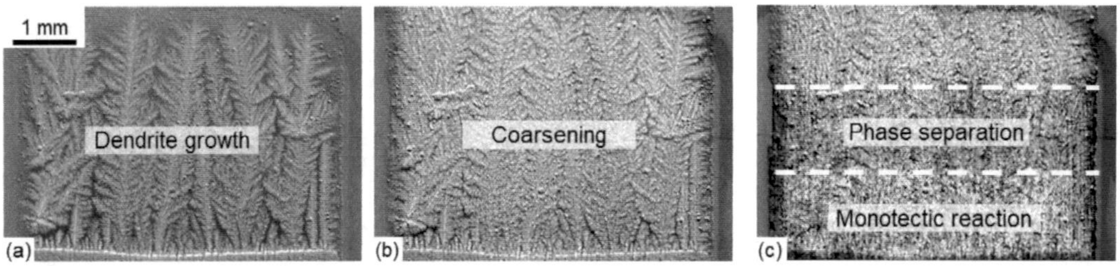

Figure 2. Sequence of radiographs showing: (a) the columnar dendritic growth ($t = t_0 + 73$ s), (b) the microstructure coarsening ($t = t_0 + 121$ s) and (c) the phase separation and monotectic reaction in the mushy zone ($t = t_0 + 162$ s). ($R = 0.45$ K/s, $G_{app} = 5.5$ K/mm).

Cooling and coarsening in the mushy zone led to the formation of thicker microstructures with time (Figure 2b). A few minutes later, the formation of a dark layer in the mushy zone is visible, that crossed the field of view from bottom to top (middle of Figure 2c). This upper boundary of the dark layer is the beginning of the liquid phase separation as expected from the theoretical solidification path, with the formation of many black droplets (a few tens of microns in size) in between the dendrites. Those droplets are Sn-rich liquid L″, having a higher X-ray absorption than the other phases. The lower boundary of the dark layer corresponds to the beginning of the monotectic reaction. During this step, a large amount of α-Al and θ-Al_2Cu are expected to form, which are mainly composed of aluminum and consequently have lower X-ray absorption coefficients. Therefore, α-Al and θ-Al_2Cu solid formations induce a brightening of the mushy zone on the radiographs. Finally, after approximately 15 minutes, the crossing of a last front induced a weak darkening of the mushy zone, unfortunately not distinguishable on static images but visible on videos. This last step of the solidification sequence is attributed to the solidification of the remaining Sn-rich liquid droplets with to the final eutectic reaction, in agreement with the theoretical solidification sequence.

3.3. Growth orientation of the dendrite arms

A careful examination of the dendrite morphology highlights that the secondary dendrite arms do not develop perpendicular to the primary trunks as usually expected in aluminium alloys, but with a lower angle (Figure 3a). The growth orientation of the dendrite arms was determined unambiguously by post-mortem analyses using EBSD (Electron Backscatter Diffraction). Orientation maps were collected using a Zeiss Gemini 500 field-emission gun scanning electron microscope (FEG-SEM) equipped with an EDAX Hikari Super acquisition system.

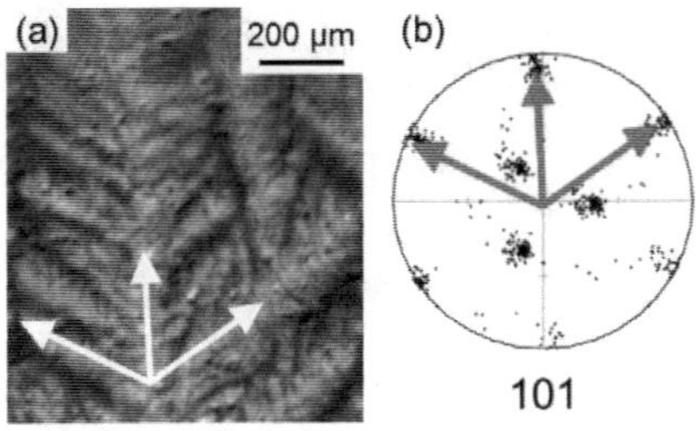

Figure 3. (a) Close-up of a radiograph showing an aluminium dendrite and (b) corresponding (101) pole figure obtained by EBSD.

The comparison between the arm growth directions (Figure 3a) and the (101) pole figure obtained for a dendritic crystal (Figure 3b) confirms that the primary trunk and secondary arms have been developing in <110> directions instead of the common <100> directions. Further experiments with various amounts of Sn and Cu are planned to determine if a dendrite orientation transition (DOT) can occur in the ternary Al-Sn-Cu system, as also reported in binary systems [22–24].

4. Conclusion

The solidification of the ternary Al-20wt.%Sn-10wt.%Cu alloy was investigated by means of in-situ X-radiography. The experimental solidification sequence agrees with calculations performed using a computational thermodynamics software. In addition, EBSD analyses showed that the dendrite arms developed along <110> crystallographic axes instead of the common <100> directions. Investigations are planned to determine how this different orientation from the standard for Al-alloys imposes trapping of tin in the form of pockets between primary and secondary dendritic arms, consequently affecting the desired properties for applications of these alloys.
Further analyses will deepen the comparison of the experimental results with the theoretical solidification path, such as the accurate determination of the phase transformation temperature by using differential scanning calorimetry (DSC). Similar investigations will also be conducted to investigate the solidification of Al-Sn-Cu alloys with various compositions and the conditions for a DOT to occur in this system. The characterisation of the droplet motion and phase distribution is also in progress.

Acknowledgement

This work is supported by the French National Space Agency (CNES). The authors would also like to thank the Swedish Space Corporation (SSC) for the development of the SFINX facility and the technical support.

References

[1] Fang X and Fan Z 2006 Rheo-diecasting of Al–Si–Pb immiscible alloys *Scr. Mater.* **54** 789–93
[2] Kong C J, Brown P D, Harris S J and McCartney D G 2005 The microstructures of a thermally sprayed and heat treated Al–20 wt.%Sn–3 wt.%Si alloy *Mater. Sci. Eng. A* **403** 205–14
[3] Bertelli F, Brito C, Ferreira I L, Reinhart G, Nguyen-Thi H, Mangelinck-Noël N, Cheung N and Garcia A 2015 Cooling thermal parameters, microstructure, segregation and hardness in directionally solidified Al–Sn-(Si;Cu) alloys *Mater. Des.* **72** 31–42
[4] Kotadia H R, Doernberg E, Patel J B, Fan Z and Schmid-Fetzer R 2009 Solidification of Al-Sn-Cu Based Immiscible Alloys under Intense Shearing *Metall. Mater. Trans. A* **40** 2202–11
[5] Kotadia H R, Das A, Doernberg E and Schmid-Fetzer R 2011 A comparative study of ternary Al–Sn–Cu immiscible alloys prepared by conventional casting and casting under high-intensity ultrasonic irradiation *Mater. Chem. Phys.* **131** 241–9
[6] Ratke L and Diefenbach S 1995 Liquid immiscible alloys *Mater. Sci. Eng. R Rep.* **15** 263–347
[7] Jiang H, Li S, Zhang L, He J and Zhao J 2019 Effect of microgravity on the solidification of aluminum–bismuth–tin immiscible alloys *Npj Microgravity* **5** 1–9
[8] Akamatsu S and Nguyen-Thi H 2016 In situ observation of solidification patterns in diffusive conditions *Acta Mater.* **108** 325–46

[9] Schaffer P L, Mathiesen R H, Arnberg L, Sabatino M D and Snigirev A 2008 In situ investigation of spinodal decomposition in hypermonotectic Al–Bi and Al–Bi–Zn alloys *New J. Phys.* **10** 053001
[10] Lu W, Zhang S, Zhang W, Kaptay G, Yu J, Fu Y and Li J 2015 Direct observation of the segregation driven by bubble evolution and liquid phase separation in Al–10 wt.% Bi immiscible alloy *Scr. Mater.* **102** 19–22
[11] Rakete C, Baumbach C, Goldschmidt A, Samberg D, Schroer C G, Breede F, Stenzel C, Zimmermann G, Pickmann C, Houltz Y, Lockowandt C, Svenonius O, Wiklund P and Mathiesen R H 2011 Compact x-ray microradiograph for in situ imaging of solidification processes: Bringing in situ x-ray micro-imaging from the synchrotron to the laboratory *Rev. Sci. Instrum.* **82** 105108
[12] Klein S, Braeuer D, Becker M, Knipstein A, Meckel S, Sondermann E and Kargl F 2016 X-RISE - A Multifunctional X-ray Radiography Device for Parabolic Flights and Laboratory Use *Int. J. Microgravity Sci. Appl.* **33** 330405
[13] Nguyen-Thi H, Reinhart G, Salloum-Abou-Jaoude G, Browne D J, Murphy A G, Houltz Y, Li J, Voss D, Verga A, Mathiesen R H and Zimmermann G 2014 XRMON-GF Experiments Devoted to the in Situ X-ray Radiographic Observation of Growth Process in Microgravity Conditions *Microgravity Sci. Technol.* **26** 37–50
[14] Ngomesse F, Reinhart G, Soltani H, Zimmermann G, Browne D J, Sillekens W and Nguyen-Thi H 2021 In situ investigation of the Columnar-to-Equiaxed Transition during directional solidification of Al–20 wt.%Cu alloys on Earth and in microgravity *Acta Mater.* **221** 117401
[15] Xavier M G C, Reyes R A V, Gomes L F, Spinelli J E, Mangelinck-Noël N, Nguyen-Thi H and Reinhart G 2020 Combined growth of α-Al and Bi in a Al-Bi-Cu monotectic alloy analyzed by in situ X-ray radiography *J. Cryst. Growth* **536** 125592
[16] Soltani H, Reinhart G, Benoudia M C, Zahzouh M and Nguyen-Thi H 2019 Impact of gravity-related phenomena on the grain structure formation: comparative study between horizontal and vertical solidification of a refined Al-20wt.%Cu alloy *IOP Conf. Ser. Mater. Sci. Eng.* **529** 012019
[17] Browne D J, Garcia-Moreno F, Nguyen-Thi H, Zimmermann G, Kargl F, Mathiesen R H, Griesche A and Minster O 2017 Overview of In Situ X-Ray Studies of Light Alloy Solidification in Microgravity *Magnesium Technology 2017* ed K N Solanki, D Orlov, A Singh and N R Neelameggham (Cham: Springer International Publishing Ag) pp 581–90
[18] Buffet A, Nguyen-Thi H, Bogno A, Schenk T, Mangelinck-Noël N, Reinhart G, Bergeon N, Billia B and Baruchel J 2010 Measurement of Solute Profiles by Means of Synchrotron X-Ray Radiography during Directional Solidification of Al-4 wt% Cu Alloys *Mater. Sci. Forum* **649** 331–6
[19] Andersson J-O, Helander T, Höglund L, Shi P and Sundman B 2002 Thermo-Calc & DICTRA, computational tools for materials science *Calphad* **26** 273–312
[20] Ansara I, Rand M and Dinsdale A 1998 COST-507 - Definition of thermochemical and thermophysical properties to provide a database for the development of new light alloys: Thermochemical database for light metal alloys vol 2 (European Commission, Directorate-General for Research and Innovation)
[21] Bertelli F, Brito C, Ferreira I L, Reinhart G, Nguyen-Thi H, Mangelinck-Noël N, Cheung N and Garcia A 2015 Cooling thermal parameters, microstructure, segregation and hardness in directionally solidified Al–Sn-(Si;Cu) alloys *Mater. Des.* **72** 31–42
[22] Gonzales F and Rappaz M 2006 Dendrite growth directions in aluminum-zinc alloys *Metall. Mater. Trans. A* **37** 2797–806
[23] Becker M, Dantzig J A, Kolbe M, Wiese S T and Kargl F 2019 Dendrite orientation transition in AlGe alloys *Acta Mater.* **165** 666–77
[24] Wang L, Hoyt J J, Wang N, Provatas N and Sinclair C W 2020 Controlling solid-liquid interfacial energy anisotropy through the isotropic liquid *Nat. Commun.* **11** 724

Microstructure evolution induced by solidification and ferrite–austenite massive-like transformation in Fe–C alloys

S Tsuji, T Nanri, T Narumi, R Katsube, H Yasuda

Department of Materials Science and Engineering, Kyoto University, Sakyo, Kyoto 606-8501, Japan

E-mail: yasuda.hideyuki.6s@kyoto-u.ac.jp

Abstract. The coarsening of austenite grains produced through a massive-like transformation from the ferrite to the austenite phase during/after solidification in Fe–0.45mass%C steel was examined through time-resolved X-ray diffraction measurements using time-resolved computed tomography. The δ phase solidification resulted in a single δ grain in the observation region (0.6 mm in diameter, 0.1 mm in height). The massive-like transformation produced multiple γ grains, and coarsening immediately followed the transformation. The coarsening in the observation region was completed even within 200 s. A correlation of crystallographic orientation between the γ grains was found in the γ grain structure. The massive-like transformation frequently produced fcc (111) twinned grains, and the twinned austenite grains tended to remain during coarsening because the twin interface energy is lower than the grain boundary energy. The massive-like transformation following the solidification contributed to the texture formation in the γ grain structure.

Introduction

Time-resolved and in-situ X-ray imaging for steel solidification has proved that ferrite (body-centred cubic, referred to as the δ phase) undercooled below the peritectic temperature massively transforms to austenite (face-centred cubic, referred to as the γ phase)[1, 2]. The transformation is referred to as massive-like transformation. Figure 1 shows a typical transformation in Fe–0.45mass%C steel, which was observed by X-ray transmission imaging [3]. γ grain coarsening was observed in the dendrite arms after the massive-like transformation. In addition, the massive-like transformation was selected even at a growth velocity of 50 μm/s during unidirectional solidification in Fe–0.3mass%C steel [4]. The critical growth velocity from the diffusion-controlled γ phase growth, which can be classified as peritectic solidification, to the massive-like transformation was as low as 10 μm/s. Thus, massive-like transformation is commonly selected in conventional solidification processes. In the Fe–18Cr–20Ni alloy (mass%), in which solidification of the γ phase (γ solidification) occurs in equilibrium, the metastable δ phase nucleated and the massive-like transformation triggered γ solidification [5, 6]. Details on the massive-like transformation can be found in [7].

An understanding of γ grain coarsening is of interest because coarse ferrite grains produced at the boundary of the coarsened γ grains in subsequent cooling after δ–γ transformation can cause defects during continuous casting. The γ grain structure was examined by quenching unidirectionally solidifying specimens [8-10]. Refined columnar γ grains always existed ahead of the coarse columnar grain region. The discontinuous changes in γ grain size was not simply explained by conventional grain coarsening.

IOP Conf. Series: Materials Science and Engineering 1274 (2023) 012055 doi:10.1088/1757-899X/1274/1/012055

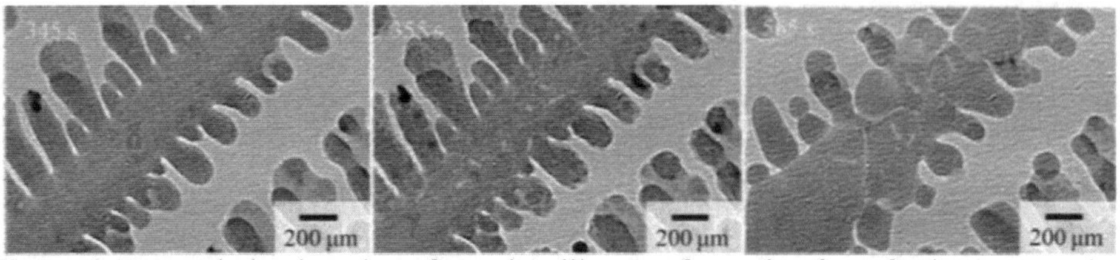

Figure 1. Transmission imaging of massive-like transformation from ferrite to austenite in Fe–0.45mass%C alloy. Blue arrows indicate grain boundaries of γ grains [3].

Our previous work using XRD [11] showed that the massive-like transformation consistently explained the formation of refined columnar grains.

Time-resolved computed tomography (referred to as 4D-CT) has been developed to observe solidification structures [12-17]. A filtering technique that used a phase field model was proposed to improve the image quality of Fe-based alloys [18, 19]. 4D-CT and XRD have been combined to measure the crystallographic orientation [18, 20]. This study used the combined technique, which allows us to evaluate coarsening kinetics. This paper demonstrates the time evolution of crystallographic orientation distribution during γ grain coarsening after massive-like transformation and the correlation of crystallographic orientation between γ grains.

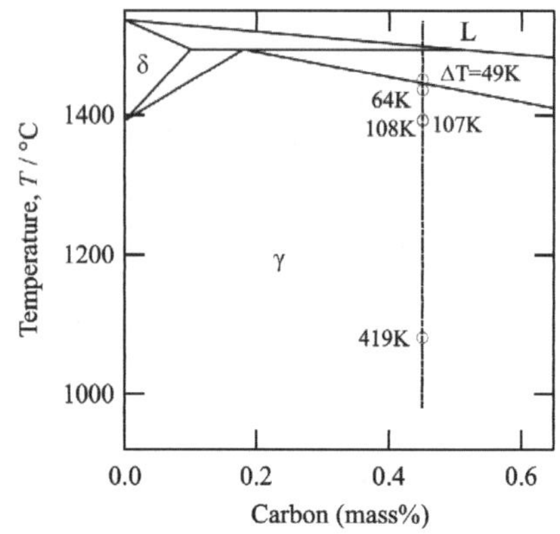

Figure 2. Fe–C binary phase diagram and nucleation temperatures of the γ phase in Figs. 3, 7 and 8.

Experiments

1.1. Specimen

Fe–0.45C–0.6Mn–0.3Si (mass%) alloys were made from pure Fe, C, Mn and Si in an arc melting furnace. Because the decrease in peritectic temperature from Mn and Si addition is less than 5 K compared with that in the binary Fe–C system [21], this study considers solidification using the binary phase diagram, as shown in Figure 2. In equilibrium, the δ phase solidifies as the primary phase. The difference between the liquidus temperature and the peritectic temperature is approximately 5–10 K, and the solid fraction expected from the phase diagram is less than 0.2 at the peritectic temperature. The dimensions of the XRD specimen were a diameter of 0.6 mm and length of approximately 2 mm. The rod-shaped specimen was inserted into an Al_2O_3 tube (inner diameter of 0.6 mm and outer diameter of 1.2 mm).

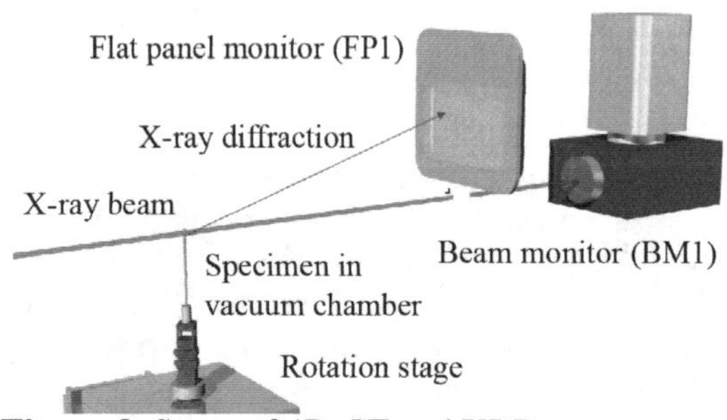

Figure 3. Setup of 4D-CT and XRD.

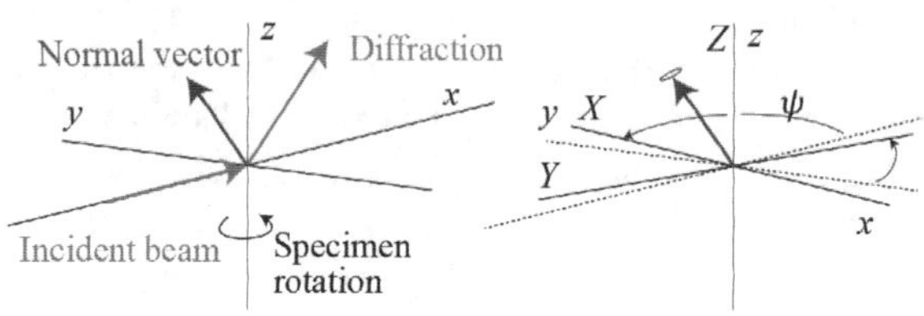

Figure 4. Coordinate systems for the X-ray beam and specimen. ψ is the specimen rotation angle.

1.2. Time-resolved X-ray diffraction

Time-resolved XRD was performed to observe the coarsening of γ grains that were produced through the massive-like transformation. Figure 3 shows the setup for time-resolved XRD using 4D-CT. The transmission images obtained by the beam monitor BM1 were used to adjust the X-ray beam position and confirm the beam height before the XRD measurement. X-ray diffraction images were observed at a frame rate of 50 fps with the flat panel monitor FP1 (observation area: 90 mm × 60 mm; pixel size: 100 μm × 100 μm). The specimen–detector distance was approximately 200 mm. X-rays with an energy of 37.7 keV were used. The beam for the XRD was 100 μm in height. Thus, XRD in the 0.6-mm-diameter and 0.1-mm-high specimen was measured by FP1. The specimen was rotated at 0.45 rps and the crystallographic orientation distribution was obtained every 4 s. During the observations, the liquid phase was cooled at a constant rate of 0.33 K/s. The cooling was suspended when the δ–γ transformation was detected by FP1, and the specimen was kept at a constant temperature to observe change in crystallographic orientation distribution due to coarsening.

1.3. Crystallographic orientation distribution

Figure 4 shows the coordinate systems of the X-ray optics (*x*–*y*–*z*) and specimen (*X*–*Y*–*Z*). The incident X-ray beam is parallel to the *x*-axis and the rotation axis is parallel to the *z*-axis. The specimen coordinate system coincides with the X-ray optics coordinate system at rotation angle $\psi = 0$. The unit vectors $\boldsymbol{S}_0$ (incident beam) and $\boldsymbol{S}_1$ (diffraction beam) are given by

$$\boldsymbol{S}_0 = (1 \quad 0 \quad 0), \boldsymbol{S}_1 = (u \quad v \quad w). \tag{1}$$

A diffraction spot on FP1 gives a normal vector $\boldsymbol{n}_s$ of the diffraction plane. The normal vector $\boldsymbol{n}_s$ is given by

$$\boldsymbol{n}_s{}^T = \begin{pmatrix} \cos\psi & -\sin\psi & 0 \\ \sin\psi & \cos\psi & 0 \\ 0 & 0 & 1 \end{pmatrix}^{-1} \left(\frac{\boldsymbol{S}_1 - \boldsymbol{S}_0}{|\boldsymbol{S}_1 - \boldsymbol{S}_0|} \right)^T . \tag{2}$$

The crystallographic orientation distribution was obtained by integrating the normal vectors over a 360° rotation. In the present setup, more than 99% of the $(111)_\gamma$ planes in the observation region were detected during the 360° rotation. Because the number and size of grains were not measured directly by time-resolved XRD, an occupancy of diffractions in the solid angle, f_d, was introduced to evaluate γ grain coarsening. The occupancy is defined by

$$f_d = \frac{\int_{\Omega_{obs}} \delta(\boldsymbol{n}_s) d\Omega}{\int_{\Omega_{obs}} d\Omega}, \tag{3}$$

where Ω_{obs} is the solid angle over which the normal vectors of the diffraction planes can be detected in the setup. If X-ray diffraction is detected at normal vector $\boldsymbol{n}_s$, then $\delta(\boldsymbol{n}_s) = 1$. If it is not detected, then $\delta(\boldsymbol{n}_s) = 0$. As mentioned in the previous paragraph, $\Omega_{obs}/(2\pi) > 0.99$. Thus, the occupancy, f_d, is an index for γ grain coarsening.

Results and Discussion

1.4. Coarsening of γ grains after massive-like transformation

Figure 5 shows X-ray diffraction spots over 360° rotation after a massive-like transformation. Each $(111)_\gamma$ plane in the observation region caused two diffraction spots on FP1 over 360° rotation in the present configuration. The massive-like transformation occurred at 64 K below the liquidus temperature. The undercooling from the liquidus temperature often exceeded 400 K. Thus, the undercooling (64 K) was relatively low. According to Figure 2, the massive-like transformation occurred during the δ solidification. Many diffraction spots of 111_γ and 200_γ indicated multiple γ grains in a single δ grain (0.6 mm in diameter, 0.1 mm in height). The number of diffraction spots decreased with an increase in holding time. The eight diffraction spots of the $(111)_\gamma$ plane (yellow arrows) at 600 s were caused by four different $(111)_\gamma$ planes, suggesting that a single γ grain remained. The crystallographic orientation distributions of the $(111)_\gamma$ and $(200)_\gamma$ planes are shown in Figure 6. As shown in the distribution at 60 s, coarsening of γ grains occurred immediately after the massive-like transformation. The normal vectors

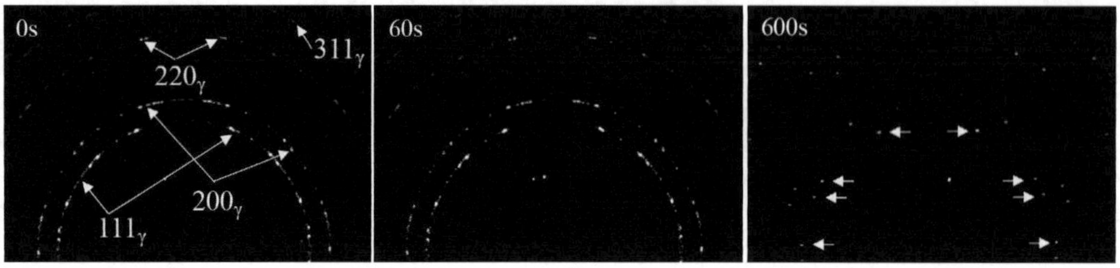

Figure 5. XRD spots of γ grains after massive-like transformation. The X-ray diffractions are integrated over 360° rotation. The arrows at 600 s indicate diffraction spots of the $(111)_\gamma$ plane.

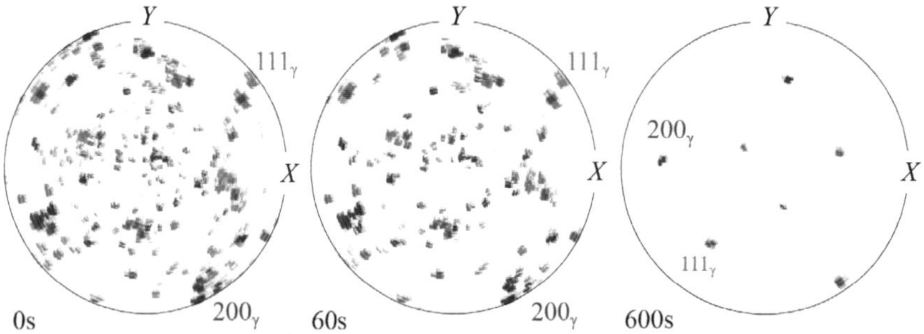

Figure 6. Crystallographic orientation distribution of γ grains after massive-like transformation. The normal vectors of the $(111)_\gamma$ and $(200)_\gamma$ planes are integrated over 360° rotation (4 s). The normal vectors of $(111)_\gamma$ and $(200)_\gamma$ are indicated by red and blue, respectively.

of the $(111)_\gamma$ and $(200)_\gamma$ planes indicated that a single γ grain remained in the observation region. Thus, the coarsening was complete before 600 s.

Figure 7 shows the occupancy defined by eq. (3) as a function of holding time. A massive-like transformation occurred in the liquid and δ phases at ΔT=49 K. Complete coarsening took approximately 300 s. However, coarsening was completed within 100 s when the undercooling exceeded 100 K. In the relatively low undercoolings, the γ grains could be isolated by the remaining liquid phase, which resulted in a reduction in the coarsening rate. The results are consistent with previous work [8-10]. On the basis of the XRD observations, γ grain coarsening could be completed within several minutes after the massive-like transformation. Thus, the γ grains are measurably coarsened in conventional casting processes such as continuous casting of steel.

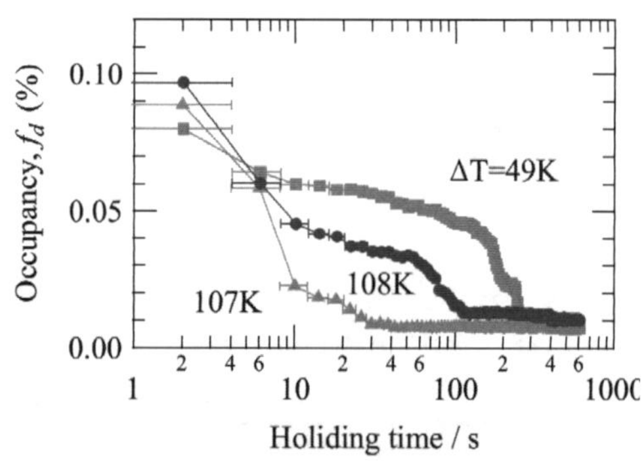

Figure 7. Occupancy of diffractions, f_d, defined by eq. (3), as a function of holding time. ΔT is an undercooling of γ nucleation from the liquidus temperature.

1.5. Formation of twinned γ grains through massive-like transformation

Figure 8 shows the dihedral angle distributions of $(111)_\gamma$ planes at 0 s and 600 s after the massive-like transformation. The undercooling from the liquidus temperature was 419 K. The blue lines correspond to the random distribution. The dihedral angle of the $(111)_\gamma$ plane in the same grain is indicated by γ_1–γ_1, and the dihedral angles of the $(111)_\gamma$ planes between the fcc (111) twinned grains are indicated by γ_1–γ_2. γ_1–γ_3 means that the γ_1 and γ_3 grains are not twinned, but they have a twin relationship with grain γ_2. The fraction of dihedral angles at 0 s exhibits several peaks, which coincide with peaks γ_1–γ_2 and γ_1–

γ_3. Thus, the massive-like transformation produced twinned grains. The formation of twinned grains could originate in γ phase nucleation (γ nucleation) because the twin variants were selected locally to reduce the elastic energy that was induced by the volume change in the δ–γ transformation.

Peaks related to the twinned grains γ_1–γ_2 at 600 s were sharpened and other peaks were negligibly small. Only the twinned grains remained in the observation region (0.6 mm in diameter, 0.1 mm high). Coarsening is generally driven by the curvature effect because the twin interface energy is relatively low compared with the grain boundary energies. Thus, the twinned grains preferentially remained. The sharpened peaks also suggest that low-angle grain boundaries that were induced by the massive-like transformation were eliminated to release the stored energy. As a result, the texture in which twinned grains are frequently observed evolved through the massive-like transformation. It should be noted that the texture is unexpected when considering common peritectic solidification.

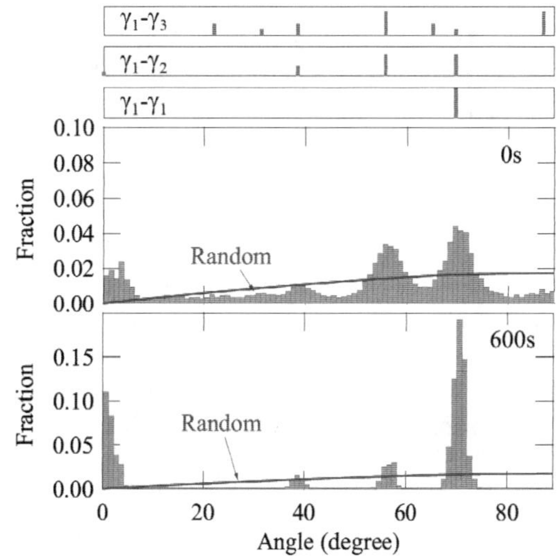

Figure 8. Angle between normal vectors of $(111)_\gamma$ planes detected by time-resolved XRD.

Summary

The evolution of a γ grain structure in Fe–0.45mass%C steel was examined through time-resolved XRD using the 4D-CT setup. Multiple γ grains produced through the massive-like transformation coarsened immediately after it. The coarsening in the observation region was nearly completed within 40 s when the massive-like transformation occurred in the highly undercooled δ phase. The massive-like transformation frequently produces fcc (111) twinned γ grains to reduce local strains in the δ–γ transformation. The twinned grains preferred to remain during coarsening because the twin interface energy was relatively low. As a result, the texture in the γ grain structure evolved. The time-resolved and in-situ XRD measurements proved that massive-like transformation influenced the γ grain structure significantly.

Acknowledgments

In situ observations using synchrotron radiation X-rays were performed as general projects (Nos. 2019B1482, 2020A1564, 2020A0769, 2021A1499, 2021B1560) at the BL20XU and the BL47XU of SPring-8 (JASRI), Japan. The study was supported by a Grant-in-Aid for Scientific Research (S) (No. 17H06155). We thank Laura Kuhar, PhD, from Edanz (https://jp.edanz.com/ac) for editing a draft of this manuscript.

References

[1] Yasuda H, Nagira T, Yoshiya M, Sugiyama A, Nakatsuka N, Kiire M, Uesugi M, Uesugi K, Umetani K and Kajiwara K 2012 Massive transformation from δ phase to γ phase in Fe-C alloys and strain induced in solidifying shell *IOP Conf Ser Mater Sci Eng* **33** 012036-

[2] Yasuda H, Nagira T, Yoshiya M, Uesugi M, Nakatsuka N, Kiire M, Sugiyama A, Uesugi K and Umetani K 2011 In-situ observation of peritectic solidification in Sn-Cd and Fe-C alloys *IOP Conf Ser Mater Sci Eng* **27** 012084-

[3] Yasuda H, Morishita K, Nakatsuka N, Nishimura T, Yoshiya M, Sugiyama A, Uesugi K and Takeuchi A 2019 Dendrite fragmentation induced by massive-like δ–γ transformation in Fe–C alloys *Nature Comm* **10** 3183-

[4] Nishimura T, Morishita K, Yoshiya M, Nagira T and Yasuda H 2020 Time-resolved and in-situ observation of δ–γ transformation during unidirectional solidification in Fe–C alloys *ISIJ Int* **60** 930-8

[5] Nishimura T, Matsubayashi R, Morishita K, Yoshiya M, Nagira T and Yasuda H 2021 Erratum to "Selection of the massive-like δ-γ transformation due to nucleation of metastable δ phase in Fe-18 Mass%Cr-Ni alloys with Ni contents of 8, 11, 14 and 20 mass%" [ISIJ International, Vol. 59 (2019), No. 3, pp. 459-465] *ISIJ Int* **61** 1053-

[6] Nishimura T, Matsubayashi R, Morishita K, Yoshiya M, Nagira T and Yasuda H 2019 Selection of the massive-like δ -γ transformation due to nucleation of metastable δ phase in Fe-18 Mass%Cr-Ni alloys with Ni contents of 8, 11, 14 and 20 mass% *ISIJ Int* **59** 459-65

[7] Yasuda H, Morishita K, Yoshiya M and Narumi T 2020 Transformation from ferrite to austenite during/after solidification in peritectic steel systems: an X-ray imaging study *ISIJ Int* **60** 2755-64

[8] Tsuchiya S, Ohno M, Matsuura K and Isobe K 2011 Formation mechanism of coarse columnar γ grains in as-cast hyperperitectic carbon steels *Acta Mater* **59** 3334-42

[9] Ohno M, Tsuchiya S and Matsuura K 2012 Austenite Grain Growth in Peritectic Solidified Carbon Steels Analyzed by Phase-Field Simulation *Metall Mater Trans A* **43A** 2031-42

[10] Ohno M, Maruyama M and Matsuura K 2013 A critical condition for the formation of a coarse columnar γ grain structure in a peritectic solidified carbon steel *Acta Mater* **61** 7334-41

[11] Yasuda H, Suga T, Ichida K, Narumi T and Morishita K 2020 In situ observation of austenite coarsening induced by massive-like transformation during solidification in Fe–C alloys *IOP Conf Ser Mater Sci Eng* **861** 012051-

[12] Ludwig O, Dimichiel M, Salvo L, Suéry M and Falus P 2005 In-situ three-dimensional microstructural investigation of solidification of an Al-Cu alloy by ultrafast x-ray microtomography *Metall Mater Trans A* **36** 1515-23

[13] Aagesen L K, Fife J L, Lauridsen E M and Voorhees P W 2011 The evolution of interfacial morphology during coarsening: A comparison between 4D experiments and phase-field simulations *Scripta Mater* **64** 394-7

[14] Sillekens W H, Casari D, Mirihanage W U, Terzi S, Mathiesen R H, Salvo L, Daudin R, Lhuissier P, Guo E and Lee P D 2016 The Use of In Situ X-ray Imaging Methods in the Research and Development of Magnesium-Based Grain-Refined and Nanocomposite Materials *JOM* **68** 3042-50

[15] Cai B, Karagadde S, Rowley D, Marrow T J, Connolley T and Lee P D 2015 Time-resolved synchrotron tomographic quantification of deformation-induced flow in a semi-solid equiaxed dendritic Al–Cu alloy *Scripta Mater* **103** 69-72

[16] Cai B, Karagadde S, Yuan L, Marrow T J, Connolley T and Lee P D 2014 In situ synchrotron tomographic quantification of granular and intragranular deformation during semi-solid compression of an equiaxed dendritic Al–Cu alloy *Acta Mater* **76** 371-80

[17] Gibbs J W, Mohan K A, Gulsoy E B, Shahani A J, Xiao X, Bouman C A, De Graef M and Voorhees P W 2015 The Three-Dimensional Morphology of Growing Dendrites *Scientific Reports* **5** 1-9

[18] Yasuda H, Hashimoto T, Sei N, Morishita K and Yoshiya M 2019 Investigation using 4D-CT of massive-like transformation from the δ to γ phase during and after δ-solidification in carbon steels *IOP Conf Ser Mater Sci Eng* **529** 012013-

[19] Yasuda H, Kawarasaki T, Tomiyori Y, Kato Y and Morishita K 2019 Characterization of dendritic growth in Fe-C system using time-resolved X-ray tomography and physics-based filtering *IOP Conf Ser Mater Sci Eng* **529** 012023-

[20] Nakano K, Narumi T, Morishita K and Yasuda H 2020 Characterization of growing dendrites in CrMnFeCoNi high-entropy alloy by time-resolved and in-situ tomography *Maater Trans* **61** 596-604

[21] Yamada H, Sakurai T and Takenouchi T 1990 Effect of Alloying Elements on the Peritectic Temperature in Low-alloy Steels *Tetsu-to-Hagane* **76** 438-45

Growth of Si particles and Al dendrites during solidification of a hypereutectic Al-Si-Cu alloy

G Zimmermann[1], L Sturz[1], E Schaberger-Zimmermann[2], Ch Pickmann[1]

[1]ACCESS e.V., Intzestrasse 5, D-52072 Aachen, Germany

[2]Foundry Institute of RWTH Aachen University, Intzestrasse 5, D-52072 Aachen, Germany

Abstract. Hypereutectic Al-Si alloys containing up to 16-20 wt% Si and often several wt% Cu are widely applied in automobile and aerospace industry, because of their high wear resistance and high-temperature resistance. The mechanical and physical properties of these alloys are strongly influenced by the size and morphology of the primary silicon particles. Here, rod-like bulk samples of hypereutectic Al-18wt%Si-10wt%Cu alloys were solidified in a Bridgman-Stockbarger furnace with different solidification parameters. Microstructure analysis of longitudinal and radial cross-sections demonstrated nearly equiaxed growth of Silicon particles in the 3D samples. At low cooling rate a few but larger Si particles exist compared to many and smaller Si particles for solidification with higher cooling rate. Also the structure of the Al dendrites is much finer at higher cooling rates. Additionally, sheet-like samples were solidified under isothermal cooling conditions. A real-time observation method using X-ray radiography allowed for the in-situ observation. During cooling down below the liquidus temperature, facetted Si particles are the first to nucleate. Reaching the nucleation temperature of Al, growth of non-facetted Al dendrites was observed simultaneously in the whole melt, generally not starting from the surface of existing Si particles. Some Al dendrites growing in the plane of the sheet-like samples develop rather rapidly in one direction, whereas other Al dendrites grow more equiaxed, which is attributed to different grain orientations of the Al-rich dendrites in quasi-2D solidification.

1. Introduction

In hypereutectic Al-Si-Cu alloys, strength is increased via a precipitation strengthening mechanism. The mechanical and physical properties of these alloys are strongly influenced by the size and morphology of the primary Si particles, which are mainly controlled by the solidification conditions [1]. Different 3D growth structures of Si particles are described [2] and experimentally reconstructed using serial sectioning [3] and synchrotron X-ray tomography methods [4]. Recently, the nucleation and growth kinetics of primary Si particles during solidification of high-purity hypereutectic Al-18wt%Si-20wt%Cu [5] and Al-22wt%Si-18.8wt%Cu alloys [6-8] was investigated using in-situ micro-focus X-radiography. As a result, the size and morphology of the growing Si particles strongly depend on the cooling rates. The quantitative experimental study of solidification of hypereutectic AlSiCu is extended to Al-18wt%Si-10wt%Cu alloy [9], containing about half the Cu content used in [5-8]. This amount of Cu is still enough to allow for good contrast in in-situ X-ray investigation and will broaden the experimental data basis.

In this work bulk samples of Al-18wt%Si-10wt%Cu alloy were solidified in a Bridgman-Stockbarger furnace with different solidification parameters to investigate the growth of Si particles in 3D environment. Additionally, sheet-like samples were solidified under cooling conditions comparable to the 3D experiments to allow for a real-time observation using in-situ X-ray radiography.

2. Solidification of rod-like Al-18wt%Si10wt%Cu samples

Thermo–Calc software [10] with TCAL4 database was used to calculate the phase diagram and phase fraction evolution as a function of temperature for the Al-18wt%Si-10wt%Cu alloy. It shows that below the liquidus temperature of T_L = 679°C, nucleation and growth of primary Si particles started, followed by growth of Al dendrites (fcc-phase) below 556°C. At the eutectic temperature of 521°C the residual melt solidifies as ternary eutectic.

For sample manufacturing high-purity Al, Si and Cu materials (better than 99.99%) were used. The measurement of the final alloy composition provided Al with 18.0 wt% Si and 10.0 wt% Cu. For directional solidification experiments, cylindrical samples with a diameter of 8 mm were machined and integrated into alumina crucibles. The directional solidification experiments were carried out in a Bridgman-Stockbarger furnace with an argon atmosphere. A constant temperature gradient of 3.0 K/mm along the sample axis was applied. Experiments with furnace movement vertically along the sample and velocities in the range between 10 µm/s and 100 µm/s were performed.

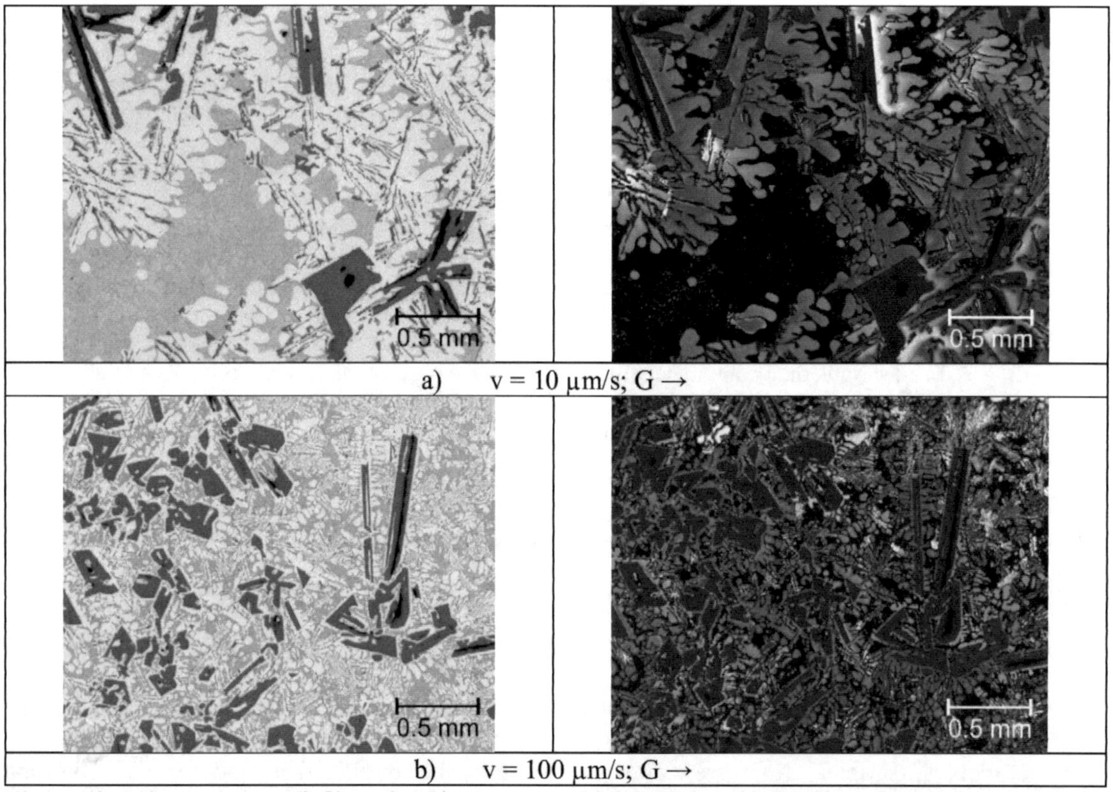

Figure 1. Microstructure (left) and grain structure (right) in longitudinal cross-sections of the Al-18wt%Si-10wt%Cu alloy samples solidified with temperature gradient G = 3 K/mm and solidification velocities a) v = 10 µm/s, and b) v = 100 µm/s (solidification direction from left to right). The microstructure shows facetted Si phase in dark grey, Al dendrites (bright) and eutectic phase in light grey. The different colors of the Al dendrites (right) represent grains with different grain orientations.

To analyze the solidified microstructure and the grain structure in the processed samples longitudinal and radial cross-sections were prepared. Electrolytically etched cross-sections were investigated in a polarized light microscope to visualize Al dendrites with different crystallographic orientations in different colors [11]. **Figure 1** shows the microstructure and grain structure in longitudinal cross-sections of Al-18wt%Si-10wt%Cu alloy samples solidified with temperature gradient G = 3 K/mm and solidification velocities of v = 10 µm/s and v = 100 µm/s. The image sizes are H x B = 2.10 mm x 2.65 mm and the solidification direction was from left to right. The microstructure shows facetted Si phase in dark grey, Al dendrites (bright) and eutectic phase in light grey. This three phases can be distinguished clearly due to the different shades of grey. The grain structure shows the different crystallographic orientations of the Al dendrites represented by different colors, and the facetted Si particles in red brown.

In **Figure 2** the microstructure and grain structure in radial cross-sections of the same parts of samples as given in Fig. 1 are shown. The size of the Si particles as well as the Al dendrites are similar in longitudinal and radial cross-sections for the same process parameters. This indicates a growth behavior nearly independent of the applied moderate temperature gradient. The number and size of the Si phases strongly depend on the cooling rate. At the low solidification velocity of v = 10 µm/s a few but larger Si particles exist compared to solidification with velocity of v = 100 µm/s showing more and smaller Si particles. Also the structure of the Al dendrites is much finer at higher cooling rates.

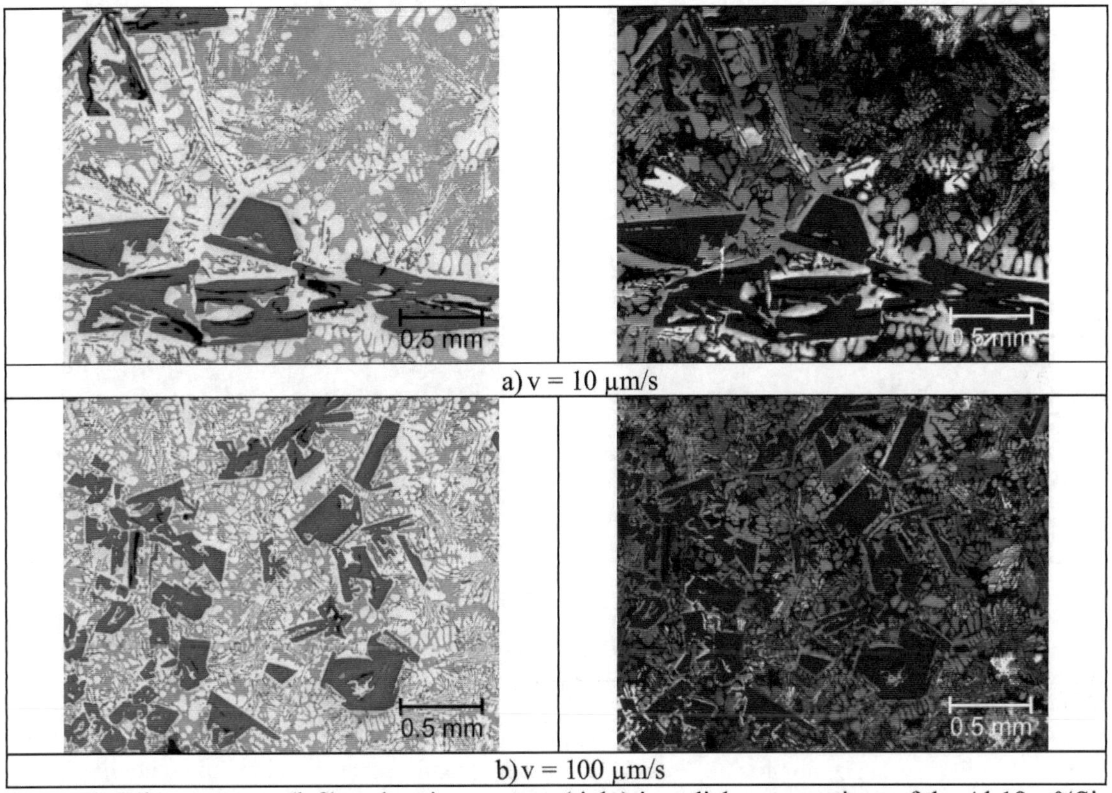

Figure 2. Microstructure (left) and grain structure (right) in radial cross-sections of the Al-18wt%Si-10wt%Cu alloy samples solidified with temperature gradient G = 3 K/mm and solidification velocities a) v = 10 µm/s, and b) v = 100 µm/s (image size: H x B = 2.10 mm x 2.65 mm; solidification direction perpendicular to image plane).

3. Solidification of sheet-like Al-18wt%Si10wt%Cu samples

From the same Al-Si-Cu alloy material as used for the bulk samples (section 2), sheet-like samples with thickness 150 µm, 50 mm in length and 5 mm in width were prepared. The polished samples were coated with boron nitride and placed between two quartz glass plates of thickness 150 µm, and mounted in a frame structure. This frame structure was placed horizontally in a Bridgman-Stockbarger type furnace, which was used in an isothermal mode and without sample movement. Therefore, isothermal solidification was induced by starting with the same heater temperatures and decreasing the heater temperatures with the same controlled cooling-rates.

For in-situ observation of the solidification behavior X-ray diagnostics was used. The setup mainly consisted of a microfocus X-ray source with a 2 µm focal spot and a high-energy tungsten target, the solidification furnace and an X-ray detector. For the experiment an acquisition rate of 0.5 frames/second was chosen to allow for a sufficient time-resolution while observing especially the growth of the Al dendrites. The magnification of about 6 resulted in a FoV of 3.12 mm x 4.47 mm, which nearly corresponds to the whole sample width of 5.0 mm.

The starting temperatures of both heaters were 800 °C for about 1 hour to fully melt the Si primary phases in the alloy and to allow for a homogenous melt composition. Then, a constant cooling rate of R = 0.027 K/s was applied to both heaters. In **Figure 3** the growth of facetted Si primary phase in Al-18wt%Si-10wt%Cu alloy is shown. The time scale starts with identification of the first Si particles appearing in the FoV.

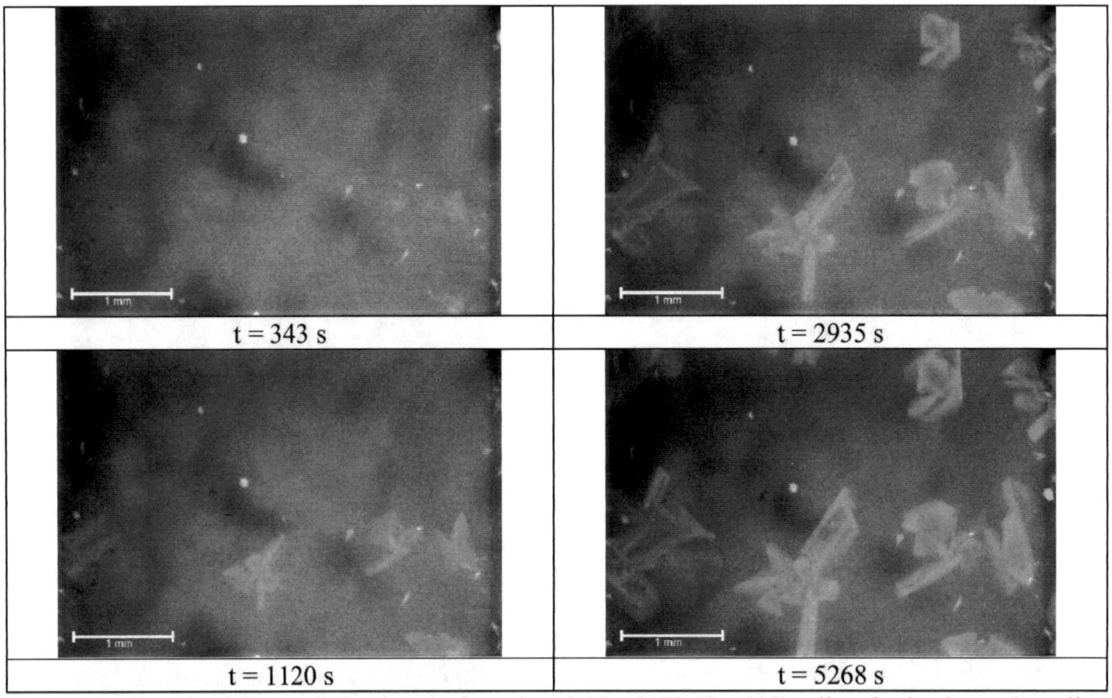

Figure 3. Growth of facetted Si primary phase in Al-18wt%Si-10wt%Cu alloy for isothermal cooling (G = 0 K/mm) with cooling rate R = 0.027 K/s. The time scale starts with identification of the first Si particle in the FoV of size 3.12 mm x 4.47 mm.

The temporal development of the number of Si particles and of the projected area fraction of Si particles is shown in **Figure 4**. The number of Si particles reaches a nearly constant value during cooling down. It is assumed that the Si concentration in the melt between already existing Si particles is reduced, because Si is segregating to existing Si particles and therefore it is not sufficient Si in the melt for nucleation of new Si particles. This behavior is confirmed by the continuously increase of the projected area fraction of Si particles (Fig. 4b).

Figure 5 shows the development of Al dendrites in the melt while cooling down below the nucleation temperature of Al, i.e. 556 °C. The size of the Si particles is nearly unchanged, and in the melt Al dendrites start growing. During cooling down to the eutectic temperature of 521°C more and more Al phase solidifies. In-situ observations (Figure 5) demonstrate that some dendrites growing in the plane of the thin experiment container develop rather rapidly in one direction, whereas other Al dendrites grow more equiaxed. To estimate the velocity of fast growing dendrites, image processing (subtracting the images from the reference image without Al dendrites, i.e. from the image taken at t = 5268 s) was used to identify just the Al phase. A maximum growth velocity of about 0.1 mm/s was measured. Maybe the grain orientation of the Al-rich dendrites play an important role in quasi-2D solidification. A dendrite nucleating with a <100> crystallographic orientation almost perfectly in the plane of the sheet-like sample, would have a growth advantage over dendrites with other orientations. A quantitative analysis of this effect has to be investigated in further experiments. The image shown in Figure 5c was taken just 18s after observation of first Al dendrites (Fig. 5a), corresponding to a temperature decrease of only 0.5

K. Within this small temperature range, Al dendrites starts growing in the whole FoV. The X-ray images demonstrate, that nucleation and early growth of Al dendrites generally not starts from the surface of existing Si particles.

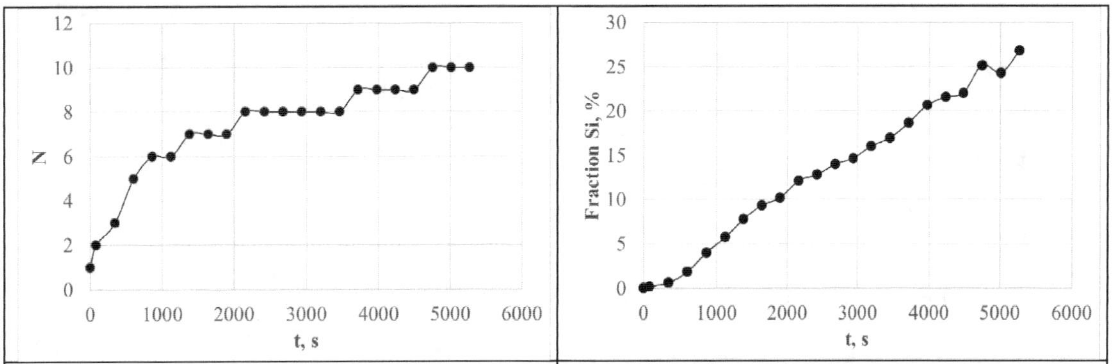

a) number of Si particles | b) area fraction of Si particles

Figure 4. Development of a) the number and of b) the projected area fraction of Si particles for the experiments shown in Figure 3, starting with the first particle being visible in the FoV (t = 0 s)

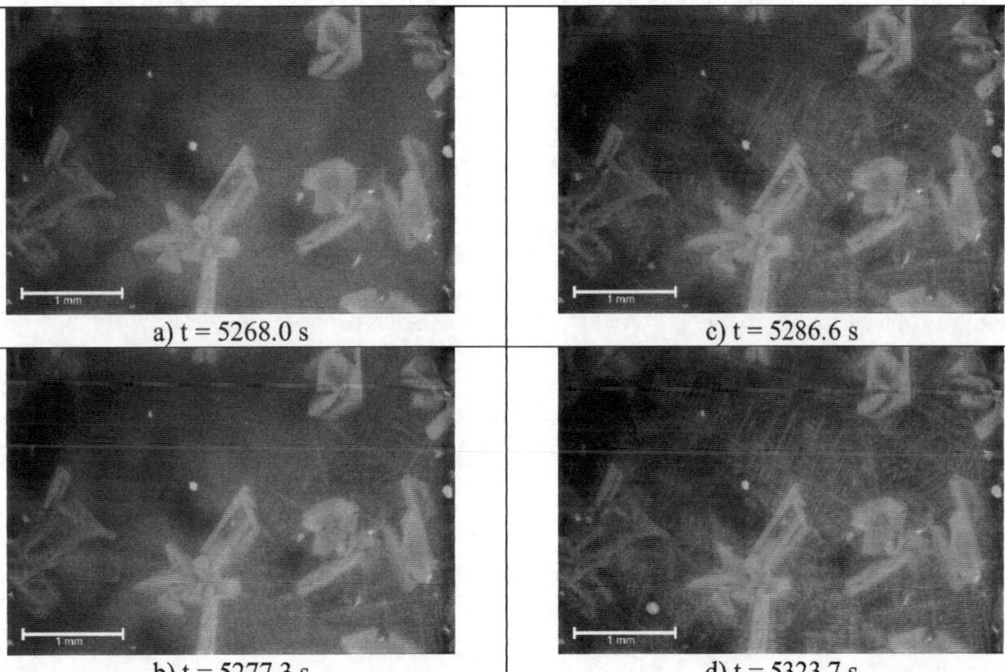

a) t = 5268.0 s | c) t = 5286.6 s
b) t = 5277.3 s | d) t = 5323.7 s

Figure 5. Development of Al dendrites (bright) in Al-18wt%Si-10wt%Cu alloy during isothermal cooling with R = 0.027 K/s. The time scale is same as given in Figure 3. (FoV = 3.12 mm x 4.47 mm)

This behavior is confirmed by microstructure analysis of the solidified sample. **Figure 6** shows the same FoV as given in Figures 3 & 5. The microstructure with facetted Si phase, Al dendrites and eutectic phases is given in Figure 6a, whereas Figure 6b shows the corresponding grain structure. Different grain orientations of the Al phase are observed from electrolytical etching combined with polarized light microscopy [11]. Therefore, different colors in Figure 6b show different grain orientations of the Al dendrites. This analysis also indicates nucleation of Al dendrites nearly independent of existing Si particles.

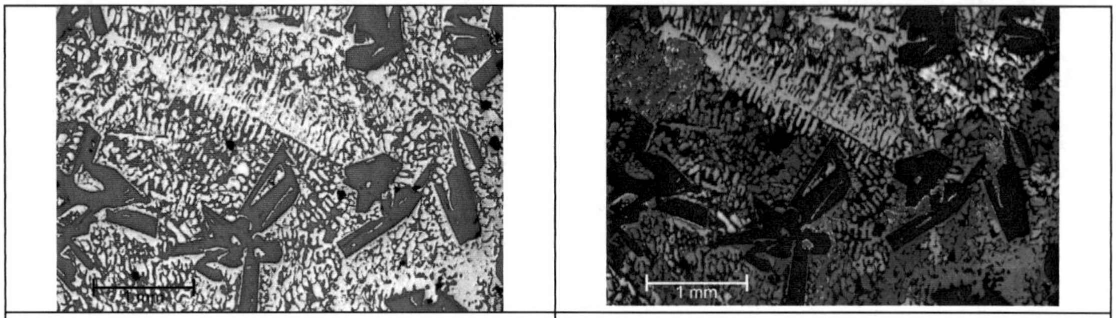

a) b)

Figure 6. Metallic preparation of the same FoV as given in the X-ray images (Figures 3 & 5), showing a) microstructure, and b) grain structure of Al dendrites with different crystallographic orientations represented by different colors.

4. Conclusion

The solidification behavior of Al-18wt%Si-10wt%Cu alloy was investigated in directionally solidified bulk samples as well as in isothermal cooling of sheet-like samples with attached X-ray diagnostics for in-situ detection of the solidification behavior. Both microstructure analysis of cross-sections of the processed 3D bulk samples and in-situ observation in sheet-like samples show nearly equiaxed growth of facetted Si particles. With increasing cooling rate the total number of Si particles increases while their sizes decrease. Also the structure of the Al dendrites is much finer at higher cooling rates. The time-dependent solidification behavior of facetted Si particles and non-facetted Al dendrites is studied in detail for isothermal cooling of sheet-like samples with cooling rate R = 0.03 K/s. In-situ observation shows that for cooling down below the liquidus temperature of Si (T_L = 679 °C), primary Si particles are the first to nucleate and grow with a facetted morphology. Reaching the nucleation temperature of Al (T = 556 °C) growth of non-facetted Al dendrites started, simultaneously in the whole melt and in general not originating from the surface of existing Si particles. Some Al dendrites growing in the plane of the sheet-like samples develop rather rapidly, whereas other Al dendrites grow more equiaxed. This finding is attributed to different grain orientations of the Al-rich dendrites in quasi-2D solidification in thin samples. Summarizing, this paper confirms previous findings concerning solidification of Si particles in sheet-like samples given in [9] and provides a more detailed investigation of nucleation and growth of Al dendrites. Furthermore, it is demonstrated that bulk samples and sheet like samples solidified with the same process parameters show comparable microstructures.

Acknowledgements

This work was financially supported by the German Space Agency DLR under BMWi grant no. FKZ 50WM2043, which is gratefully acknowledged.

References

[1] Xu X L, Jiang Q C 2006 *Materials Science and Engineering* **A437(2)** 451
[2] Wang R, Lu W, Hogan L M 1999 *J. Cryst. Growth* **207** 43
[3] Schaberger-Zimmermann E, Mathes M, Zimmermann G 2016 *JOM* **68/8** 2073
[4] Wang J, Guo Z, Song J L, Hu W X, Li J C, Xiong S M 2018 *Materials & Design* **137** 176
[5] Xu Y J, Casari D, Mathiesen R H, Liu X F, Li Y J 2017 *Proc. 6th Decennial Inter. Conf. on Solidification Processing*, 78
[6] Xu Y, Deng Y, Casari D, Mathiesen R H, Liu X, Li Y 2020 *J. Mater Science* **55** 15621
[7] Xu Y, Deng Y, Casari D, Mathiesen R H, Li Y 2020 *J. Alloys and Compounds* **832** 54948
[8] Xu Y, Deng Y, Casari D, Mathiesen R H, Liu X, Li Y 2021 *J. Alloys and Compounds* **854** 155323
[9] Zimmermann G, Pickmann Ch, Schaberger-Zimmermann E 2022, *J. Cryst. Growth* **585** 126589
[10] Thermo-Calc Software, Computational Materials Engg., https://thermocalc.com
[11] Schaberger E, Grote F, Schievenbusch A 2000 *Prakt. Metallographie* **37** 419

Effect of Zn addition and natural aging on the microstructure and mechanical properties of secondary AlSi7Cu2 alloys

G Scampone[1], O Gursoy, S Brutti and G Timelli

Department of Management and Engineering, University of Padova, Vicenza, I-36100, Italy

[1] Email: giulia.scampone@phd.unipd.it

Abstract. The effect of different Zn contents on the microstructure of a secondary AlSi7Cu2 alloy has been studied. The mechanical properties of the alloy during natural aging has been investigated too. The Zn additions range from 0.7 to 8 wt.%, and the tensile testing has been carried out at the as-cast condition and after 3, 7, 12 and 21 days of natural aging. Metallographic and image analysis techniques have been used to characterize the microstructural changes of the alloy. The results indicate that the alloy is strengthened as a function of the aging time and Zn addition. Both yield stress and elongation to fracture tend to stabilize over long aging times, independently of Zn level. The steady-state tensile properties are achieved earlier by increasing the Zn content. Furthermore, high Zn contents promote the formation of eutectic structure and the precipitation of needle-like Fe-rich compounds, which decrease the ductility of the material.

1. Introduction

Excellent castability, low density, high mechanical properties and good corrosion resistance are the main features of the Al-Si foundry alloys, which are widely used in the transportation industry. These alloys can reduce the vehicle's weight without compromising the safety of the cars, improving the fuel economy and the environmental sustainability [1].

In order to maximise the strength of heat-treatable Al-Si alloys, thermal heat treatments can be performed to dissolve coarse intermetallic phases, spheroidize the eutectic Si and promote the precipitation of fine strengthening particles. However, the industrial costs and the energy consumption of these post-casting heat treatments are not negligible. Therefore, self-hardening Al alloys are now increasingly attracting the attention of industrial companies, which are considering these alloys as alternative to the standard heat-treated AlSiMg or AlSiCuMg alloys to produce automotive components [2,3].

Zn-containing Al-Si alloys can naturally age after casting process, and they can strengthen over time without any further heat treatment. Tillová et al. [4,5] demonstrated how the mechanical properties of an AlZn10Si8Mg casting alloy can be maximized after 7÷10 days of storage at room temperature. Moreover, further straightening can be induced by high cooling rates during solidification and through the addition of 3 wt.% Mg [6].

In general, Zn shows great solubility in the α-Al matrix (up to ~83 wt.% at 381°C [7]). Further, after the addition of Zn to Al-Si alloys, the system behaves as though the eutectic point is shifted to a higher silicon content and the eutectic temperature is depressed [8]. The volume fraction of the eutectic increases as well as the size of the eutectic Si particles [8,9]. From the mechanical point of view, the hardness, ultimate tensile strength and yield stress increase after Zn addition, while the elongation to fracture, as well as the impact toughness, is reduced by increasing the Zn amount [8–11].

To the best of authors' knowledge, even though the performance of Al-Zn alloys by varying the aging parameters has been deeply investigated [12,13], the evolution of the mechanical and microstructural properties during the natural aging of Zn-containing Al-Si-Cu alloys has not been studied yet.

In this study, the influence of Zn content and natural aging time on the microstructure and mechanical properties of secondary AlSi7Cu2 casting alloy was investigated.

2. Experimental Procedure

A secondary AlSi7Cu2 cast alloy (EN AC-46600 [14]), provided in the form of ingots, was selected as baseline for the casting trial. The alloy was grain refined and eutectic modified by adding Al-5Ti-1B and Al-10Sr master alloys, respectively. The chemical composition is shown in Table 1.

In order to study the effect of Zn addition on the microstructural and mechanical properties of the alloy, three additional levels of Zn were selected for the analysis. Weighted amounts of pure Zn in the form of waffle-shaped were added to achieve the nominal Zn contents of 3, 5 and 8 wt.%.

Table 1. Chemical composition (wt%) of the base AlSi7Cu2 alloy.

	Al	Si	Fe	Cu	Mn	Mg	Zn	Ti	Sr	B
AlSi7Cu2	bal.	6.451	0.665	2.20	0.374	0.423	0.687	0.165	0.019	0.005

The supplied AlSi7Cu2(Fe) alloy was melted at 780 ± 5 °C into a SiC crucible in an electric resistance furnace. The melt temperature was then decreased to 750 ± 5 °C for the Zn addition and the following degassing phase. Pure Argon was blown for 30 mins with a graphite lance and the molten metal was then skimmed according to standard foundry practice. After degassing, Al-5Ti-1B and Al-10Sr master alloys were added to the molten metal, which was then held at 720 ± 5 °C for about 30 min to ensure a complete homogenisation of the bath.

The metal was then poured at 735 ± 5 °C into a permanent steel mould preheated at 250 ± 3 °C and covered with a semi-permanent layer of Foseco Dycote® F34 coating. The die allowed the production of flat tensile specimens whose gauge length, width and thickness were 35, 9.5, and 6.1 mm, respectively. In order to continuously monitor the casting temperature and ensure good reproducibility of the tests, a thermocouple was embedded in the die at 2 mm from the inner surface of the gauge length. A 30 ppi SiC filter was used and located before the die cavity, in order to reduce the entrapment of oxides and inclusions inside the specimens. A batch of at least 15 specimens was produced for each alloy. The as-cast specimens were stored at room temperature and tested after 3, 5, 7, 12 and 21 days of natural aging and were therefore similar to a T1-condition. Generally, this temper designation applies to products that are cooled from an elevated-temperature shaping process, and for which mechanical properties have been stabilised by room-temperature aging.

Tensile testing was performed at room temperature according to ISO 6892-1:2019 standard. An electromechanical MTS Criterion Series 40 system equipped with a 50 kN load cell was used to carry out the tests. At least 3 specimens for each condition were tested, by applying a strain rate of 10^{-3} s^{-1}. The strain was measured using a 25 mm extensometer.

In order to assess the influence of Zn additions on the hardness of the material, Brinell hardness tests were performed on the cross section of the specimens after 21 days of natural aging. A load of 62.5 kgf was applied, using an indenter of 2.5 mm diameter. The dwell time was 15 seconds according to the ASTM E10-18 standard. Finally, in order to evaluate the strengthening of the α-Al matrix as a function of the Zn level, Vickers microhardness tests were carried out in the aforementioned specimens by applying a load of 0.005 kgf and a dwell time of 10 seconds.

Some samples were selected for the microstructural investigation. A cross-section in the gauge length was drawn and mechanically prepared according to standard metallographic techniques. The microstructure was investigated using an optical microscope equipped with an image analyser to measure the secondary dendrite arm spacing (SDAS), the eutectic fraction, and the precipitation of Fe-rich compounds. In each sample, a total area of more than 17 mm^2 was investigated by analysing progressive micrographs of 1.2 × 0.9 mm^2. The polished specimens were etched in a 70°C heated solution of 10% sulphuric acid in water, to easily measure the area fraction of Fe-bearing compounds by image analysis software.

3. Results and discussion

3.1. Microstructure

Figure 1 shows the typical microstructure of the experimental AlSi7Cu2 alloy. It consists mainly of a primary α-Al phase, a eutectic mixture of Al and Si, Fe- and Cu-rich compounds. Due to the high cooling rate during solidification (~40 °C/s) and the Sr addition, the eutectic Si particles are fine and fibrous, ranging in size from 0.6 to 3 μm. The Fe-rich compounds appear mainly with a needle-like shape and are uniformly distributed throughout the microstructure.

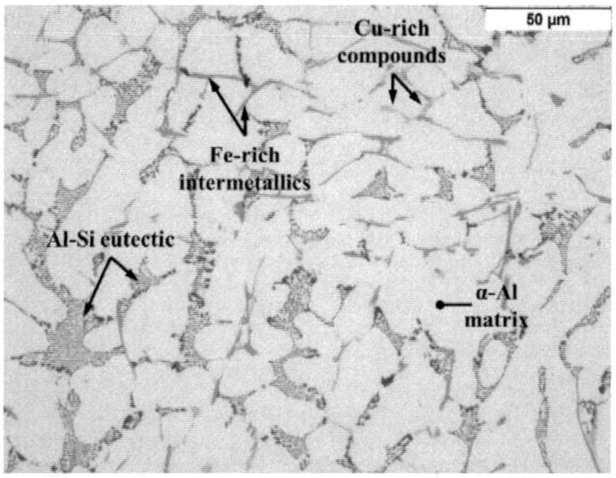

Figure 1. Typical microstructure of as-cast secondary AlSi7Cu2 alloy.

Regardless of the Zn content, the SDAS values were measured to be constant and equal to 12 ± 2 μm. Upon increasing the Zn concentration, some microstructural changes can be observed in the AlSi7Cu2 alloy (Figure 2). While the average size of the eutectic Si particles remains almost constant, the eutectic fraction greatly increases with Zn addition. Moreover, the precipitation of Fe-rich compounds, mostly in the needle-like shape, is promoted.

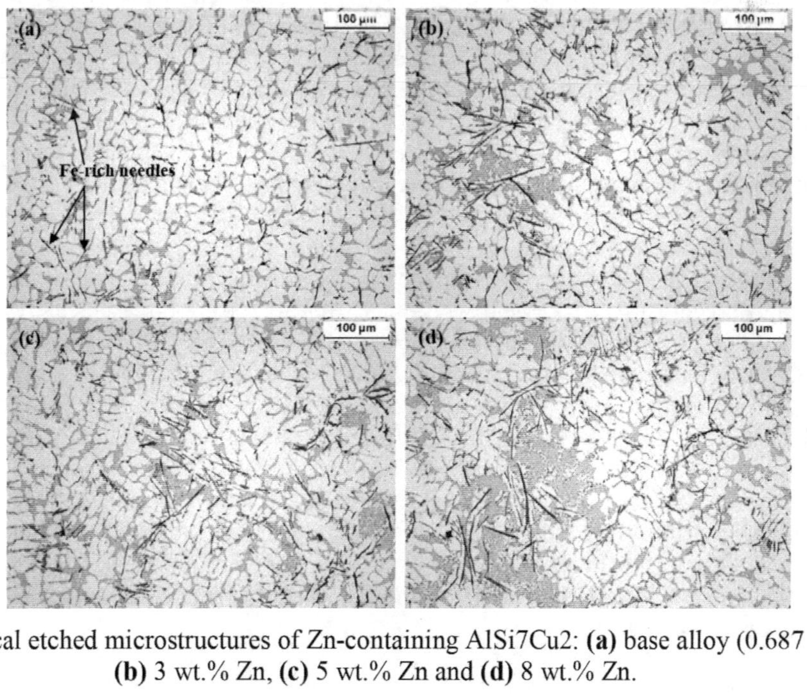

Figure 2. Typical etched microstructures of Zn-containing AlSi7Cu2: **(a)** base alloy (0.687 wt.% Zn), **(b)** 3 wt.% Zn, **(c)** 5 wt.% Zn and **(d)** 8 wt.% Zn.

Figure 3 shows the evolution of the eutectic fraction and the area fraction of Fe-rich compounds as a function of the Zn level. Both these microstructural features show an increasing trend, but with different slopes according to the Zn content. The eutectic fraction increases significantly from about 0.26 to 0.36 by increasing the Zn level from 0.7 to 3 wt.%. Further addition of Zn slightly affects the formation of the eutectic structure; upon increasing the Zn concentration from 3 to 8 wt.%, the eutectic fraction increases by 3.3%. This trend can be explained by considering the solidification path of Al-Si and Al-Si-Zn alloys. Upon increasing the Zn level in the Al-Si alloys, the eutectic point shifts toward higher Si content, resulting in an increase of the size and volume fraction of silicon particles, as reported by Alemdağ et al. [8]. In the present work, even though the dimension of the eutectic Si particles remains unchanged, the eutectic fraction increases. The high cooling rate during solidification and the Sr-modification prevented here coarsening phenomena of the eutectic Si particles.

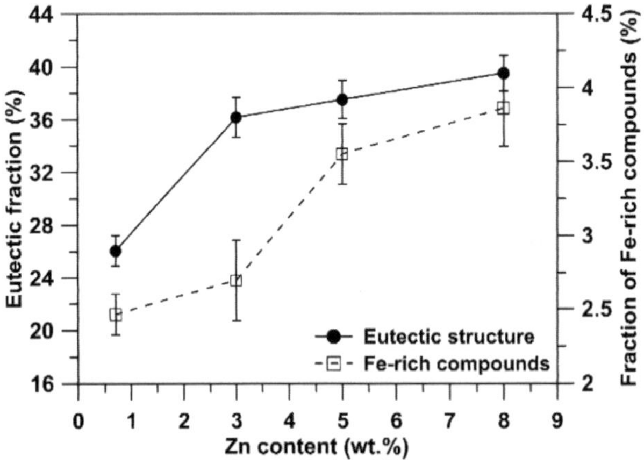

Figure 3. Evolutions of the amount of the eutectic and Fe-rich compounds as a function of Zn content in the AlSi7Cu2 alloy. Standard deviations are given as error bars.

The Zn addition facilitates significantly the precipitation of Fe-rich compounds (Figure 3). When the Zn content is about 8 wt.%, the area fraction of Fe-bearing compounds is about 3.9%, which is much higher with respect to the base alloy (~2.5%). The increase is particularly emphasized when the Zn concentration varies from 3 to 5 wt.%; in this range, the fraction of Fe-rich particles increases from about 2.7 to 3.6%. Most of the Fe-rich particles exhibit a needle-like morphology, which is consistent with previous findings by Sanna et al. [15]. From the thermodynamic point of view, high Zn concentrations increase the precipitation temperature of specific Fe-rich phases, promoting the formation of acicular Fe-rich compounds. As demonstrated in [16], this morphology can negatively affect the mechanical properties of the alloy, especially the ductility and the fatigue behaviour.

3.2. Mechanical properties

Figure 4 shows the evolution of the tensile properties of the AlSi7Cu2 alloy as a function of the aging time and Zn content. In agreement with other studies [8,10,11], the yield stress (YS) increases with the Zn addition, while the elongation to fracture decreases. The yield stress shows the same trend as a function of the aging time, independently of the Zn concentration, that is a rapid increase at the beginning of the aging time and a further stabilization (Figure 4a). Different Zn levels affect mainly the YS of the as-cast alloy, the initial increase of the YS during early aging time as well as the stabilization time (see Figure 4a). Upon increasing the Zn content, the YS increases rapidly, and the material reaches a stable condition faster. The yield stress increases by about 15 MPa during the first three days of aging in the alloys with 0.75, 3 and 5 wt.% Zn, while, after the addition of 8 wt.% Zn, it increases from 144 to 180 MPa in the same aging time. Furthermore, at the highest Zn addition, the YS is stabilised at ~190 MPa after 7 days of aging, while longer times are necessary to stabilize the alloy at lower Zn concentrations.

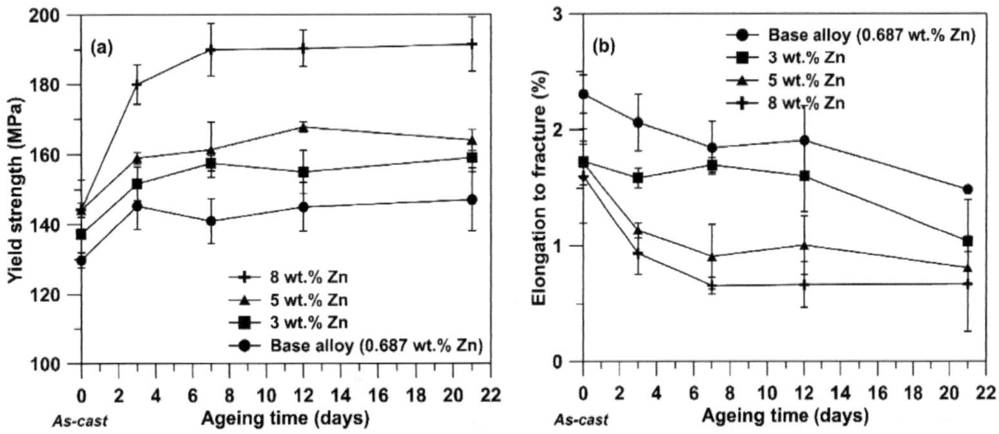

Figure 4. Evolution of **(a)** the yield stress and **(b)** the elongation to fracture as a function of the natural aging time and the Zn content in the AlSi7Cu2 alloy.

Figure 4b shows the variation of the elongation to fracture over aging time. After 7 days of aging, the ductility of the 5- and 8-wt.% Zn containing alloys is almost stabilised around 0.8%. At lower Zn levels, the elongation to fracture still decreases after 12 days of aging. This behaviour can be explained by considering the strengthening mechanisms of the alloys. Due to the high solubility of Zn in the α-Al matrix, solid solution strengthening contributes to the hardening of the alloy. Moreover, the addition of this element allows the spontaneous formation of precipitates even at room temperature. Therefore, during natural aging, the YS increases with time due to the gradual formation of Zn-rich precipitates. Higher Zn concentrations result in a greater tendency to form precipitates and require shorter aging times to maximize the strength of the alloy. Therefore, after the addition of 5 and 8 wt.% Zn, the trends of yield stress and elongation to fracture tend to stabilize soon with respect to the base alloy (Figure 4). Furthermore, it worth to be mentioned how higher Zn additions promote the precipitation of Fe-rich needles, which negatively affects the ductility of the alloy. The rapid decrease of the elongation to fracture by increasing the Zn level may also be related to greater precipitation of Fe-rich phases.

The hardness of AlSi7Cu2 alloy and the microhardness of α-Al phase are also affected by Zn addition. The results of the hardness testing after 21 days of natural aging show that the alloy strengthens with increasing Zn content. The hardness increases linearly from 86±2 to 102±1 HB by adding 8 wt.% of Zn to the base alloy. The microhardness of the α-Al phase exhibits the same trend, increasing linearly from 65±1 to 85±2 HV with varying Zn content from 0.7 to 8 wt%. Therefore, as the Zn content increases, a strengthened α-Al phase is formed and the mechanical properties of the alloy are improved through the combined effect of solid solution and precipitation hardening mechanisms.

4. Conclusions

In the present work, the influence of different Zn additions on the microstructure and mechanical properties of a secondary AlSi7Cu2 alloy was analysed. The following conclusions can be drawn.

- The formation of the eutectic structure and the precipitation of Fe-rich compounds are affected by the Zn concentration. Higher Zn amounts increase the eutectic fraction formed during the solidification of the alloy and promote the formation of acicular Fe-rich phases, which affect negatively the ductility of the alloy.
- During natural aging, the yield stress increases with time until a stable condition is reached. Similarly, the elongation to fracture decreases over time until it stabilizes over long aging times. The Zn addition affects the stabilization time, which occurs earlier when the Zn content is high.

- The yield stress and hardness of the alloy increases with the Zn content in the alloy. The α-Al matrix results harder due to the presence of Zn as solid solution or strengthening precipitates.

Acknowledgments
The present work was developed with the financial support of Fondazione Cassa di Risparmio Padova e Rovigo (CARIPARO).

5. References

[1] Hirsch J 2011 Aluminium in innovative light-weight car design *Mater. Trans.* **52** 818–24
[2] Peter I, Rosso M, Castella C and Molina R 2014 Self-hardening alloys for automotive application *Mater. Sci. Forum* **794–796** 1221–6
[3] Vicen M, Fabian P and Tillová E 2017 Self-hardening AlZn10Si8Mg aluminium alloy as an alternative replacement for AlSi7Mg0.3 aluminium alloy *Arch. Foundry Eng.* **17** 139–42
[4] Tillová E, Ďuriníková E and Chalupová M 2011 Structural analysis of secondary AlZn10Si8Mg cast alloy *Acta Metall. Slovaca* **17** 4–10
[5] Tillova E, Durinikova E and Chalupova M 2011 Characterization of phases in secondary AlZn10Si8Mg cast alloy *Mater. Eng.* **18** 1–7
[6] Rosso M, Peter I, Castella C and Molina R 2015 *Mater. Today: Proc.* vol 2 (Elsevier Ltd.) pp. 4949 – 56
[7] Murray J L 1983 The Al-Zn (Aluminum-Zinc) System *Bull. of Alloy Phase Diagr.* **4** 55–73
[8] Alemdağ Y and Beder M 2014 Microstructural, mechanical and tribological properties of Al-7Si-(0-5)Zn alloys *Mater. Des.* **63** 159–67
[9] Alemdag Y and Beder M 2019 Effects of zinc content on strength and wear performance of Al−12Si−3Cu based alloy *T. Nonferr. Metal. Soc.* **29** 2463–71
[10] Zhu X, Liu F, Wang S and Ji S 2021 The development of low-temperature heat-treatable high-pressure die-cast Al–Mg–Fe–Mn alloys with Zn *J. Mater.* **56** 11083–97
[11] Li L, Ji S, Zhu Q, Wang Y, Dong X, Yang W, Midson S and Kang Y 2018 Effect of Zn concentration on the microstructure and mechanical properties of Al-Mg-Si-Zn Alloys processed by gravity die casting *Metall. Mater. Trans. A Phys. Metall. and Mater. Sci.* **49** 3247–56
[12] Pezda J 2012 Heat treatment of AlZn10Si7MgCu alloy and its effect on change of mechanical properties *Arch. Foundry Eng.* **12** 135–8
[13] Abubakar M and Usman M 2020 Influence of ageing time on mechanical properties and weibull probability distribution of tensile strength of ternary Al–Zn–Mg alloy produced from aluminium and zinc scrap *Transactions of the Indian Institute of Metals* **73** 1827–36
[14] EN 1706:2020, Aluminium and aluminium alloys - Castings - Chemical composition and mechanical properties
[15] Sanna F, Fabrizi A, Ferraro S, Timelli G, Ferro P and Bonollo F 2013 Multiscale characterisation of AISi9Cu3(Fe) die casting alloys after Cu, Mg, Zn and Sr addition *Metall. Ital.* **105** 13–24
[16] Závodská D, Kuchariková L, Tillová E, Guagliano M, Chalupová M, Uhríčik M and Belan J 2019 The effect of iron content on fatigue lifetime of AlZn10Si8Mg cast alloy *Int. J.Fatigue* **128** 1–8

IOP Conf. Series: Materials Science and Engineering 1274 (2023) 012058 doi:10.1088/1757-899X/1274/1/012058

Microstructural study of containerless solidification of Al-5wt%Ce alloy

J Valloton[1,a], S C Vogel[2] and H Henein[1]

[1] Advanced Materials and Processing Laboratory, University of Alberta, Edmonton, Canada
[2] Materials Science & Technology Division, Los Alamos National Laboratory, Los Alamos, NM, USA
[a] Corresponding author: valloton@ualberta.ca

Abstract. Containerless solidification of Al-5wt%Ce is investigated experimentally using the electromagnetic levitation technique. The processed samples show small primary undercooling and minimal eutectic undercooling. Phase fractions determined by neutron diffraction are similar to the values obtained from a Scheil-Gulliver solidification simulation. Significant solidification shrinkage at the top of the samples indicate that solidification started at the bottom and the front propagated upwards. X-ray microscopy results show multiple nucleation events throughout the samples.

1. Introduction

Aluminum alloys are heavily used within the transportation sector, namely automotive and aerospace industries, due to their excellent mechanical properties, high strength to weight ratio, corrosion resistance and high thermal conductivity. The strengthening of these alloys typically occurs through precipitation of intermetallic either directly during the solidification process or during subsequent heat treatment. However, degradation of mechanical properties occurs at high temperatures due to coarsening or dissolution of the strengthening phases, such that operating temperatures should remain lower than the heat treatment temperature (100-200°C [1]). The development of aluminum alloys with improved performances at high temperature focuses mainly on systems forming Al_3X precipitates, such as Al-Sc, Al-Zr or Al-V [2]. These intermetallics have a coherent lattice with fcc aluminum, which increases thermal stability. But this coherence is lost at high temperatures (e.g. ~300°C for Al-Sc) [3], resulting in performance loss. In contrast, Al-Ce based alloys, which consist primarily of fcc aluminum and $Al_{11}Ce_3$, remain thermodynamically stable and retain their mechanical properties at high temperatures [4]. This is due to the near zero solubility of cerium in fcc aluminum, i.e. less than 0.005 wt% at the eutectic temperature [5]. This is orders of magnitude lower than even Sc, which currently leads to some of the most competitive high-temperature alloys (0.2 wt% Sc solubility at 600°C [6]). The diffusion coefficient of Ce in aluminium is also lower than that of other alloying elements (5.7×10^{-18} m^2s^{-1} for Ce vs 6.0×10^{-14} m^2s^{-1} for Cu at 500°C [2]) due to the large size of Ce atoms (182 pm for Ce vs 128 pm for Cu [7]). Thus, the strengthening $Al_{11}Ce_3$ intermetallic is far more stable against high temperature dissolution than precipitates in typical Al-alloys. For example, Al-Ce alloys retain over 60% of their room-temperature yield and more than 40% of their ultimate tensile strength (UTS) at 300°C. On the other hand, traditional casting aluminum alloys such as 240 or 512 do not exceed 45% yield and 50% UTS retention values at 300°C [2]. Finally, while cerium is the most abundant of the rare earth elements (as abundant as copper), it does not have any significant high-

IOP Conf. Series: Materials Science and Engineering 1274 (2023) 012058 doi:10.1088/1757-899X/1274/1/012058

volume applications. As a result, it is considered a by-product of rare earth mining and is so far relatively inexpensive as an alloying element [8]. In this work, solidification of hypoeutectic Al-5wt%Ce is carried out using electromagnetic levitation (EML) under terrestrial conditions. By using this containerless processing method, undercooling of bulk samples could be achieved since heterogeneous nucleation on crucible walls is avoided. Microstructures are analyzed using neutron diffraction, X-ray and electron microscopy. Preliminary results are presented here.

2. Experimental method

This study was carried out using an Al-5wt%Ce alloy obtain from American Elements (5N purity). Al-5Ce is of hypo-eutectic composition (see phase diagram in Figure 1). Upon cooling, solidification of Al-5Ce starts at 652°C with the crystallization of α-Al solid solution followed by an α-Al–$Al_{11}Ce_3$ eutectic at 640°C. α-Al has a face-centered cubic structure with the Pearson symbol cF4 and space group number 225 while $Al_{11}Ce_3$ has an orthorhombic structure with the Pearson symbol oI28 and space group number 71.

Electromagnetic levitation is a containerless solidification technique for the processing of metallic and semiconductor samples with a large range of undercoolings. A schematic view of the apparatus is shown in Figure 2. An alternating current flowing through a water-cooled levitation coil produces an alternating electromagnetic field. A conducting sample placed within this field is levitated by the Lorentz force F_L which compensates for the gravitational force F_G. Simultaneously, the eddy currents induced in the sample heat and melt the sample by resistive heating. To solidify the sample, cooling jets of inert gas are used. The temperature of the sample is monitored continuously with a two-color pyrometer with an accuracy of ±5 K. This allows for a direct measurement of cooling rates and undercoolings. Detailed information on the EML technique can be found in [9].

The levitated samples (~6 mm in diameter) were first analysed using non-destructive methods. Phase fractions were determined by neutron diffraction at Los Alamos Laboratory [10, 11]. The surface of the samples was observed with a scanning electron microscope, while the bulk microstructure was characterized using X-ray microscopy. Neutron bulk texture, SEM and EBSD analysis are still to follow.

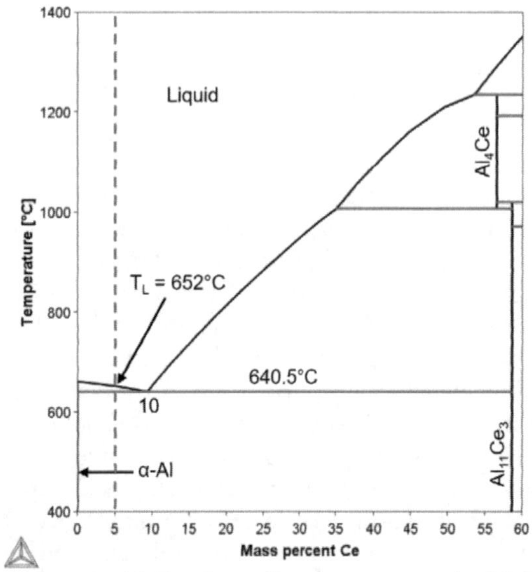

Figure 1: Al-Ce phase diagram computed with Thermo-Calc using the TCAL7 database.

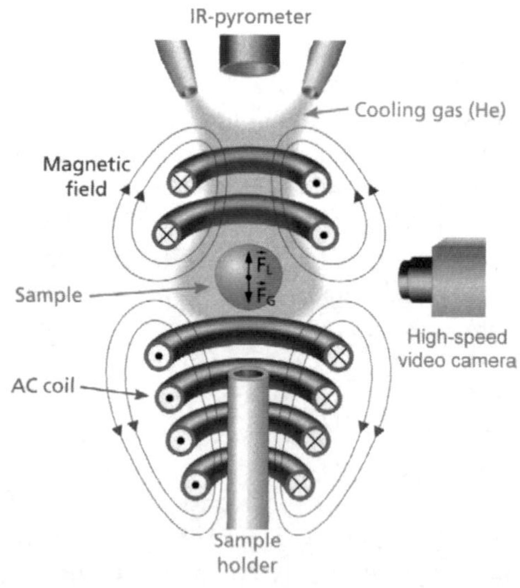

Figure 2: Schematic view of an electromagnetic levitation apparatus.

3. Results and discussion

Figure 3 shows a typical temperature-time profile obtained during EML solidification of an Al-5Ce sample. Two solidification events can be identified. First, primary solidification of α-Al is observed after an undercooling of ΔT_p = 26 K. Following recalescence, solidification of α-Al proceeds at a slower cooling rate (due to the release of latent heat) until the eutectic point is reached. A slight eutectic undercooling ΔT_e = 1 K is observed. This is followed by a temperature plateau during which the remaining interdendritic solidifies as α-Al–$Al_{11}Ce_3$ eutectic. Once solidification is complete, the sample cools down at a rate similar to when it was fully liquid. Table 1 shows the primary and eutectic undercoolings as well as the liquid cooling rate of the three Al-5Ce samples processed in EML. All samples present similar solidification conditions. Thus it is expected that the samples should show similar microstructures. Note that in all cases, the primary undercooling is greater than the freezing range of the alloy, i.e. the primary nucleation event occurred below the eutectic temperature.

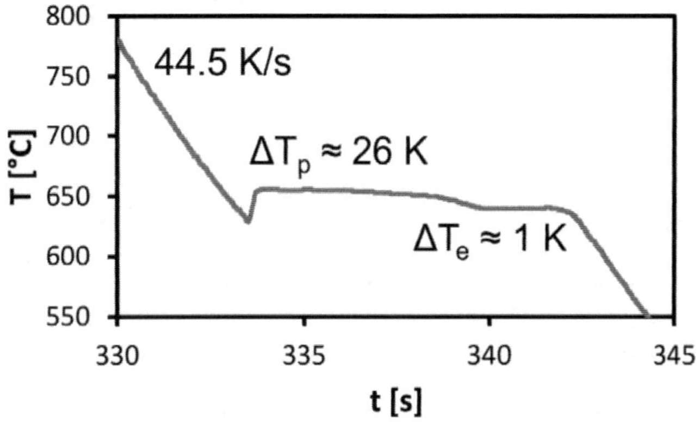

Figure 3: Typical temperature-time profile obtained during EML solidification (Al-5Ce sample #3). Primary and eutectic solidifications are identified by the corresponding recalescence events of undercoolings ΔT_p and ΔT_e respectively. The liquid cooling rate prior to solidification is also identified.

Table 1: Primary and eutectic undercoolings (ΔT_p and ΔT_e) and liquid cooling rate ($\dot{T}$) of the samples.

Al-5Ce	ΔT_p [K]	ΔT_e [K]	$\dot{T}$ [K/s]
1	33	5	37.4
2	19	7	42.9
3	26	1	44.5

Two levitated samples and a piece of as-received material were analyzed using neutron diffraction. A typical spectrum is shown in Figure 4. The main diffraction peaks of α-Al and $Al_{11}Ce_3$ are highlighted. Rietveld refinement was performed on the spectra to determine the mass fraction of the two phases. Results are presented in

Table 2. Scheil-Gulliver solidification simulation was performed using Thermo-Calc (TCAL7 database). The predicted mass fraction of $Al_{11}Ce_3$ is 8.5wt%. Both the as-received material and the EML-processed samples show close values. As the solubility of Ce in α-Al is minimal and the undercooling values are low, no supersaturation of Ce is observed in the EML samples. Solidification proceeded following the non-equilibrium Scheil equation.

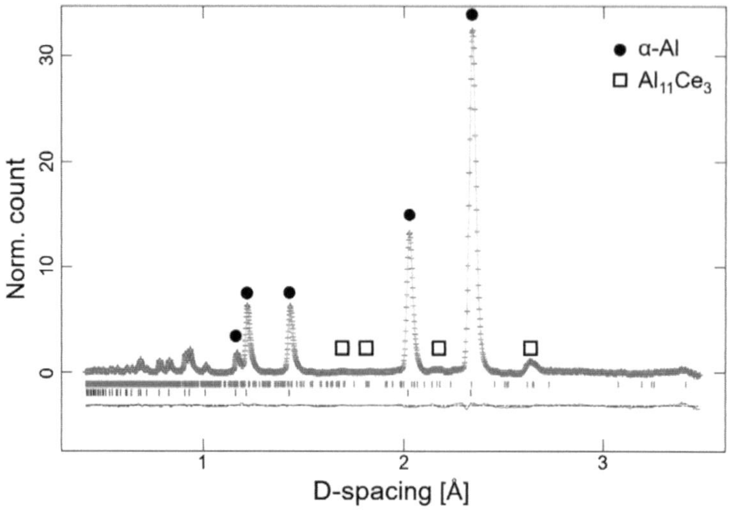

Figure 4: Neutron diffraction spectrum of sample #3 (red +) and Rietveld refinement (green solid line).

Table 2: Mass fractions of α-Al and $Al_{11}Ce_3$ obtained from Rietveld refinement of neutron diffraction spectra, as well as from a Scheil-Gulliver simulation.

Al-5Ce	α-Al [wt%]	$Al_{11}Ce_3$ [wt%]
As-received	91.7	7.3
2	91.3	8.9
3	91.1	8.7
Scheil simulation	91.5	8.5

Figure 5(a) and (b) show SEM micrographs of the top and bottom of sample #3, respectively. Significant solidification shrinkage can be observed in Figure 5(a), indicating that solidification started at the bottom of the sample and the front propagated upwards. The image in Figure 5(c) is a 2D slice reconstructed from X-ray microscopy analysis. It shows the peculiar shape of the solidified sample. Internal porosity due to solidification shrinkage can be seen at the top of the sample (black patches). It is clear that the microstructure is not homogeneous throughout the sample. Figure 5(d) is a close-up of the highlighted area in (c). Though the resolution is not enough to fully resolve the features, some dendrites with well-developed secondary arms (dark regions) are observed in the top left corner, surrounded by the α-Al–$Al_{11}Ce_3$ eutectic (light region). Coarse dendrites or cells of α-Al can be seen in the center of (d). Thus, despite the liquid being undercooled below the eutectic temperature, the first solid to form is the primary α-Al phase. The circular patch in the bottom right corner seems to be composed of much finer dendrites or so-called "seaweed" microstructure. Their smaller scale indicates that their growth velocity was higher. They were likely the first solid to form at the onset of solidification while the larger ones grew post-recalescence. The presence of several circular patches suggests that multiple nucleation events took place. This will be further investigated with SEM and EBSD analysis of cross-sections of the samples.

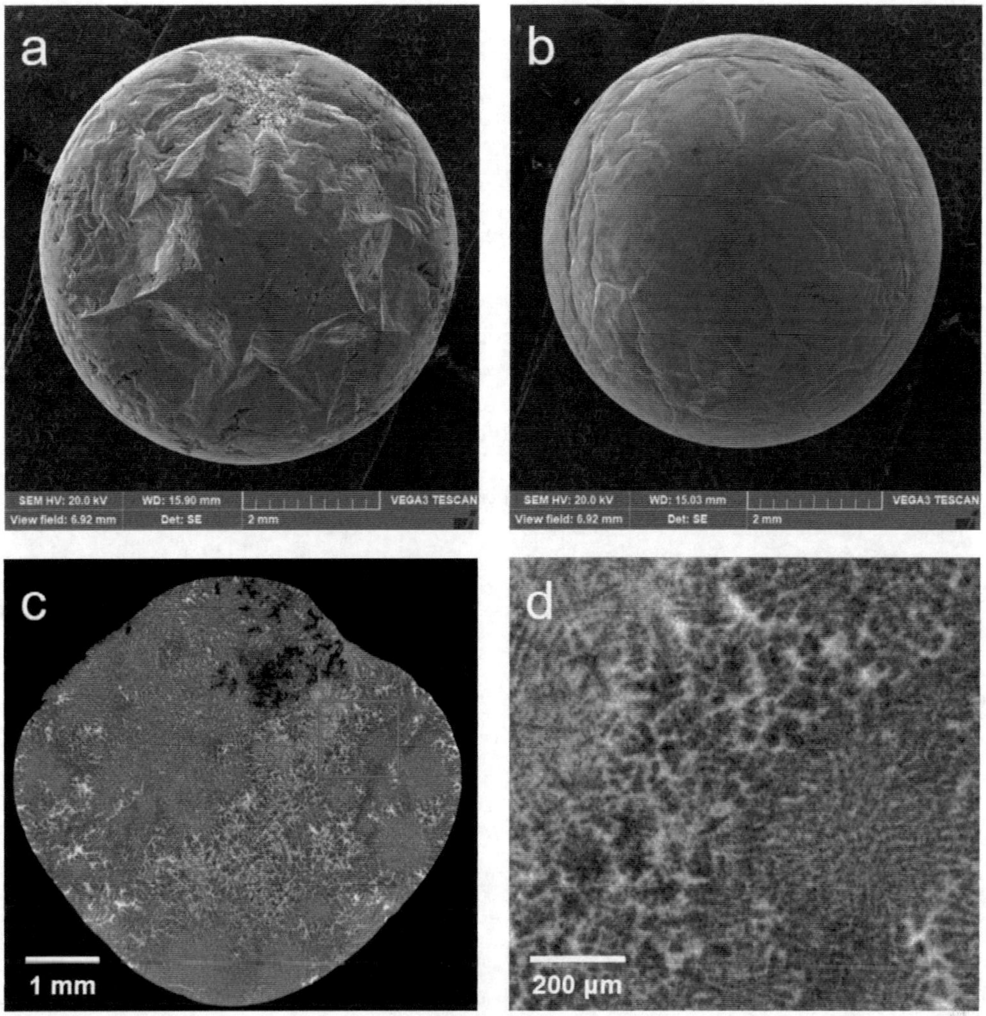

Figure 5: SEM micrographs of the top (a) and bottom (b) surfaces of sample #3. XRM image of the center of the sample (c) and close-up of the highlighted area (d).

4. Conclusions

Samples of hypo-eutectic Al-5wt%Ce were solidified using the electromagnetic levitation technique. The processed samples show small primary undercooling and minimal eutectic undercooling. Rietveld refinement of neutron diffraction spectra give phases fractions that are similar to the values obtained from a Scheil-Gulliver solidification simulation, assuming all Ce remains in the intermetallic phase. No supersaturation of Ce in α-Al is thus expected. Significant solidification shrinkage at the top of the samples indicate that solidification started at the bottom and the front propagated upwards. X-ray microscopy results seem to show multiple nucleation events throughout the samples. Cross-sections of the samples will be prepared and investigated with SEM and EBSD to further analyze the obtained microstructures.

Acknowledgments

The financial support of the Natural Sciences and Engineering Research Council of Canada (NSERC) is gratefully acknowledged. The authors are grateful to Los Alamos National Laboratory for beam time and expert support.

References

[1] Glazoff MV, Zolotorevsky VS and Belov NA 2010 *Casting Aluminum Alloys* (New York: Elsevier)

[2] Sims ZC, Rios OR, Weiss D, Turchi PEA, Perron A, Lee JRI, Li TT, Hammons JS, Bagge-Hanse M, Willey TV et al. 2017 *Mater. Horiz.* **4** 1070

[3] Marquis EA and Seidman DN 2001 *Acta Mater.* **49** 1909

[4] Weiping C 1997 *J. Mater. Sic. Lett.* **16** 1824

[5] Gschneidner Jr KA and Calderwood FW 1986 *Handbook on the Physics and Chemistry of Rare Earths* **8** 1

[6] Drits ME, Kadaner ES and Shoa ND 1969 *Russ. Metall* **1** 113

[7] Wells AF 1984 *Structural Inorganic Chemistry* (Oxford: Clarendon Press) 1288

[8] Plotkowski A, Rios OR, Sridharan N, Sims ZC, Unoci K, Ott RT, Dehoff R and Babu SS 2017 *Acta Mater.* **126** 507

[9] Herlach DM and Matson DM 2012 *Solidification of Containerless Undercooled Melts* (Weinheim: Wiley-VCH) 9

[10] Wenk HR, Lutteroti L and Vogel S 2003 *Nucl. Instrum. Meth. A* **515** 575

[11] Wenk HR, Lutteroti L and Vogel S 2010 *Powder Diffr.* **25** 283

Solidification of Al-Mn-based alloys under various processing parameters

B Markoli[1], B Leskovar[1], A Zaky[1], I Naglič[1]

[1]University of Ljubljana, Faculty of Natural Sciences and Engineering, Department of Materials and Metallurgy, Aškerčeva cesta 12, 1000 Ljubljana, Slovenia

Abstract. A number of Al-Mn-based alloys were investigated with a view to their commercialization. Various casting techniques were used to affect the constitution of the alloys studied from crystalline to quasi-crystalline as well as other metastable states. Variations in cooling rates and chemical composition, along with other processing parameters, have been shown to allow us to control the final constitution and properties of what we believe could be the next generation of commercial Al-based alloys.
Our research has shown that the solidification process of Al-Mn based alloys with additions of other metallic elements can lead to the formation of very complex final states, opening up possibilities for designing and controlling the properties of alloys.
The solidification process in the alloys studied proved to be very complex, but can be explained by basic knowledge of the crystallisation of metallic melts. We have discovered that by changing the processing parameters and the chemical composition of the alloys in a controlled way, it is possible to tailor the final properties of the alloys, making them attractive for commercial purposes.

Keywords: casting, process, Al-alloys; quasicrystals; commercialization.

1. Introduction

Aluminium-based alloys with high-strength and light-weight have attracted rapidly increasing interest because of the increase in the energy and environmental problems on the earth. Since the discovery of an age-hardening phenomenon [1], various kinds of high-strength/high-performance Al-alloys have been produced. It is known that these Al-alloys have been developed by use of the following strengthening mechanisms, i.e. (I) solid solution strengthening, (II) precipitation strengthening, (III) grain size refinement strengthening, (IV) dispersion strengthening, (V) work strengthening and (VI) fibre reinforcement. However, the use of conventional strengthening mechanisms for wrought and cast alloys leads to an upper limit of tensile strength ranging from 500 to 600 MPa at room temperature when elongation at break is higher than 10 %. Consequently, in order to overcome these inherent limitations, the use of a completely different strengthening mechanism with metastable phases shows the best results. For the last decade, one can notice that high mechanical strength exceeding 1000 MPa is achieved by the formation of a metastable structure, amorphous or quasicrystal (QC) phases, in the aluminium matrix [2].
First discovered icosahedral (i)QC (by Shechtman) in the Al–Mn alloys system were thermodynamically metastable and formed only when the melt was rapidly cooled (~10^6 K/s) [3]. A few years later, it was shown that less rapidly solidified Al-Mn alloys forms a metastable decagonal (d)QC. Only in the case of even slower cooling rates a stable phase forms, which is then periodic [4,

5]. Annealing of the alloys with primary metastable iQC phase showed that decomposition of the iQC phase into the thermodynamically more stable crystalline phase, the so-called approximant, occurs [3, 6, 7].

From the field of microstructural engineering by solidification processing, composite-like materials containing QCs are the subject of various attempts for application in view of enhanced mechanical performance, whether they are prepared by blending with polymers or light metals or by in-situ reinforcement produced by nanoprecipitation of particles [8]. However, extremely high-strength values reaching over 1000 MPa were obtained for the melt-spun ribbon alloys and extruded atomized powders where high cooling rates are obtained (~10^6 K/s) [9]. For practical applications as high-strength Al-based alloys, it is essential to produce an Al-based alloy with metastable phase in a bulk form.

The most promising process to produce composite like material reinforced with iQC particles in a bulk form is to alloy the aluminium melt with elements that reduced the cooling rate needed for primary metastable iQC formation. In reports by Zupanič *et al.* [10-12], Bončina *et al.* [13], Markoli *et al.* [14, 15], Naglič *et al.* [16], and Stan-Głowińska *et al.* [17, 18] the effect of different alloying elements on the formation of the primary metastable iQC phase in Al–Mn–X alloys (X = B, Be, C, Ca, Cu, Cr, Co, Ni, Ti, Sr, Fe, Si and Mg) was evaluated. They discovered that additions of boron, carbon, calcium, chromium, cobalt, titanium and strontium could not be proven to be effective in promoting the formation of iQCs in cast Al–Mn-based alloys. In contrast, beryllium, iron, copper, nickel, magnesium and silicon proved to be highly effective in promoting iQC formation at cooling rates between 500 and 1350 K/s.

Novel research show that the combination of promoting elements in Al–Mn-based alloys can significantly reduce the cooling rate needed for metastable iQC phase. Critical cooling rates vary around a few 100 K/s and can be achieved by simple casting into a permanent mould [13, 19-23] or even, by industrial high pressure die casting procedure, as shown in this work. High-pressure die casting (HPDC) is the process of filling a cavity at a high speed and pressure in a preheated, usually steel (hot working tool steel) mould or tool, which enables the production of thin-walled castings and short production cycles. The cooling rates are generally around a few 10 to a few 100 K/s, largely depending on the wall thickness.

2. Materials and Methods

Aluminium alloy containing manganese, silicon, copper and magnesium was prepared in an industrial furnace. We used technically pure aluminium (99.8 wt.%), Al-Mn master alloy (75 wt.% Mn), copper (99.9 wt.%) and commercial EN AC-51200 in EN AC-46000 aluminium alloys. The experimental casting was carried out on an industrial HPDC machine. Part with a wall thickness of 3 mm was cast with different carefully selected parameters and conditions. The alloy was also gravity cast into a sand measuring cell, steel mould and copper mould to compare different solidification states. The chemical compositions of the prepared alloy are presented in table 1 and is further refer as QC alloy. The castings were cut, mount and prepared by grinding and polishing for characterization of microstructure.

Table 1. Chemical compositions of the QC alloy in wt.% with corresponding experimental uncertainties.

Alloy	Al	Mn	Si	Cu	Mg	Fe
Al–Mn–Si-Cu-Mg	88.53 ± 0.5	4.5 ± 0.4	1.76 ± 0.3	2.76 ± 0.2	2.23 ± 0.2	0.22 ± 0.05

Characterization of microstructure was preformed using a light optical microscope (LOM), ZEISS Axio Imager A1m, equipped with a digital camera AxioCam ICc 3 (3.3 million pixels) and AxioVision software. A field-emission-gun scanning electron microscope JEOL JSM-7600F equipped with energy-dispersive x-ray spectroscopy (EDS) and electron backscatter diffraction (EBSD) was used for the microstructural characterization. The analyses were performed using Oxford Instruments INCA Microanalysis Suite, with an X-Max 20 SDD-EDS detector and CHANNEL5 EBSD software with a Nordlys detector. The crystallinity of the samples was studied by EBSD analysis using a 20 kV.

In order to identify and evaluate the variables that have the most influence on the amount of the metastable phase in investigated alloys we employed several different approaches. First method used was to assess the solidification rate for non-equilibrium solidification conditions by measuring the secondary dendrite arm spacing (SDAS) of α-Al. We executed measurements using LOM and backscattered electron (BE) images. Our SDAS measurements were used to make an estimate of the solidification rate for QC alloy based on SDAS-cooling rate relation devised by Cho *et al.* [24]. In the study by Cho *et al.*, cylindrical bar castings with different diameters were cast to acquire the cooling curves in the solidification range from 15°C/s to 100°C/s, and then the microstructures were compared to estimate the cooling rate in HPDC. The solidification characteristics, including liquidus/solidus temperature, dendrite arm spacing, and cooling rate, were analysed. The results showed a strong proportional (linear) relationship between dendrite arm spacing (SDAS) and cooling rate in HPDC. This approach provided us with the average solidification rate according to Cho *et al.* and will be referred to hence forth as the Cho model.

For calculating the mean solidification rate we applied also a modified Chvorinov's rule [25, 26]. This rule enables us to calculate solidification time for melt cast into a simple cylindrical mould by equation:

$$t = B \times \left(\frac{A_s}{A}\right)^{1.31} \times V^{0.67} \qquad (1)$$

In Equation 1 t represents the solidification time, B is the mould constant, A_s is the surface area of the sphere having the same volume as the casting, A is the surface area of the casting that is in contact with the mould and V is the volume of the casting. The mould constant is defined as:

$$B = \left[\frac{\rho_m L}{(T_l - T_s)}\right]^2 \times \left[\frac{\pi}{4k\rho c}\right] \times \left[1 + \left(\frac{c_m \Delta T_s}{L}\right)^2\right] \qquad (2)$$

where T_l is liquidus temperature of the alloy (K), T_s, is solidus temperature of the alloy (K), ΔT_s, is superheat (K), L is heat of fusion or heat of solidification ($J \cdot kg^{-1}$), k is a thermal conductivity of the mould ($W \cdot m^{-1} \cdot K^{-1}$), ρ is a density of the mould ($kg \cdot m^{-3}$), c is specific heat of the mould ($J \cdot kg^{-1} \cdot K^{-1}$), ρ_m is the density of the alloy ($kg \cdot m^{-3}$), and c_m is the specific heat of the alloy ($J \cdot kg^{-1} \cdot K^{-1}$). The simple thermal analysis (STA) was performed to define characteristic solidification temperatures for the studied alloy. The cooling curves of the Al-Mn base alloy were obtained by casting into a sand measuring cell. The temperature was measured using a K-type thermocouple. The values obtained by thermal analysis of the studied alloy were later used to calculate the solidification time. The average solidification rate was determined from the calculated solidification time and the obtained solidification interval. The values used in equations are: T_l=953 K, T_s=763 K, ΔT_s=200 K, L=360000 $J \cdot kg^{-1}$, k=398 $W \cdot m^{-1} \cdot K^{-1}$(copper), ρ=8960 $kg \cdot m^{-3}$(copper), c=385 $J \cdot kg^{-1} \cdot K^{-1}$(copper), k=45 $W \cdot m^{-1} \cdot K^{-1}$(steel), ρ=7850 $kg \cdot m^{-3}$(steel), c=460 $J \cdot kg^{-1} \cdot K^{-1}$(steel) ρ_m=2800 $kg \cdot m^{-3}$, c_m=1080 $J \cdot kg^{-1} \cdot K^{-1}$.

MAGMASOFT® ((MAGMA) Giessereitechnologie GmbH, Aachen, Germany) is a finite volume difference based metal casting process simulation and optimization software. The simulations were performed with the objective of determining the average solidification rate in HPDC castings. The databases covering the thermodynamic properties of the material were selected from the standard material databases containing the HPDC - cold chamber process mode and the aluminium material type. Simulation was performed to simulate the entire HPDC process with defined process parameters and cycle conditions.

The main differences between the approaches are that (I) the Cho model estimates the solidification rate determined during solidification of the EN AC-46000 alloy by the SDAS measurements of α-Al, which show a linear dependence between solidification rate and SDAS; (II) the Chvorinov rule uses the thermodynamic properties of the studied alloy defined by the thermal analysis and the thermal properties of the mould material, while (III) the MAGMA simulation uses the EN AC-46000 alloy from the standard databases to calculate the solidification rates using different industrial casting and process parameters. The solidification rate obtained was used as an estimate for the alloy under study since the casting parameters of the new alloy are not yet defined.

3. Results and discussion

The LOM images presented in figure 1 show the microstructure of the QC alloy as a function of mould composition and casting cavity diameter. Figure 1a and 1b show the microstructure of the sample cast into the sand measuring cell with a diameter of 36 mm. Due to the slow cooling of the alloy, the solidification follows the thermodynamically stable phase diagrams. The phase diagram of solidification for this system is well explained by Chang *et al.* [27] and will not be discussed in detail here. The primary phase formed during solidification is hexagonal β-AlMnSi and was confirmed by the EBSD technique. The obtained EBSD pattern is shown as an inset in figure 1a. Figure 1c and 1d show the microstructure obtained when casting into a steel mould with a circular casting cavity of 10 mm diameter. The amount of metastable phase has greatly increased. EBSD analysis confirms that the primary phase is predominantly cubic α-AlMnSi. The corresponding EBSD pattern of α-AlMnSi is shown as an inset in figure 1c. The microstructure obtained by casting into a copper mould with a circular casting cavity of 5 mm diameter is shown in figure 1e and 1f. The EBSD analysis shows that the primary phase is predominantly composed of iQC. The corresponding EBSD pattern of iQC is shown as inset in figure 1c.

For the HPDC process, we chose a casting with a wall thickness of 3 mm. By changing the process parameters and conditions, we obtained different metastable states. Figure 2 shows HPDC microstructures cast under typical casting conditions (figure 2a and 2b), under slow cycle conditions (figure 2c and 2d), and under fast cycle conditions (figure 2e and 2f). Direct process parameters are not reported in this work as they are under consideration for patent application to produce novel alloys. Solidification rates obtained using STA, Chvorinov's rule, MAGMA simulations, and SDAS are shown in Table 2. Since the casting parameters of the novel alloy are not yet defined, we have to use the Al-Si based alloy from the MAGMA databases to calculate the solidification rates using different casting and process parameters. The estimation of solidification rates using SDAS was also focused on the results obtained when casting Al-Si based alloys by Cho *et al.* [24].

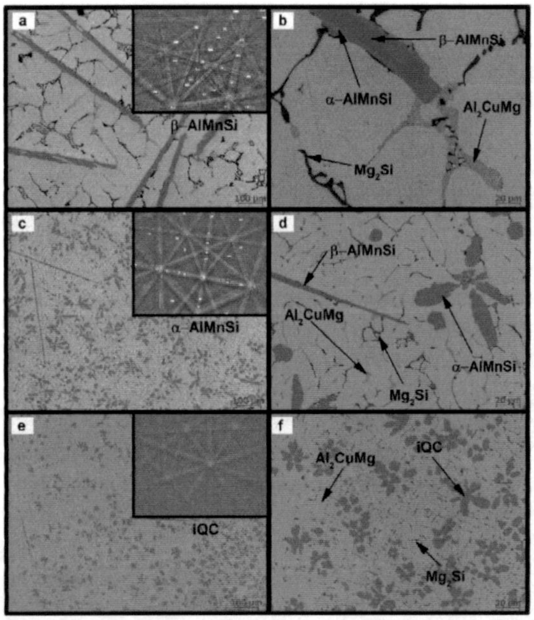

Figure 1. Microstructure of Al-Mn-Si-Cu-Mg alloy cast into the sand measuring cell, figure 1a and 1b, cast into the steel mould, figure 1c and 1d, and cast into the copper mould, figure 1e and 1f. The EBSD patterns attached as insets in figure 1 are showing the dominated primary phase which is hexagonal β-AlMnSi, figure 1a,

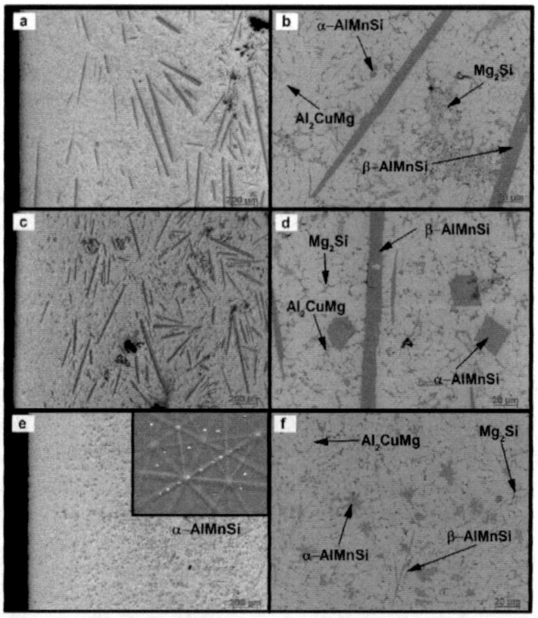

Figure 2. Microstructure of Al-Mn-Si-Cu-Mg alloy cast by different casting processes and conditions. Figure 2a and 2b showing microstructure obtained by typical HPDC parameters used for the commercial Al-alloys castings. Figure 1c and 1d showing microstructure obtained by slow cycle casting

cubic α-AlMnSi, figure 1c, and icosahedral iQC, figure 1e.

and figure 1e and 1f by fast cycle casting. The EBSD pattern attached as inset in figure 1e is showing that the primary phase, for fast cycle casting procedure, is predominantly cubical α-AlMnSi.

Table 2. Solidification rates obtained by thermocouple measurement, different simulation methods and based on SDAS (Cho model).

Method	STA	Steel mould	Copper mould	HPDC (typical)	HPDC (slow)	HPDC (fast)
Thermocouple measurement	0.3 K/s	-	-	-	-	-
Chvorinov's rule	-	10 K/s	190 K/s	-	-	-
MAGMA	-	-	-	70 K/s	40 K/s	200 K/s
Cho model	0.2 K/s	20 K/s	200 K/s	100 K/s	90 K/s	150 K/s

From table 2, it can be concluded that the solidification rates obtained with HPDC are high enough to obtain the metastable phases. However, a large hexagonal β-AlMnSi phase is present in parts produced under typical casting conditions (figure 2a and 2b) and under slow cycle conditions (figure 2c and 2d). The reason for their formation is the excessively long filling and dosing time, which causes a sharp drop in the alloy melt temperature near the contact surface with the shot sleeve wall. MAGMA simulations show that the temperature can even drop by more than 100 K, falling below the liquidus temperature of the alloy under investigation. This can also be clearly seen from figure 2a and figure 2c, where the highest amount of β-AlMnSi is present under slow cycle conditions. In the case of HPDC under fast cycle conditions (figure 2e and 2f), it can be seen that only a small amount of β-AlMnSi phase is formed. For the Cho model and MAGMA simulation, the average solidification rate for the QC alloy was estimated based on results from a commercial casting of high-pressure EN AC-46000 alloy. The Chvorinov equation, on the other hand, refers to the thermodynamic properties of the alloy under study. In the case of gravity casting, small differences occur between the solidification rate according to Chvorinov's rule and Cho's model. This could be related to the changing thermal conductivity of the metal during solidification, but also to the accuracy of measurement of the SDAS. Since the thermal properties of the studied alloy are not yet included in the MAGMA database, there are also differences in the solidification rate in HPDC between the Cho model and the MAGMA simulation.

We strongly believe that the process parameters and conditions for HPDC can be further modified to obtain a QC reinforced alloy. Based on the results obtained by gravity and HPDC, we have outlined a temperature-time-transformation (TTT) diagram for the alloy under study. In further work, we will calculate the TTT diagrams exactly as prepared by Gillessen *et al.* [4, 5] for the alloy $Al_{88}Mn_{12}$. Figure 3 shows the difference between the formation of metastable phases for $Al_{88}Mn_{12}$ and the QC alloy.

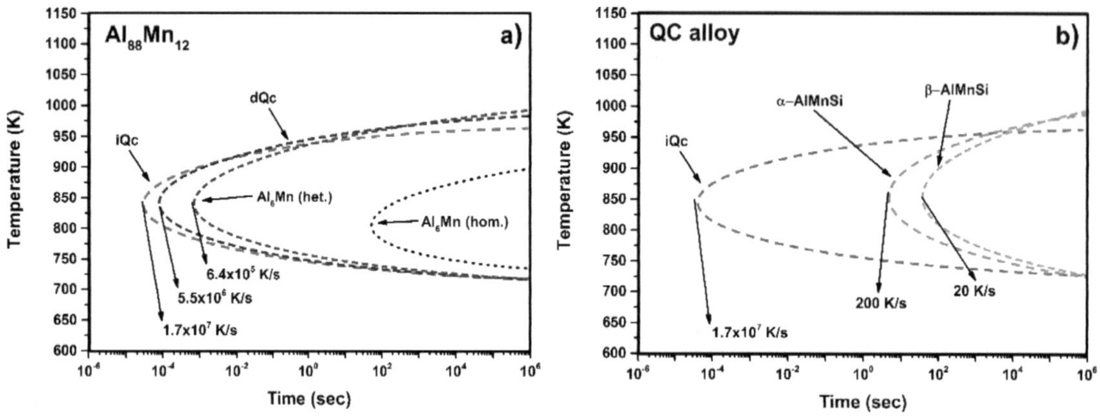

Figure 3. Calculated temperature-time-transformation (TTT) diagrams for $Al_{88}Mn_{12}$ alloy based on Al_6Mn-phase, dodecagonal dQC and icosahedral iQC phase according to Gillessen *et al.* [4, 5] figure 3a, and sketch of (TTT) diagrams for QC alloy based on primary α-AlMnSi, β-AlMnSi and icosahedral iQC phase, figure 3b.

4. Conclusions

It was found that the Al-Mn-based alloy can have at least three different solidification paths depending on the cooling rate. The thermodynamically stable primary β-AlMnSi phase can form up to 20 K/s. The thermodynamically metastable α-AlMnSi phase forms in the range of 20 K/s to 200 K/s. The thermodynamically metastable iQC phase, on the other hand, forms at solidification rates greater than 200 K/s. Solidification rates of more than 100 K/s can be achieved with the HPDC process, but the process parameters and conditions must be properly selected to avoid the formation of the primary β-AlMnSi phase.

Acknowledgements
The authors would like to thank the Slovenian Research Agency (Research Core Funding No. P1-0195) and Iskra ISD - Livarna d.o.o. for technical and financial support.

References

[1] Wilm A 1911 *Metallurgie: Zeitschrift für de gesamte Hüttenkunde* **8** 225-7
[2] Inoue A and Kimura H 2001 *J. Light Met.* **1** 31-41
[3] Shechtman D, Blech I, Gratias D and Cahn J W 1984 *Phys. Rev. Lett.* **53** 1951-3
[4] Battezzati L, Antonione C and Marino F 1989 *J. Mater. Sci.* **24** 2324-30
[5] Gillessen F and Herlach D 1991 *Mater. Sci. Eng.* **134** 1220-4
[6] McAlister A, Bendersky L, Schaefer R and Biancaniello F 1987 *Scripta metallurgica* **21** 103-6
[7] Bancel P A, Heiney P A, Stephens P W, Goldman A I and Horn P M 1985 *Phys. Rev. Lett.* **54** 2422-5
[8] Dubois J-M 2012 *Chem. Soc. Rev.* **41** 6760-77
[9] Inoue A and Kimura H 2000 *Mater. Sci. Eng.* **286** 1-10
[10] Zupanic F, Boncina T, Rozman N, Anzel I, Grogger W, Gspan C, Hofer F and Markoli B 2008 *Z Kristallogr Cryst Mater,* 735
[11] Zupanič F, Wang D, Gspan C and Bončina T 2015 *Mater. Charact.* **106** 93-9
[12] Zupanič F, Lojen G, Barba L and Bončina T 2012 *Mater. Charact.* **70** 48-54
[13] Bončina T, Markoli B and Zupanič F 2009 *J. Microsc.* **233** 364-71
[14] Markoli B, Delijic K, Strekelj N and Naglic I 2014 *Contemp. Mater.* **5** 30-6
[15] Markoli B, Boncina T and Zupanic F 2015 *Recent Pat. Mater. Sci.* **108** 109-18
[16] Naglič I, Samardžija Z, Delijić K, Kobe S, Dubois J-M, Leskovar B and Markoli B 2017 *J. Mater. Sci.* **52** 13657-68

[17] Stan-Głowińska K, Rogal Ł, Góral A, Wierzbicka-Miernik A, Wojewoda-Budka J, Schell N and Lityńska-Dobrzyńska L 2017 *J. Mater. Sci.* **52** 7794-807
[18] Stan-Głowińska K and Lityńska-Dobrzyńska L 2017 *Mater. Charact.* **128** 203-8
[19] Tsai A-P 2013 *Chem. Soc. Rev.* **42** 5352-65
[20] De Laissardière G T, Nguyen-Manh D and Mayou D 2005 *Prog. Mater. Sci.* **50** 679-788
[21] Chang H-J, Fleury E, Song G, Lee M, Kim W-T and Kim D 2004 *Mater. Sci. Eng.* **375** 992-7
[22] Rozman N, Bončina T, Anžel I and Zupanič F 2008 *Mater. Tehnol.* **42** 65-8
[23] Kang H, Li X, Su Y, Liu D, Guo J and Fu H 2012 *Intermetallics* **23** 32-8
[24] Cho J-I and Kim C-W 2014 *Int. J. Met.* **8** 49-55
[25] Jelinek P and Elbel T 2010 *Arch. Foundry Eng.* **10** 77-82
[26] Tiryakioglu M, Tiryakioglu E and Askeland D 1998 *Trans. Am. Foundrymen's Soc.* 907-14
[27] Chang K, Liu S, Zhao D, Du Y, Zhou L and Chen L 2011 *Thermochim. Acta* **512** 258-67

Microstructural Characterization and Mechanical Properties Investigation of Nemalloy HE700 alloy: Effect of Pre-Solidified Grains

T. Aziz[1], K. Sadayappan[3], A. Phillion[1], S. Shankar[2,*]

[1] Department of Materials Science and Engineering, McMaster University, Hamilton, ON, Canada

[2] Light Metal Casting Research Centre (LMCRC), Department of Mechanical Engineering, McMaster University, Hamilton, ON, Canada

[3] Canmet MATERIALS, Hamilton, ON, Canada

E-mail:: shankar@mcmaster.ca

Abstract. Structural plate castings of Nemalloy HE700 alloy, an Al-Fe eutectic alloy system with Zn and Mg as precipitation strengtheners were produced through high vacuum high pressure die casting (HVHPDC) and designed as an alternative to existing structural automotive materials. Cast materials are subject to the existence of pre-solidified grains (PSG) within the microstructure that originate from external solidification in the liquid melt and cause defects in the final cast components. In the case of HPDC, these PSG emerge in the shot sleeve and are injected through the gating system, influencing liquid metal flow behaviour and solidification characteristics. The extent of PSG and their distribution were manipulated through the altering of process route parameters such as shot delay intervals. Quantitative metallography and the effect of PSGs on the uniaxial tensile properties for the Nemalloy HE700 alloy are presented in this publication.

1. Introduction

Throughout the last few decades, increased fuel, and electric vehicle (EV) ranges and reduced CO_2 emissions in automotive vehicles have gained an upsurge in traction in the automotive industry. These goals can be realized through implementing lighter weight material alternatives such as various metallic alloys, metal-matrix composites, laminates, or sandwich structures to traditional advanced high strength steel (AHSS) counterparts which are commonly used for structural automotive components [1]. Czerwinski from CanmetMATERIALS reports that alternative materials such as aluminum have grown in use in the automotive industry from 154 kg per vehicle in 2010 to 208 kg in 2020 and are predicted to grow to 233 kg by the year 2026 [1]. Joost from The U.S. Department of Energy noted that by replacing standard steel and cast-iron components in a vehicle with aluminum, an approximate mass reduction of 40-60% can be achieved [2]. Additionally, Czerwinski also states that the demand for EVs will increase by 30-35% by the year 2030 and that there will be a rise in lighter weight material development and selection for these vehicles as electrification introduces additional loads to the vehicle as compared to traditional internal combustion engine vehicles (ICE) [1]. However, in the case of light-weighting, it is not simply about weight reduction; there must also be an emphasis on factors such as structural performance and economical and environmental aspects of the components and vehicles in general, especially in the automotive sector [1].

As the automotive industry shifts its focus towards lighter weight vehicles while maintaining modern day strength and safety requirements, an endeavour to develop a new family of Al-Fe eutectic-based alloy system named the Nemalloy HE700 alloy with Zn and Mg additions for strengthening was undertaken and the alloy was validated to be used for structural automotive components. The alloy system was fabricated using high pressure die casting (HPDC) as this process route allows for increased economical production and the ability to produce near-net shaped castings which is beneficial for manufacturing complex parts and geometries. In high pressure die casting, molten metal is poured into a shot sleeve chamber which is injected at high pressures and velocities through a gating system into a die mould where it then solidifies. However, typical components produced via HPDC are subject to casting defects such as hot tearing, shrinkages and porosity formation which hinder cast product performance and quality. From this, the high vacuum high pressure die casting (HVHPDC) process route was chosen for the manufacturing of structural components using HPDC. HVHPDC operates in a similar manner as HPDC, however, it has an added inclusion of a high-powered vacuum which is utilized to evacuate the die cavity allowing for a reduction in defect formation such as porosity in the cast products which invites the idea that components for structural applications can be manufactured using this process route.

With the reduction of porosity defects by adapting HVHPDC and other process route variations alike, another notable defect known as externally solidified crystals (ESC) or pre-solidified grains (PSG) rose in prominence. ESCs or PSGs are irregular shaped grains that are considerably larger in size than those normally found in cast microstructures and are a product of external solidification from outside of the die cavity. An example of an AlSi4MgMn cast microstructure produced via HPDC investigated by Otarawanna et al. can be found in Figure 1. The arrows found in the images identify PSGs.

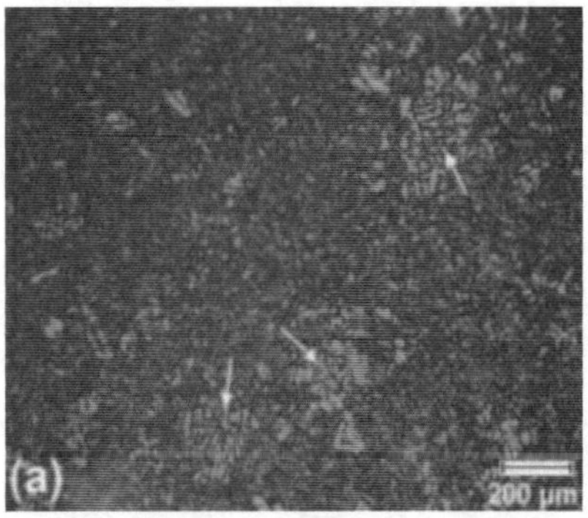

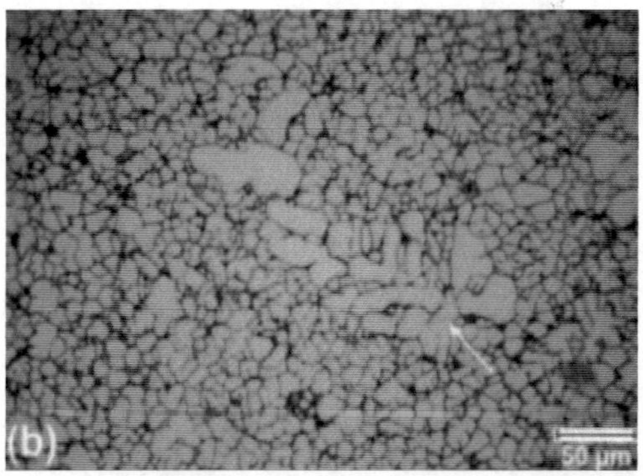

Figure 1. Typical micrographs of an AlSi4MgMn HPDC alloy with (a) crystals formed in the shot sleeve and (b) in-cavity solidified grains [3].

Although PSGs have begun to attract attention in recent years, they are not unfamiliar to cast microstructures. In experiments and simulations performed by Bi et al., the distribution and amount of PSGs or ESCs within an HPDC component were determined to have a sizable impact on both the microstructure and its mechanical properties [3]. In the micrographs shown in Figure 1 above, Otarwanna et al. encountered two distinct primary grain sizes, one being ESCs originating from the shot sleeve as large as 200 μm and the other only ~10 μm in size within the cavity [4]. Touching on the size of primary grains, Otarawanna et al. state that the difference in size can be attributed to the varying cooling rates experienced between the shot sleeve and the die cavity itself [4]. Another notable discovery by Otarawanna et al. through experiments was that ESCs were commonly found to drift towards the centers of the casting cross-sections as early as the filling stage of the HPDC process [5].

In this paper, the effects varying shot delay intervals of 1, 4, 7, and 10 s (referred to as D1, D4, D7 and D10 in latter sections) in both the Nemalloy HE700 alloy and AlSi7Mg (known as Aural-5 or A5 in some notations) have been explored to investigate the effect of interval delay on the microstructures and mechanical properties of the cast components. Grain size measurements of in-cavity solidified grains and PSGs within the various shot delay intervals were determined using ImageJ Fiji to understand microstructural evolution of samples with varying delay intervals and their mechanical properties.

2. Experimental Procedure.

The HPDC samples investigated were fabricated in one of many series of casting trials using the Bühler Carat type 1200-ton high vacuum pressure die casting press in conjunction with CanmetMATERIALS. Alloy compositions in weight percent for the Aural-5 Al-Si-Mg system are given in Table 1. The 3-step plate geometry component is shown in Figure 2 (a) where the 3-step plate section of the castings were separated from the rest of the geometry for analysis. Additionally, typical shot profile velocities and metal pressures for each delay interval are presented in Figure 2 (b) and (c), respectively, demonstrating that the casting conditions were nearly identical to one another, except for the delay time. The castings were subjected to a waiting period (1, 4, 7 and 10 s) in the shot sleeve before being injected into the die cavity, to ensure formation of PSG; in Figure 2 (b) and 2 (c), the delays are shown as D1 to D10, respectively, for the four cases. Figure 2 (a) also illustrates the shot direction and sample locations (red boxes) used for metallographic analysis.

Table 1. Aural-5 alloy composition in weight percent

Alloy	Si	Cu	Mg	Zn	Fe	Ti	Mn	Sr
Aural-5	8.04	0.02	0.16	0	0.19	0.052	0.53	0.02

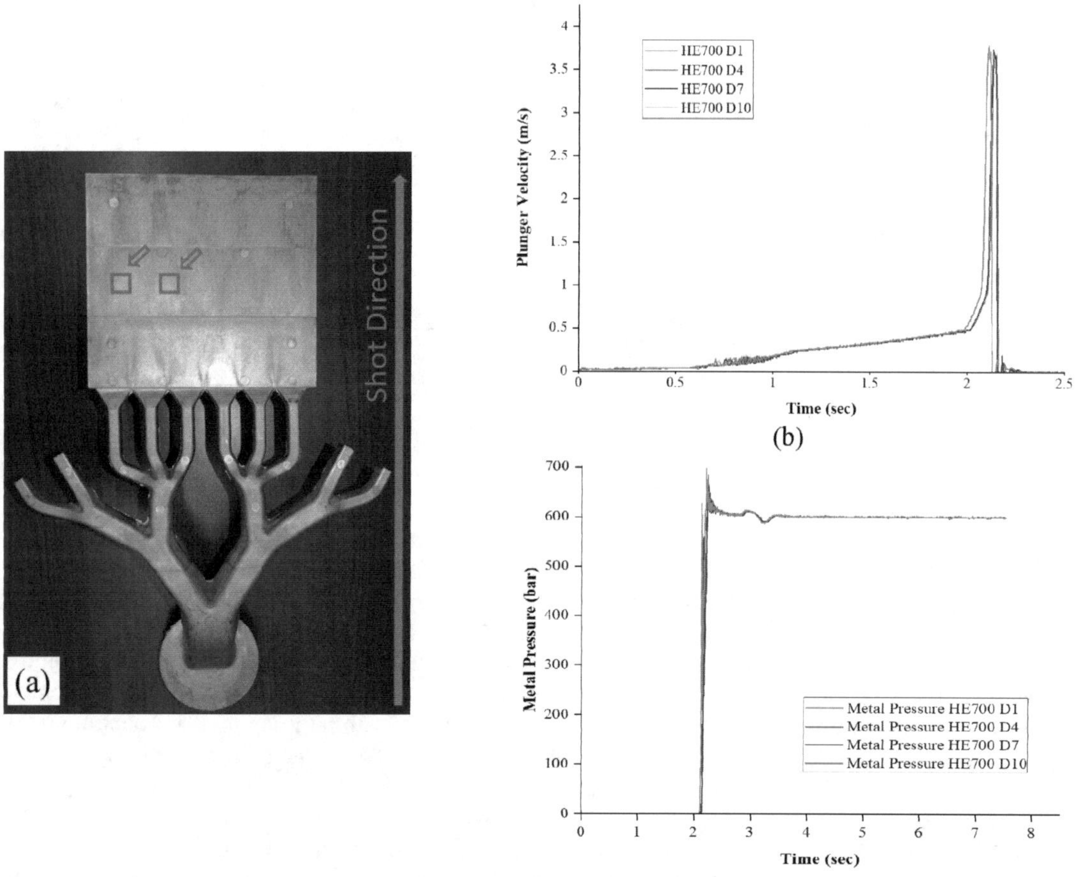

Figure 2. (a) Photograph of typical cast geometry with sample locations used for analysis for both alloys, (b) Typical shot casting plunger velocities and (c) Typical metal pressures for the HE700 castings

The dimensions for the 3-step plate sections are shown in Figure 3 with each plate adhering to nominal thicknesses of 3mm, 2.5mm, and 2.3mm. Tensile bars following the ASTM E8 standard as shown in the schematic representation were machined through water jet machining such that the mechanical properties of both alloys could be quantified. It is to be noted that the orientation of these tensile specimens is perpendicular to the shot fill direction. In this paper, quantitative metallographic analyses and mechanical testing were performed only on the 2.5mm plates of both the Nemalloy HE700 and Aural-5 alloys.

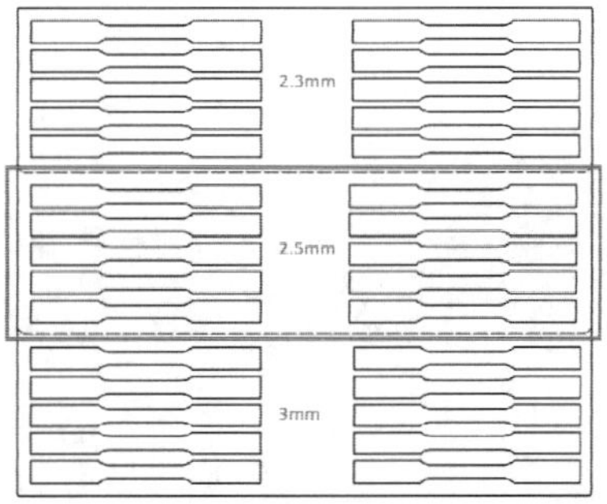

Figure 3. Schematic representation of typical tensile bar plate design used for mechanical testing for 2.5mm plate sections.

Metallographic investigations of the samples shown in Figure 2 (a) were performed using standard grinding, polishing and etching steps. The cross-sectional faces of the samples were chosen for viewing under a light optical microscope and the samples were cold mounted using a fast-curing epoxy and hardener. Kroll's reagent was chosen as the etchant for exploration of the samples as the distinction between primary Al grain structure and the intermetallic phases were more defined.

3. Results

In Figure 4 below, light optical images of the center specimen locations of both alloys are presented with 1 and 7 s delay times. Within the micrographs, primary α-Al grains are identified to be white in colour and dark regions are the intermetallic or eutectic phases. The mean grain size of the cast component for the in-cavity solidified grains were 10 µm in the Aural-5 alloy and 8 µm in the Nemalloy HE700 alloy. The size of PSG in both the alloys ranged between 100-200 µm, which, is similar to those reported by Gourlay et al. [5] in their metallographic investigation of AlSi7Mg HPDC samples. Furthermore, from the optical images it can be seen that the agglomeration of PSGs was more significant in the centers of the cast specimens collectively and the content was higher in Aural-5 than in HE700. This PSG drift phenomenon was noted by Otarawanna et al. [4] and Bi et al. [3], both through experiments and simulations, alike.

The measured uniaxial tensile properties for each shot delay time for both alloys can be found in Table 2. The values show a minimal variability in the mechanical properties of both alloys despite the increased interval times between the cast specimens. The resultant tensile properties of Nemalloy HE700 alloy were determined to be higher than those seen in the Aural-5 counterpart with an approximate ultimate tensile strength (UTS) of 330 MPa, 190 MPa yield stress (YS) and elongation of 9-11%. Notably, the cast plates for Nemalloy HE700 and Aural-5 were fabricated approximately 500 and 365 days before testing, respectively. As such, both alloys were subject to natural ageing; the stress-strain graphs from the tensile testing is shown in Figure 5; which represented typical tensile test curves for each delay time shown in Table 2. Figure 5 shows that there is a decrease in elongation and toughness of the castings with delay time increases, however, the yield strength (0.2% proof stress) does not seem to vary much with presence of PSG, except that the 10s delay sample show some decrease in the same.

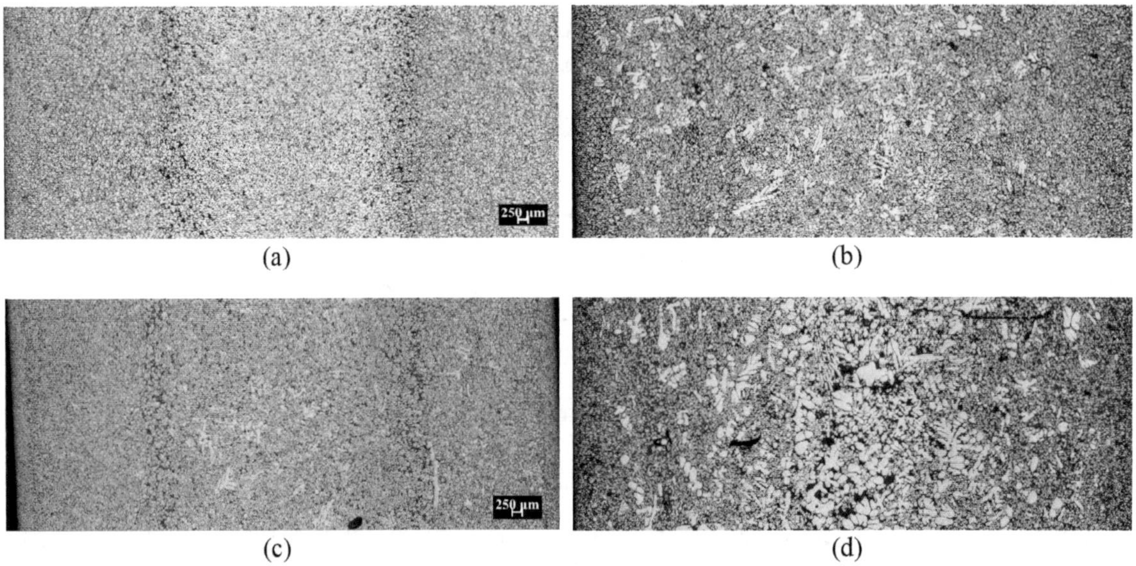

Figure 4. Light optical microscopy image sections of the center locations of (a) HE700 1s, (b) HE700 7s, (c) Aural-5 1s delay and (d) Aural-5 7s delay.

Table 2. Average ultimate tensile strength (UTS), yield stress (YS), and elongation results with standard deviation for HE700 and Aural-5 2.5mm plate thicknesses for each delay time

Sample #	UTS (MPa)	YS (MPa)	Elongation (%)	Sample #	UTS (MPa)	YS (MPa)	Elongation (%)
HE700 D1	332 ± 17	197 ± 3	11.5 ± 3	**A5 D1**	262 ± 10	134 ± 3	8.2 ± 2
HE700 D4	330 ± 8	196 ± 3	9.3 ± 2	**A5 D4**	261 ± 3	132 ± 3	8.0 ± 1
HE700 D7	308 ± 7	190 ± 4	6.3 ± 1	**A5 D7**	259 ± 4	135 ± 4	7.0 ± 1
HE700 D10	327 ± 5	191 ± 4	9.3 ± 1	**A5 D10**	257 ± 1	126 ± 3	7.8 ± 1

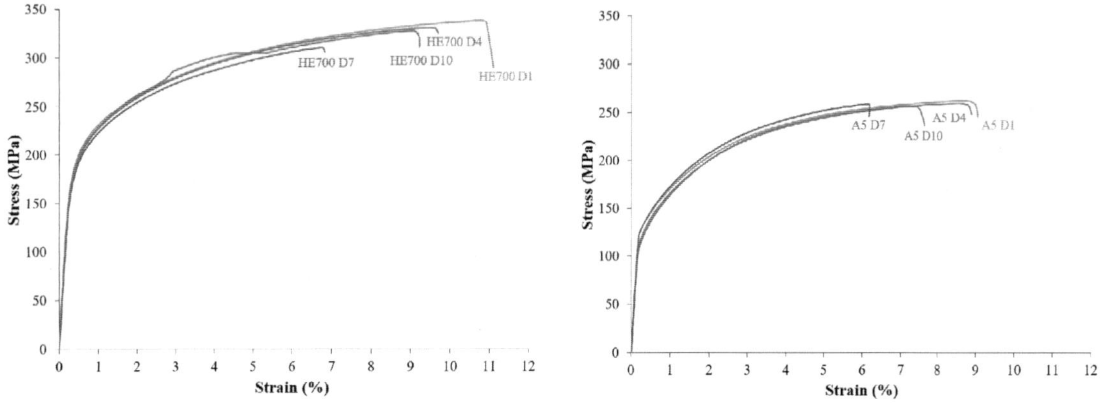

(a) (b)

Figure 5. Typical stress vs strain curves for select specimens of all shot delay intervals for (a) HE700 and (b) Aural-5 2.5mm plate thickness tensile specimen castings

A notable observation in the microstructures of these samples (Figure 4) is that there is an increased incidence of shrinkage voids at the locations of the PSG; both, within the dendritic structure of the PSG and at the interface between the PSG and the small grains solidified within the cavity. Since, there is a typical skin layer (100 to 200 μm) of very compact primary Al grains at the die tool interface of the castings, the effect of shrinkage increases within the center of the castings due to the increase in PSG, arising from the imposed delays in the shot sleeve is not readily reflected in the yield strength of the casting locations, However, at larger delays such as 7 and 10 s, there seem to be a notable effect on the yield strength due to the increased incidence of abnormal grain sizes (PSG) and the associated shrinkage cavities from them.

4. Summary

The study has shown that HVHPDC provides a viable process route for manufacturing structural automotive components. Although the presence of PSGs pose a critical defect in cast materials produced via HPDC, the effects seen in the 3-step geometry 2.5 mm plate castings of both Nemalloy HE700 alloy and Aural-5 showed minimal variation in their uniaxial tensile properties, on gauge lengths perpendicular to the melt flow direction in the die cavity. Moreover, the Nemalloy HE700 alloy was determined to have higher uniaxial tensile properties over its Aural-5 counterpart as it had higher strength and ductility values.

References.

[1] F. Czerwinski: *Materials (Basel).*, DOI:10.3390/ma14216631.
[2] US DOE: 2011, p. 30.
[3] C. Bi, S. Xiong, X. Li, and Z. Guo: *Metall. Mater. Trans. B Process Metall. Mater. Process. Sci.*, 2016, vol. 47, pp. 939–47.
[4] S. Otarawanna, C.M. Gourlay, H.I. Laukli, and A.K. Dahle: *Trans. Indian Inst. Met.*, 2009, vol. 62, pp. 499–503.
[5] C.M. Gourlay, H.I. Laukli, and A.K. Dahle: *Metall. Mater. Trans. A Phys. Metall. Mater. Sci.*, 2004, vol. 35 A, pp. 2881–91.

AUTHOR INDEX